GROWTH OF THE BACTERIAL CELL

GROWTH OF THE BACTERIAL CELL

John L. Ingraham
UNIVERSITY OF CALIFORNIA, DAVIS

Ole Maaløe
UNIVERSITY OF COPENHAGEN, DENMARK

Frederick C. Neidhardt
UNIVERSITY OF MICHIGAN, ANN ARBOR

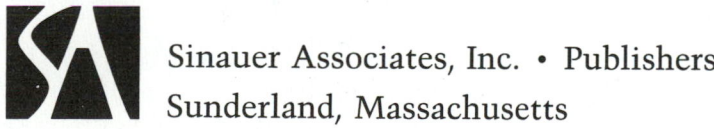

Sinauer Associates, Inc. • Publishers
Sunderland, Massachusetts

We wish to pay tribute to
Jacques Monod
and
Roger Y. Stanier
for their unique contributions
to the study of the bacterial
cell and its growth.

GROWTH OF THE BACTERIAL CELL

Copyright © 1983 by Sinauer Associates, Inc.

Sinauer Associates Inc.
Sunderland, MA 01375
U.S.A.

Library of Congress Cataloging in Publication Data

Ingraham, John L.
  Growth of the bacterial cell.

  Includes bibliographical references and index.
  1. Bacterial growth. 2. Escherichia coli—Growth.
I. Maaløe, Ole. II. Neidhardt, Frederick C.
III. Title.
QR86.I53 1983   589.9'031   83-496
ISBN 0-87893-352-2

Printed in U.S.A.

9 8 7 6 5 4 3 2

Contents

vi Contents

viii   *Contents*

Introduction

The paramount evolutionary accomplishment of bacteria as a group is rapid, efficient cell growth in many environments. Bacteria grow and divide as rapidly as the environment permits. This simple lifestyle has been maintained through a long period of evolution, with the result that bacteria grow and reproduce faster and more efficiently than any other type of organism. This is not just because they are small and their enzymes are effective, but also, and perhaps more, because a network of control mechanisms, the complexity of which we only begin to appreciate, has evolved to balance the many biosyntheses and polymerizations, the net result of which is growth.

The perception that growth is the core of bacterial physiology has led us to produce a text in which the results of structural, biochemical, genetic, and physiological studies on bacteria are presented in the context of cell growth and division. In this way we have hoped to provide a simple and meaningful framework to guide the reader in approaching the vast amount of information known about the bacterial cell.

We have presumed that the reader has some background in general biochemistry and genetics. Previous study in bacteriology is not prerequisite—we hope in fact that this book will serve to introduce the advanced biochemistry student (undergraduate and graduate) to the excitement of examining how biochemical processes come together to produce the balanced and efficient synthesis of a whole cell. On the other hand, those specializing in microbiology are likely to bring to our book a knowledge of the microbial world that should enrich their appreciation of our approach, and they can test its usefulness as a conceptual framework.

To the extent possible, we have treated the processes of growth and division as they apply to bacteria in general, and we have omitted specialized cases such as the growth cycles of sporulating or stalked bacteria; also, important processes such as photosynthesis and nitrogen fixation are described only briefly. Our decisions not to broaden the treatment of bacterial growth and not to deal with the fascinating diversity of bacteria were made in order to present and analyze some experiments in considerable detail. In this way we hope to have provided a basic text that will be useful not only to students and their teachers but to many microbiologists engaged in basic or in applied research with bacteria.

Escherichia coli is our model organism, and this bias has been unavoidable because many key experiments (involving the use of sophisticated genetic techniques, including the isolation of mutants and their use to analyze the functioning of such structures as promoters and operators) have been done only with this organism or its close relative *Salmonella typhimurium*. What indeed is noteworthy about the study of *E. coli* is that from our current knowledge one can now formulate quite precisely a large number of fundamental questions about cell growth.

Our book is certainly not the first about bacteria to emphasize the process of growth. Suffice it to mention those of Henrici (1928), Rahn (1932), Monod (1941), and Maaløe and Kjeldgaard (1966). During the half-century since Henrici's book appeared, experimental biology has developed very rapidly, and even since the latest book of this sort appeared in 1966 the introduction of gel electrophoresis, DNA/RNA sequencing, and new in vitro and in vivo genetic tools has enormously increased the scope of experiments. In writing this book we have tried to give due weight to past as well as recent developments.

Some housekeeping notes are in order. Throughout the text we refer to books and journal articles by the first author's (or editor's) name and the date of publication. The references are collected at the end of each chapter. We have not provided encyclopedic documentation for each statement but have instead cited some key journal articles and reviews that should provide appropriate entrée to the literature of this field. Sources of figures and tables are abbreviated by the same convention; full identification of these sources are listed at the end of the book in Credits for Figures and Tables.

Each individual chapter was initially conceived and drafted by a different one of the three authors. A significant collaboration was then undertaken to produce a harmonious text. The reader can judge whether this effort was successful, but the authors can testify that it was a stimulating and pleasurable task. Each chapter in the final draft truly has three authors, and the three of us contributed equally to the overall conception of the book. The listing of the authors on the title page is, therefore, deliberately alphabetical.

J.L.I., O.M., and F.C.N.
Winters, California
June, 1982

Henrici, A. T. 1928. *Morphologic Variation and the Rate of Growth of Bacteria.* Microbiology Monographs. Baillière, Tindall and Cox, London.

Maaløe, O. and N. O. Kjeldgaard. 1966. *Control of Macromolecular Synthesis.* W. A. Benjamin, Inc., New York.

Monod, J. 1941. Recherches sur la Croissance des Cultures Bacteriennes. Doctoral Thesis, University of Paris. Hermann & Cie, Paris.

Rahn, O. 1932. *Physiology of Bacteria.* P. Blakiston's Son & Co., Philadelphia.

Acknowledgments

Skillful and dedicated individuals in Davis, California, in Copenhagen, Denmark, and in Ann Arbor, Michigan typed parts of the many drafts of our manuscript: Alice Henin, Emmy Sidenius and Lauren Gagnon with an assist from Nan Plummer. Lauren Gagnon was responsible for assembling and typing the separately produced parts of the manuscript and preparing them for final editing. Marjorie Ingraham provided valuable typing. The complete text was read by our friend and counselor, Moselio Schaechter, who kept us honest to our purpose and to the theme presented in the Introduction, but who cannot be held responsible for errors that remain. H. Edwin Umbarger, who inspired much of our treatment of metabolism, helped remove errors, but likewise cannot be blamed for those that persist. During the conception and production of this book Andy Sinauer became a respected advisor and a friend—a tribute to his personal qualities as well as his professional competence. Not the least of his contributions was the assignment to the project of a masterly copy editor, Jodi Simpson.

To all these people, and to the many colleagues who provided generous help of various kinds we express our deep gratitude.

J.L.I., O.M., and F.C.N.

Chapter One

Composition, Organization and Structure of the Bacterial Cell

INTRODUCTION

One of the many ways to observe the growth of bacteria is to spread some on the surface of a nutrient agar block mounted on the stage of a light microscope. Individual cells can be seen to increase their volume and mass by invisible means until a division produces two offspring, each of which proceeds directly to repeat the cycle.

In a pure culture the cells produced in this manner are quite homogeneous with respect to viability and growth rate. Direct microscopy shows that if a large number of cells from a liquid culture are spread on nutrient agar, very few fail to divide, and microcolonies of remarkably uniform size develop. In fact, for hundreds of testable phenotypic and genotypic traits, the cells (except for rare mutants) are equally homogeneous. Thus, bacterial growth and division must be organized so as to ensure genetic continuity and phenotypic homogeneity; and the overall process, in spite of the short time (20–30 minutes) required, must be a display of the fundamental property of living systems—self-replication.

Bacteria are ideal subjects with which to study self-replication—or cell growth, as we shall call it—because as a group they emerge from over two billion years of evolution as masters of rapid growth in many environments. Their existence today is a function largely of their ability to grow rapidly when conditions permit and to survive when conditions prevent growth.

Multicellular forms of life have evolved specialized structures that generate the myriad of survival mechanisms we associate with plants, animals, and the higher protists. In many organisms these are so highly developed that the generation time (or the mass doubling time) must be measured in years or decades rather than minutes. Bacteria, too, have special survival mecha-

1

nisms, including both offensive and defensive weaponry, but these do not obscure the bacterial cell's display of its central accomplishment. It is the master of rapid, independent cellular growth.

Overall in this book we shall want to discuss how bacteria can be used to study growth and to assess where we are in understanding this process.

We begin in this chapter by asking what a bacterium is built of. The answer will include many details about molecules and structures and how they are organized. These details may at first sight appear overwhelming, but they are necessary to our story, and they can in fact be rather easily managed because of their direct relevance to the central act of growth. We shall omit much biochemical detail that can be found in any of the several excellent texts (Stanier, 1976; Gottschalk, 1979) dealing with microbial chemistry; we choose rather to emphasize how these details relate to growth.

CHEMICAL COMPOSITION OF BACTERIAL PROTOPLASM

By centrifugation or filtration, one can collect the bacterial cells from a pure culture growing in liquid medium. Assume that filtration of a culture has yielded 100 mg of packed bacterial mass. This biomass can be analyzed by any of a number of methods to learn its chemical composition. An aliquot can be dried to learn its water content, and a portion of the dried material used to ascertain its elemental composition as though it were a single substance. Typically one finds that approximately 70% of the packed cell mass is water—somewhat less than the 90% found in cells of higher organisms.

Elemental assay of the dry mass of *Escherichia coli* cells reveals a fairly typical composition of protoplasm: approximately 50% carbon, 20% oxygen, 14% nitrogen, 8% hydrogen, 3% phosphorus, 1% sulfur, 2% potassium, 0.05% each of calcium, magnesium, and chlorine, 0.2% iron, and a total of 0.3% trace elements including manganese, cobalt, copper, zinc, and molybdenum (Luria, 1960). From these data one can anticipate that most of the bacterial mass consists of organic compounds—a correct guess—but otherwise the elemental analysis is too unsophisticated for our purpose, and we must turn to the analytical techniques of organic chemistry.

The second column of Table 1 shows the results of organic analysis of the dry mass of cells harvested by rapid filtration from a bacterial culture of a given strain in a steady-state growth under specified conditions of culture medium and temperature.

Right at the start it should be noted that biochemical data are meaningful only if attention has been given to specify (1) the organism, (2) the growth environment, and (3) the state of growth. These parameters have a profound effect on biochemical results but often are not adequately documented in the reports of experiments.

1. *The organism.* Although all bacteria, as PROCARYOTES, share a basic cellular organization (to be examined later in this chapter), the over-5000 species of bacteria are distinguished by many individual characteristics;

Table 1. Overall macromolecular composition of an average E. coli B/r cell[a]

Macromolecule	Percentage of total dry weight	Weight per cell (10^{15} × weight, grams)		Molecular weight	Number of molecules per cell	Different kinds of molecules
Protein	55.0	155.0		4.0×10^4	2,360,000	1050
RNA	20.5	59.0				
23 S rRNA			31.0	1.0×10^6	18,700	1
16 S rRNA			16.0	5.0×10^5	18,700	1
5 S rRNA			1.0	3.9×10^4	18,700	1
transfer			8.6	2.5×10^4	205,000	60
messenger			2.4	1.0×10^6	1,380	400
DNA	3.1	9.0		2.5×10^9	2.13	1
Lipid	9.1	26.0		705	22,000,000	4[b]
Lipopolysaccharide	3.4	10.0		4346	1,200,000	1
Peptidoglycan	2.5	7.0		$(904)_n$	1	1
Glycogen	2.5	7.0		1.0×10^6	4,360	1
Total macromolecules	96.1	273.0				
Soluble pool	2.9	8.0				
building blocks			7.0			
metabolites, vitamins			1.0			
Inorganic ions	1.0	3.0				
Total dry weight	100.0	284.0				
Total dry weight/cell		2.8×10^{-13}g				
Water (at 70% of cell)		6.7×10^{-13}g				
Total weight of one cell		9.5×10^{-13}g				

[a] In balanced growth at 37°C in glucose minimal medium, mass doubling time, g, of 40 minutes. The data are assembled from Dennis (1974), Maaløe (1979), F. C. Neidhardt (unpublished), Roberts (1955), and Umbarger (1977).

[b] There are four classes of phospholipids, each of which exists in many varieties as a result of variable fatty acyl residues.

including chemical composition. An average composition could perhaps be roughly estimated (e.g., 50% protein, 25% RNA, 3% DNA), but this would be too imprecise for our purposes and would, in fact, not necessarily be the composition of any real cell. We must, right at the start, specify our cell.

This is done first by naming the GENUS, SPECIES, and STRAIN; for example, *Escherichia coli* strain B. Some clarification of these terms is in order. The concept of species is rooted in the sexual reproduction by which most organisms other than bacteria increase their numbers. A population of sexually reproducing organisms that are free to interbreed will share its total gene pool and assort new mutations among its members. If one portion of the population becomes geographically isolated and cannot interbreed with the other, the two subpopulations are likely to evolve along different lines, leading eventually to their physiological incapacity to interbreed. At that point, speciation is said to have occurred, and the decision whether any two subpopulations are the same or are different species is made on the basis of this criterion. Of course many other characteristics help distinguish one group from the other.

For bacteria, which are haploid and reproduce only asexually, the very concept of species seems inapplicable. Furthermore, their short generation time and the ease with which they can be found to evolve genetically in the laboratory does not lead one, *a priori*, to expect that any large, random population of bacteria in nature would consist of identifiable clusters of closely similar biotypes. But such, indeed, is the case. Bacteria can be ordered in taxonomic units, unfortunately called SPECIES, each of which consists of a large number of independently isolated STRAINS (also called ISOLATES) that share the vast majority of their observable characteristics. Furthermore, species can be arranged according to their degree of similarity (and presumed relatedness) into GENERA (singular: GENUS). Similar genera are clustered into FAMILIES, and for some kinds of bacteria, families into ORDERS (Buchanan, 1974). The name *Escherichia coli*, therefore, refers to a group of bacteria that are independently obtained from nature as separate strains, or isolates, and that share a large number of qualitative and quantitative characteristics (genus: *Escherichia*; species: *coli*).

One strain of *E. coli* is called K12; others are called B, ML30, C, and so on. By appropriate biochemical tests, these strains can easily be recognized as *E. coli* and not strains of the related species *Escherichia freundii*. Strains (isolates) of the same species usually share 80% or more genetic homology. Of course, this means they can differ in very many genes. Unfortunately, it has become customary to designate as "strains" mutant variants of a given isolate. A histidine-requiring derivative of *E. coli* strain K12 is usually given a "strain number." No term has come into common use to designate the tens of thousands of genetic variants of *E. coli* strain K12 other than the word "strain" itself. The resulting ambiguity (strain for isolate and strain for variant) has fostered an unfortunate sloppiness in reporting genealogies of variants. At one point a considerable effort had to be expended to clarify the confusing state of *E. coli* strain K12 genealogies (Bachmann, 1972).

For many laboratory strains (original sense) of *E. coli*, long cultivation and storage under different conditions in different laboratories has led to the inadvertent selection of variants that have different growth rates and different chemical composition in a given medium. For these strains it is necessary to identify the laboratory source of the strain. This problem is greatest for *E. coli* strain K12, less so for strain B and for bacteria of other genera and species. The problem of strain variation among laboratories can be virtually eliminated by the modern practices of lyophilization or of low temperature ($-80°C$) storage of strains.

2. *The medium.* Bacteria have control devices that bring about adaptive adjustment of the levels of their major constituents in response to the chemical and physical environment. These adjustments are an important aspect of the cell's mastery of rapid growth in nature. Bacteria have little

ability to modify their environment to advantage, and they rely mostly on internal adjustment of their own growth machinery. As we shall see in Chapter 6, many aspects of cell composition vary with growth rate, including the total content of ribonucleic acid. In addition, at the same growth rate in different media, an organism will produce different sets of proteins and other specific components. Therefore, meaningful and reproducible chemical measurements on bacteria must involve careful definition of the growth conditions.

A simple, yet very effective, check to help assure that one has used the desired strain of bacterium and has constructed the desired medium in any experiment is to measure the (exponential) growth rate of the cells (Chapters 5 and 6). This parameter should be invariant from experiment to experiment and from laboratory to laboratory. Its value should always be included in the other data reported on experiments in bacterial physiology.

3. *The state of growth.* When bacterial cells of a single strain have been growing in a medium for sufficient time to complete their chemical adjustments to that growth condition, the exponential increase in population mass occurs at an extraordinarily constant and reproducible rate characteristic of the strain and growth condition. For a time—until the medium is changed as a result of the growth of the bacteria—a truly steady-state situation exists, one in which cell growth is said to be BAL-ANCED because every extensive property (i.e., every component) of the cell culture increases by the same constant factor per unit time. It is this growth state that in practice is the *only* reproducible state of these cells, and every effort must be made to achieve it before commencing serious measurements. Why balanced growth is important, and how it is achieved in practice is dealt with in detail in Chapter 6.

Now, with these three elements (strain, medium, state of growth) specified, we can examine the data of Table 1 with some assurance of their validity and with some understanding of the requirements for obtaining similar data for other bacteria in which we may be interested.

Four general observations can be gleaned from the data of Table 1, column 2, that pertain to our overall analysis of bacterial growth: (1) macromolecules constitute the preponderance (96%) of the cellular dry mass; (2) proteins constitute over half of this amount; (3) there is a high content of RNA compared to that in most cells of higher organisms; (4) there are many familiar substances, but two that are unique to bacteria: peptidoglycan and lipopolysaccharide.

DYNAMIC ASPECTS OF PROTOPLASMIC COMPONENTS DURING GROWTH

A simple experiment can tell us much of the dynamic relationships of these substances in growing cells. Adding glucose uniformly labeled with ^{14}C to a culture growing with this sugar as its sole organic source of carbon enables one to follow the flow of this element through the various classes of molecules in the cells. This is done by sampling at intervals after addition of the labeled glucose to a culture already in a steady state of growth. The samples may then be treated to separate molecules that are soluble from those insoluble in boiling water, that is, roughly separating low-molecular-weight substances from macromolecules. Biochemical methods allow fractionation into general classes, such as proteins, nucleic acids, and fat-soluble molecules, or further into subclasses of these. The fractions can then be assayed for radioactivity.

Eventually every organic molecule in the cell will approach a specific ^{14}C content equal to that in the glucose substrate, but the time course of labeling will be governed by the metabolic characteristics of each molecule. As a class the small quantity of soluble, low-molecular-weight molecules attain values near their maximum specific activity within a small fraction of a generation time—a few minutes. This behavior, shown idealized as curve A in Figure 1, reveals that they are rapidly formed and consumed during growth; that is, there is a rapid flow through the pool of these substances in the cell. This is the behavior of molecules that include building blocks that are destined to be polymerized into macromolecules, metabolic intermediates on their way to becoming such building blocks, and molecules on their way to being excreted as waste products. Macromolecules, on the other hand, as a class become labeled *pari passu* with net growth: after one doubling of bacterial mass, their specific radioactivity is roughly half that of the glucose substrate; after two doublings (a fourfold increase in mass), it is 75% that of the glucose; and so on. This labeling pattern, idealized as curve B in Figure 1, is characteristic of substances that are stable components of the cell. They include most proteins, DNA, some RNA species, and other large molecules.

Some compounds acquire label with a time course that is apparently a composite of curves A and B. Such behavior is usually seen for molecular species that label quickly because they are present in small quantities and are converted into other, stable products but also are formed by degradation of macromolecules. Pools of ribonucleotides exhibit this behavior as a result of

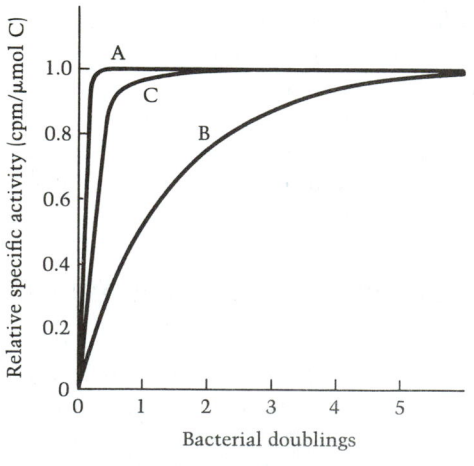

Figure 1.

Idealized time course of labeling of organic molecules following addition of [^{14}C]glucose to a culture of bacteria in steady-state growth in glucose minimal medium. Curves A, B, and C are idealized data for different classes of macromolecules described in the text.

mRNA degradation. Curve C in Figure 1 represents one of a large family of complex labeling patterns observed for different compounds.

Of special interest are the large molecules, such as messenger RNA (mRNA), that label by some time course C, indicating metabolic lability, and small molecules, such as the coenzyme NADP, that label by time course B, indicating metabolic stability, which is a property of molecules with catalytic functions in cell growth.

It is clear that this simple experiment can reveal much about the general role of compounds in cellular growth. The complexity of labeling of the nucleotide pool, however, serves as a cautionary note to avoid both simple expectations and unwarranted conclusions from such experiments. Later, in Chapters 3 and 6, we shall return to this point when we discuss precise measurements of rates of synthesis of cellular components.

CHEMICAL COMPOSITION OF AN AVERAGE BACTERIAL CELL

We have more to discuss concerning the chemical makeup of bacteria, and it is now appropriate to refine our analysis by considering not a 100-mg lump of bacterial mass, but rather the individual cells, the real biological units constituting that mass.

How many cells are in our sample? Several techniques can tell us. We can carefully suspend a measured portion in a suitable solution and count the particles with an electronic particle counter (or count them more laboriously by inspection under a microscope with the aid of a chamber of measured volume). Bacteriologists frequently use a simpler method that relies on the fact that normally greater than 99% of the cells taken from growing cultures are viable, that is, can themselves grow and divide to form a colony of descendants. Spreading a measured volume of a suitably diluted suspension of cells over an agar plate results, after incubation of 10–18 hours, in macroscopic colonies that are easy to count. (A colony of 10^7 cells is easily seen with the unaided eye.) With the use of appropriate duplicate and control samples, a tally can be obtained that is adequate for many purposes.

In the example we have been using (a culture of *E. coli* strain B grown in glucose minimal medium at 37°C), we would find that 100 mg of bacterial protoplasm consists of roughly 100 billion cells (10^{11} cells). The precise value obtained as an average of many measurements of this sort is that there are 1.05×10^{12} cells per gram of biomass in such a culture. The average cell in this population must therefore have a wet mass of 9.5×10^{-13} g, or 2.9×10^{-13} g dry mass.

What is the meaning of the term "average cell," in the sense we have been using it? Is it a cell midway in age and size between those in the population just formed by cell division and those just about to divide? Consider that upon division (in *E. coli* and most other bacteria, this occurs by binary fission) one old cell becomes two young ones. In a random population proceeding regularly through the cell cycle, therefore, there are twice as many just-born cells as there are cells ready to give birth. The idealized frequency distribution of cell age is therefore given by

$$f(x) = 2^{1-x}$$

where x is the age of the cells from 0 to 1 cell generation time (Powell, 1956). This is illustrated in Figure 2. The cell described by dividing the total biomass, or any measured component of our randomized, steady-state population, by the total number of cells in that population is one that is approximately 44% along in age in the cell cycle and, if individual cells increase in mass exponentially, is approximately 33% larger than when it was born. This is the average cell presented in Table 1.

The amount of each chemical component of the average cell presented in Table 1 has been carefully determined by direct as well as indirect measurements. These values will be used by us throughout our analysis of growth. They enable us to calculate how much of each component must be made by the cell to reproduce itself and what this biosynthesis costs (Chapters 2 and 3). They provide a framework for describing the apparatus of macromolecular synthesis (Chapter 3) and how it must be regulated (Chapters 6, 7, and 8). And they provide us with an inventory of molecules to be assigned to the various structures and compartments of the cell (remainder of Chapter 1).

For these purposes it will be necessary to complete two tasks: we must express the amount of each chemical class of molecule in terms of number of molecules per cell, and we must then ascertain how many different molecular species are represented in each of the major chemical classes.

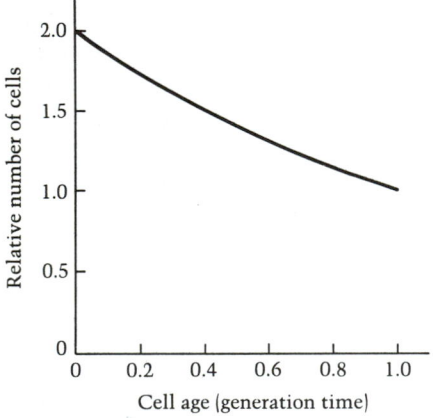

Figure 2.

Idealized frequency distribution of cell age in a steady-state culture increasing by binary fission.

VARIETY AND NUMBERS OF MOLECULES

Proteins constitute 55% of the cell's dry mass, so our average cell contains 155 femtograms (1 femtogram = 10^{-15} g) of protein. How many molecules does this represent? How many kinds of proteins are there in the cell? And what is their range of molecular abundance?

We can give reasonably accurate answers to each of these questions from analysis of two-dimensional gels that resolve total cell protein (O'Farrell gels). In this technique, proteins are separated on the basis of their net charge by means of ISOELECTRIC FOCUSING in one dimension and on the basis of their apparent size by SDS discontinuous slab gel electrophoresis in the second dimension. Figure 3 depicts autoradiograms of dried O'Farrell gels prepared from a reference culture identical to the one used for the chemical analyses given in Table 1, except for the presence of $^{35}SO_4$ in the medium to label cell proteins. Figure 3A is an autoradiogram on which approximately 850 radioactive spots are visible on the original film. With few exceptions each spot is a different polypeptide chain—the product of a different gene. The pH range of the gradient set by the ampholines in the gel tube of the first dimension is not sufficiently alkaline to permit entry of the many basic proteins of the cell. Figure 3B is an autoradiogram of a gel prepared by a modified method to circumvent this problem; it reveals the approximately 200 major basic proteins of the cell.

There is some evidence that the 1050 proteins (polypeptides) displayed by the O'Farrell technique is a close approximation of the number of proteins present in significant amounts (30 or more molecules) in an *E. coli* cell growing in this medium. (You might consider what sort of evidence might support this contention.) Because a rough approximation of the molecular weight of each of these 1050 polypeptides is given by its vertical position on the gel, the O'Farrell display permits an estimate of the average molecular weight of *E. coli* polypeptides: approximately 40,000. Our reference cell must therefore contain on the order of 2.4 million protein molecules.

For the third question, a glance at Figure 3 is sufficient to indicate an enormous range of molecular abundance among these different protein species. Optical methods for measuring the intensity of individual spots, confirmed by measurement in a scintillation counter of their radioactivity in the original gel, reveal that they vary in their content of molecules over a 10^5-fold range. Evidence indicates that the relative amount of individual proteins in the spots on the gels reflects in most instances their in vivo abundance. We, therefore, face the interesting fact that some proteins are present as over 100,000 molecules per cell and others as only a handful. As we shall see, most proteins are remarkably stable in *E. coli*, so we can anticipate powerful mechanisms for controlling the expression of individual genes.

These control mechanisms not only set the rate of synthesis of individual proteins at appropriate values with respect to each other but also adjust these rates in response to the environment. The two autoradiograms reproduced in Figure 4 illustrate the magnitude of these adjustments. They show amounts of synthesis of individual proteins in one strain of *E. coli* during (A) steady-state growth in acetate minimal medium and (B) steady-state growth in glucose-rich medium. The adjustments of individual proteins under these circumstances is readily apparent visually.

Approximately 200 of the protein spots on these gels have already been identified as known enzymes, factors, and structural proteins (Phillips, 1980). A few of these identifications are given in Table 2.

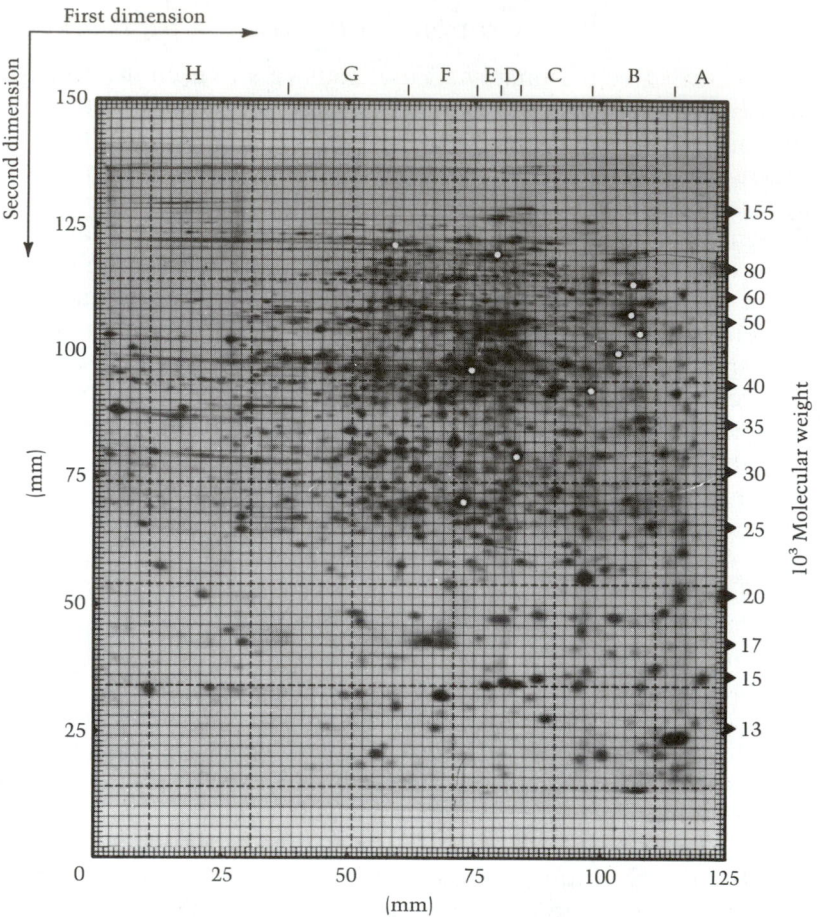

Figure 3. Autoradiograms of two-dimensional gels of extracts of *E. coli* B/r strain NC3 grown in glucose minimal medium at 37°C and labeled with $^{35}SO_4$. A grid overlay provides coordinates for individual spots. The letter designations A to I, from the acidic to the basic end of the first dimension, indicate zones containing proteins of increasing isoelectric points. The approximate molecular weight scale from the top to the bottom of the second dimension was determined by using the average migration distance on these two gels of proteins of known molecular weight. These parameters are used to designate proteins; for example, a protein with coordinates 15,81 in panel A that is in zone H and is 35,000 molecular weight is called protein H35.0. A. Isoelectric focusing to equilibrium. B. Nonequilibrium isoelectric focusing. Very basic proteins do not enter the isoelectric focusing gel in the equilibrium method. In the nonequilibrium method, electrofocusing is stopped before these proteins migrate out of the gel and the proteins of zone A have not entered the gel. Small white dots placed over the same 10 proteins in the two autoradiograms show the different migration in the two systems. (From Bloch, 1980.)

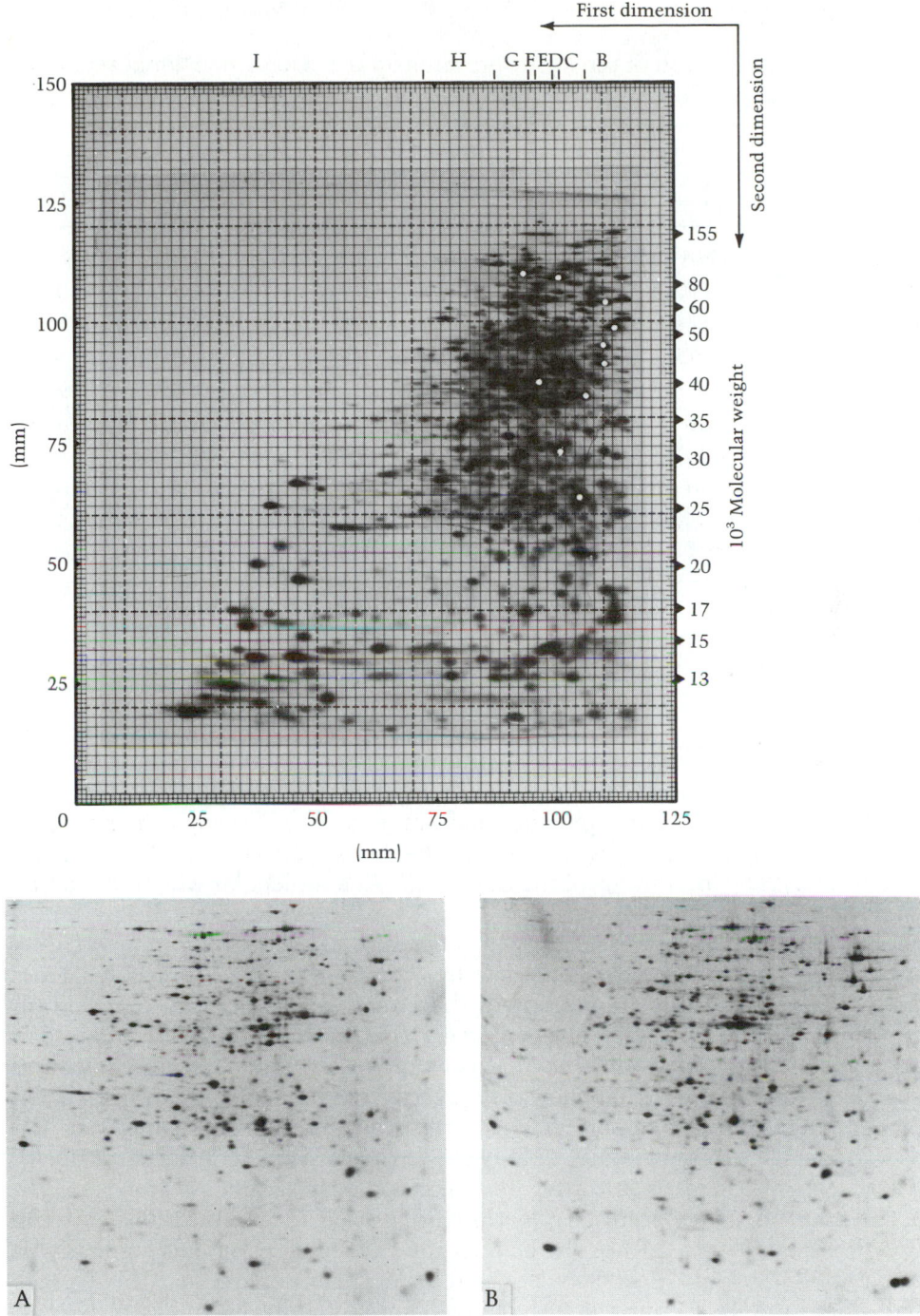

Figure 4. Comparison of proteins in cells of *E. coli* strain B/r in two different conditions. The two-dimensional gels were prepared in the same manner as that in Figure 3A. A. Steady-state growth in acetate minimal medium. B. Steady-state growth in glucose-rich medium. (F. C. Neidhardt, unpublished.)

11

Table 2. Identification of ten of the proteins on the two-dimensional reference gels of *E. coli* B/r (Figure 3)

Alphanumeric designation	Protein name	Location[a]	
		Gel A	Gel B
B40.7	RNA polymerase, α subunit	98, 92	106, 84
B46.7	ATPase, β subunit	103, 99	110, 91
B65.0	Ribosomal protein S1	106,113	111,104
D40.7	Leucine-isoleucine-valine binding protein	81, 93	—
D84.0	Protein synthesis elongation factor G	80,119	99,109
E42.0	Protein synthesis elongation factor Tu	74, 96	96, 88
E106	Valyl-tRNA synthetase	84,123	—
F24.5	Outer membrane protein A	73, 70	100, 73
I21.4	Cyclic AMP receptor protein	—	65, 57
I23.0	50 S Ribosomal subunit protein L3	—	40, 62

[a] Numbers represent x and y coordinates (in that order) of individual proteins.

Bacterial cells are rich in RNA. One-fifth of the dry mass of our reference cell growing in glucose minimal medium is RNA. The preponderance (81%) of this is ribosomal RNA, comprising three molecular species (23 S, 16 S, and 5 S rRNA) that are present in equimolar amounts—18,700 copies per cell.

Transfer RNA constitutes most of the remainder (15%). There are approximately 8.5 femtograms of tRNA per cell. At a molecular weight of 25,000 each, there are somewhat over 200,000 tRNA molecules per cell—10 times the number of ribosomes. The total structure is known for many tRNA molecules; Figure 5 illustrates the features common to these molecules. How many species of tRNA are present in the *E. coli* cell is not yet known with certainty, but the number is estimated to be near 60. Figure 6 shows an autoradiogram of a two-dimensional gel that resolves at least 40 molecular species of tRNA. Note that they vary greatly in abundance, a fact supported by considerable independent evidence about their in vivo abundance.

Messenger RNA amounts to only 2.3 femtograms (4% of the total RNA). The experiment described earlier, in which radioactive glucose was added to a culture in a steady state of growth, can be used to illustrate the extreme metabolic lability of mRNA. Its acquisition of label corresponds generally to curve C in Figure 1. More sophisticated experiments reveal that the average half-life of mRNA in our reference cell is approximately 1.3 minutes. Not all molecular species are equally labile (the range is from 0.5 to 10 minutes), and the reason for this variation is not yet known. A further contrast between rRNA and tRNA is the structural heterogeneity of mRNA. The actual number of kinds of mRNA at any one time in the cell is not known with certainty, but one arrives at an estimate of 350–400 by assuming that the genes for the

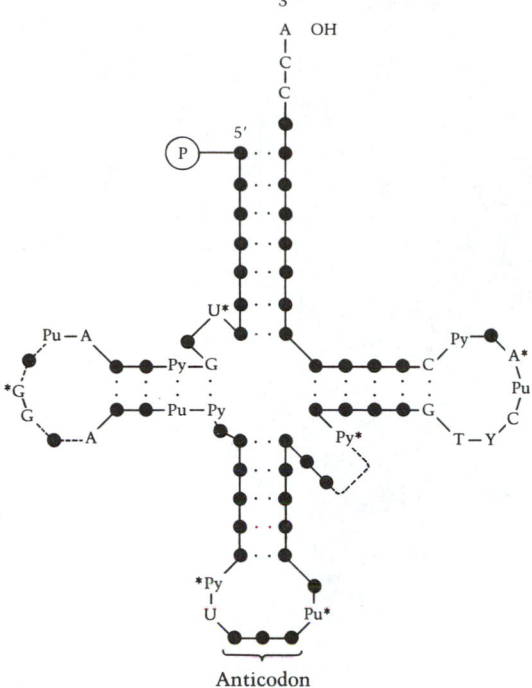

Figure 5. Generalized cloverleaf structure of tRNA. The primary structure is shown by heavy solid lines connecting nucleosides (A, adenosine; C, cytidine; G, guanosine; T, thymidine; U, uridine; Pu, purine nucleoside; Py, pyrimidine nucleoside; Y, pseudouridine; solid circles, variable residues in different tRNAs). Asterisks indicate where modifications of nucleosides are found. Dotted lines show Watson-Crick base pairing in the four stems of the cloverleaf. Dashed lines indicate regions of variable length in different tRNAs. (Modified from Clark, 1980.)

1050 proteins in the cell are arranged on the average in operons (gene clusters that cotranscribed into a single mRNA) of two to three genes each.

DNA constitutes approximately 9 femtograms of our standard cell. The complete, single chromosome of *E. coli* is known, with fair precision, to have a molecular weight of 2.5×10^9. It is a circular, covalently closed, double-stranded molecule of approximately four million nucleotide base pairs. Our reference cell, caught in the act of replicating its DNA, has slightly over two complete chromosomes (see Chapter 6).

The lipid of *E. coli*, exclusive of lipid A of the lipopolysaccharide, is entirely phospholipid. Bacteria do not accumulate neutral fats as reserve material; rather, lipids are a functional part of their membranes (cf., Goldfine, 1972). The major phospholipids in *E. coli* are phosphatidylethanolamine (75%), followed by phosphatidylglycerol (18%), cardiolipin (5%), and traces of phos-

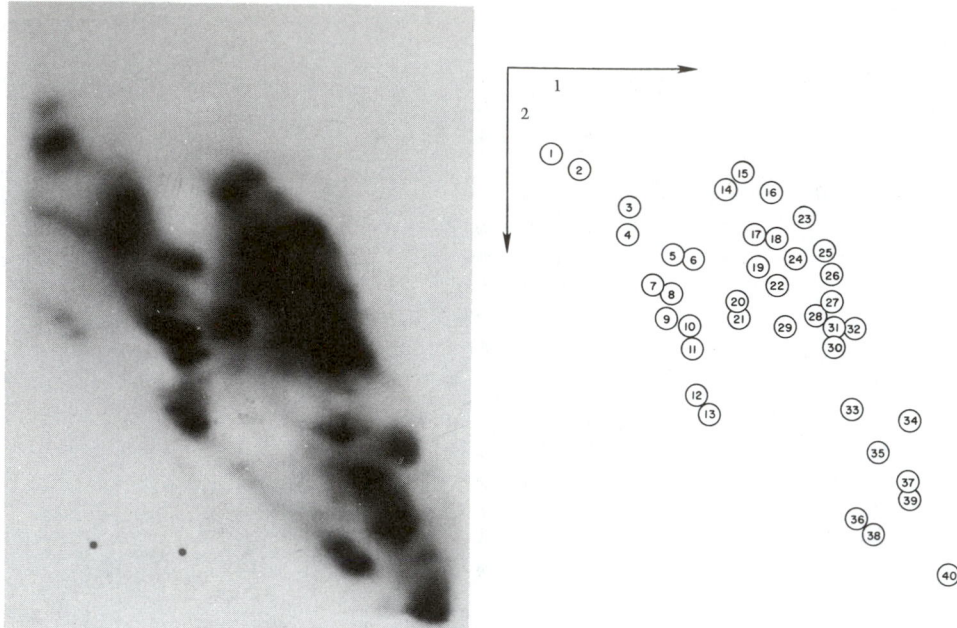

Figure 6. Partial resolution of transfer RNA molecules by electrophoresis on a two-dimensional gel. *E. coli* cells were labeled with ^{32}P-phosphate during growth in glucose minimal medium. An extract of the cells was passed through a DEAE-cellulose column to produce a fraction enriched for tRNA. Electrophoresis in the first dimension was performed through a 10% polyacrylamide, 7 *M* urea gel in a Tris-borate buffer, containing EDTA, at pH 8.3. For the second dimension the polyacrylamide was raised to 20% and the urea concentration lowered to 4 *M*. In contrast to the two-dimensional gels of total cell protein (Figures 3 and 4), resolution of tRNA species is not sharp. Several factors account for the lesser clarity. The small size of tRNA molecules and the long duration (3–4 days) of electrophoresis lead to diffusional spread of each species. Also, the two dimensions do not fractionate by independent parameters as for the protein gels (size and charge), and therefore there is a clustering of species along the diagonal. Nevertheless, with prolonged exposure autoradiograms reveal 50–60 spots; lesser exposures show clearer separation but do not reveal species of minor abundance. In the analysis shown, 40 of the abundant spots (indicated by the tracing on the right) were measured. (Courtesy of D. McKay and M. J. Fourner, manuscript in preparation.)

phatidylserine (Figure 7). The major fatty acids found in these molecules are 43% palmitic (16:0), 32.6% palmitoleic (16:1), and 25.2% *cis*-vaccenic acids (18:0) (Figure 8). Together the membrane phospholipids amount to 26 femtograms in our average cell.

The lipopolysaccharide (LPS) of *E. coli* is an extremely complex molecule,

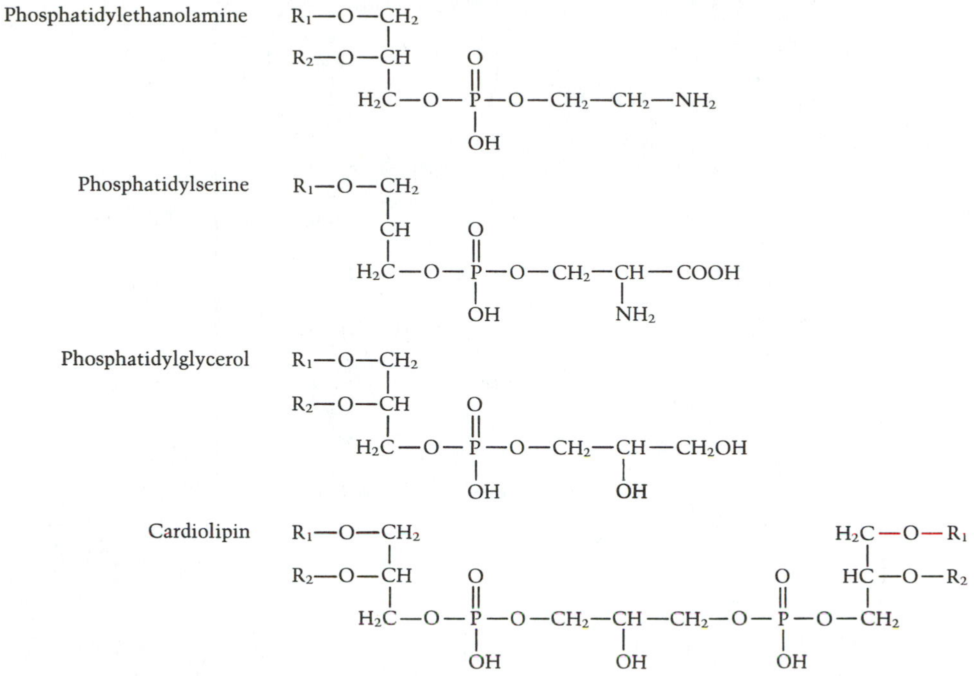

Phosphatidylethanolamine

$$R_1-O-CH_2$$
$$R_2-O-CH$$
$$H_2C-O-\overset{\overset{\displaystyle O}{\|}}{\underset{\underset{\displaystyle OH}{|}}{P}}-O-CH_2-CH_2-NH_2$$

Phosphatidylserine

$$R_1-O-CH_2$$
$$CH$$
$$H_2C-O-\overset{\overset{\displaystyle O}{\|}}{\underset{\underset{\displaystyle OH}{|}}{P}}-O-CH_2-\underset{\underset{\displaystyle NH_2}{|}}{CH}-COOH$$

Phosphatidylglycerol

$$R_1-O-CH_2$$
$$R_2-O-CH$$
$$H_2C-O-\overset{\overset{\displaystyle O}{\|}}{\underset{\underset{\displaystyle OH}{|}}{P}}-O-CH_2-\underset{\underset{\displaystyle OH}{|}}{CH}-CH_2OH$$

Cardiolipin

$$R_1-O-CH_2$$
$$R_2-O-CH \qquad\qquad\qquad H_2C-O-R_1$$
$$H_2C-O-\overset{O}{P}-O-CH_2-CH-CH_2-O-\overset{O}{P}-O-CH_2 \qquad HC-O-R_2$$
$$OH \qquad OH \qquad OH$$

Figure 7. Structures of common bacterial phospholipids. The four types shown are all found in *E. coli.* R_1 and R_2 = fatty acyl residues.

$CH_3(CH_2)_{12}COOH$	Myristic acid
$CH_3(CH_2)_{10}CHOHCH_2COOH$	β-Hydroxymyristic acid
$CH_3(CH_2)_{14}COOH$	Palmitic acid
$CH_3(CH_2)_5CH=CH(CH_2)_9COOH$	Palmitoleic acid
$CH_3(CH_2)_5CH-CH(CH_2)_7COOH$ with CH_2 bridge	*cis*-9,10-Methylenehexadecanoic acid
$CH_3(CH_2)_5CH=CH(CH_2)_7COOH$	*cis*-Vaccenic acid

Figure 8. Structures of common fatty acids in bacteria. The six molecules shown are all found in *E. coli,* but only palmitic, palmitoleic, and *cis*-vaccenic acids are prevalent.

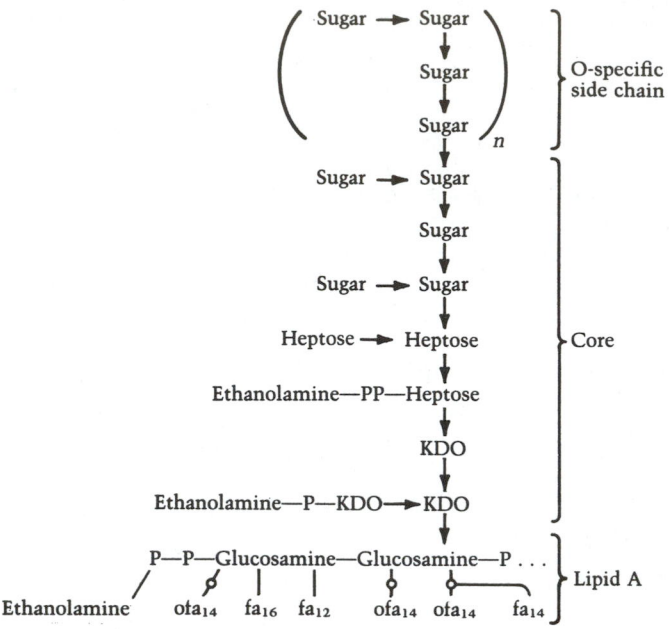

Figure 9. General diagram of a subunit of lipopolysaccharide. Molecules of lipopolysaccharide contain, on average, three of these subunits. The structure shown is based on analyses of the envelope of enteric bacteria, particularly species of *Salmonella*. LIPID A is a phosphorylated glucosamine disaccharide esterified with fatty acids (fa), commonly including dodecanoic (fa$_{12}$), tetradecanoic (fa$_{14}$), 3-hydroxytetradecanoic (ofa$_{14}$), and hexadecanoic (fa$_{16}$) acids. The CORE is an oligosaccharide that commonly includes heptose (L-glycero-D-mannoheptose) and KDO (3-deoxy-D-mannooctulosonic acid). The O-SPECIFIC SIDE CHAIN is usually longer than the core oligosaccharide and consists of many repeating tri-, tetra-, or pentasaccharide subunits, including (in different species) a variety of unusual sugars. Lipid A and the core are quite similar among all enteric species, but the O-specific side chain is highly strain specific. (Modified from Nikaido, 1973.)

the elucidation of which is a triumph of structural chemistry and microbial genetics. A diagram of the general structure of the LPS is shown in Figure 9. All 9.7 femtograms of LPS are found in the outer layer of the wall of this cell. The structure of LPS varies among species of bacteria, and even among strains, but all Gram-negative bacteria possess it in some form or other.

Peptidoglycan is found in all bacteria, Gram-positive and Gram-negative alike, except for the recently recognized archaebacteria. In Gram-positive bacteria, as we shall see, it forms the bulk of the cell wall. Though present in lesser amount in Gram-negative cells such as *E. coli*, it nonetheless plays a significant role in maintaining the structural integrity of the cell. How it does so can perhaps be intuitively understood from its structure (Figure 10); the lattice of peptidoglycan forms a covalent shell (or better, hauberk) of great strength around the entire cell. In many Gram-positive cells (and never in Gram-negative cells), teichoic acids are associated with the peptidoglycan. Teichoic acids (Figure 11) are polymers of glycerol or ribitol joined through phosphodiester bonds and carrying one or more amino acid or sugar substituents.

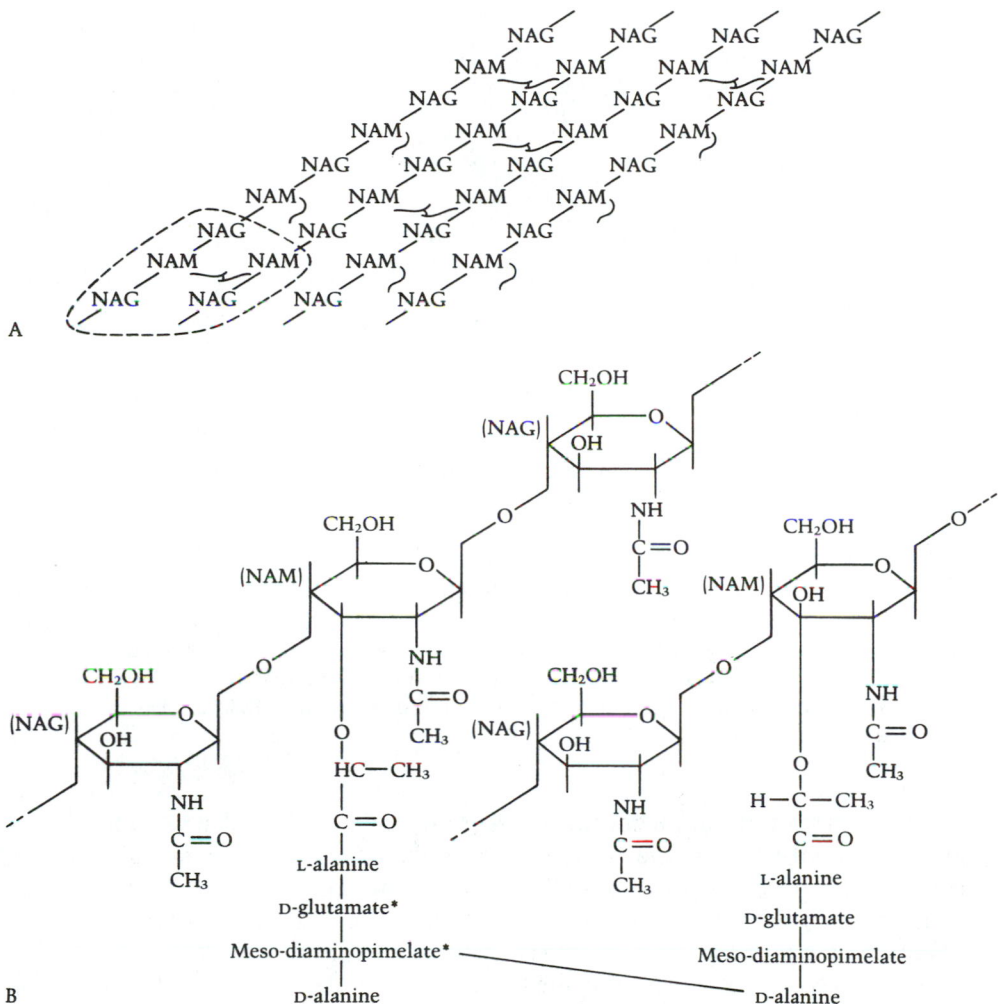

Figure 10. Diagrammatic representation of the peptidoglycan of *E. coli.* A. The arrangement of *N*-acetylglucosamine (NAG) and *N*-acetylmuramic acid (NAM) residues linked by β-1,4 glycosidic bonds (diagonal lines). B. The portion of A enclosed within the dotted line is enlarged to show the cross linkage between tetrapeptides attached to NAM residues of adjacent chains. The peptidoglycan of other species may differ in the nature of the amino acids at positions 2 and 3 of the tetrapeptide and in the nature and frequency of the cross link (noted by asterisks in the figure). (Modified from Ghuysen, 1968.)

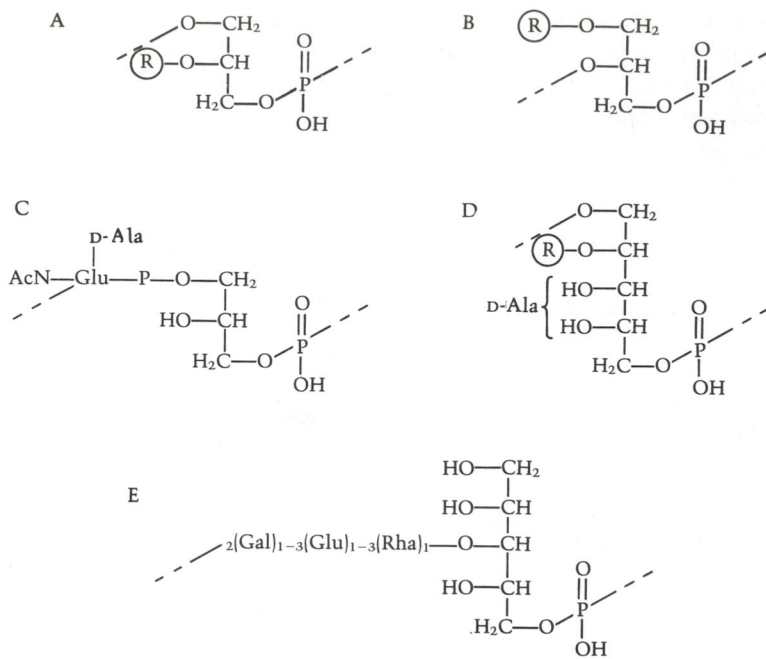

Figure 11. Structure of teichoic acids. Repeat units of five teichoic acids from different Gram-positive species are shown. A. Glycerol teichoic acid of *Lactobacillus casei* 7469 (R = D-alanine). B. Glycerol teichoic acid of *Actinomyces antibioticus* (R = D-alanine). C. Glycerol teichoic acid of *Staphylococcus lactis*; D-alanine occurs in the 6-position of *N*-acetylglucosamine. D. Ribitol teichoic acids of *Bacillus subtilis* (R = glucose) and *Actinomyces streptomycini* (R = succinate). (The D-alanine is attached to position 3 or position 4 ribitol.) E. Ribitol teichoic acid of the type 6 pneumococcal capsule. (From Stanier, 1976.)

During growth on glucose, a small proportion of the substrate molecules are connected to form the polysaccharide glycogen, with a reserve function believed to be analogous to that in higher organisms. Our reference cell contains approximately 7 femtograms of glycogen. This tiny amount is insufficient to support the carbon needs of the cell for growth for more than a minute or so in the event the medium becomes depleted of substrate. Nor would the energy made available by metabolizing this amount of glycogen serve the maintenance energy (Chapter 5) needs for very long in the absence of growth. The real significance of the glycogen reserve seems to be that it can become enormously increased when growth is restricted by other than the depletion of the carbon and energy source. Under these conditions the cell converts available substrate into large quantities of intracellular glycogen, sequestering it from competing organisms in a form readily used when growth is again possible.

The small portion of the cell that consists of molecules of low molecular weight—approximately 8.3 femtograms—consists of many hundreds of different kinds of molecules. In general, they are (1) precursors (building blocks) of macromolecules (amino acids, nucleotides, sugars, fatty acids), (2) metabolic intermediates on the way to becoming precursors or to being excreted, (3) cofactors of enzymes, and (4) polyamines bound to DNA. Many—perhaps most of them—are known (Table 3).

Inorganic ions complete our inventory of small molecules found in bacterial protoplasm. The major anions depend to some extent on the anionic composition of the medium, but they always include phosphate, and usually sulfate and chloride. Potassium (K^+) is the major cation in most cases (many bacteria seem able to dispense with sodium) and is involved in active adjustment of the internal osmotic pressure. Ammonia (NH_4^+), magnesium (Mg^{2+}), calcium (Ca^{2+}), and iron (Fe^{2+}, Fe^{3+}) are present in significant amounts; manganese (Mn^{2+}), molybdenum (Mb^{2+}), cobalt (Co^{2+}), copper (Cu^{2+}), and zinc (Zn^{2+}) are found in quite small amounts, bound to a few enzymes. Altogether the inorganic molecules of the cell amount to only 1% of the dry mass, but their functions (indicated in Table 4) are essential for growth.

One comes away from this overview of the chemical composition of the bacterial cell with a number of impressions. First, for all its smallness, the bacterial cell is chemically complex. This complexity is of a manageable order; techniques now exist that enable one to resolve and study complex mixtures of molecular species, and one can foresee the complete definition of the molecules present in a bacterial cell in the near future. Second, the composition of the bacterial cell is extremely orderly. A given strain growing in a given environment achieves a predictable steady-state composition and readjusts that composition in a predictable and precise manner when the environment changes. Finally, though there are some molecules unique to bacteria (lipopolysaccharide, peptidoglycan, and teichoic acid), on the whole there is a

Table 3. Small molecules present in a bacterial cell growing in glucose minimal medium

Molecule	Approximate number of kinds
Amino acids, their precursors and derivatives	120
Nucleotides, their precursors and derivatives[a]	100
Fatty acids and their precursors	50
Sugars, carbohydrates, and their precursors	250
Quinones, polyisoprenoids, porphyrins, vitamins, other coenzymes and prosthetic groups, and their precursors	300

[a] Bochner (1982) has detected 55 nucleotides (42 identified, 13 unknown) in significant amounts in acid-soluble extracts of *Salmonella typhimurium* that were resolved by a new system of two-dimensional chromatography; these do not include many of the precursors of nucleotides.

Table 4. Inorganic ions present in the bacterial cell growing in glucose minimal medium

Ion	Function
K^+	Principal cation, cofactor for certain enzymes
NH_4^+	Principal form of inorganic N for assimilation
Mg^{2+}	Cofactor for a large number of enzymes
Ca^{2+}	Cofactor for certain enzymes
Fe^{2+}	Present in cytochromes and other enzymes
Mn^{2+}	Cofactor for several enzymes
Mo^{2+}	Present in several enzymes
Co^{2+}	Present in vitamin B_{12} and its coenzyme derivatives
Cu^{2+}	Present in several enzymes
Zn^{2+}	Present in several enzymes
Cl^-	Not required for many bacteria
SO_4^{2-}	Main source of S in most media
PO_4^{3-}	Participant in many metabolic reactions

high degree of commonality between the biochemistry of bacteria and of other organisms—a fundamental observation often referred to as the "unity of biochemistry."

We shall have frequent occasion to refer to these three aspects of bacterial cell composition: its *complexity, orderliness,* and *commonality.*

ORGANIZATION OF THE PROCARYOTIC CELL

It was only with the development in the 1950s of techniques for preparing very thin sections of bacteria (100 nm or less in thickness) and the availability of the electron microscope that the ultrastructure of bacterial cells was revealed (cf., Costerton, 1979). A properly prepared and stained section, such as the one through a cell of *E. coli* shown in Figure 12A, can reveal almost all of the important cellular structures and organelles. When such micrographs became available, it was quickly recognized that the internal structure of bacterial cells was strikingly different from that of other cells.

The PROCARYOTIC body plan of bacterial cells is a unique design that bears little similarity to the EUCARYOTIC plan shared by all cells of animals, plants, algae, fungi, and protozoa. Procaryotic cells lack any internal UNIT MEMBRANES, that is, phospholipid bilayer membranes, inside the cell membrane, and their genetic material is never assembled into the complex structures—chromosomes—found in eucaryotic cells. As a consequence, procaryotes contain no endoplasmic reticulum, golgi apparatus, lysosomes, mitochondria, chloroplasts, mitotic spindle fibers, or membrane-bounded nuclei. Their overall intracellular appearance is that of densely packed cytosol surrounding an amorphous, rather fibrous-appearing, less electron-dense NUCLEAR REGION. Functional as well as organizational features distinguish procaryotic cells from eucaryotic cells, and many of these are listed in Table 5.

Being procaryotic means being small, lacking internal membranes and membrane-bounded compartments, having intimate contact between DNA and the abundant protein-forming machinery, lacking complex organelles, concentrating organized electron transport systems at the cell surface, having highly developed transport mechanisms that actively concentrate a variety of nutrient solutes, and possessing genetic material, a cell division apparatus, and organelles of locomotion of minimal structural complexity. The ensemble of these features creates an unmistakable picture: the procaryotic cell shown in Figure 12 is a cell that appears customized for a high metabolic rate leading to rapid cell growth and division. Indeed, the metabolic rate (amount of substrate or oxygen consumed per hour per unit cell mass) of procaryotic cells is typically 10- to 100-fold higher than that of eucaryotic cells.

Table 5. Comparison of procaryotic and eucaryotic cell organization[a]

Structural/functional feature	Procaryotes	Eucaryotes
Endoplasmic reticulum	Absent	Present
Nuclear membrane	Absent	Present
Golgi apparatus	Absent	Present
Lysosomes	Absent	Present
Mitochondria	Absent	Present
Chloroplasts	Absent	Present in plants
Glyoxosomes	Absent	Sometimes present
Microtubules	Absent	Present
Cell wall with peptidoglycan	Present, except mycoplasma and archaebacteria	Absent
Ribosomes	70 S structure	80 S structure
Number of chromosomes	1	> 1
Chromosome with histones	Absent	Present
Nucleolus	Absent	Present
Genetic exchange by conjugation	Plasmid-mediated, unidirectional	By gamete fusion
Phagocytosis	Absent	Sometimes present
Pinocytosis	Absent	Sometimes present
Cellular endosymbionts	Absent	Sometimes present
Ameboid motion	Absent	sometimes present
Cytoplasmic streaming	Absent	Present
Site of electron transport	Cell membrane	Organellar

[a] Adapted from Stanier (1976).

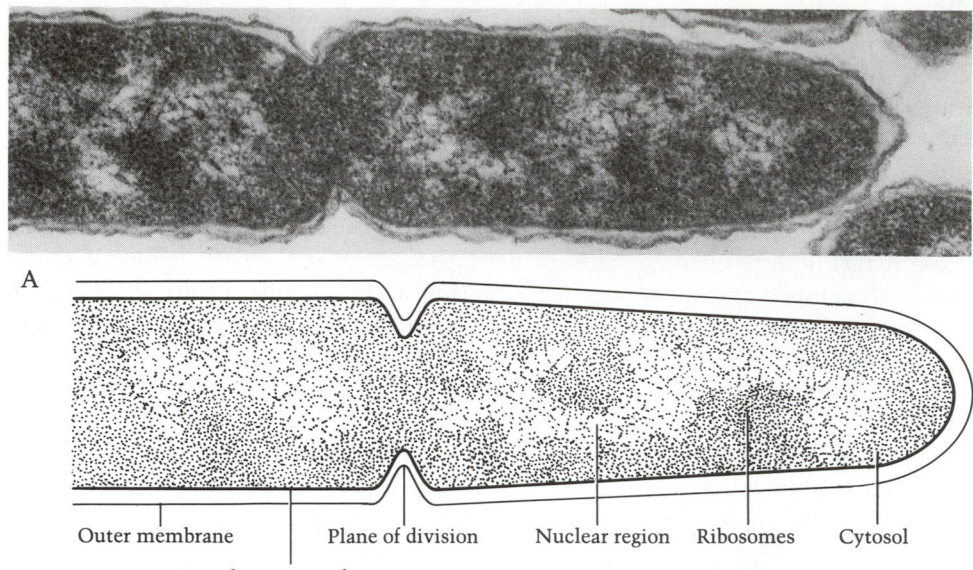

A

| Outer membrane | Plane of division | Nuclear region | Ribosomes | Cytosol |

Cytoplasmic membrane

Figure 12. A. Transmission electron micrograph of a thin section of a dividing cell of *E. coli* B/r. The locations of some important subcellular structures are indicated on the accompanying diagram. The thin peptidoglycan layer that lies between the outer membrane and the cytoplasmic membrane is not revealed. The section is approximately 50 nm thick. ×23,500. (Courtesy of Jack Pangborn.) B. Shadowed electron micrograph of *Vibrio alginolyticus* showing the pattern of insertion of flagella at the cell surface. This organism, like *E. coli*, produces numerous unsheathed peritrichously arranged flagella; unlike *E. coli*, it produces, in addition, a single, sheathed, polar flagellum (arrow). ×12,500. (From de Boer, 1975.) C. The pili of *E. coli*. The bacterial cell that is covered with numerous appendages is a genetic donor connected to a recipient cell (without appendages) by an F pilus, which is necessary for the transfer of genes to the recipient (see Chapter 4). The F pilus is labeled along its length by special bacteriophage particles that infect donor cells through the F pilus. The numerous other appendages on the donor are called Type 1 pili and have no role in conjugation, but they are required for *E. coli* to colonize the intestine and to cause diarrhea. This class of bacterial pili are sometimes called somatic pili. The specimen was negatively stained with 1% phosphotungstic acid. (Courtesy of Charles C. Brinton, Jr.) D. Nucleoids in *E. coli*. Cells in exponential growth in tryptone broth were rapidly deposited onto silver membrane filters under constant supply of oxygen; they were then rapidly frozen. The ice (at −50°C) was substituted with acetone containing OsO_4. Samples were then imbedded in Epon, thin-sliced, and stained with uranyl acetate. The lower copy of the micrograph has been labeled to delineate the two compact nucleoids (N) surrounded by the ribosome-rich cytosol (C) of this cell. Freeze-substitution leads to preparations with more compact nucleoids than those in specimens fixed by aldehyde, and reveal changes in nucleoid appearance with different physiological conditions. (Micrograph supplied by J. Hobot, W. Villiger, J. Escaig and E. Kellenberger, manuscript in preparation.)

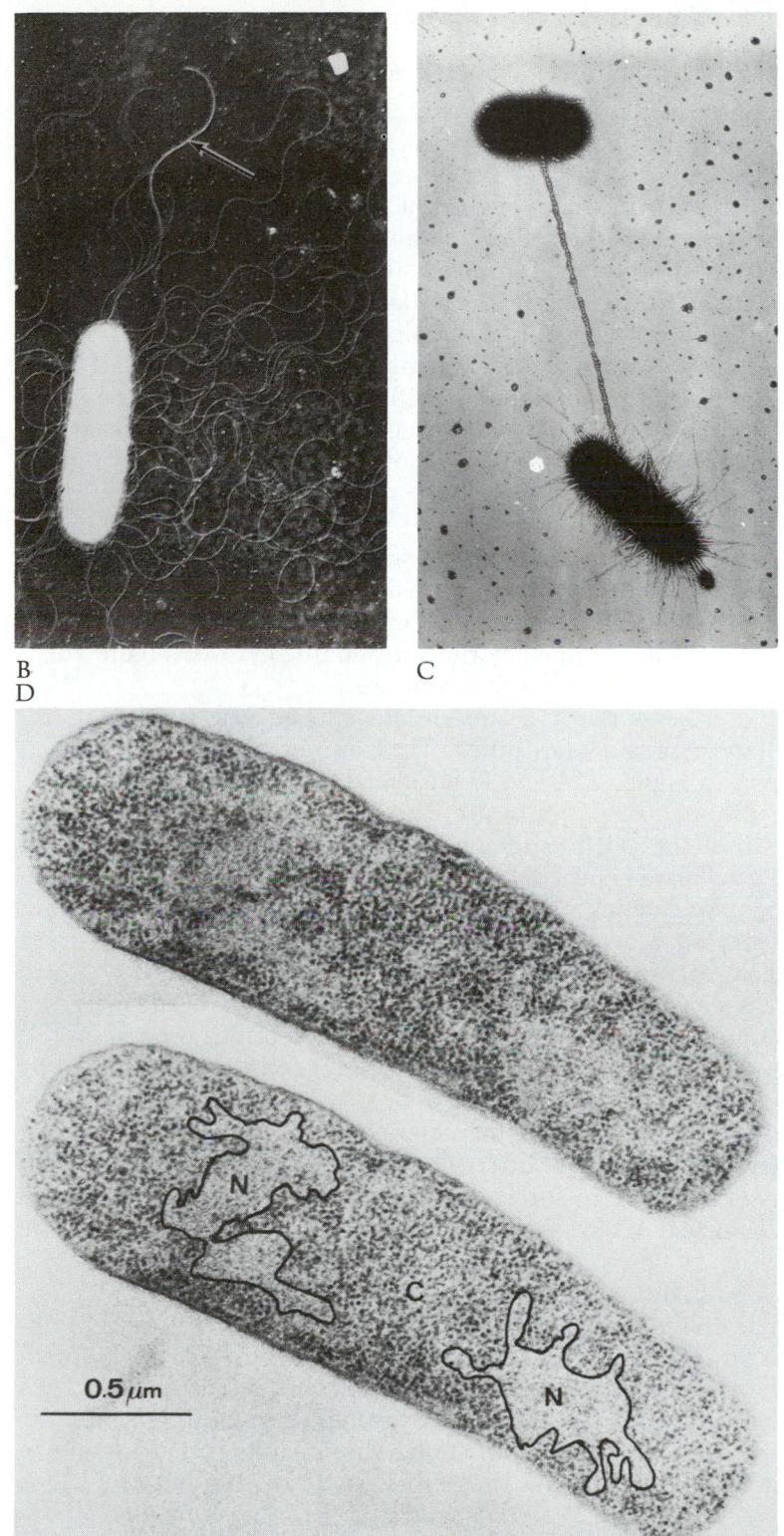

B

C

D

0.5 μm

A closer look at Figure 12A reveals the important macromolecular structures. The ENVELOPE that bounds the cell can be seen to be composed of an inner CELL MEMBRANE surrounded by a WALL. The cell membrane appears as a double layer, a pattern that is typical for the phospholipid bilayers or unit membranes common to all cells. The wall is more complex. Its innermost portion is a thin layer of peptidoglycan—the saclike macromolecule that confers structural strength and helps determine cellular shape. The outermost portion of the wall resembles the cell membrane in appearance and in certain chemical and physical properties; it is therefore commonly called the OUTER MEMBRANE. In contrast to the peptidoglycan layer, the outer membrane is a significant barrier to the passage of macromolecules and certain small molecules. Thus, the region between the cell membrane and the outer wall layer has a unique solute composition differing significantly from that of the cytosol and that of the surrounding medium. This region is called the PERIPLASMIC SPACE and is found (as we shall shortly see) only in the envelope of Gram-negative bacteria like *E. coli*.

Interior to the cell membrane, the dense CYTOSOL appears to be granular because it is densely packed with RIBOSOMES, which at this magnification can be individually resolved as discrete oval-shaped bodies. The NUCLEAR REGION or NUCLEOID appears to ramify through the central region of the cell almost as though its shape were determined by the surrounding cytosol (Figure 12D). Very little additional internal structure can be seen.

Typically, *E. coli* bears two types of surface appendages, FLAGELLA (singular: flagellum) and PILI (singular: pilus). They do not happen to appear in the sections shown in Figure 12A and D but are shown in Figure 12B and C. Flagella and pili, although structurally similar, serve quite different functions. They both arise from the cell membrane and in the main are aggregates of single proteins. Flagella are helical-shaped organelles of locomotion. The function of the straight organelles, the pili, is less clear. Certain types of pili, termed sex pili, play an essential role in the conjugal transfer of DNA from one cell to another. Others play essential roles in the adhesion of bacteria to surfaces, a process considered to be of considerable selective advantage in certain natural environments and essential to some infectious processes.

Although not shown in Figure 12, many bacterial cells are surrounded by an extensive layer of polysaccharide, termed a CAPSULE. This loose layer may extend beyond the cell for many cell diameters and undoubtedly is essential to the cell's ability to compete in a variety of natural environments. It provides one of the modes of adhesion to surfaces so the bacterium can be established in a favorable ecological niche. Sometimes such adhesive interactions are remarkably specific, as when certain bacteria adsorb only to the cells of a particular tissue of a higher animal.

Bacteria can be divided into two broad classes—Gram-positive and Gram-negative—on the basis of their differential abilities to retain a crystal violet–iodine stain when treated with organic solvents (alcohol or acetone). In 1884 a Danish microbiologist, Christian Gram, discovered this staining technique, which has become known as the Gram stain. Those bacteria that retain the stain are termed Gram-positive and those that lose it, Gram-negative. This staining property depends on the morphology and composition of the bacterial envelope, even though Gram-positive and Gram-negative bacteria differ in a number of important aspects in addition to envelope structure. Some bacterial species possess the key structural and compositional features of Gram-positive cells but exhibit a weak or variable ability to stain Gram-positive; such bacteria are considered to be Gram-positive.

The ultrastructure of the *E. coli* wall is characteristic of Gram-negative bacteria: a very thin peptidoglycan layer, the periplasm, and the outer membranelike layer. The structure of *Bacillus subtilis*, a typical Gram-positive bacterium (Figure 13), differs strikingly. Outside the cell membrane is a single, thickened wall layer that is composed largely of peptidoglycan, together with TEICHOIC ACID. The wall is not bounded by the outer membranous layer that can be seen in Gram-negative bacteria, and there is, therefore, no periplasmic space. The diagram in Figure 14 compares the architecture of the Gram-positive and Gram-negative cell envelopes. An excellent overview of the bacterial envelope has been presented by Leive (1980).

The Gram-positive–Gram-negative distinction is a profound one. It reflects not only a fundamental difference in wall structure but other important biochemical, physiological, and genetic differences as well. The majority of, but not all, bacteria are members of these two groups (Schleifer, 1972). The suggestion has been made (Gibbons, 1978) that the remaining groups ought also to be divided on the basis of wall structure. They fall into two classes: (1) those that completely lack walls, the MYCOPLASMAS, and (2) those that produce walls that do not contain peptidoglycan, the ARCHAEBACTERIA.

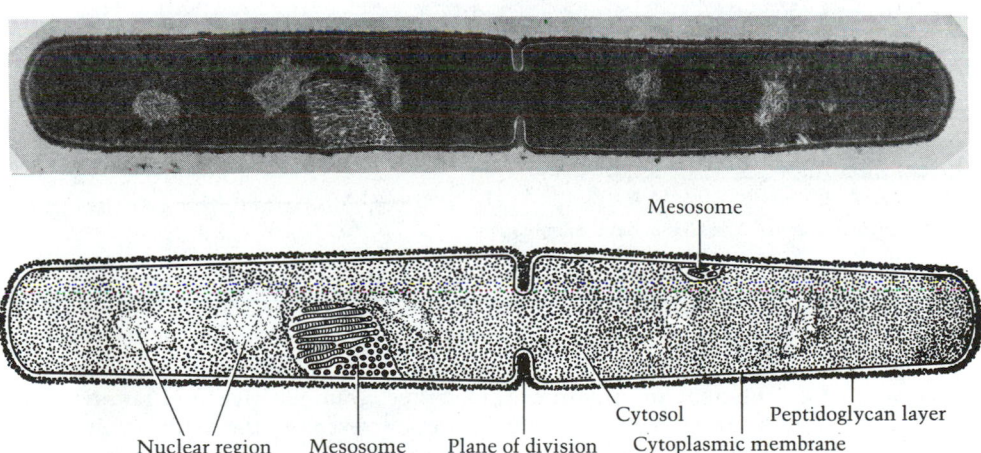

Figure 13. Electron micrograph of a thin section through a dividing cell of *Bacillus subtilis.* Significant intracellular structures are indicated on the accompanying diagram. Mesosomes are tubular involutions of the cytoplasmic membrane; their function is not completely clear but they often are associated with cell division. (Courtesy of Jack Pangborn.)

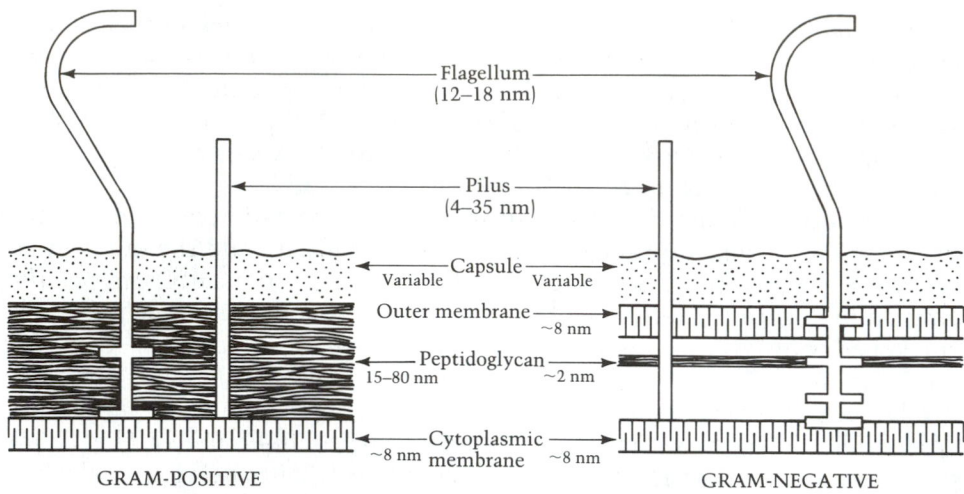

Figure 14. Comparison of the structure of the Gram-positive and Gram-negative cell envelopes showing the major molecular components and their approximate dimensions. The region between the outer membrane and the cytoplasmic membrane of the Gram-negative envelope is called the periplasmic space.

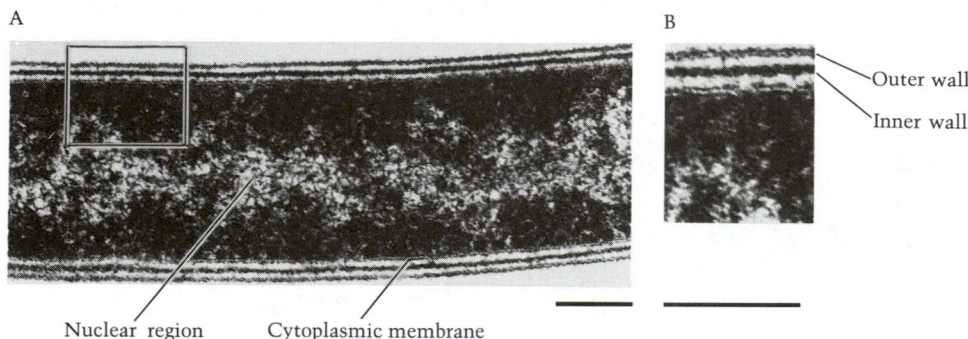

Figure 15. Transmission electron micrograph of a section of the archaebacterium *Methanospirillum hungatii.* Although the variation in ultrastructure among archaebacteria is considerable and no single one can be said to be representative of the group, most can clearly be seen to be quite different in structure from either Gram-positive or Gram-negative bacteria. A. A longitudinal section revealing the cytoplasmic membrane and the double-track appearance of the cell wall. Bar represents 0.13 μm. The nuclear region (N) is similar to that seen in both Gram-positive and Gram-negative bacteria. B. High magnification of the bracketed area of A showing the distinctive structure of the outer wall and inner regions of the double-track wall. Bar represents 0.06 μm. (Courtesy of J. G. Zeikus; Zeikus, 1975.)

The archaebacteria are particularly interesting. They are clearly procaryotic, but on the basis of the sequence of bases in their ribosomal RNA, the chemical composition of their lipids, and their wall structure, the archaebacteria seem to be only quite distantly related to other bacteria (Woese, 1978). Indeed, it appears they are no more closely related to other bacteria than they are to plants or animals. Archaebacteria constitute a small group of procaryotes that are found in specialized or extreme habitats, including the hot acidic niches of the thermoacidophiles, the nearly saturated salt solutions of the extreme halophiles, and the highly reducing (low redox potential) conditions required for the growth of methane producers. Wall composition and ultrastructure is quite variable among archaebacteria (Kandler, 1978). Some contain a heteropolymer, termed PSEUDOMUREIN, that resembles peptidoglycan but lacks two of its characteristic components, muramic acid and diaminopimelic acid. Other archaebacteria have walls composed of protein subunits with or without traces of glucosamine; still others have heteropolysaccharide walls. Because of such chemical diversity, it is not possible to select one organism as having a typical archaebacterial ultrastructure. A thin section of the methanogen *Methanospirillum hungatii* (Figure 15) reveals a typical procaryotic appearance: the lack of internal membranes and the dense ribosomal packing of the cytoplasm. However, the wall, which is composed of protein and is surrounded by an external sheath, differs dramatically in appearance from either a Gram-positive or a Gram-negative cell.

MOLECULAR STRUCTURE OF CELL PARTS

Cell wall

Much of the vigor and toughness of bacteria is related to their cell wall. This structure provides rigid mechanical support, preventing turgor pressure from bursting the cell (osmotic lysis). But the Gram-negative wall of *E. coli* contributes in other ways to the cell's ability to grow rapidly and to survive inhospitable environments (DiRienzo, 1978; Osborn, 1980). Because of its outer membrane layer, it is a barrier to hydrophobic and amphipathic (meaning "having polar and nonpolar regions") molecules (including many antibiotics, detergents, and other toxic molecules), and it protects the inner layers of the envelope, especially its own peptidoglycan, from exposure to hydrolytic enzymes such as lysozyme and lipases. This exclusion function is accomplished without sacrifice of ready permeability to nutrient solutes. The high rate of metabolism of *E. coli* requires that substrates enter the cell rapidly. In fact, the wall not only is freely permeable to small hydrophilic molecules (up to 800–900 molecular weight), but it also contains specific mechanisms that aid in the entry of selected larger molecules. In addition, the wall confers on these cells some resistance to engulfment by phagocytes of an animal host and is an important determinant of social interactions between cells, including the exchange of genetic material.

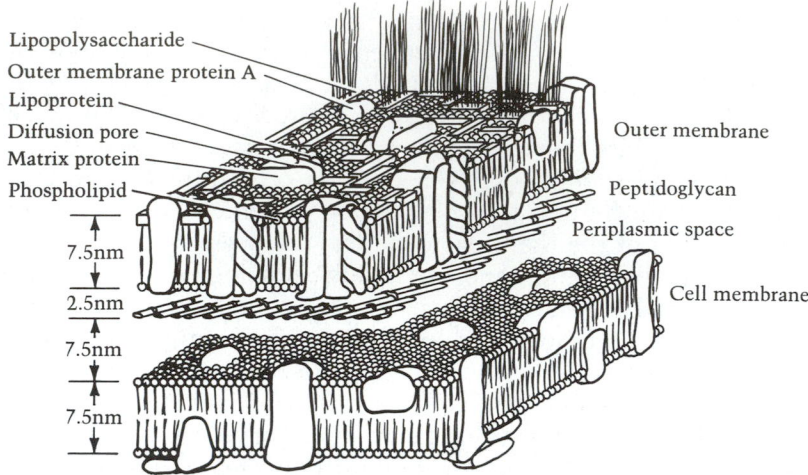

Lipopolysaccharide

Outer membrane protein A

Lipoprotein

Diffusion pore

Matrix protein

Phospholipid

Outer membrane

Peptidoglycan

Periplasmic space

Cell membrane

7.5nm

2.5nm

7.5nm

7.5nm

Figure 16. Diagrammatic representation of the possible molecular architecture of the *E. coli* cell envelope. Only some of the polysaccharide side chains are shown on lipopolysaccharide molecules. (Modified from DiRienzo, 1978.)

The molecular architecture of the Gram-negative wall is summarized in Figure 16; its relationship to other layers of the cell envelope has already been shown in Figure 14.

The peptidoglycan layer is one giant molecule. In *E. coli* it is a single two-dimensional sheet of the general structure shown in Figure 10. All the N-acetylmuramic acid residues are substituted with a tetrapeptide, approximately 25% of which are cross-linked by a direct peptide bond between the free NH_2 group of the diaminopimelic acid residue of one tetrapeptide and the COOH group of the terminal D-alanine of another. There is diversity among different bacteria in the nature and frequency of the cross-linking bridge as well as in the nature of the amino acids at positions 2 and 3 of the tetrapeptide. The same cross-links serve also to join together sheets of peptidoglycan in species that have multiple sheets. The mechanical strength of the peptidoglycan layer derives largely from the fact that it is a single molecule (just as was the fantasy cloth made by Alec Guinness in the film *The Man in the White Suit*); however, also contributing to its strength are the β-1,4 bonds of the polysaccharide backbone, the alternation of D- and L-amino acids in the tetrapeptide, and the possibility of extensive internal hydrogen bonding.

The peptidoglycan layer is not believed to be attached covalently to the underlying cell membrane (except possibly at zones of adhesion that we shall discuss in Chapter 2), but it is firmly bound to the outer membrane of the wall by covalent linkage of some molecules of the outer membrane protein (called MUREIN LIPOPROTEIN) to occasional tetrapeptide residues. Noncovalent bonds between peptidoglycan and certain other outer membrane proteins (called matrix proteins) also contribute to the attachment.

Outer membrane of the Gram-negative wall

The outer membrane consists of two opposed leaflets forming the familiar lipid bilayer structure common to all biological membranes. The inner leaflet consists of phospholipid molecules (Figure 7) of the variety found in the cell membrane. The outer leaflet also may contain some standard phospholipids, but in large measure they are replaced by the molecule unique to the outer membrane, the LIPOPOLYSACCHARIDE (Figure 9). As a result, this is a membrane of unusual asymmetry. The lipid moiety of the lipopolysaccharide, called LIPID A, forms the hydrophobic portion of the leaflet; the CORE polysaccharide with its attached O POLYSACCHARIDE SIDE CHAINS projects outward, and the fatty acid residues project toward the center. Mg^{2+} ions stabilize the lipopolysaccharide molecules. Mutational loss of the ability to make the polysaccharide side chains (the O-antigen) diminishes the cell's resistance to phagocytosis but not its growth; loss of lipid A, including the first sugar residue of the core polysaccharide, however, is lethal. Lipid A is toxic to animals, including humans; it is called ENDOTOXIN, and it plays a significant and still poorly understood role in the infectious diseases caused by Gram-negative bacteria. Depending on dosage and other circumstances, injections of lipid A into experimental animals can cause hemorrhage, fever, abnormal numbers of circulating leukocytes, hemorrhagic necrosis of tumors, circulatory malfunction, abortion, and fatal shock. Septicemias caused by many Gram-negative bacteria (ranging from *E. coli* to the causative agent of bubonic and pneumonic plague, *Yersinia pestis*) owe most of their characteristic and life-threatening symptoms to the biological properties of endotoxin (cf. Davis, 1980).

Were the outer leaflet composed only of lipopolysaccharide, few kinds of molecules could penetrate it to reach the cell membrane. Interspersed throughout the outer membrane, however, are 20–30 proteins that considerably modify its permeability properties (cf., Osborn, 1980). In *E. coli*, more than half of this protein consists of three or four major species. One is called MATRIX PROTEIN or PORIN. There are two—OmpC and OmpF—in the K12 strain of *E. coli*; our reference strain, B/r, makes only OmpF (Omp stands for *o*uter *m*embrane *p*rotein). Porins are unusual proteins. They resist denaturation by sodium dodecyl sulfate (SDS) except at near-boiling temperature, and they also resist enzymatic proteolysis. They have a high content of β structure. Upon isolation they exhibit strong self-association as well as binding to peptidoglycan, forming ordered structures of characteristic form when analyzed by electron microscopy. Studies on these in vitro complexes suggested that porins have largely a trimer structure (three polypeptide subunits of 36,000 molecular weight each) and that the trimers form channels of sufficient diameter to pass molecules of up to 800–900 molecular weight. Because the porins span the membrane, these channels serve as water-filled pores for the rapid transmembrane diffusion of small hydrophilic nutrients, including sugars and disaccharides, amino acids and di- and tripeptides, β-lactams (e.g., penicillin), and inorganic ions.

The outer membrane therefore acts as a molecular sieve, but not all molecules that pass through it do so by diffusing through these nonspecific pores. Many of the "minor" proteins of the outer membrane are specific TRANSPORT PROTEINS or RECEPTORS that facilitate the entry of molecules too large for the regular pores. These proteins are responsible for the entry of iron chelates (Fe^{3+} complexed with ferrichrome or enterochelin), vitamin B_{12}, maltose oligosaccharides, and degradation products of nucleic acids. How these transport proteins function is not known. Perhaps they form specialized pores. Many of them share the physical properties of OmpC and OmpF. Interestingly, the specific transport proteins are made in increased amount when their substrate is present in the medium; the content of individual proteins in the outer membrane can vary considerably.

Other proteins of the outer membrane have important structural functions. One of these, the OmpA protein, plays a role in the integrity of the outer membrane, possibly through interaction with the peptidoglycan. Finally, the most prevalent protein of the outer membrane (indeed, on a numerical basis, the most prevalent protein of the cell) is a small polypeptide, molecular weight 7200, with a lipid substituent on its NH_2-terminal cysteine. This MUREIN LIPOPROTEIN is found in two forms. Both are imbedded in the outer membrane, but one is covalently attached through the ϵ-amino group of the COOH-terminal lysine and the COOH of the diaminopimelic acid residue of peptidoglycan. Lipoprotein helps stabilize the outer membrane as well as bind it to the underlying peptidoglycan.

There is evidence that optimal growth of Gram-negative bacteria requires that both the cell membrane and the outer membrane be in a semifluid rather than crystalline state. Growth stops if more than half of the phospholipid is in an ordered (crystalline or solid) state; growth is possible even if 90% of the lipid is in the disordered, fluid state. One adjustment *E. coli* makes to a lowering of the growth temperature is the synthesis and incorporation of increased amounts of unsaturated fatty acids into its phospholipids (Chapter 5).

The outer membrane, as we have seen, is an effective barrier against the passage of many molecules, and so is the cell membrane. The space between the two is only partially occupied by peptidoglycan; most of it is a compartment called the PERIPLASMIC SPACE. Sequestered within it are a variety of proteins—perhaps 30–50—with three known functions. Some are hydrolytic enzymes (proteases, RNA and DNA nucleases, phosphatases, phosphodiesterases, lactamase) and serve obvious roles in nutrient acquisition. Some are BINDING PROTEINS (for sulfate, galactose, maltose, glutamine, and many other amino acids) that act in cooperation with the cell membrane in ACTIVE TRANSPORT of their ligands. The periplasmic binding proteins have turned out to be also the specific CHEMORECEPTORS that enable *E. coli* to sense nutrients in the environment and move toward them (CHEMOTAXIS). Periplasmic proteins are usually released into the medium when the cell is subjected to osmotic shock (Heppel, 1969) or has its wall removed by EDTA–lysozyme treatment.

The Gram-positive wall, as noted before, is essentially a multisheet peptidoglycan structure. In *B. subtilis* (Figure 13) the thickness of the wall suggests a structure of 40 sheets. Variation in peptidoglycan structure is particularly marked among Gram-positive species, but it is the nature of the teichoic acids present that really confers individuality on Gram-positive strains. As indicated previously (Figure 11), teichoic acids can appear in great variety depending on the nature of the repeating unit of the backbone (ribitol, glycerol, and more complex sugars), the position of the phosphodiester linkage, the nature and position of various substituents (D-alanine, glucose, succinate) as well as the length of the chain. Up to 50% of the wall mass may be teichoic acid, most of which is covalently linked to the 6-OH of occasional *N*-acetyl-muramic acid residues of the peptidoglycan. The rest, called LIPOTEICHOIC ACID, is linked not to the wall but to a glycolipid in the cell membrane. Teichoic acids are the principal surface antigens of Gram-positive cells. Their function is not known. They do bind Mg^{2+}, and it has been suggested that they play a role in supplying this ion to the cell.

Capsule

Many bacteria, Gram-positive and Gram-negative alike, secrete a slime of amorphous structure. In most cases it is polysaccharide, either heteropolymeric (i.e., containing more than one kind of monosaccharide) or homopolymeric. Although its synthesis is optional, depending on the growth medium, and although growth occurs readily without capsule formation, this substance is frequently the major determinant of a bacterial cell's ability to colonize a given niche (such as *Streptococcus salivarius* on your teeth) or avoid phagocytosis (such as *Streptococcus pneumoniae* in your lungs).

Cell membrane

All bacteria possess a cell membrane of the usual bilayer, "unit-membrane" structure (cf., Salton, 1976). That of *E. coli* consists of the phospholipids shown in Figure 7 and proteins. The proteins account for approximately 70% of the mass of the membrane. Sterols are absent, except in certain mycoplasma. Basically it is an osmotic barrier (phosopholipid function) modified by the presence of specific transport systems (protein function). Because all important macromolecular syntheses and virtually all metabolic reactions occur within or internal to the cell membrane, it is the true boundary between the cell interior and the outside. Except for its high content of protein, there is little in its composition to reveal the uniqueness of function of the bacterial cell membrane: its involvement in a multitude of complex metabolic processes to an extent unknown in the rest of the living world.

A diagram of the current view of bacterial cell membrane structure is shown in Figure 16. As already mentioned, the membrane lipids are mostly in a fluid state. Because individual lipid molecules are free to exchange their contacts with each other within each leaflet of the membrane, the whole structure resembles a two-dimensional fluid. Nevertheless there must be considerable order within the membrane with respect to its individual protein components. The membrane proteins are enzymes and carrier molecules that perform highly complex tasks, frequently requiring the coordinated actions of several individual proteins: (1) electron transport and oxidative phosphorylation by the chemiosmotic mechanism; (2) synthesis of complex lipids; (3) synthesis of the components of the wall; (4) replication of DNA; (5) specific transport of nutrients and ions; (6) transduction of sensory signals and energy for motility and chemotaxis; and (7) secretion of proteins.

In many bacteria, especially Gram-positive species, the cell membrane forms large, convoluted invaginations called MESOSOMES (cf., Greenawalt, 1975). It is suspected that mesosomes located near the area where cross wall formation occurs preparatory to cell division—so-called septal mesosomes—participate in some way in DNA replication and cell division. There is evidence that mesosomes located elsewhere—lateral mesosomes—may give rise in some specialized instances to secretory vesicles.

Flagella

The flagellum (plural: flagella) is an organelle of locomotion (cf., Silverman, 1977). Driven by a motor at its base, the helical filament rotates, thereby propelling the cell through the medium. The flagellum is composed of three parts, which have different molecular complexities. Outermost is the long helical FILAMENT, which extends 15 to 20 μm into the medium—many times the length of the cell (Figure 17). Connecting the filament to the BASAL STRUCTURE that anchors the flagellum to the cell envelope is a relatively short (less than 1/100 the length of the filament) curved region termed the HOOK.

The filament is composed of a single type of protein building block termed FLAGELLIN. (The filament of the flagellum of *Caulobacter* is an exception; it is a copolymer of two types of flagellin.) The aggregation properties of flagellin determine the structure of the filament: its diameter (approximately 20 nm in *E. coli*) and the wavelength of its helical turns (approximately 2.3 μm in *E. coli*). Isolated filaments can be dissociated into a solution of flagellin in vitro, and, if the ionic environment is properly adjusted, they spontaneously reaggregate to form filaments virtually indistinguishable from the natural product. The pattern of aggregation of flagellin molecules in the filaments of *E. coli* and the closely related organism *Salmonella typhimurium* is known in some detail. Eleven longitudinal rows of flagellin form the wall of the cylindrical filament. These rows are arranged in a helical pattern with just under 11 subunits being required to make two complete turns.

The helical filament serves the purpose of a propeller. The function of the hook is less clear. Possibly it serves as a universal joint between the motor in the basal structure and the filament. The molecular structure of the hook is quite simple. Like the filament, it is an aggregation of a single type of protein subunit. It is slightly larger in diameter than the filament and has a determined length—approximately 90 nm.

The basal structure, although comparatively small—it constitutes approximately 1% of the mass of the flagellum—is complex. It is composed of 10 to 13 different kinds of protein subunits that aggregate to form a rodlike structure to which (in the case of Gram-negative bacteria) four rings are

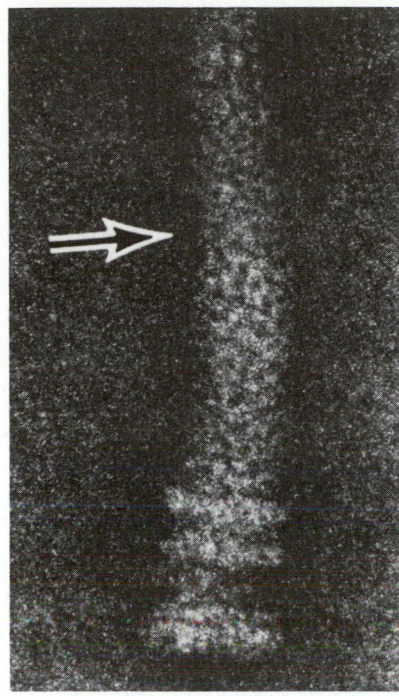

A

Figure 17.

The bacterial flagellum. Fine details of the basal body of the flagellum of *E. coli* are shown by negative staining in A: four rings and the hook (arrow). (From DePamphilis, 1971a.) B is a diagram of the basal body inferred from such photographs. (From DePamphilis, 1971a.) C shows the relationship between the four rings and the envelope of *E. coli*. (From DePamphilis, 1971b.)

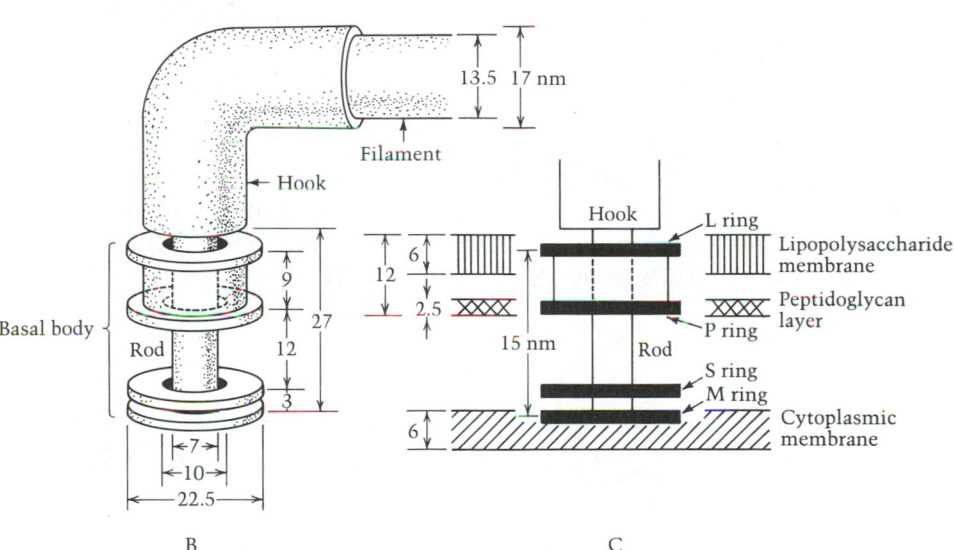

B   C

attached. The rings are associated with, or imbedded in the various layers of the cell envelope. The innermost, the M ring, lies in the cell membrane. The next, the S ring, sits on the cell membrane; the P ring lies in the peptidoglycan layer; and the L ring lies in the lipopolysaccharide-containing outer membrane.

As might be expected, the difference in envelope structure between Gram-positive and Gram-negative bacteria is reflected in a difference of the basal structure of their flagella. Gram-positive bacteria have only two rings—one imbedded in the cell membrane and another associated with the teichoic acid component of the wall.

Flagellar motors are part of the complex behavioral system that enables motile bacteria to direct their net movement toward environments favorable to growth. Flagellar motors alternate between clockwise and counterclockwise rotation. At least in *E. coli*, counterclockwise rotation leads to the formation of a stable bundle of flagella that propels the bacterium smoothly forward; clockwise rotation disperses the bundle and leads to a nonproductive tumbling of the cell (Figure 18). Addition of a chemical ATTRACTANT induces counterclockwise rotation by all motors: a process called EXCITATION. Excitation invokes binding of the attractant molecule to a specific receptor (located in Gram-negative cells either in the periplasm or the cell membrane) in the envelope and transduction of the information that the receptors are occupied to the flagellar motors. Excitation cannot accomplish chemotaxis, but only smooth, undirected swimming.

Next, however, a process called ADAPTATION restores the initial pattern of alternating rotation. Adaptation involves a change in the signal transduced to the motors, and its biochemical nature is just becoming understood. The transducers turn out to be cell membrane proteins that can assume different conformations depending on their state of METHYLATION. In *E. coli* there are three of these METHYL-ACCEPTING CHEMOTAXIS PROTEINS: MCPI, MCPII, and MCPIII are the products of genes called *tsr*, *tar*, and *trg*, respectively. Demethylation of glutamic acid residues of MCP molecules occurs when attractants bind to their specific acceptors. For some attractants, such as aspartate and serine, binding occurs directly to one of the MCP proteins; for others, such as glucose and maltose, binding is mediated by a specific periplasmic binding protein. In either case, binding changes the conformation of the MCP, producing a signal that excites the flagellar motors into their counterclockwise mode. Adaptation results from a gradual enzymatic remethylation of the MCP

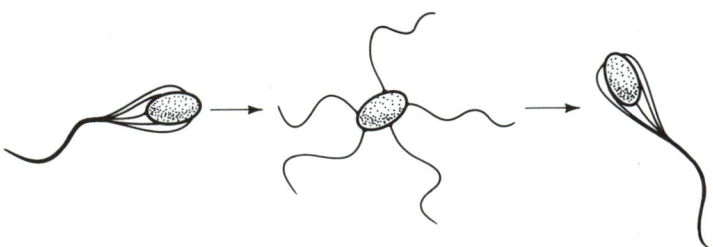

Figure 18. Tumbling caused by dispersion of flagellar bundles. On left, bacterium swims smoothly with all flagellar filaments rotating counterclockwise and forming a stable bundle. In the middle, reversal of rotation disperses the bundle and causes tumbling. On right, return to counterclockwise rotation causes reformation of bundle and smooth swimming again. (After Koshland, 1980.)

molecules. (The methylation is accomplished by a methyltransferase that uses *S*-adenosylmethionine as substrate; an esterase removes the methyl group.)

The principle of directed motility can now be understood. When swimming in an environment with a CHEMICAL GRADIENT of an attractant solute, all cells will alternate between smooth swimming and tumbling because adaptation acts to keep excitation a transient state. However, cells that by chance are swimming in the direction of increased concentration will be subject to increased stimulus and excitation, and cells swimming in the opposite direction will more quickly become adapted to the (decreasing) stimulus. The finite time lag built into the adaptation process makes it possible for cells to compare the concentration of stimulant at one time with what it had been a short time before. In other words, bacteria sense a chemical gradient by a memory rather than, for example, detecting concentration differences between their head and tail.

Figure 19 presents a schematic view of many of the elements of the *E.*

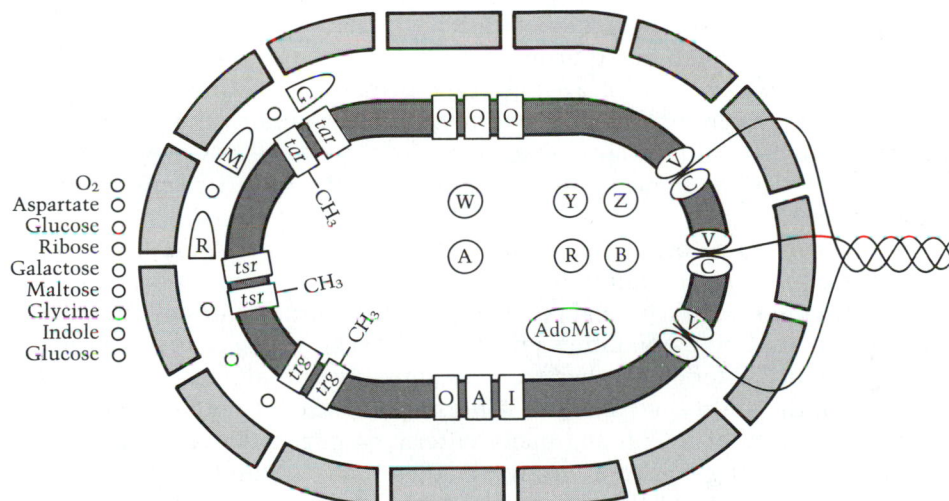

Figure 19. Schematic summary of chemotactic system in *E. coli.* On the left are shown small molecules (O₂, aspartate, etc.) that act as chemoeffectors by passing through the pores of the outer membrane to interact with specific receptor proteins in the periplasmic space (G, M, R, etc.) and in the cell membrane (O, A, I, etc.). Signals from many of these receptors are focused through the *tar*, *tsr*, and *trg* proteins (the so-called methyl-accepting chemotaxis proteins). Other signals are processed by separate pathways (Q). The level of methylation of the *tar*, *tsr*, and *trg* proteins is determined by the level of the chemoeffector, which determines the balance between methylation from *S*-adenosylmethionine (AdoMet) and demethylation by an esterase. The methylation-related signals are processed in some way by the cytoplasmic proteins A, W, R, Y, B, and Z and alter the level of some response regulator that interacts with proteins C and V, which are in intimate contact with the flagellar motors. (After Koshland, 1980.)

coli motility–chemotaxis system. There are many fascinating features of bacterial motility and taxis, including the nature of this molecular sensing and behavioral system, the response to chemical repellents, and to temperature, pH, light, and the gravitational field of the earth. The reader is urged to consult the several fine reviews that cover different aspects of the work on motility and chemotaxis; one of the most recent is by Hazelbauer (1982), and it provides a guide to earlier reviews. Koshland (1980) has written a small but comprehensive book about chemotaxis in *E. coli*, with many references to other organisms. Perusal of the lists of attractants and repellents in Koshland's book provides convincing evidence that the bacterial motility–chemotaxis system has evolved under the selective pressure of moving toward growth-favorable environments, though the role of chemotaxis in facilitating colonization of the gut is interesting because it involves movement toward a surface for attachment rather than toward a nutrient (Freter, 1981).

Pili

Pili (singular: pilus) are a second type of proteinaceous structure that extend beyond the bacterial cell surface. They originate from the cell membrane and extend 10 or more micrometers into the medium (Figure 12C). They are composed of a single type of protein subunit termed PILIN. These units are arranged helically to form a straight cylinder. In some respects, the structure of a pilus is similar to that of a flagellum; however, a pilus is straight rather than helical, its diameter invariant over its length, and it doesn't rotate.

Pili appear to be organelles of attachment and exhibit remarkable specificity in this respect. Certain pili, termed SEX PILI, are produced by male cells and play a vital role in bacterial conjugation; they form the initial attachment between the mating pairs. Other pili, termed TYPE 1 PILI, are involved in the attachment of bacteria to other surfaces, notably the surface of eucaryotic cells. Thus, the ability of *Neisseria gonorrhea* to attach to membranes of the urinary tract, and of *E. coli* and many enteric pathogens to attach to cells of the intestinal lining is essentially the result of specific binding between the pili of these bacteria and components—probably glycoproteins—of the host cell surface. The term ADHESIN has been given to pili that play this role in host-parasite interactions.

Polysomes

The RIBOSOMES that appear to fill the cytosol are complex organelles that catalyze the process of translation of information encoded in messenger RNA (mRNA) into protein. At any moment, 80–90% of ribosomes in the cell are attached to mRNA and are actively engaged in protein synthesis. This structure for synthesizing protein—mRNA with attached ribosomes—is termed a POLYSOME. Remarkable electron micrographs by O. Miller enable one to visualize the role of polysomes in protein synthesis and gene expression. The one shown in Figure 20 illustrates the expression of a single gene. A set

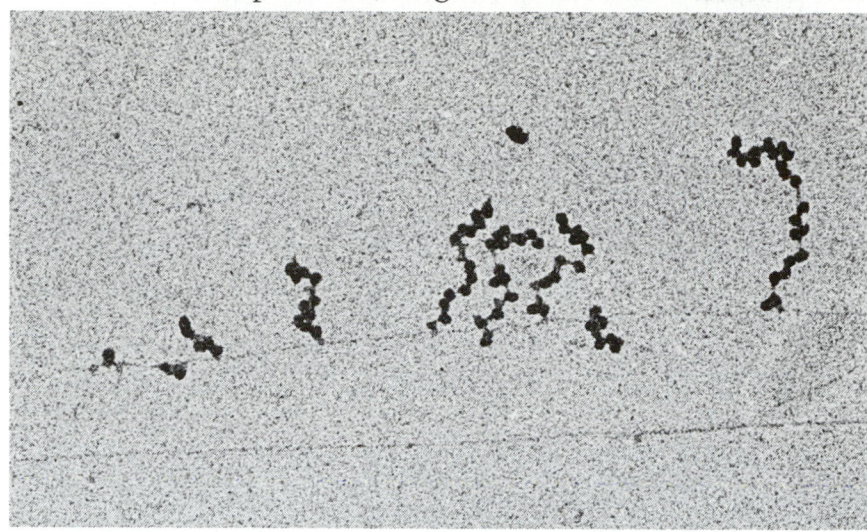

Figure 20. Electron micrograph of polysomes. The micrograph (×78,200) shows the expression of a single gene. The two thin horizontal lines are strands of DNA. Polysomes are attached to the upper strand. Their length increases from left to right; this observation indicates that transcription occurred in the same direction. RNA polymerase molecules can be seen as small objects at the intersection of the polysomes and the DNA. Nascent protein molecules that must be attached to the ribosomes cannot be resolved. (From Miller, 1970.)

of polysomes of increasing length can be seen attached to a segment of DNA. The increasing length of the polysomes (from left to right in the picture) indicate that transcription of DNA into mRNA is proceeding in that direction. Indeed, RNA polymerase (RNA-P), which is the enzyme that catalyzes the process of transcription, can just be discerned as a small object assumed to be at the mRNA–DNA junction. The picture shows that polysomes form (i.e., ribosomes attach to mRNA) while mRNA is still being synthesized. Because it is known that ribosomes attach near the end of the mRNA molecule, the picture establishes that they move down the mRNA molecule, synthesizing protein during this same period. The nascent molecules of protein attached to the individual ribosomes are not visible in the micrograph.

As noted earlier, it is the procaryotic body plan that permits translation and transcription of a gene to occur simultaneously. In eucaryotic cells, transcription occurs within the membrane-bounded nucleus, whereas translation occurs largely on the endoplasmic reticulum in the cytoplasm. We shall see later how the procaryotic arrangement combined with mRNA degradation permits speedier adjustments of gene expression.

An individual procaryotic ribosome sediments in a centrifugal force field at a velocity of 70 Svedberg units (S) and on this basis is designated a 70 S ribosome. (Eucaryotic ribosomes are 80 S.) It is a complex ribonucleoprotein

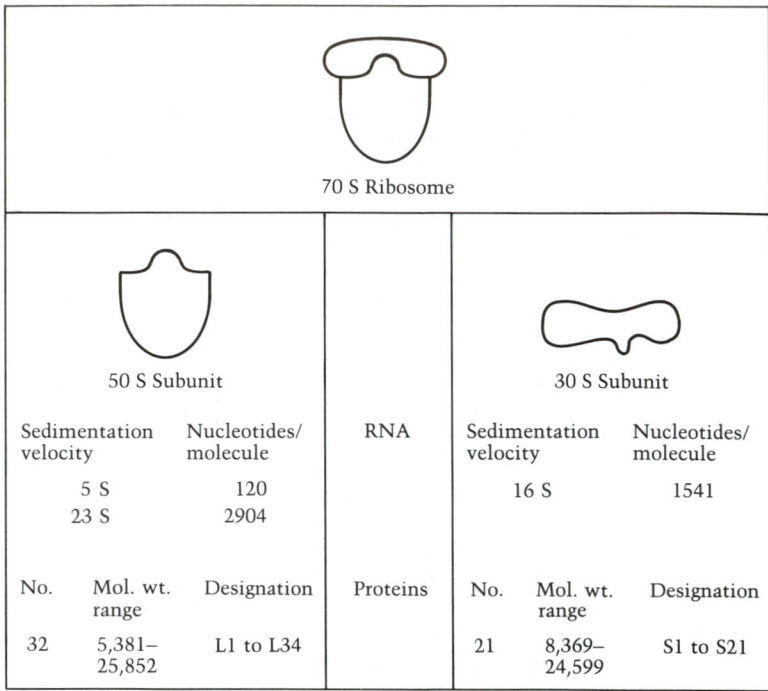

| 70 S Ribosome | | | | | |

50 S Subunit		RNA	30 S Subunit		
Sedimentation velocity	Nucleotides/ molecule			Sedimentation velocity	Nucleotides/ molecule
5 S	120			16 S	1541
23 S	2904				

No.	Mol. wt. range	Designation	Proteins	No.	Mol. wt. range	Designation
32	5,381– 25,852	L1 to L34		21	8,369– 24,599	S1 to S21

Figure 21. The molecular structure of ribosomes from *E. coli*: based on the molecular weights of 19 proteins from the small subunit and 29 proteins from the large subunit, the primary structures of which are known. The discrepancy between numbers of proteins (32) in the large subunit and their designation (up to L34) derives from the fact that one protein, L8, which was thought to be unique when numbered as a spot on a chromatogram, is now known to be an association of L7, L12, and L10 and that another, L26, is identical to S20. Only small amounts of L26 remain associated with the large subunit upon dissociation of the 70 S ribosome.

particle, the molecular composition and structure of which are now known in some detail (Figure 21). It can be dissociated in vitro into two subunits—a 50 S unit and a 30 S unit—by reducing the concentration of Mg^{2+}. This dissociation occurs in vivo each time the synthesis of a protein is completed. Each subunit is composed of RNA and protein. The smaller one contains a single molecule of RNA (16 S) and 21 different proteins, which are designated S (for small subunit) 1 through 21. The large subunit contains two RNA molecules (5 S and 23 S) and 32 different proteins, designated L (for large subunit) 1 through 34 (Figure 21). With the exception of L7 and L12, which differ only by the presence in L7 of an acetyl group of the NH_2-terminus, all ribosomal proteins are encoded by different genes and thus differ in primary structure. A number of proteins are modified posttranscriptionally by acetylation or methylation.

Some aspects of the relative position of the molecular components in the ribosomal subunits are now understood in some detail as a result of studies in which adjacent proteins are chemically joined by bifunctional reagents and of studies in which the binding of specific protein components to particular regions of RNA is measured. Moreover, the complete sequence of nucleotides in the 5, 16, and 23 S RNA molecules is now known, data that enables informed speculation about the degree of secondary structure (double-strandedness) in the molecules. Certain facts are clear. The RNA molecules contain significant double-stranded regions; they serve as an extended core to which specific proteins are bound at specific regions; the relative positions of proteins is definite and fixed.

Although most studies on the bacterial ribosome have been on those from *E. coli*, sufficient information on ribosomes from other bacteria is now available to reveal a remarkable similarity among all procaryotic ribosomes. This similarity is dramatically illustrated by the finding that proteins taken from 30 S ribosomes of *E. coli* will bind specifically to 16 S RNA of other enteric bacteria, of species of *Bacillus*, and of such diverse bacteria as *Anacystis nidulans* (a cyanobacterium), *Clostridium pasteurianum* (a strict anaerobe), *Chromatium vinosum* (a photosynthetic bacterium of the purple sulfur group), and even certain archaebacteria.

Many details of ribosome structure are summarized in a recent symposium report (Chambliss, 1979).

Nuclear body

The DNA of the bacterial cell is contained in a central region termed a NUCLEUS, NUCLEAR BODY (or NUCLEAR REGION), or NUCLEOID. These terms emphasize the fundamental structural distinctions between this central region and the membrane-bounded nucleus of eucaryotic cells. The terms are not strictly synonymous but unfortunately are used so loosely as to be currently interchangeable. A brief discussion here may help the reader to cope with the confusion. The advent of useful mutant methodology in the 1950s (Chapter 4) led to a great interest in verifying that the genetic apparatus of bacteria was similar to that of other cells because of the hope that exploring physiological genetics in bacteria would yield information of general validity. Hence, the term *nucleus* was optimistically adopted and is still widely used even though there is known to be no structural justification for such a broad use of an otherwise quite definitive term. (The careful reader may even find nucleus used in parts of this book.) *Nucleus* is poor usage, but probably not eradicable. *Nuclear body* (or *nuclear region*) is the preferable term for what one sees in situ with preparations stained to reveal DNA or with thin sections viewed in the electron microscope. *Nucleoid* could, on linguistic grounds, refer to the same structure, but it was first applied to the structure one obtains after gentle lysis of the cell; the nucleoid is a folded DNA molecule complexed with RNA and protein. As information about nucleoids increases, it is likely that this term will come to refer to both the in vivo and in vitro structure, as we shall presently see.

The shape of the nuclear body in an electron micrograph (Figure 12A, D) appears amorphous, an appearance that belies its complex molecular structure. It can also be visualized by light microscopy under appropriate conditions. If the RNA is removed from the cytoplasm by treatment with either HCl or RNase, the nuclear region seems to consist of discrete bodies when stained with basic dyes and can be shown to be composed of DNA because they stain with the DNA-specific Feulgen reagent. They can also be detected by conventional phase-contrast microscopy, and they become very distinct if the cells are suspended in a concentrated protein solution that matches the refractive index of the cytoplasm, thereby enhancing contrast between the nuclear region and its surroundings.

Bacteria contain from one to several nuclear bodies, depending on the rate at which the cells have been grown (Chapter 6). On the basis of DNA content, it is assumed that each nuclear body is a bacterial chromosome in a replicating or nonreplicating state. It has been known for some time that all bacterial genes are linked in a single structural unit—the CHROMOSOME—and that bacteria behave genetically as HAPLOID organisms. Thus, the several nuclear bodies seen in certain bacterial cells must each contain the same haploid chromosome. Much evidence indicates each chromosome is attached to the cell membrane (Leibowitz, 1975).

Since the early 1960s it has been known that bacteria must contain a single circular chromosome, because extensive genetic studies on *E. coli* had established that each known bacterial gene was genetically linked on both sides to other known genes. In 1963 John Cairns succeeded in establishing the chromosome's circularity by direct observation, a remarkable feat when one considers that the width of the chromosome has molecular dimensions and its length is macroscopic. Cairns accomplished this by autoradiography. He labeled the chromosome by growing a culture in a medium containing [^{3}H]thymine. After lysing the cells under conditions that minimize turbulence and thereby minimize shearing forces on the fragile molecule, proteins were dissociated from the chromosome and it was allowed to settle on a cellulose nitrate (membrane) filter. Placement of the dried filter against X-ray film and subsequent development of the film revealed the form of the chromosome by the pattern of dots caused by disintegration of ^{3}H during the period of exposure. The circular chromosome, thus seen, was approximately 1 mm long, a value that corresponds closely to the average amount of DNA in a nucleoid, corrected for the added amount of DNA present as a consequence of the chromosome being partially replicated. Indeed, the chromosome depicted in Figure 22 has a theta (Θ) shape, showing it to be partially replicated.

It was immediately recognized that the molecular organization of the chromosome within the nuclear region must be complex: the length of the chromosome is approximately a thousand cell diameters. The studies of D.E. Pettijohn (1976) and A. Worcel (1972) have revealed certain aspects of the organization of the chromosome within the nucleoid. By gentle lysis of cells, nuclear structures (Figure 23) can be isolated that contain approximately 80% DNA, 10% RNA (mostly nascent), and 10% protein (mostly RNA polymerase).

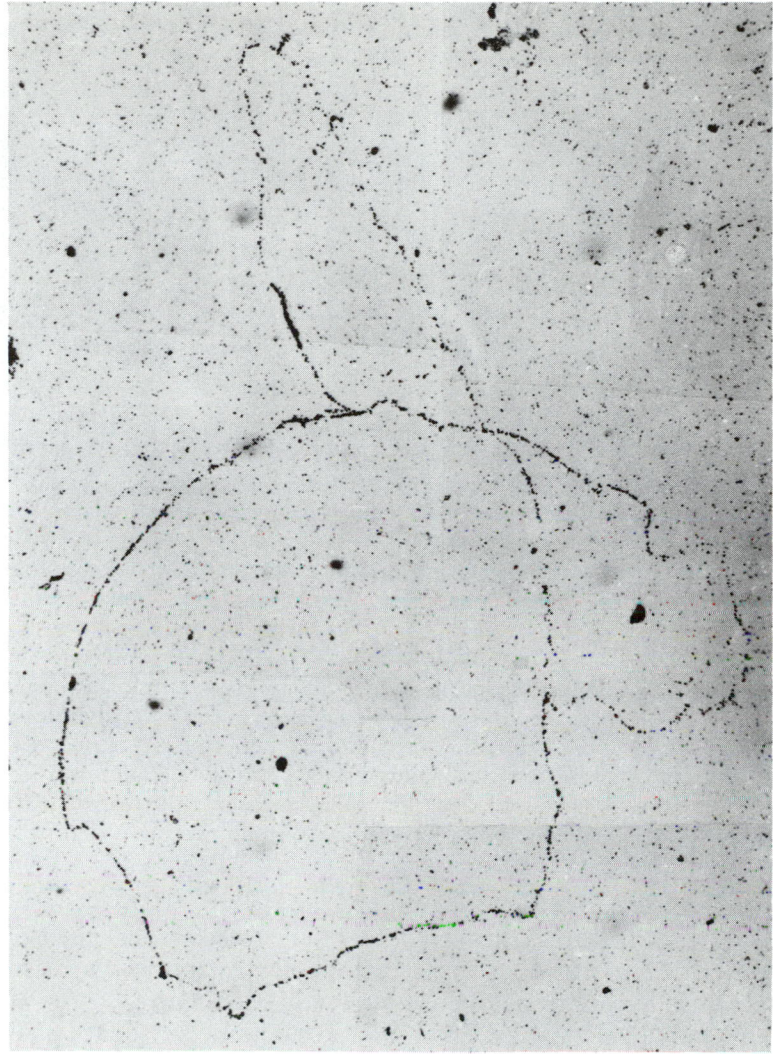

Figure 22. The bacterial chromosome. This autoradiograph of a molecule of *E. coli* DNA was prepared from a culture that had been labeled for approximately 1.8 generations with [³H]thymidine. (From Cairns, 1963.)

Figure 23 shows an isolated intact nucleoid and one with its DNA partially released. Measurements were made to determine how many single-strand breaks (nicks) had to be introduced to relax all the supercoiling of the DNA in the isolated structure. The numbers of such nicks define regions of topological constraint on the rotation of the double helix, and each of these domains can be imagined to correspond to the folds of DNA shown in Figure

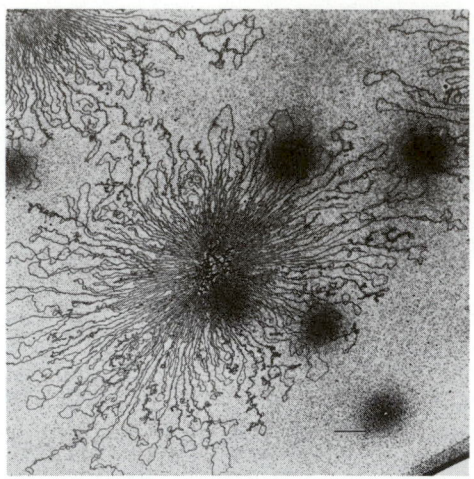

Figure 23. The bacterial nucleoid. On the left are three nucleoids isolated from *E. coli* and visualized by scanning electron microscopy (Appendix A). The preparation was made by gentle lysis of cells under conditions chosen to minimize the possibility of disrupting the non-covalent bonds that hold the nucleoid components in a compact structure. ×11,600. (From Van Ness, 1979; micrograph courtesy of D. Pettijohn.) On the right is a nucleoid (viewed by transmission electron microscopy) from which the DNA has been partially released. The nucleoid was surface spread with cytochrome *c*, stained with uranyl acetate and rotary shadowed with platinum-palladium to reveal the loops of DNA released from the nucleoid by this treatment. The bar represents 1 μm. (From Kavenoff, 1976; micrograph courtesy of D. Pettijohn.)

23. These measurements and other studies suggest that the isolated structures contain 12 to 80 loops of supercoiled DNA, forming a reasonably compact structure by additional helical coiling of each loop.

Sinden and co-workers (1981) have obtained evidence about the nature of the *E. coli* nucleoid without removing it from the cell. They made use of the fact that trimethylpsoralen photobinds to double-stranded DNA at a rate proportional to its torsional tension and that this agent can be used to measure such tension in DNA in the living cell. Nicks can be introduced by γ irradiation of the whole cells, thereby making it possible to learn the physical nature of DNA in the nucleoid in situ. The number of nicks per genome equivalent length of DNA required to relax 95% of the tension is consistent with there being 43 ± 10 domains of tension, or folds, in this length of DNA. Nucleoids with partially replicated chromosomes in rapidly growing cells could have three times this number of folds. The number of domains increased with growth rate; and, in contrast to results obtained in vitro, the domains did not depend on the existence of nascent RNA in the intact cell.

The implications of nucleoid structure and supercoiling domains for DNA replication and function during growth are discussed in Chapters 2, 3 and 6.

Inclusion bodies

Frequently, granular INCLUSION BODIES are seen in thin sections of procaryotes (cf., Shiveley, 1974). In the case of *E. coli*, they are electron-dense bodies of uneven appearance, ranging from 20 to 100 nm in diameter and consisting of GLYCOGEN, which is a highly branched polymer of glucose. Glycogen granules accumulate in response to excess carbon in the medium when the source of nitrogen, sulfur, or phosphorus is limiting or when the pH is low. Upon carbon starvation, these granules disappear. Thus, glycogen granules are a storage form of carbon. A number of bacteria, including many enteric bacteria, the spore-formers, and the cyanobacteria, store carbon as glycogen. Others, including pseudomonads and rhizobia, store carbon in granules of POLY-β-HYDROXYBUTYRATE. Still others, including certain photosynthetic bacteria, have the capacity to store carbon in either of these forms. Some bacteria, for example, *Acinetobacter*, appear to be incapable of storing carbon in insoluble granules.

Carbon-rich compounds are not the only nutrients that bacteria store. Some, but not *E. coli*, store phosphate as granules of polyphosphate and sulfur as elemental sulfur.

OTHER ORGANIZATIONAL VARIATIONS

In addition to the profound differences in the organization of procaryotic cells that distinguish the three major classes—Gram-positive bacteria, Gram-negative bacteria, and the archaebacteria—a number of other important, but less fundamental, structural variations occur among procaryotes. A study of them constitutes the richly detailed field of general microbiology or bacterial diversity—not the topic of this book, but one extensively considered in others (e.g., Stanier, 1976). The more obvious variations in procaryotic cell organization that reflect adaptations to specialized environments or modes of metabolism are briefly summarized here.

Shapes of procaryotic cells include round ones [coccus (singular), cocci (plural)], rod-shaped ones [bacillus (singular), bacilli (plural)], helical or spiral-shaped ones, and comma-shaped ones (vibrios). Some bacteria are filamentous and quite similar in appearance to fungi. Certain procaryotic cells have tube-like appendages extending from their surfaces (prostheca). The selective or adaptive advantage of one or another cell shape is by no means clear, although there are suggestions that prosthecate bacteria, for example, are exclusively associated with aquatic environments. On the other hand, it is quite clear that cell shape is determined by the molecular architecture of the cell wall (cf., Henning, 1975). In all cases except the archaebacteria and those cells devoid of a wall (the mycoplasma group), shape is determined by the peptidoglycan layer. Mutations affecting synthesis of peptidoglycan alter cell shape. Enzymatic (with lysozyme) or other treatments (e.g., exposure to penicillin) that remove the peptidoglycan layer obliterate normal cell shape and generate spherical structures, which, if osmotically protected, continue to grow and divide.

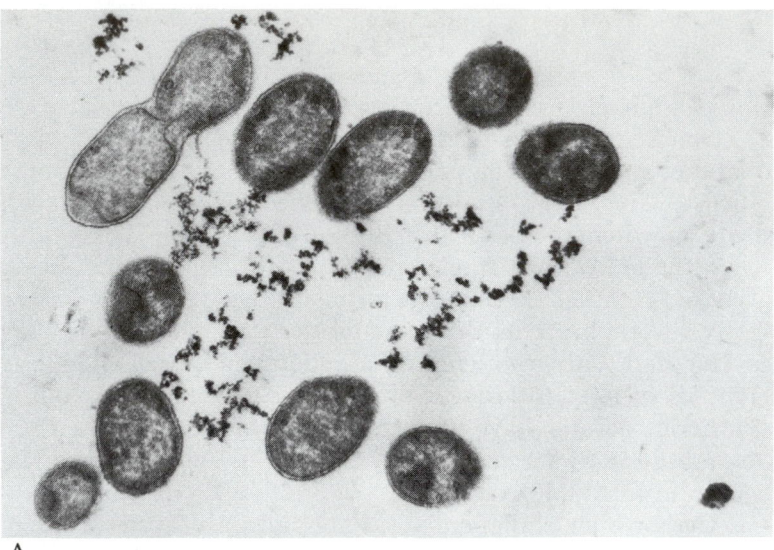

A

Figure 24. Elaboration of the cell membrane. A. Thin section through a group of cells of the photosynthetic bacterium *Rhodopseudomonas sphaeroides*. Invaginations of the cell membrane on which the photosynthetic pigments are located are seen as circular objects near the envelope. These are continuous with the cell membrane, but the connecting membranes are only rarely seen in thin sections. Cells were grown anaerobically with dim (200 foot-candles) illumination,

Flagella, the organelles of locomotion for all motile bacteria except those that can glide by an unknown mechanism over solid surfaces, are quite uniform in structure but are arranged on the cell surface in a variety of patterns. They may occur in large numbers uniformly distributed over the cell surface, an arrangement termed peritrichous and typical of *E. coli* and *Bacillus subtilus*. They may be restricted to the ends of rod-shaped cells (lophotrichous or polar) or may be inserted near the ends (subpolar). They may occur singly on a cell, in small numbers or in quite large numbers. In all cases, they are helical in shape and propel the cell by rotating at the point of insertion in the envelope.

Certain procaryotic cells have specialized organelles within them. Gas-filled vacuoles provide flotation for some aquatic forms. Highly heat-resistant bodies (actually specialized cells) termed endospores are formed by several groups. The endospore has been the object of intensive genetic, chemical, and physiological investigations because of its extreme resistance to heat (most survive extended periods of boiling) and because of its attractiveness as a model system in which to study morphogenesis.

The cell membrane, which is the cell's osmotic barrier and site of ATP generation by electron transport, is quite elaborate in form and exaggerated in amount in certain procaryotes. In photosynthetic procaryotes, where the

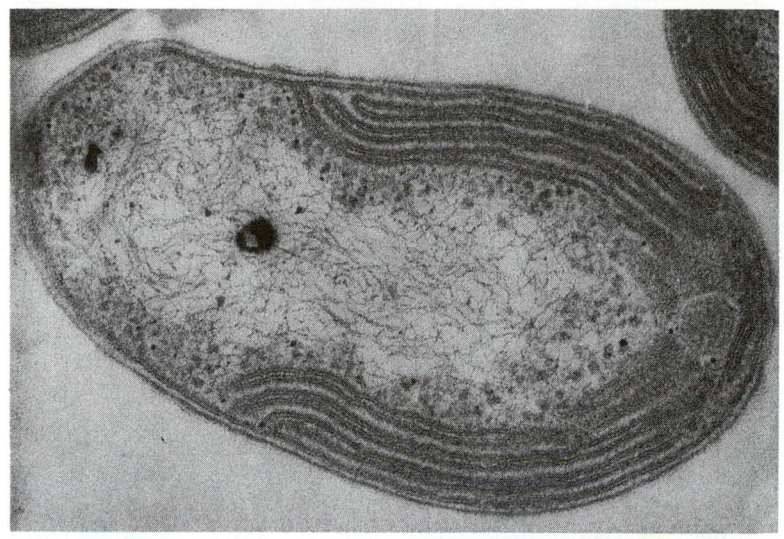

B

fixed in 2% glutaraldehyde with 2% unbuffered $KMnO_4$, and stained with lead citrate and uranyl acetate. (Courtesy Anthony Macaluso.) B. Thin section through a single cell of the autotrophic bacterium *Nitrobacter winogradskyi.* Invaginations of the cell membrane form a laminated cap at one pole. The three-layered appearance of the membrane is clearly seen. (Courtesy of S. W. Watson; from Watson, 1971.)

cell membrane has the added role of accommodating photosynthetic pigment and associated components of photophosphorylation, the area of the membrane is greatly enlarged, a circumstance that causes invaginations of it into the cytoplasm. In one photosynthetic procaryote (*Rhodopseudomonas sphaeroides*), transmission electron micrographs of thin sections reveal them to be somewhat regularly arranged hollow vesicles (Figure 24A). Similarly, some organisms that generate ATP by respiration of inorganic substrates by reactions with low negative free energy yields possess elaborately extended folds of membrane into the cytoplasm. *Nitrobacter winogradskyi*, which generates ATP by oxidizing nitrite to nitrate ion, contains folded extensions of the cytoplasmic membrane that run the length of the cell (Figure 24B).

SUMMARY

1. Bacteria generally have been selected in nature for high metabolic rates and rapid growth. Their ease of cultivation and the ready availability of powerful techniques for biochemical and genetic analyses make bacteria an excellent paradigm for the study of cell growth.
2. Bacteria are chemically complex. Over 95% of their mass is macromole-

cular. Most of their molecules resemble those of other cells (exceptions: lipopolysaccharide, peptidoglycan, teichoic acid); the "unity of biochemistry" includes bacteria.

3. Bacteria are small and their genomes are the smallest known for any cell. The maximum number of kinds of proteins (~ 3000) that can be made by a bacterium such as *E. coli* is small enough that they can be resolved and cataloged by modern methods.

4. Bacteria display biochemical orderliness. Their chemical composition is reproducible under any given growth condition, and the adjustments they make when growth conditions change are similarly predictable.

5. The bacterial cell body plan differs in fundamental ways from that of other cells. The procaryotic organization of bacteria helps make their rapid growth possible.

6. The interior of the bacterial cell is plain; its envelope, fancy. Great biological individuality is expressed in the molecular variations of the enveloping layers of the cell. This individuality is a basis for grouping bacteria into major classes (mycoplasma, archaebacteria, Gram-positive bacteria, Gram-negative bacteria), as well as for distinguishing individual strains (by antigenic type) within species.

7. Many cellular functions are concentrated at the periphery of bacteria. Besides conventional functions (selective permeability, secretion, active transport, and osmotic balance), the bacterial envelope is involved in electron transport, oxidative phosphorylation, DNA replication, motility and chemotaxis, and the synthesis of selected proteins.

Many features of the composition, organization, and ultrastructure of the bacterial cell remain to be elucidated. We know of one-third, at most, of the possible gene products of *E. coli*, fewer in other bacteria. Details must still be supplied about the structural interaction between chromosome and cell membrane, the architecture of outer membrane pores, and the arrangement of proteins involved in specific transport, the structure of chemiosmotic pumps involved in energy transduction, and the spatial organization of proteins involved in chemoreception and motility. Nevertheless, the progress of just the last decade leaves one with the impression that near-total elucidation of the structure of a complete cell is achievable with reasonable effort. The benefits to general cell biology of this accomplishment will be considerable.

REFERENCES

Bachmann, B. J. 1972. Pedigrees of some mutant strains of *Escherichia coli* K-12. Bacteriol. Rev. 36:525.

Buchanan, R. E. and N. E. Gibbons (eds.). 1974. *Bergey's Manual of Determinative Bacteriology*, 8th Edition. Williams and Wilkens, Baltimore.

Cairns, J. 1963. The bacterial chromosome and its manner of replication as seen by autoradiography. J. Mol. Biol. 6:208.

Chambliss, G., G. R. Craven, J. Davies, K. Davis, L. Kahan and M. Nomura (eds.). 1979. *Ribosomes: Structure, Function, and Genetics*. University Park Press, Baltimore.

Costerton, J. W. 1978. How bacteria stick. Sci. Am. 238:86.

Costerton, J. W. 1979. The role of electron microscopy in the elucidation of bacterial structure and function. Ann. Rev. Microbiol. 33:459.

Cronan, J. 1978. Molecular biology of bacterial membrane lipids. Ann. Rev. Biochem. 47:163.

Davis, B. D., R. Dulbecco, H. N. Eisen and H. S. Ginsberg. 1980. *Microbiology*, 3rd Edition. Harper & Row, Hagerstown, MD.

DiRienzo, J. M., K. Nakamura and M. Inouye. 1978. The outer membrane proteins of Gram-negative bacteria: biosynthesis, assembly, and functions. Ann. Rev. Biochem. 47:481.

Freter, R., P. C. M. O'Brien and M. S. Macsai. 1981. Role of chemotaxis in the association of motile bacteria with intestinal mucosa: in vivo studies. Inf. and Immun. 34:234.

Gibbons, N. E. and R. G. E. Murray. 1978. Proposals concerning the higher taxa of bacteria. Int. J. of Systematic Bacteriol. 28:1.

Goldfine, H. 1972. Comparative aspects of bacterial lipids. Adv. Microb. Physiol. 8:1.

Gottschalk, G. 1979. *Bacterial Metabolism*. Springer-Verlag, New York.

Greenawalt, J. W. and T. L. Whiteside. 1975. Mesosomes: membranous bacterial organelles. Bacteriol. Rev. 39:405.

Hazelbauer, G. L. and S. Harayama. 1982. Sensory transduction in bacterial chemotaxis. Int. Rev. Cytol. In press.

Henning, U. 1975. Determination of cell shape in bacteria. Ann. Rev. Microbiol. 29:45.

Heppel, L. A. 1969. The effect of osmotic shock on release of bacterial proteins and on active transport. J. Gen. Physiol. 54:953.

Kandler, O. and H. Konig. 1978. Chemical composition of the peptidoglycan-free cell walls of methanogenic bacteria. Arch. Microbiol. 118:141.

Koshland, D. E., Jr. 1980. *Bacterial Chemotaxis as a Model Behavioral System*. Raven Press, New York.

Leibowitz, P. J. and M. Schaechter. 1975. The attachment of the bacterial chromosome to the cell membrane. Int. Rev. Cytol. 41:1.

Leive, L. and B. D. Davis. 1980. Cell envelope; spores, Chp. 6. In *Microbiology*, 3rd edition, B. D. Davis, R. Dulbecco, H. N. Eisen, and H. S. Ginsberg. Harper & Row, Hagerstown, MD.

Luria, S. 1960. In *The Bacteria*, Vol. I, I. C. Gunsalus and R. Y. Stanier, eds. Academic Press, New York.

Miller, O. L. Jr., B. A. Hamkolo, and C. A. Thomas, Jr. 1970. Visualization of bacterial genes in action. Science 169:392.

Osborn, M. J. and H. C. P. Wu. 1980. Proteins of the outer membrane of Gram-negative bacteria. Ann. Rev. Microbiol. 34:369.

Pettijohn, D. E. 1976. Prokaryotic DNA in nucleoid structure. CRC Crit. Rev. Biochem. 4:175.

Phillips, T. A., P. L. Bloch and F. C. Neidhardt. 1980. Protein identifications on O'Farrell two-dimensional gels: Locations of 55 additional *Escherichia coli* proteins. J. Bacteriol. 144:1024.

Powell, E. O. 1956. Growth rate and generation time of bacteria, with special reference to continuous culture. J. Gen. Microbiol. 15:492.

Salton, M. R. J. and P. Owen. 1976. Bacterial membrane structure. Ann. Rev. Microbiol. 30:451.

Schleifer, K. H. and O. Kandler. 1972. Peptidoglycan types of bacterial cell walls and their taxonomic implications. Bacteriol. Rev. 36:407.

Shiveley, J. M. 1974. Inclusion bodies of procaryotes. Ann. Rev. Microbiol. 28:167.

Silverman, M. and M. I. Simon. 1977. Bacterial flagella. Ann. Rev. Microbiol. 31:397.

Sinden, R. R. and D. E. Pettijohn. 1981. Chromosomes in living *Escherichia coli* cells are segregated into domains of supercoiling. Proc. Natl. Acad. Sci. USA 78:224.

Stanier, R. Y., E. A. Adelberg and J. Ingraham. 1976. *The Microbial World*, 4th Edition. Prentice-Hall, Englewood Cliffs, NJ.

Woese, C. R., L. J. Magrum and G. E. Fox. 1978. Archaebacteria. J. Mol. Evol. 11:245.

Worcel, A. and E. Burgi. 1972. On the structure of the folded chromosome of *Escherichia coli*. J. Mol. Biol. 81:127.

Chapter Two

Chemical Synthesis of the Bacterial Cell: Assembly

INTRODUCTION

Return now to the observation with which we began our study of the bacterial cell in Chapter 1. Through the light microscope we can actually watch *E. coli* cells make faithful copies of themselves from glucose and some salts. The whole process takes just 40 minutes at 37°C.

In this chapter and the next we shall address the chemistry of this process. Later, in Chapters 6, 7, and 8, we shall discuss how the individual chemical events are brought into balance. We shall concentrate on one specific bacterium in order to present as complete a picture as possible of how a cell grows. The choice of *E. coli* is made simply on the basis of the wealth of information we have about it; the selection is otherwise arbitrary. Because different bacteria exhibit great similarity in many of the biochemical processes of growth, our treatment is not as restricted as might at first appear. We shall highlight their metabolic unity and indicate the areas of metabolism in which diversity is the rule.

OVERVIEW OF METABOLISM AND SOME GENERALIZATIONS

The enormity of the problem of describing the individual chemical reactions necessary to transmute glucose and inorganic ions into a living cell has until recently discouraged all but the most superficial or partial of treatments of the chemistry of cell growth. This situation has changed with the introduction of powerful biochemical and genetic techniques that greatly enhance our ability to gain new information. Also, the recognition of certain patterns and principles considerably simplifies the conceptual task of describing cell growth, and it will be useful, therefore, to precede our analysis of the

49

chemistry of growth with some of these generalizations that help organize the myriad biochemical details.

Evidence presented in Chapter 1 indicated that the cell consists of structures formed by specific associations of macromolecules; that the latter are made from small, rapidly used pools of low-molecular-weight compounds; and that these pools are constantly replenished by chemical synthesis from precursors derivable ultimately from glucose or certain other carbon sources. Accordingly, on the basis of their primary function in growth, the 2000–3000 individual reactions in the cell can be usefully organized as (1) assembly, (2) polymerization, (3) biosynthetic, and (4) fueling reactions (Figure 1).

1. ASSEMBLY REACTIONS involve the chemical modification of macromolecules, their transport to prespecified locations in the cell, and their association to form cellular structures: envelope, appendages, nucleoid, polysomes, inclusions, and enzyme complexes. In some cases structures form by the spontaneous association of molecules with mutual affinity (SELF-ASSEMBLY); in other cases special devices must aid the process (DIRECTED ASSEMBLY).

2. POLYMERIZATION REACTIONS consist of directed, sequential linking of activated molecules into long (in some cases, branched) polymeric chains. All the macromolecules are produced from a moderately large number of BUILDING BLOCKS that include 20 amino acids, 8 nucleotides, and numerous sugar derivatives and fatty acids. Polymerization of protein, RNA, DNA, and glycogen occur within the cell, whereas lipopolysaccharide, capsular polysaccharide, and peptidoglycan undergo their final polymerization steps on the outer surface of the cell.

3. BIOSYNTHETIC REACTIONS produce the building blocks needed for the polymerization reactions; they produce also coenzymes and related metabolic factors, including signal molecules called ALARMONES. There are hundreds of biosynthetic reactions. They occur in functional units called BIOSYNTHETIC PATHWAYS, each consisting of from one to a dozen sequential reactions that produce one or more building blocks. Pathways are easily recognized. They are usually controlled *en bloc* and in some cases consist of enzymes made from a single mRNA transcribed from a set of contiguous genes. Interestingly, all the cellular biosynthetic pathways begin with one or another of a small, elite group of compounds called PRECURSOR METABOLITES. Some pathways begin directly with these metabolites, others indirectly by branching from an intermediate or end product of a related pathway. There are just 12 of these precursor metabolites, yet they lead to something on the order of 75 building-block and coenzyme products. With few exceptions, the routes from precursor metabolites to end products are the same in all bacteria, but species differ in the number of pathways they possess.

4. FUELING REACTIONS produce the 12 key precursor metabolites, the ATP, and the reducing power needed for biosynthesis. They also produce the ATP needed for polymerization and all other endergonic processes of the cell (transport, motility, repair, etc.). These fueling reactions include all

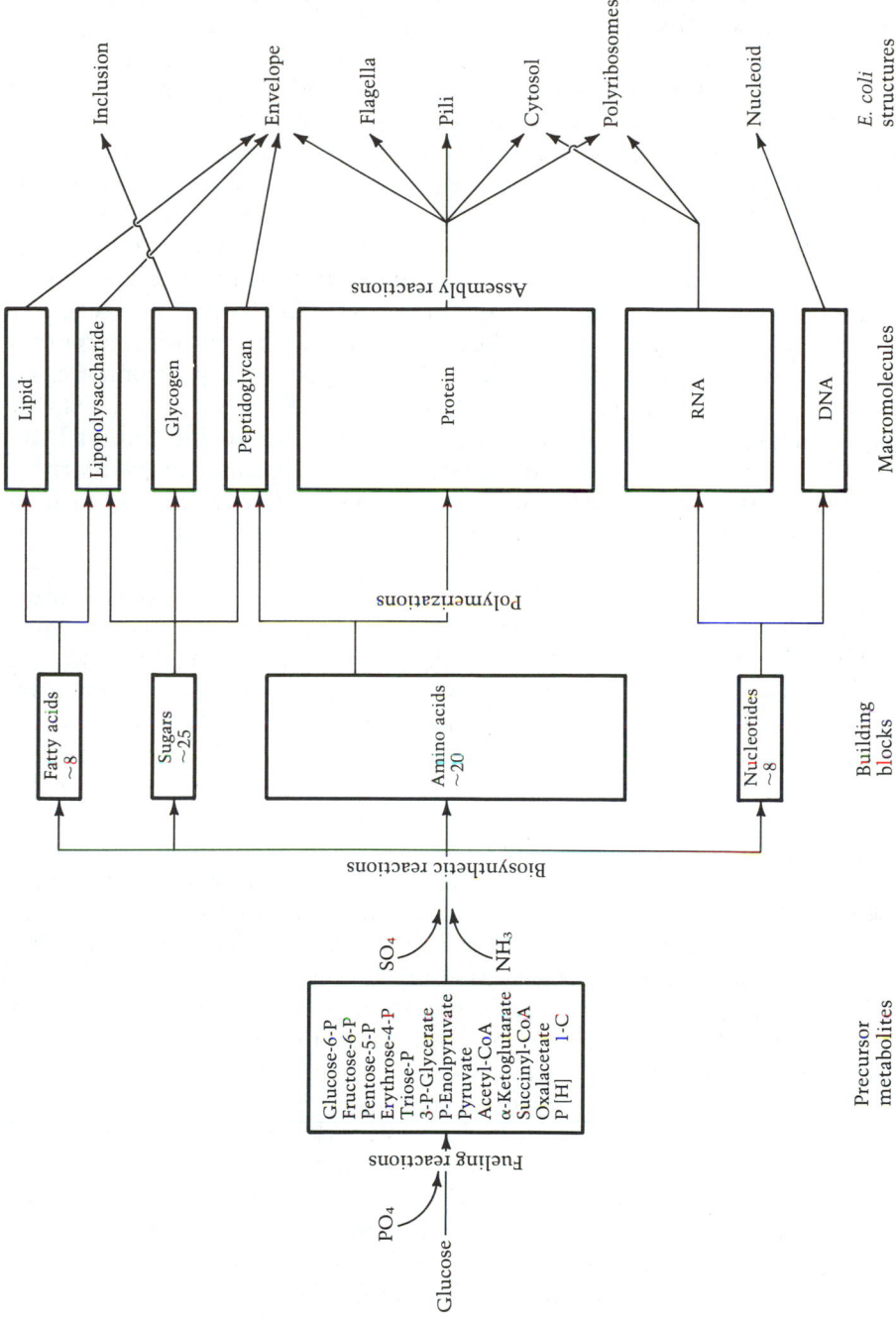

Figure 1. Overall view of metabolism leading to chemical synthesis of *E. coli* from glucose. Boxes are proportional to need in making *E. coli*.

those pathways referred to as CATABOLIC (degrading and oxidizing substrates) and AMPHIBOLIC (serving to produce both energy and precursor metabolites). Certain of these reactions, notably the CENTRAL PATHWAYS (glycolysis, TCA cycle, pentose pathway) are well nigh universal, but otherwise there is a rich diversity of fueling reactions among different bacteria. Even in a single species such as *E. coli*, the reactions used when the cell is presented with glucose are somewhat different than when it is presented with a different substrate, such as acetate or succinate. And there are important differences in the fueling reactions when *E. coli* grows with glucose in the absence of O_2 compared with those reactions occurring in its presence. Some organic compounds not usable by *E. coli* as a source of carbon and energy may be used by certain other species of bacteria that have evolved to occupy special ecological niches. Finally, in some cases evolution has led to alternative pathways for handling the same substrate (e.g., glucose by either Embden-Meyerhof glycolysis or the Entner-Doudoroff pathway). The sole (but significant) biochemical unity in these fueling reactions lies in their function for growth: the provision of carbon-containing precursor metabolites, reducing power, and ATP.

The assignment of certain cellular reactions to one or another of these four metabolic processes is somewhat arbitrary. REPAIR REACTIONS (which involve chiefly DNA) can be grouped with polymerization reactions. SOLUTE TRANSPORT REACTIONS enable uptake and retention of nutrients. Those involving transport of building blocks, precursor metabolites, CO_2 and inorganic forms of sulfur, phosphorus, and nitrogen can be grouped with biosynthetic reactions because of their primary function in these pathways. Transport reactions of K^+ and other cations for purposes of osmotic balance might also be grouped as biosynthetic, largely on the basis of their being energy-requiring. ASSIMILATION of sulfur and nitrogen occurs in certain biosynthetic reactions; that of phosphorus occurs as an integral part of the fueling reactions. METABOLIC TURNOVER REACTIONS degrade proteins, nucleic acids, and other macromolecules, particularly during starvation conditions; these also may be viewed as fueling reactions. Finally, there are several reactions not directly involved in growth but clearly of survival value in hunting nutrients and avoiding toxins: MOTILITY and CHEMOTAXIS.

It has been customary to trace the metabolism of glucose (or another substrate molecule) through the central pathways, into biosynthesis, and thence into macromolecules. The reverse order makes more sense because it is the formation of macromolecules and their assembly into cellular structures that define for us (and the cell) the nature and magnitude of the biosynthetic demands for building blocks and energy. From these demands, in turn, are derived the requirements that must be met by the fueling reactions.

ASSEMBLY: INTRODUCTION

All the chemical reactions that transmute glucose into macromolecules proceed silently and invisibly. What one actually sees while watching a bacterial cell grow is cellular structures assembling from macromolecules. It is the condensation of individual molecules into highly ordered planar and globular assemblages that makes the chemical process of growth visible to us.

Though grossly visible under the microscope, assembly is the least understood step of metabolism. Complete analysis of how structures such as the wall and membrane become assembled must wait until the complex molecular composition and anatomy of these structures have been learned. Fortunately, physical and genetic tools appropriate to investigating assembly are becoming available. Rapid progress can be expected in the 1980s.

Loosely put, cell structures form by the condensation of macromolecules. In the broadest sense this is correct, but it begs two important questions: (1) Do structures emerge simply from the spontaneous aggregation of their molecular parts, or does special construction machinery guide and energize the process? (2) Are macromolecules made at the sites where they assemble, or must they be transported to these sites? These questions are, of course, interrelated.

It turns out that the answers to these questions are unique for each specific structure, and the alternatives posed—self-assembly versus guided assembly and on-site construction versus delivery of prefabricated units—may not be exclusive alternatives for any structures. Assembly involves a combination of these modes. Self-assembly, wherein the macromolecules to be assembled provide sufficient information and energy for the process, is a major element in the morphogenesis of most cellular structures. Many assembly processes, however, are guided and/or energized by preexising structures or by special scaffolding or other morphogenetic elements. In certain cases large structures are constructed, not from macromolecules, but directly from precursors too small to be termed macromolecules; in other cases elaborate channels and vectorial reactions carry ready-made macromolecules (usually proteins) from their sites of synthesis to their site of assembly.

It is convenient to organize assembly into four parts: (1) the envelope with its appendages, (2) the nucleoid, (3) the polysomes, and (4) the rest of the cytosol. This grouping is based on the nature of the assembly processes in each, as well as on the nature of certain global control systems governing these structures (see Chapters 6, 7, and 8). The macromolecules needed for each of the cell's component parts were described in Chapter 1; these requirements are summarized here (Figure 2) for easy reference.

ASSEMBLY: THE ENVELOPE

Assembling the envelope is a difficult and exciting problem. This structure (Figure 3) is extremely complex in procaryotes because of its many specialized functions (mechanical support, protection against phagocytosis, cell–cell recognition, motility, chemosensing, selective permeability, solute transport, electron transport, oxidative phosphorylation, secretion, and DNA segregation). Yet, some general features of bacterial envelope assembly are thought to apply to all living cells. The possibility of combining powerful genetic analysis with biochemical and morphological studies has made bacterial envelope assembly an active area of contemporary research.

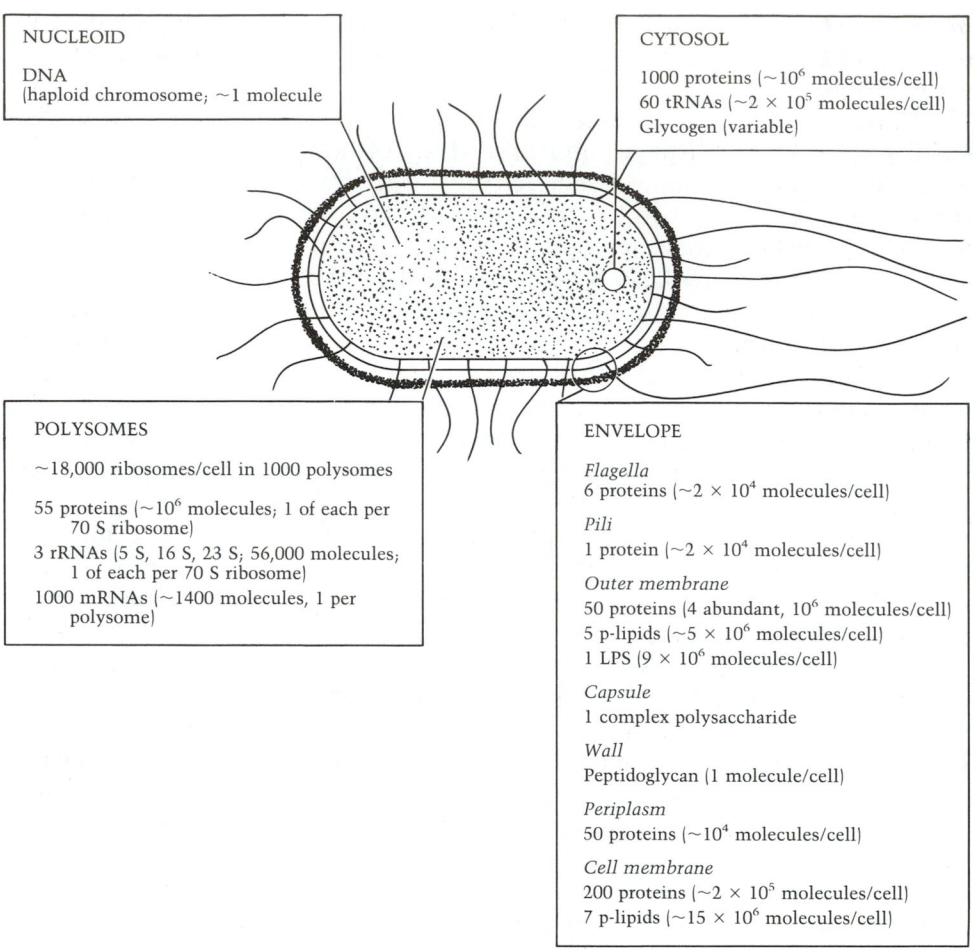

NUCLEOID

DNA
(haploid chromosome; ~1 molecule

CYTOSOL

1000 proteins (~10^6 molecules/cell)
60 tRNAs (~2×10^5 molecules/cell)
Glycogen (variable)

POLYSOMES

~18,000 ribosomes/cell in 1000 polysomes

55 proteins (~10^6 molecules; 1 of each per
   70 S ribosome)
3 rRNAs (5 S, 16 S, 23 S; 56,000 molecules;
   1 of each per 70 S ribosome)
1000 mRNAs (~1400 molecules, 1 per
   polysome)

ENVELOPE

Flagella
6 proteins (~2×10^4 molecules/cell)
Pili
1 protein (~2×10^4 molecules/cell)
Outer membrane
50 proteins (4 abundant, 10^6 molecules/cell)
5 p-lipids (~5×10^6 molecules/cell)
1 LPS (9×10^6 molecules/cell)
Capsule
1 complex polysaccharide
Wall
Peptidoglycan (1 molecule/cell)
Periplasm
50 proteins (~10^4 molecules/cell)
Cell membrane
200 proteins (~2×10^5 molecules/cell)
7 p-lipids (~15×10^6 molecules/cell)

Figure 2. Macromolecular composition of structures of an average *E. coli* cell.
Based on data presented in Chapter 1, Table 1.

The individual problems are fascinating. Orderliness of membrane and wall structure combined with their chemical complexity imply a highly programmed mode of formation. The envelope is multilayered, and all but the innermost layer (the cell membrane) would seem to be assembled, technically, outside the cell. Topologically the layers are closed surfaces; for cell integrity and viability they must to some degree always remain closed and yet still be able to expand through addition of new material. The cell membrane, wall, and (in Gram-negative cells) outer membrane must grow coordinately. Proteins must be incorporated into their correct final location: the cell membrane, the periplasm, the outer membrane, or the external environment. Each generation the assembly of the envelope must in some way facilitate division of the cell into two viable offspring.

How all this is brought about with speed, accuracy, and adaptive flexibility is not fully understood. But it is possible to construct a reasonable picture of some of the larger aspects of envelope growth and to point in what directions some of the unknown answers must lie. The weaknesses and vagueness of parts of this story constitute challenges for current research.

An overview of growth of the envelope is as follows: the cell membrane largely self-assembles, growing by intercalation of newly made components at special growth regions. The cell membrane directs cell wall formation. Precursors of the cell wall peptidoglycan are transported through the cell membrane on a carrier lipid, undecaprenol phosphate (formerly called bacto-prenol). Polymerization into peptidoglycan chains occurs outside the membrane, and these polymers become cross-linked and attached to the existing peptidoglycan sac. New peptidoglycan can perhaps be added to the existing wall at many sites, but most addition is particularly active at the equatorial zone. Certain proteins in the cell membrane catalyze the reactions that elongate the peptidoglycan polymer, cross-link it, and determine formation of the cross wall preparatory to cell division. The cell membrane participates also in assembly of the outer membrane. At certain places the cell membrane adheres tightly to the outer membrane. At these zones of adhesion—known as BAYER'S JUNCTIONS—the inner leaflet of the cell membrane seems to be in contact with the outer leaflet of the outer membrane; no periplasmic space and no peptidoglycan intervene. Through these regions, precursors of the

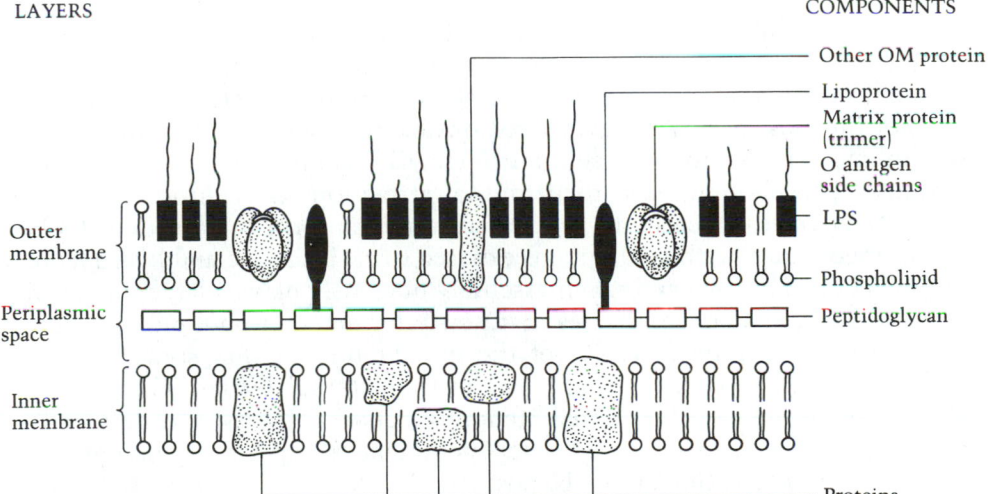

LAYERS   COMPONENTS

— Other OM protein

— Lipoprotein
— Matrix protein (trimer)
— O antigen side chains
— LPS

Outer membrane

— Phospholipid

Periplasmic space

— Peptidoglycan

Inner membrane

— Proteins

Figure 3. Diagram of a Gram-negative cell envelope. Components are listed on the right. The trimers of matrix protein of the outer membrane (OM) are associated with lipoprotein and with LPS (of variable polysaccharide length), and lipoprotein is covalently bound to peptidoglycan. The diagram also illustrates some general properties of membranes (see Chapter 1, Cell Membrane). Phospholipid molecules are illustrated with a *circle* for the polar groups, and a *line* for each fatty acid acyl moiety. (From Davis, 1980.)

lipopolysaccharide are conveyed by undecaprenol phosphate, to be assembled at the outer membrane. Polypeptides reach the outer membrane by growing through these patches as well, and then they fold into their appropriate configuration, guided by the lipopolysaccharide as in the case of the matrix porins that form the transmembrane pores. Capsular polysaccharide as well seems to be made at Bayer's junctions. Proteins destined for the periplasm are secreted through the cell membrane, probably all over the cell. Three features known to contribute to the correct delivery of proteins to the outer membrane, periplasm, and cell membrane are (1) the location of the polysomes that make them, (2) the presence of a signal sequence at the NH_2-terminus of the protein, and (3) the primary structure and folding of the protein itself.

It is already clear from this preview of envelope assembly that Bayer's junctions are critical. More information is needed about them. Many interesting calculations about the dynamics of envelope assembly await careful estimation of the number of junctions per cell, their average size, their topological distribution in the envelope, and their precise molecular architecture. Early estimates place their number at 200–400 per cell; their diameters range from 25 to 50 nm (Bayer, 1968).

The individual processes of envelope growth require some detailed discussion, particularly with regard to the evidence for this somewhat hypothetical picture of envelope assembly.

Cell membrane

Making the cell membrane means forming a phospholipid bilayer containing up to seven kinds of phospholipids and approximately 200 different proteins in amounts characteristic of the specific growth conditions.

For the most part the enzymes responsible for phospholipid synthesis are within the cell membrane itself (see review by Cronan, 1978). The bipolar nature of phospholipids and their mutual attraction for each other leads to their ready coalescence in an aqueous medium to form the familiar double leaflet membrane, with polar groups exposed to each surface and hydrophobic residues buried in the center. Self-assembly obviously plays a large role in the process. Some evidence indicates that newly formed phospholipid molecules appear first in the inner leaflet of the membrane and that some are then rotated ("flip-flopped") to become part of the external leaflet. The different phospholipids are not distributed randomly between the two leaflets or between the inner membrane and outer membrane, so specific lipid–lipid and lipid–protein interactions must be postulated.

Where does membrane growth occur? Interest in the topology of membrane assembly has been high because of the postulated role of the membrane in assisting segregation of replicated chromosomes into daughter cells at the time of cell division. The attachment of the chromosome to the envelope provides a plausible model for segregation, a model in which intercalation of newly made envelope gradually forces apart the two chromosomes as they are replicated (see also Chapter 6).

The fluidity of membrane lipids—their ability to exchange contacts and migrate within the individual leaflets of the bilayer—makes it extremely difficult to learn whether the cell membrane grows by intercalation of phospholipids throughout the structure or whether there are special growing regions. Radioactive precursors of lipids (glycerol, fatty acids) quickly appear distributed throughout the cell membrane. But interpretation of these results is uncertain; the fluidity of the membrane can obscure evidence of a localized region of assembly.

In fact, evidence using a different sort of membrane marker—an inducible protein that facilitates lactose entry into the cell—indicates that intercalation occurs at localized sites (Képès, 1972). A culture of *E. coli* induced for lactose permease by growth in lactose was shifted to a noninducing growth medium. Samples of the culture taken during subsequent growth were analyzed to detect the first appearance of cells lacking lactose permease in their membrane. Such cells are resistant to penicillin killing in a lactose medium because only growing cells are killed by penicillin. A large portion of the population was found to become permease negative between the second and third generation of growth. The diagram shown in Figure 4 shows why the observed result would be expected if membrane grows by assembly of new components in the central region of the cell. Other models of regional growth could also account for these results, but generalized intercalation all over the membrane could not.

Another kind of experiment suggests limited growing zones in the membrane. The synthesis of membrane phosphatidylethanolamine and phosphatidylglycerol was followed by measuring the incorporation of [32P]phosphate and [3H]palmitate (Joseleau-Petit, 1980) or [14C]serine and [14C]acetate (Pierucci, 1980) in cultures growing with division synchrony (Chapter 5). The rates of synthesis remained constant per cell throughout most of the division cycle and then suddenly doubled. This result is inconsistent with a generalized, diffuse growth of the membrane, although it does not prove that growth occurs in the equatorial zone. Of some interest is the finding of Pierucci (1980) that the time of doubling of the rate of phospholipid incorporation coincided with the initiation of replication of the chromosome. It is tempting to visu-

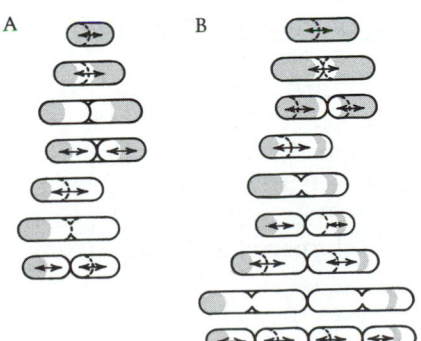

Figure 4.

Theoretical distribution of parental membrane (shaded area) in descendants of a bacterium with a single median growing zone. Arrow indicates elongation of new membrane. A. A newly formed daughter cell. B. A cell midway between two divisions. (From Képès, 1972.)

alize a process by which duplication of the origin of replication of the chromosome (see Chapter 3, DNA Polymerization) triggers the activation of a new zone of membrane growth between the attachment points of the replicating chromosomes.

Wherever new cell membrane is assembled, the insertion of proteins is a major part of the process. Much has been learned of how certain proteins become assembled into the cell membrane and the outer membrane, how others become translocated into the periplasmic space, and how still others are secreted from the cell. It will be best to examine this information before proceeding to describe assembly of the rest of the cell envelope.

Assembly of proteins into the envelope

All membrane proteins possess hydrophobic domains imbedded in the hydrocarbon core of the lipid bilayer and polar regions exposed to the aqueous phase. Transmembrane proteins have polar regions exposed to both the interior and exterior of the cell; other membrane proteins may be exposed at only one of the surfaces of the membrane. Some protein molecules exist singly in the phospholipid leaflets, but many are found associated with other protein molecules, in some cases as quite elaborate structures (e.g., the membrane ATPase responsible for the proton translocation that is coupled to oxidative phosphorylation of ADP).

The force of attraction between hydrophobic molecules in an aqueous medium is considerable. Membrane proteins are notoriously insoluble in water (at least in their mature configuration, as explained later). Unless a detergent is added to coat their hydrophobic surface, membrane proteins rapidly aggregate into large complexes, stick to glass and other surfaces, denature, and even precipitate. Hydrophobicity is clearly sufficient to provide much of the energy for assembly of proteins into membranes, and, grossly, to provide a distinction between proteins destined for the membrane and those that will remain in the cytosol. The primary structure (amino acid sequence) of a polypeptide therefore ordains the ultimate cellular location of a protein by determining the nature and extent of its hydrophobic surfaces when folded.

This can be only part of the story. It leaves unexplained the orderliness of the membrane. It does not take into account the necessarily enormous diversity of structure found among the 200 or so proteins constituting the envelope. It fails to specify how proteins are partitioned between the outer membrane and the cell membrane, how some find their way into the periplasmic space and still others exit from the cell entirely. It is difficult to imagine how self-assembly based on hydrophobicity alone could determine these different ultimate destinations, and it is particularly troublesome to account for the entry of a protein into the cell membrane, as the result of hydrophobic attraction, followed by an exit on the periplasmic side as a water-soluble molecule.

Clearly, special assembly devices are at work to guide this process. Some of these have been elucidated greatly since 1975, and many ideas have come from the study of protein secretion and membrane assembly in eucaryotic cells. This extensive work can be only briefly summarized here. The reader is encouraged to consult the excellent reviews and critical discussions prepared by Wickner (1979, 1980), Osborn (1980), Inouye (1980a), and Davis (1980).

There are four or five special aspects of protein assembly into membrane (and protein secretion) that go beyond the primal force of hydrophobicity. First, many proteins that are to pass into and through the cell membrane are made on polysomes that become attached by the nascent polypeptide to the inner surface of the cell membrane. These polysomes accompany the cell membrane during centrifugal sedimentation in sucrose gradients. Direct labeling of nascent polypeptide chains protruding through the cell membrane has been achieved through the use of nonpenetrating reagents that react with certain amino acid residues. This demonstrates that at least some secreted proteins are translocated as they are translated. Because the ribosomes are freed from the membrane by the drug puromycin, which dissociates ribosomes and their nascent polypeptides, the polysomes may be attached to the membrane solely by the polypeptides they are synthesizing. The force of chain elongation, therefore, seems not to be responsible for pushing the peptide chain through the membrane; we shall discuss an alternative force in a moment.

A second discovery helps explain how polysomes making proteins that must enter the cell membrane become attached to the cell membrane. A large number of secreted proteins have been found to be made with an extra sequence of amino acids at their NH_2-terminus. These SIGNAL SEQUENCES differ from protein to protein but share certain characteristics, including length (15–30 amino acids, but most commonly approximately 20), the presence of one or two basic amino acids (e.g., lysine) near the NH_2-terminus of the peptide followed by a stretch of apolar amino acids that usually includes two glycine or proline residues, and the absence of acidic amino acids (Figure 5). The last residue has an α-carbon with either no side chain or at most a side chain with only one carbon (glycine, alanine, serine). It is thought that the synthesis of a membrane or secretory protein begins in a normal way on unattached polysomes. The formation of the signal peptide causes the binding of the growing chain to the cell membrane at a region containing a channel, or a potential channel, through which the polypeptide is threaded as it is made.

One view of how the signal peptide guides the rest of the molecule through to the periplasm is shown in Figure 6. According to the BENT LOOP MODEL, the signal peptide remains in the membrane and is hydrolytically cleaved from the secreted body of the protein by a SIGNAL PEPTIDASE, the existence of which has already been demonstrated. It is not yet known how many signal peptidases exist in the cell. Specific membrane proteins that might bind the signal peptides and/or form pores for the peptide chains are being sought.

The diagram in Figure 7A illustrates how this mechanism could be envisioned to function for proteins to be inserted in the cell membrane, secreted into the periplasm, or assembled into the outer membrane. This scheme incorporates a third device important in envelope assembly—the existence of

Figure 5.

Precursor of	Localization	Amino Acid sequence
Lipoprotein	OM	Met Lys Ala Thr Lys Leu Val Leu ⟨Gly⟩ Ala Val Ile Leu ⟨Gly⟩ Ser Thr Leu Leu Ala ⟨Gly⟩ Cys
λ Receptor	OM	(Met) Met Ile Thr Leu Arg Lys Leu ⟨Pro⟩ Leu Ala Val Ala Val Ala Ala ⟨Gly⟩ Val Met Ser Ala Gln Ala Met [Ala] Val
Maltose binding protein	PS	Met Lys Ile Lys Thr Gly Ala Arg Ile Leu Ala Leu Ser Ala Leu Thr Thr Met Met Phe Ser Ala Ser Ala Leu [Ala] Lys
β-Lactamase	PS	Met Ser Ile Gln His Phe Arg Val Ala Leu Ile ⟨Pro⟩ Phe Phe Ala Ala Phe Cys Leu ⟨Pro⟩ Val Phe [Ala] His
Arabinose binding protein	PS	Met Lys — Thr Lys Leu Val ⟨Gly⟩ Ala Val Ile Leu Ser; — Gly Ala —— [Ala] Glu
fd phage major coat protein	CM	Met Lys Lys Ser Leu Val Leu Lys Ala Ser Val Ala Val Ala Thr Leu Val ⟨Pro⟩ Met Leu Ser Phe [Ala] Ala
fd phage minor coat protein	CM	Met Lys Lys Leu Leu Phe Ala Ile ⟨Pro⟩ Leu Val Val ⟨Pro⟩ Phe Tyr Ser His Ser [Ser] Ala

—— Hydrophilic basic region ——   —— Hydrophobic region ——

Figure 5.

Signal peptide sequences of proteins that enter the *E. coli* envelope. Basic amino acid residues are underlined, glycine and proline residues are circled, and the residue at the cleavage site (arrowhead) is boxed. The sequences are arranged to emphasize homology by being aligned at the last basic amino acid of the hydrophilic region. Blanks in the arabinose binding protein indicate residues of uncertain identity. OM, Outer membrane; PS, periplasmic space; CM, cell membrane. (After Osborn, 1980.)

Figure 6.

Bent loop model for the translocation of secretory proteins across the membrane. An arrow designates the cleavage site in the precursor molecule. In this illustration ribosomes are bound to the membrane by interaction with a hypothetical ribophorin-like protein in the cytoplasmic membrane. (After Inouye, 1980.)

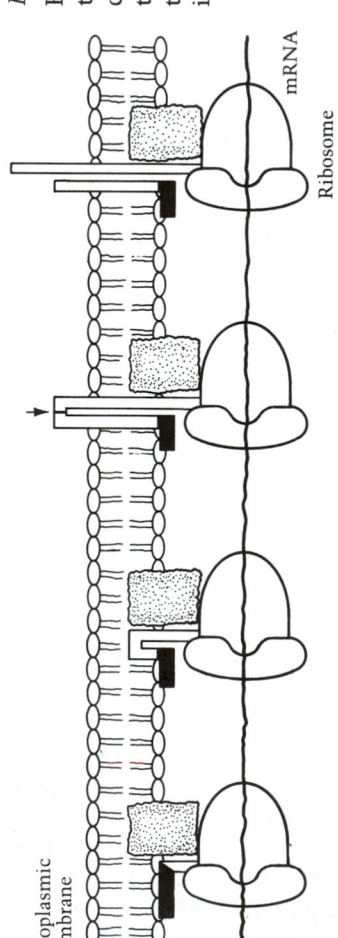

Cytoplasmic membrane

mRNA

Ribosome

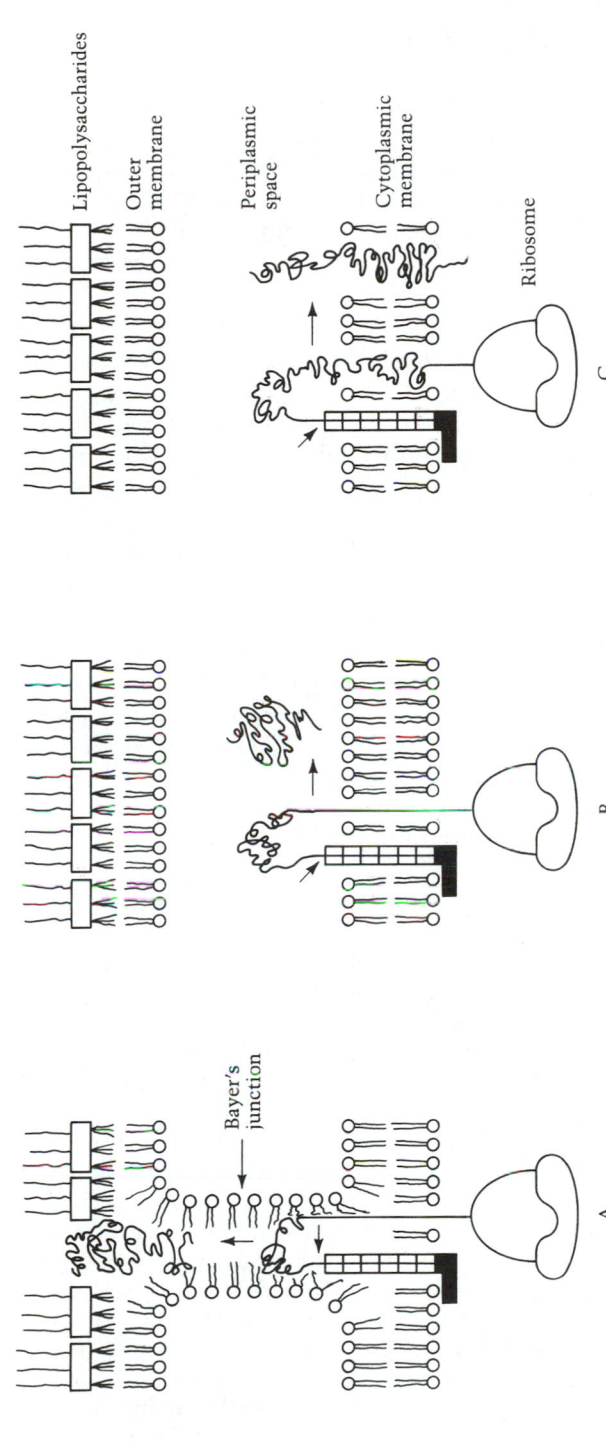

Figure 7.

Possible mechanism of secretion and final localization of envelope proteins in *E. coli*. A. Outer membrane proteins. When the cleavage site is exposed to the outside of the cytoplasmic membrane, the peptide extension is cleaved off before the completion of the chain. The resultant NH_2-terminal part of the nascent peptide of the outer membrane protein begins folding and then interacts specifically with an outer membrane component such as the lipopolysaccharide in a junction between the cytoplasmic and outer membranes (Bayer's junction); this process leads the protein to the outer membrane. B. Periplasmic proteins. When the cleavage site is exposed to the outside surface of the cytoplasmic membrane, the peptide extension is cleaved off and the resultant nascent peptide begins to fold. Because of the hydrophilic property of the proteins, the part that extrudes to the outside of the cytoplasmic membrane remains soluble in the periplasmic space. C. Cytoplasmic membrane proteins. After the peptide extension is cleaved off, the new NH_2-terminal part of the protein stays in the periplasmic space, and the other region assembles in the lipid bilayer of the cytoplasmic membrane because of the hydrophobic property of this portion of the protein. The peptidoglycan layer is not illustrated between the outer membrane and the cytoplasmic membrane for simplicity. (After Inouye, 1980b.)

61

zones of adhesion (Bayer's junctions) that make possible direct delivery of polypeptides and other components from the cytosol to the outer membrane without transversing the periplasmic space and cell wall (Figure 7A).

Much evidence supports the signal peptide mechanism for the entry of many proteins into the membrane. Proteins with a mutationally altered signal sequence—with only a single misplaced amino acid—cannot be transported normally. Fusion of signal peptide sequences to proteins not normally exported has shown that these sequences are necessary, though not in all cases sufficient, to direct the export of the proteins.

A second model, the MEMBRANE TRIGGER HYPOTHESIS, makes use of the fact that some proteins can exist in alternative conformations: a water-soluble one and a hydrophobic or lipid-soluble one. Contact with a membrane can trigger a change from the first to the second conformation and, driven by the energy of hydrophobic attraction with the phospholipids of the membrane, this refolding is accompanied by entry of the polypeptide into the lipid bilayer (Figure 8).

Proteins of this sort can be made on unattached polysomes in soluble form and later assembled into the cell membrane. In some cases a leader peptide might still be involved, not solely as a membrane recognizer, but as a modifier of the folding pathway of the protein, in a sense activating it for assembly. Attachment of polysomes to the membrane and the presence of signal peptides are regarded as features of assembly necessary for some classes of proteins and not others, in acknowledgment of the great diversity of structure and function among envelope proteins.

It is clear now that assembly of proteins into the membranes of the cell envelope and into the periplasm takes many routes. The mechanisms we have

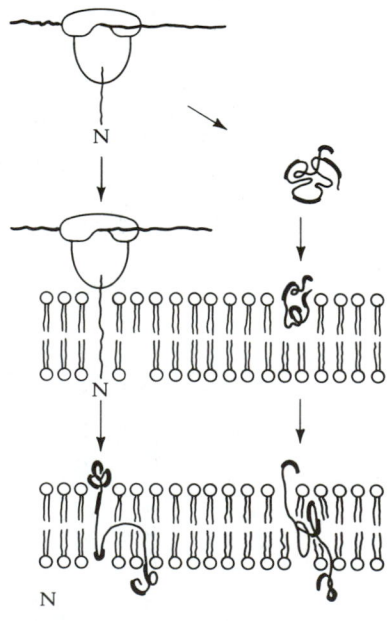

Figure 8.

A schematic presentation of the membrane trigger hypothesis. Thick lines represent polar regions, thin lines represent hydrophobic regions. A soluble complex of nascent peptide, tRNA, ribosome, and mRNA is shown at the top of the figure. On the left, a protein is inserted into the membrane during synthesis; the function of the leader peptide in this case is to alter the folding pathway. On the right, a protein is fully made before insertion. Its initial folding, with polar regions exposed, is compatible with its aqueous environment. On encountering a specific membrane region, it assumes a second conformation with its hydrophobic regions imbedded in the hydrophobic regions of the lipid bilayer. (After Wickner, 1979.)

examined here (signal peptides, Bayer's junctions, membrane-attached ribosomes, triggered folding) facilitate the specific assembly of different proteins into the envelope. Other devices, including posttranslational addition of components to proteins, probably are important as well, and many must be awaiting discovery. An excellent illustration that there is not a unique way for proteins to assemble in membranes comes from the observation that the bacterial proteins normally found in the cell membrane (for which information is available) are made without a signal peptide, and yet two bacteriophages, fd and M13, make coat proteins that insert asymmetrically into the *E. coli* cell membrane with the NH$_2$-terminus on the inner surface and the COOH-terminus on the external surface, and these proteins are each made with a classic signal peptide.

No important differences in cell membrane assembly have been discovered between Gram-positive and Gram-negative bacteria. Of course, features of the membrane in Gram-negative cells that relate specifically to formation of the periplasm and outer membrane are absent from the cell membrane of Gram-positive bacteria; and features of the latter dealing specifically with the assembly and turnover of the peptidoglycan and teichoic acids of the thick Gram-positive wall can be expected to be different or absent in Gram-negative species.

Assembly of the remaining structures of the envelope, as we shall now see, all involve the cell membrane in both Gram-positive and Gram-negative species.

Periplasm

The periplasm is the space between the two membranes of Gram-negative cells and consists, as far as is known, of an aqueous solution of approximately 50 different kinds of proteins.

Whatever the mechanisms for assembling proteins into the cell membrane, there is excellent evidence that signal peptides are involved in translocation of proteins into the periplasm, as shown in Figure 7. Some of these proteins have been studied in great detail, including alkaline phosphatase, β-lactamase (an enzyme that hydrolyzes penicillin), and binding proteins for maltose, arabinose, leucine-isoleucine-valine, and leucine.

Although each of these is made with signal peptides, certain of them (β-lactamase and the maltose-binding protein) are initially synthesized as soluble, cytoplasmic proteins by free polysomes and only later secreted across the membrane. They are, therefore, good candidates for the membrane trigger mechanism of translocation.

Elegant experiments involving gene fusions (Chapter 4, In Vivo Genetic Engineering) have allowed an assessment of the significance of signal sequences and protein structure in secretion of periplasmic proteins. In particular, a protein containing the signal peptide and almost all the maltose-binding protein fused to β-galactosidase (a typical cytoplasmic enzyme) was found to enter the cell membrane, but not the periplasm. Apparently a sequence of amino acids in β-galactosidase stops passage through the secretory channel.

Interestingly, these stuck hybrid molecules interfere with translocation of normal outer membrane proteins; the latter accumulate in the cytosol in precursor form, an indication that polypeptides destined for the periplasm and those destined for the outer membrane share some early sites critical for translocation (Bassford, 1979).

Gram-positive cells have no periplasm because they have no outer membrane.

Wall

The wall in *E. coli* consists of a single, gigantic, sac-shaped molecule of peptidoglycan one layer thick (Figure 10 of Chapter 1). Its assembly occurs concomitantly with the final states of polymerization of the peptidoglycan.

In the cytosol the peptidoglycan precursor, UDP-N-acetylmuramic acid-pentapeptide is synthesized by numerous steps (Chapter 3) and then attached to the cell membrane carrier lipid, undecaprenol phosphate. In the cell membrane, further additions take place; then disaccharide residues that are still attached to the carrier are polymerized to form a glycan chain. When the latter has reached 10–50 disaccharide units in length, they are released from the periplasmic side of the cell membrane and a battery of seven or more proteins catalyze covalent reactions that result in the extension, morphogenesis, and eventual septation of the peptidoglycan sacculus (Figure 9). These proteins have the unique property of binding penicillin covalently (Spratt, 1977).

Penicillin-binding protein 1A (PBP 1A) has no known function but seems to be able to substitute for PBP 1B, which has transglycosylase and transpeptidase activities. These collectively seem responsible for cross-linking glycan

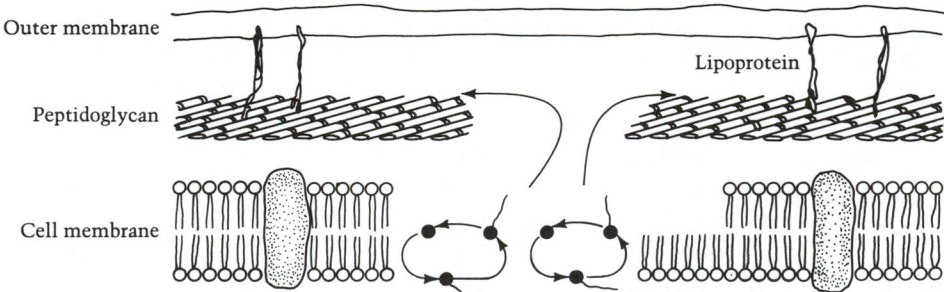

Figure 9. Wall assembly in Gram-negative bacteria. The diagram depicts a region where peptidoglycan precursors are shuttled through the cell membrane on undecaprenol phosphate, polymerized into glycan chains, intercalated with covalent links to the existing peptidoglycan sac, and finally attached to the outer membrane by reaction with the lipoprotein. The number of such growing regions, their distribution in the envelope, and their architecture are unknown.

chains (and perhaps polymerizing them) and therefore are involved in growth-by-extension of the existing peptidoglycan sacculus. The enzymatic activities of PBP 2 and 3 are not yet known, but the former is necessary for *E. coli* to grow rod-shaped and the latter is essential for septum formation. The remaining proteins—PBP 4, 5, and 6—all have one or another type of carboxypeptidase activity capable of attacking various bonds in the peptides of peptidoglycan (for references, consult Buchanan, 1981).

For simplicity one can say that wall assembly involves polymerization of peptidoglycan precursors (by PBP 1A and 1B), cutting and shaping the sacculus at growing points (by PBP 4, 5, and 6), initiating a septum, or cross wall, preparatory to cell division (by PBP 3) and some reaction that leads to rod-shaped extension of the sacculus (by PBP 2).

As in the case of the outer membrane, attempts to demonstrate zonal or regional growth (as, for example, at the equator) have not completely settled the topology of wall synthesis. Some early reports indicated that incorporation of [^{3}H]diaminopimelic acid (DAP), which is an amino acid in *E. coli* found only in peptidoglycan, occurred equatorially, but more refined measurements seem consistent with diffuse growth throughout the cell—though growth is more intense at the equator where septum formation is occurring (reviewed in Verwer, 1980).

Many of the fundamental assembly reactions by which the peptidoglycan sacculus is formed from precursors in *E. coli* have their counterparts in Gram-positive cells and indeed were discovered there first. Nevertheless, wall growth in Gram-positive bacteria displays interesting differences from that in *E. coli*, partially reflecting the much thicker wall structure (40 layers in *Bacillus subtilis* compared to 1 in *E. coli*).

Newly synthesized peptidoglycan precursors are continually added to the existing wall at the innermost surface, and at the external surface peptidoglycan is continually hydrolyzed and lost to the medium (Figure 10). The Gram-positive wall is, therefore, a dynamic structure in which peptidoglycan moieties moved outward as a result of new growth and turnover (see references in Frehel, 1979). Because the wall retains a constant thickness in all but the most contrived of experimental stresses, there must be mechanisms that keep wall assembly and turnover in balance.

The anionic polymers called teichoic acids, which are found with the peptidoglycan, become linked to the glycan strands during assembly at the inner surface, though some may be secreted through the wall without becoming attached to glycan. Assembly of the teichoic acids appears to take place on a membrane carrier, possibly undecaprenol phosphate. The membrane form of teichoic acid, called lipoteichoic acid, is attached to some glycolipid in the cell membrane rather than to the wall peptidoglycan; its metabolic relationship to the wall teichoic acid is unclear.

Autoradiography to reveal incorporated wall precursors in *Bacillus subtilis* (a Gram-positive, rod-shaped species) indicate there may be limited wall insertion sites: as few as two or three per cell during slow growth and as many as a dozen during rapid growth (Pooley, 1978). In Gram-positive species with spherical shape (e.g., *Streptococcus pyogenes*), labeling with fluorescent anti-

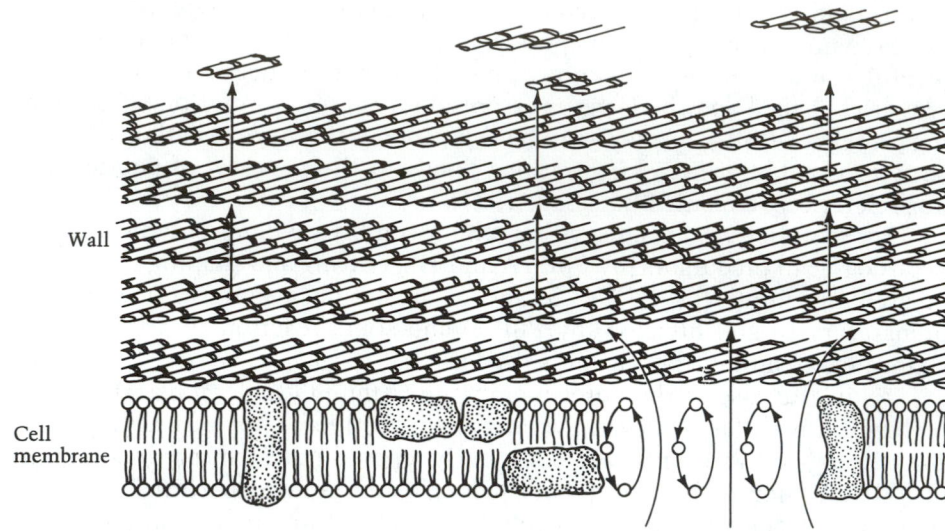

Wall

Cell
membrane

Figure 10. Wall assembly and turnover in Gram-positive cells. The dynamic nature of the wall is illustrated by the arrows, which show the progression of newly formed peptidoglycan from its site of synthesis at the cell membrane to the periphery of the wall where it is sloughed off.

body directed against the wall has provided clear evidence of an equatorial growth zone (Cole, 1962).

Outer membrane

The outer membrane consists of phospholipids, lipopolysaccharide, and a moderately large array of different proteins. The highly ordered complexity of this specialized Gram-negative structure (Figure 11) is produced by independent assembly of each component rather than by formation of an intermediate subunit that contains all components. Although these separate assembly reactions proceed in concert rather than in sequence, formation of the outer membrane can more readily be understood by visualizing the individual steps (Figure 12). An excellent and detailed review of this subject has been prepared by Osborn (1980).

Step 1. Assembly of the lipopolysaccharide. Lipopolysaccharide (LPS) subunits are synthesized in the cell membrane in two parallel processes (see Figure 13). One process produces the repeating polysaccharide side chain built up on the lipid carrier, undecaprenol phosphate, and the other process produces the core polysaccharide built up on lipid A (which functions both as primer and carrier). The subunits are made at the inner surface of the cell membrane, then translocated by their respective carriers to the outer surface, presumably through Bayer's junctions. Translocation is thought to be driven by the protonmotive force operating across the cell membrane. At the outer surface, a

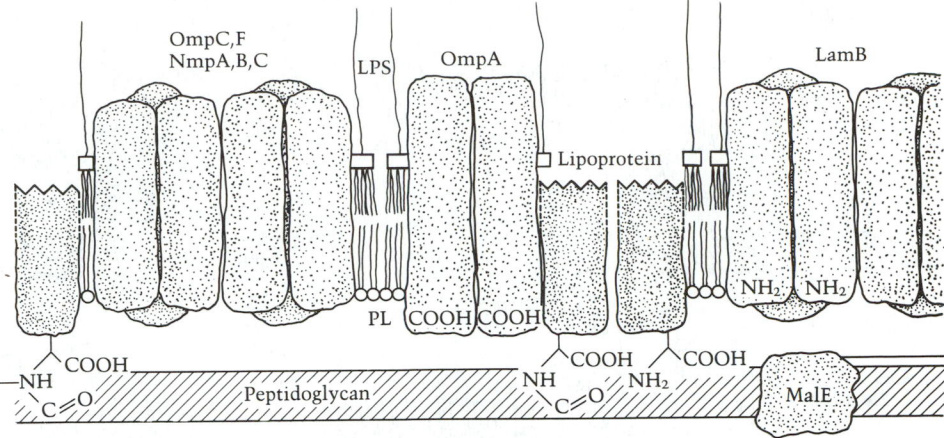

Figure 11. Schematic representation of the molecular organization of major proteins in the outer membrane of Gram-negative bacteria. The diagram
shows the amide linkage of lipoprotein to the peptidoglycan, the
transmembrane nature of OmpA and the porins (OmpC, OmpF,
NmpA, NmpB, and NmpC). The LamB protein is an outer membrane
protein that functions as a porin for the transport of maltose and
maltodextrins [it is also a receptor for bacteriophage lambda (λ)]; it is
shown in proximity to the MalE protein, a periplasmic maltose-
binding protein. LPS, Lipopolysaccharide; PL, phospholipid. (From
Osborn, 1980.)

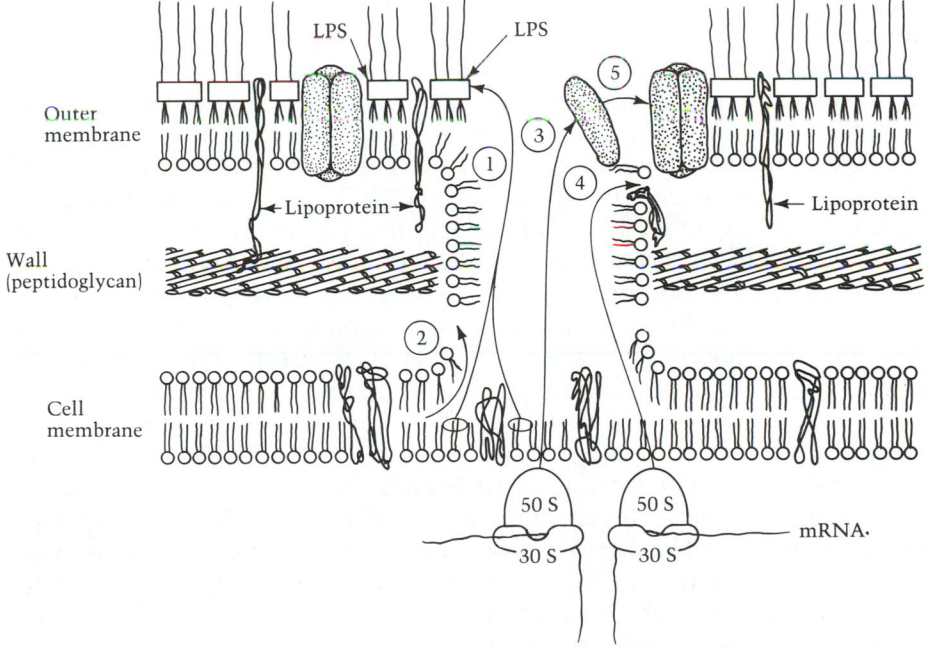

Figure 12. Assembly reactions of the outer membrane. See text for details on
assembly steps 1–5.

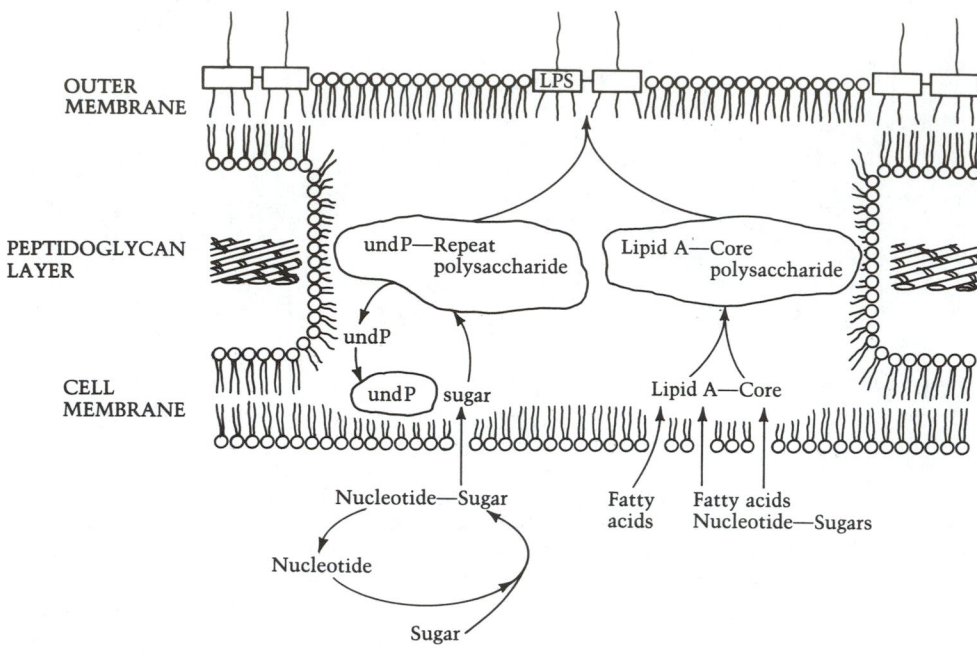

Figure 13. Assembly of the lipopolysaccharide. See text for details. und-P, Undecaprenol-PO₄.

transfer enzyme attaches the completed side chain to the core polysaccharide on lipid A. The individual LPS molecules condense by mutual attraction into the two-dimensional array of the outer leaflet of the outer membrane.

Step 2. Assembly of phospholipids. As is true for lipopolysaccharide, synthesis of phospholipids occurs in the cell membrane, and their translocation is by an equally obscure mechanism that is driven by the electrochemical potential (protonmotive force) across the membrane rather than by ATP. A reasonable suggestion (Donohue-Rolfe, 1980) is that newly synthesized lipid molecules in the inner leaflet of the cell membrane are flip-flopped into the outer leaflet of this structure and then migrate by lateral diffusion at the region of Bayer's junctions to the (possibly) contiguous inner leaflet of the outer membrane. After translocation, hydrophobic interactions with lipid A and with neighboring phospholipids largely determine the self-assembly of phospholipids to form the inner leaflet of the outer membrane.

Step 3. Assembly of outer membrane proteins. There are, as we have seen, approximately three or four major outer membrane proteins, and as many as 50 less abundant ones. Ribosomes attached to the inner surface of the cell membrane make the polypeptides that are to be transported to the outer membrane. The translocation involves two steps: (1) initiation of the process by interaction of the signal sequence with the membrane, followed by (2) interaction of the nascent peptide with LPS. The bent loop model of membrane protein synthesis accounts for this process for most outer membrane proteins. The LPS already aggregated in the outer membrane participates in the process by interacting with portions of the NH_2-terminal end of the mature polypeptide after the signal peptide has been removed. This interaction guides the folding of the nascent polypeptide chains, promoting their translocation from the polysome to their close association with the LPS in the outer membrane (Figure 14).

An interesting possibility for control of the synthesis of outer membrane proteins is suggested by this image. If LPS synthesis were to decline, or outer membrane growth to stop for whatever reason, interaction of nascent chains of outer membrane proteins with LPS could not occur and translocation would halt. If for some reason translocation were essential for a normal rate of synthesis of the protein from mRNA, a coupling can be imagined that would coordinate synthesis of outer membrane proteins with LPS availability (Datta, 1976). Such coupling could even extend to the mRNA by, for example, an attenuation mechanism (Chapter 7). A coordination of this sort has been postulated by Beher (1981) to account for the lack of accumulation of OmpA protein (or precursors) in cells defective in making a normal LPS.

The role of LPS in creating assembly sites and the plasticity of outer membrane composition are illustrated by the study of certain mutants that form neither the OmpC nor the OmpF general pore proteins. These porin-deficient cells grow poorly, and secondary mutants can be isolated as faster growers that now produce a new membrane protein that self-assembles in the LPS to form functional pores (Foulds, 1978a, b).

Step 4. Assembly of the lipoprotein. The outer membrane contains a small protein that is the most abundant protein (numerically, not by weight) in *E. coli*. This protein is made with a signal peptide of 20 amino acids at its NH_2-terminus (Figure 5). Shortly after cleavage of this leader, the NH_2-terminal cysteine of the mature polypeptide becomes modified by the addition of glycerol in thioether linkage, followed by fatty acid acylation of the two remaining glycerol hydroxyls and the α-NH_2 group of the cysteine. Presumably the lipid end of the modified protein interacts with the LPS and other lipids of the outer membrane, facilitating an intercalation in which this protein

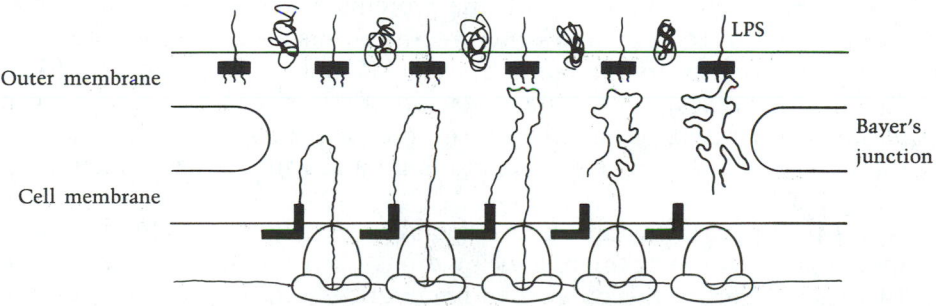

Figure 14. Model for the assembly of proteins into the outer membrane. The interaction between a nascent protein chain and the LPS of the outer membrane at one of the junctions of Bayer is depicted as guiding the final location of protein in the outer membrane. (After Osborn, 1980.)

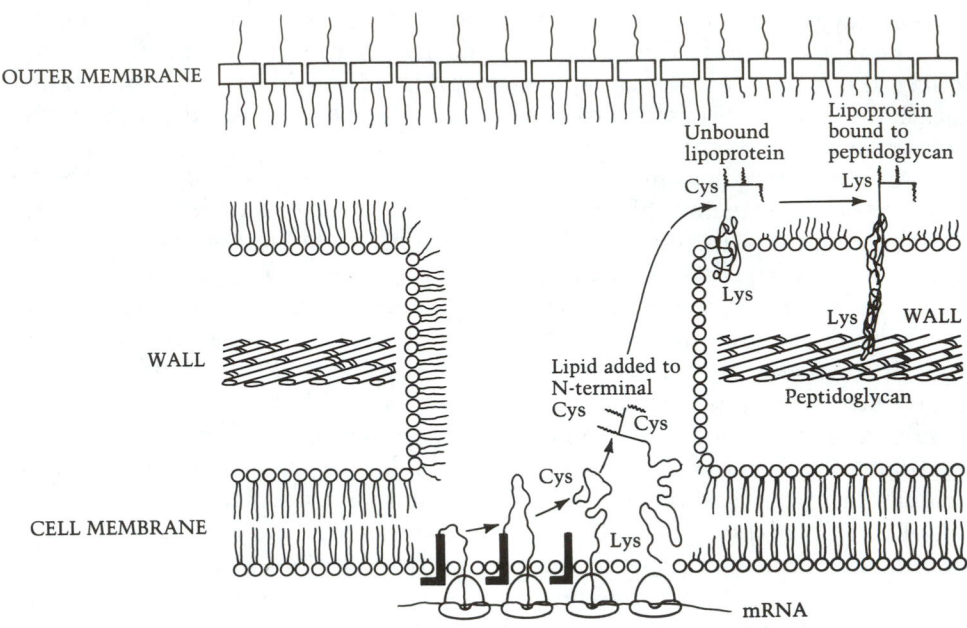

OUTER MEMBRANE

Unbound lipoprotein

Lipoprotein bound to peptidoglycan

Cys

Lys

Lys

WALL

Lys WALL

Lipid added to N-terminal Cys

Cys

Peptidoglycan

Cys

Cys

CELL MEMBRANE

Lys

mRNA

Figure 15. Assembly of the lipoprotein. See text for details. (After Osborn, 1980; Wu, 1980.)

spans the membrane. The COOH-terminal lysine of the mature lipoprotein projects inward into the periplasm. At some point in the process, this residue reacts with the diaminopimelic acid residue of side chains of the peptidoglycan, thereby anchoring the lipoprotein and the whole outer membrane to the underlying wall (Figure 15).

Step 5. Self-assembly of the outer membrane. The LPS not only helps translocate the outer membrane proteins, it also guides the overall morphogenesis of this structure. An organized hexagonal lattice strongly resembling the outer membrane can be produced spontaneously by simply mixing LPS and one or more major outer membrane proteins, such as the porins OmpF and OmpC. In this reconstituted structure, the porins are correctly condensed as trimers that form pores extending through LPS matrix. Furthermore, if LPS and outer membrane proteins are mixed in the presence of peptidoglycan containing covalently attached lipoprotein, the membrane that is formed covers the surface of the peptidoglycan and is attached to it by the lipoprotein (Yamada, 1978).

The same uncertainty about the topology of growth exists for the outer membrane as for the cell membrane. Various clever approaches have been used to learn whether the outer membrane grows diffusely or in restricted zones. Bayer (1975) studied the synthesis of antigenically novel LPS induced by phage ε15 in *Salmonella anatum*. He concluded that LPS synthesis occurred at 20–30 sites, corresponding to zones of adhesion located evenly over the cell surface. A different study (Begg, 1977) employed an *E. coli* mutant in which synthesis of the outer membrane receptor for phage T6 could be induced by a temperature shift. It was decided that new receptors were preferentially produced at one pole of the cell, or at both poles in the case of large doublet cells.

Flagella

At least 40 gene products are necessary for the assembly and function of flagella in *E. coli* (Appendix B). Known flagellar proteins are listed in Table 1.

How the 10 or so proteins that make up the complex basal structure and hook come to be assembled is not yet understood, but this structure must form before the long filament of the flagellum itself can grow. It should be recalled that the rings of the basal structure (four in Gram-negative cells, two in Gram-positive cells) are embedded in the various layers of the envelope (see Figure 17 in Chapter 1). It seems a fair guess that self-assembly plays a major role. Once formed, the basal body remains intact during purification. No pool of unassembled basal body proteins can be found in the cytosol or elsewhere. It is reasonable to assume, therefore, that synthesis occurs at the cell membrane by attached ribosomes and that assembly occurs as the nascent polypeptide chains grow. The main protein—flagellin—that constitutes the

Table 1. Flagellar components

Components (mol wt)	Location
54,000 (flagellin)	filament
42,000	hook
60,000	basal structure
39,000	basal structure
31,000	basal structure
27,000	basal structure
20,000	basal structure
18,000	basal structure
13,000	basal structure
11,000	basal structure
9,000	basal structure
31,000	inner membrane
39,000	inner membrane
76,000 66,000	cytoplasm
12,000	cytoplasm
60,000 61,000 63,000	inner membrane
38,000	cytoplasm
38,000	cytoplasm
8,000	cytoplasm
24,000	cytoplasm
64,000 65,000	inner membrane

After Silverman, 1977.

filament plays no role in assembly of the basal body and hook. Can the basal structures be made anywhere at already existing envelope, or must assembly occur at the sites of membrane, wall, and outer membrane assembly? We do not yet know, but the answer should be forthcoming soon.

Assembly of the main filament of the flagellum is somewhat better understood. Neither wild-type cells nor mutants defective in various flagellar genes accumulate detectable pools of flagellin, a finding that suggests that assembly occurs during synthesis and that the two processes are tightly coupled. Growth of the filament requires the presence of the basal structure and hook and begins at the tip of the hook. Flagellin molecules self-assemble into the helical filament, but this process continues to occur at the tip rather than at the base of the growing filament. This remarkable feat is accomplished by the threading of flagellin, as it is made, through the hollow core of the filament; upon reaching the tip each molecule spontaneously condenses with its predecessor and thus elongates the filament.

In Figure 16 we have presented a partially hypothetical model of flagellum assembly. As far as is known, assembly of flagella occurs similarly in all bacteria. Obviously the details of basal body formation will be somewhat different in Gram-positive and Gram-negative cells.

Assembly of the other filamentous appendages of bacteria—the pili—has not been studied in detail. Molecules of pilin, the structural protein of pili, have the property of spontaneously aggregating into filaments.

Capsule

Though perhaps not a structure in the formal sense, the amorphous outer layer of polysaccharide presents an assembly problem similar in some respects to that of LPS. (See review by Troy, 1979.) In both cases a large hydrophilic polymer must somehow be placed outside the cell membrane. And in both cases the solution is to build subunits on a lipid carrier, translocate them across the membrane, and assemble the subunits on the outer surface, presumably at Bayer's junctions (Bayer, 1977). Again, undecaprenol phosphate is believed to be the membrane carrier in some strains of *E. coli*. At the inner surface of the cell membrane, a repeating unit of three sugars is built up on the carrier by transfer from nucleoside diphosphate sugar precursors. At the outer face (and this probably must be at the outer face of the outer membrane), the trisaccharides are linked together by head addition, that is, by the transfer of the growing polymer from its carrier to the next trisaccharide–carrier complex. (This is diagrammed in Figure 13 for the repeating side chains of LPS.)

Formation of capsular material proceeds more or less continuously, and the completed product sloughs off into the medium. The thickness of this "structure" is, therefore, not fixed. It varies in abundance with the nature of the growth medium and the growth temperature.

With the exception of a few *Bacillus* species that produce a capsule of poly-D-glutamate, bacterial exopolymers are usually polysaccharides and include cellulose, glucose, colanic acid, colominic acid, polyuronides, and the

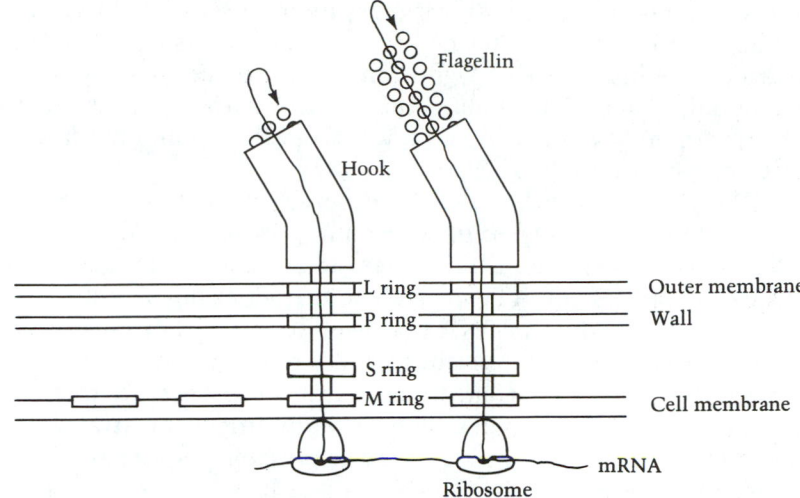

Figure 16. Assembly of the flagellum in a Gram-negative cell. (Drawn partially from the structural studies of DePamphlis, 1971a, b.) This model is based solely on evidence about filament growth.

many types of pneumococcal polysaccharides. One important group, the dextrans and levans, are assembled outside the cell membrane directly from exogenously supplied sucrose.

ASSEMBLY: THE NUCLEOID

In a sense, the newly replicated DNA needs no special assembly process to become a chromosome. It is already attached to the bacterial membrane by means that assure appropriate segregation of the chromosomes into daughter cells as they form. Also, as the DNA is produced, it attracts all the many regulatory proteins that bind to their appropriate recognition sites and modulate the initiation of RNA synthesis. Nevertheless, the newly replicated DNA must undergo at least three further processes (METHYLATION, SUPERCOILING, and FOLDING) in its maturation as a chromosome.

Many species and strains of bacteria have enzymes called RESTRICTION ENDONUCLEASES. These enzymes, of which over 200 are known, recognize specific nucleotide sequences in DNA and produce double-strand scissions, which lead subsequently to the complete degradation of the DNA (see Table 9 in Chapter 4). Each bacterial strain that possesses a restriction system is able also to disguise these recognition sites in its own DNA by methylating the adenine or cytosine residues they contain. Type I restriction endonucleases require Mg^{2+}-ATP and *S*-adenosylmethionine as cofactors and exhibit many puzzling biochemical properties. Enzymes of this type and of the biochemically simpler Type III have both endonuclease and methylation activity. Type II restriction enzymes have only endonuclease activity, and cells with Type II enzymes possess separate methylating enzymes. In either case, a bacterial cell with a restriction–modification system "marks" its DNA by methylating the cognate restriction sites and thereby renders its own DNA insensitive to restriction. This device is significant as a protection system against foreign (e.g., viral) DNA. The modification occurs with *S*-adenosylmethionine as a methyl donor. Several excellent reviews of restriction–modification have been written (e.g., Modrich, 1979; Yuan, 1982).

The DNA of the *E. coli* chromosome exists as negatively supertwisted circular DNA. The degree of supercoiling is controlled by a set of enzymes called TOPOISOMERASES. One of these enzymes, topoisomerase II, also called GYRASE, catalyzes an ATP-dependent introduction of negative superhelical turns into closed duplex DNA that is either relaxed or positively supercoiled. Positive supercoiling occurs ahead of the replication forks as the double helix untwists during replication; unless removed by gyrase, this positive supercoiling will eventually block replication by impairing strand separation. Not only DNA synthesis, but also gene expression, is affected by supercoiling, because synthesis of RNA from DNA (see later) requires local strand separation ("melting"). Transcription is aided by negative twisting, which tends to separate strands, and is hindered by positive twisting. We shall see evidence (Chapter 6) that transcription occurs at the interface between nucleoid and cytosol. This process, therefore, presents a topological problem for the release of mRNA from the DNA template. Newly made mRNA would remain twisted around the duplex unless the latter were free to swivel. Transcription as well as replication is, therefore, dependent on a gyraselike activity. Topoisomerase I, also called ω (omega) protein, relaxes negatively supercoiled DNA and therefore opposes the ATP-dependent effects of gyrase. The situation is even more complex, however, in that gyrase in the absence of ATP can relax negative supercoils, and a third enzyme, topoisomerase II', can relax either positively or negatively supertwisted DNA. This complexity is understandable; because supercoiling of DNA affects both gene expression and DNA synthesis, there must be a very effective means to adjust the degree of supercoiling in all regions of the very long DNA molecule.

It will be recalled from Chapter 1 that the chromosome is divided into approximately 40 domains of supercoiling. Nicking the duplex to permit uncoiling leads to a relaxation of the DNA only within the domain in which the nick occurs. It is not known what structure segregates the domains. Perhaps the domains represent folds of DNA stabilized by (40) patches of binding to the cell membrane—or by some other structure.

Speculation about the assembly reaction would be premature, but one of its most perplexing features deserves special note—the daughter strands of newly replicated molecules must somehow be told apart so that lengths of one duplex condense only with themselves and not with those of the sister duplex. This problem is discussed in Chapter 6.

It has been suggested that the cell might adjust supercoiling of different domains of the chromosome to different extents and thereby regulate the level of gene expression locally. Genes normally expressed at high rates might be clustered in a domain where negative supercoiling is greater than average.

There is no evidence for this idea, but there is evidence for another level of control. Polyamine auxotrophs of *E. coli*, that is, mutant cell lines that cannot make putrescine and spermidine, grow at only one-third the normal rate unless provided with spermidine. Unlike cells growing slowly with other nutritional deficiencies, these deprived cells have reduced chain elongation rates for protein, RNA, and DNA. Effects of this sort are rare but are seen in mutants defective in DNA gyrase and in normal cells in the presence of novobiocin, which is an antibiotic inhibitor of DNA gyrase. Could spermidine be involved in DNA supercoiling? In vitro, the polyamines bind to DNA (and RNA); however, their effect seems to be to strengthen the duplex and counteract the effect of strand-separating agents, so it is unclear how the absence of polyamine in vivo would slow replication, transcription, and translation. A more intriguing possibility is suggested by another property of spermidine: it interacts with both topoisomerases, stimulating gyrase and inhibiting ω protein. The hypothesis has been suggested (Morris, 1981) that spermidine is an effector that modulates DNA supercoiling. In support of this hypothesis is the observation that the supercoiling of the DNA of a particular plasmid in the spermidine auxotroph decreases from an average of -28 turns to -21 turns upon spermidine deprivation (Morris, 1981). Evidence is still lacking, however, that changes (if they occur) in the intracellular spermidine concentration in the normal life of a bacterial cell serve to modulate chromosome conformation.

ASSEMBLY: THE POLYSOMES

The polysomes constitute the most abundant structure of the growing bacterial cell. Our reference *E. coli* cell (Table 1 in Chapter 1) has approximately 1000 polysomes, each consisting on the average of approximately 20 ribosomes attached to mRNA, together with accessory factors, enzymes, and tRNA cycling between the soluble phase and the polysomes. And, of course, each polysome will have a spectrum of nascent polypeptide chains averaging half the size of whatever proteins it is making.

Polysomes are formed by the association of ribosomal subunits, aided by protein initiation factors, with nascent mRNA. This association is actually the initiation step of translation (protein synthesis), and we have chosen to describe this final stage of polysome assembly in the section on protein synthesis. Here we shall deal with the assembly of the two major polysome components, the 30 S and 50 S ribosomal subunits.

The growing *E. coli* cell contains no significant pools of either rRNA or ribosomal proteins (r-proteins). So once again we meet an assembly process that is rapid and efficient. In fact, it will be helpful to keep in mind as we review the complex process of ribosome assembly that the whole job takes 2–5 minutes in the growing cell.

Processing of rRNA

There are seven gene sets (*rrnA* to *rrnG*) that code for rRNA in *E. coli*. Their location on the chromosome is given in Figure 17. The general structure of each of these transcriptional units is:

16 S gene–spacer tRNA–23 S gene–5 S gene–[distal tRNA]

Not all of the seven gene sets have a distal tRNA gene. Table 2 describes which tRNA genes are associated with each of the units.

This family of genes is believed to be required for the very high rate of rRNA synthesis needed during rapid growth. The PROMOTER regions recognized by RNA polymerase have been sequenced for five of these units, yielding the surprising result that each has two promoters in tandem: the first approximately 310 nucleotides and the second approximately 190 nucleotides from the start of the 16 S rRNA gene. In addition, there may be other, weaker, binding or storage sites for RNA polymerase. Presumably these all help guarantee a great transcription potential for rRNA.

It is not known why a few tRNA genes are included in the operons coding for rRNA. The existence of one gene per operon for each rRNA species would seem to ensure their equimolar synthesis, and therefore the meaning of two 5 S rRNA sequences in the *rrnD* operon is unclear.

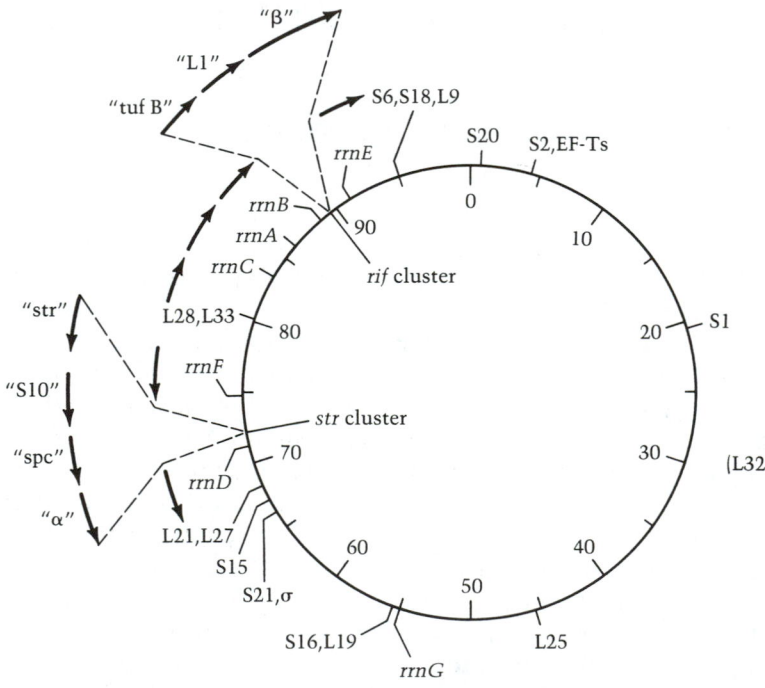

Figure 17. Genetic map of *E. coli* K12 chromosome showing locations of genes for rRNA, r-proteins, RNA polymerase subunits, and peptide elongation factors. Where it is known, the direction of transcription is indicated by arrows. Gene sets of rRNA are indicated by *rrn"x"*. Genes coding for proteins are indicated by the names of their products. Transcription units in the *str* and *rif* clusters are identified by the operon "names." (From Lindahl, 1982.)

Table 2. Ribosomal RNA transcription units of *E. coli* K12: chromosomal locations and associated tRNA genes[a]

rRNA operon	Map position (minutes)	Spacer tRNA genes	Distal tRNA genes
rrnA	86	tRNA^{Ile}, tRNA^{Ala}	—
rrnB	89	tRNA^{Glu}	—
rrnC	84	tRNA^{Glu}	tRNA^{Asp}, tRNA^{Trp}
rrnD	72	tRNA^{Ile}, tRNA^{Ala}	tRNA^{Thr}
rrnE	90	tRNA^{Glu}	—
rrnF	74	[tRNA^{Ile}, tRNA^{Ala}]	[tRNA^{Asp}]
rrnG	56	tRNA^{Glu}	—

[a]The brackets around the genes of *rrnF* indicate that the information is tentative. From Lindahl, 1982.

At any rate, simple transcription of these operons would yield a rather useless hybrid giant. This does not occur. A group of at least four ribonucleases (RNases) operate with interesting site specificity on the transcript as it is being produced. These four enzymes (RNase III, RNase P, RNase F, and RNase E) perform PRIMARY PROCESSING, cutting the transcript into pieces that are precursors of the various RNA species (thus, one p16 S rRNA, one p23 S rRNA, one p5 S rRNA, and one or more ptRNA molecules). These endonucleolytic cuts occur so closely on the heels of transcription by RNA polymerase that an intact transcript of the operon is not normally produced.

Ribosomal proteins begin to bind to the rRNA sequences as they are transcribed. Therefore, ribonucleoprotein particles form containing the p16 S rRNA and the p23 S rRNA plus p5 S rRNA. At this point SECONDARY PROCESSING begins, utilizing at least three additional RNases: RNase M16, RNase M23, and RNase M5, for the final trimming to size of the 16 S, 23 S, and 5 S rRNA precursors, respectively.

A summary of these primary and secondary processing reactions is presented in Figure 18. Additional details are presented in the review by Apirion (1980).

In addition, specific enzymes catalyze the transfer of methyl groups from *S*-adenosylmethionine to certain bases in each of the rRNA molecules.

Assembly of r-proteins

There are approximately 20 r-protein operons, only some of which are shown in Figure 17. The task of coordinating the synthesis of r-proteins is therefore far from simple, and the rationale for the particular organization of r-protein genes in *E. coli* has not been discovered. Later, in Chapter 7, we shall examine in detail what is known about the regulation of

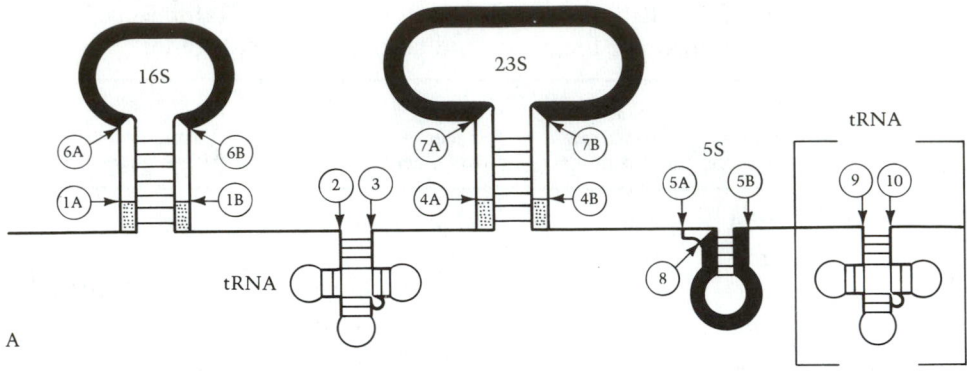

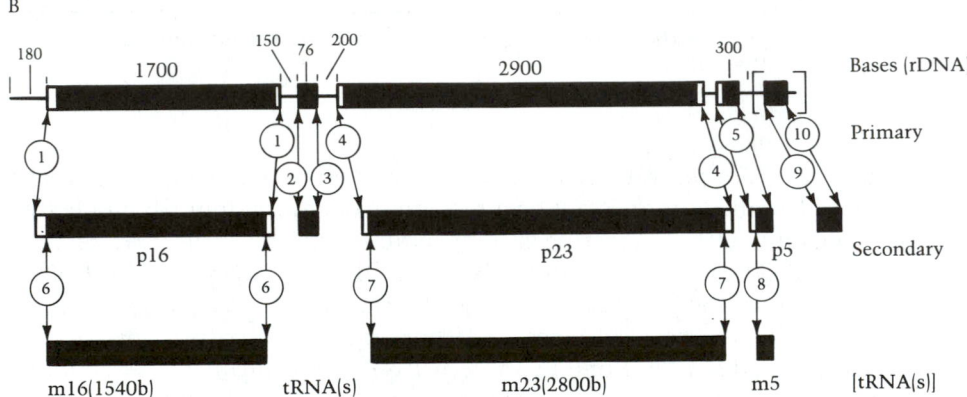

Figure 18. Processing map of ribosomal RNA. A. The primary transcript from an *rrr* gene is shown with its hypothetical secondary structure. The arrows with circled numbers indicate cleavage sites for the processing enzymes: Cuts 1,4, RNase III; 2,9, RNase P; 3,10, RNase F; 5, RNase E; 6, RNase M16; 7, RNase "M23"; 8, RNase "M5". B. The sequence of processing steps in wild-type strains is illustrated, but without indicating the binding of r-proteins to the precursor rRNAs. (From Apirion, 1981.)

r-protein synthesis. For our current purpose we need only assume an efficient and rapid synthesis of all 53 polypeptides that must be assembled to yield the ribosomal subunits.

Each subunit assembles by a series of successive reactions. The binding of 10 particular r-proteins to the p16 S rRNA leads to an intermediate that can be detected in vivo—apparently the next step is a rate-limiting one. This intermediate has a sedimentation value of 20 S and is called the p30 S. After whatever change must occur (probably a conformational one), the remaining proteins are added, the p16 S rRNA is processed completely, and active 30 S ribosomal subunits are produced.

From in vitro studies on the reconstitution of ribosomes from their purified components (work done largely by Nomura and his colleagues in the late 1960s), the following general sequence of assembly steps, starting with mature 16 S RNA, has been developed (Held, 1973):

$$\text{16 S RNA} + \begin{bmatrix} S4,S5,S6,S7/S8,S9, \\ S11,S13,S15,S16, \\ S17,S18,S19,S20 \end{bmatrix} \xrightarrow{0°C} \text{RI}_{30} \text{ particles} \qquad \text{(Step 1)}$$

$$\text{RI}_{30} \text{ particles} \xrightarrow{37°C} \text{RI}_{30}^* \text{ particles} \qquad \text{(Step 2)}$$

$$\text{RI}_{30}^* \text{ particles} + \begin{bmatrix} S1,S2,S3, \\ S10,S14,S21 \end{bmatrix} \xrightarrow{0°C} \text{30 S particles} \qquad \text{(Step 3)}$$

The precise nature of the heat-induced conversion of the intermediate ribosome particle (RI_{30}) to its activated form (RI_{30}^*) is unknown. A complete assembly map showing more detail is given in Figure 19A.

Solution of the in vitro assembly of the 50 S subunit has been more difficult. The general steps observed in vitro can be summarized in an analogous way as follows (Spillman, 1977):

$$\text{28 S RNA} + \text{5 S RNA} + \begin{bmatrix} L1,L2,L3, \\ L4,L5, \\ L8/9,L10, \\ L11,L13, \\ L19–L24, \\ L29,L33 \end{bmatrix} \xrightarrow[\text{4 mM Mg}^{2+}]{0°C} \text{RI}_{50}\ (1) \text{ particles} \qquad \text{(Step 1)}$$

$$\text{RI}_{50}\ (1) \text{ particles} \xrightarrow{44°C} \text{RI}_{50}^*\ (1) \text{ particles} \qquad \text{(Step 2)}$$

$$\text{RI}_{50}^*\ (1) \text{ particles} + \begin{bmatrix} L6,L15,L16, \\ L25,L27,L28, \\ L30,L32 \end{bmatrix} \xrightarrow[\text{4 m}M^{2+}]{44°C} \text{RI}_{50}\ (2) \qquad \text{(Step 3)}$$

$$\text{RI}_{50}\ (2) \xrightarrow[\text{20 m}M \text{ Mg}^{2+}]{50°C} \text{50 S particles} \qquad \text{(Step 4)}$$

The position of six proteins (L7/L12, L14, L18, L26, L31, and L34) have not been rigorously assigned. Conversion of the two intermediate particles, $\text{RI}_{50}(1)$ and $\text{RI}_{50}(2)$, to their activated forms, $\text{RI}_{50}^*(1)$ and $\text{RI}_{50}^*(2)$, is not understood in detail.

It has long been suspected that some morphogenetic factors might operate in the cell to accomplish the steps in ribosome biogenesis that require heating in vitro, but none have been found.

In Figure 19 additional details of the in vitro assembly processes are given. It should be kept in mind that the assembly of ribosomes in the cell can be expected to follow different, though similar paths.

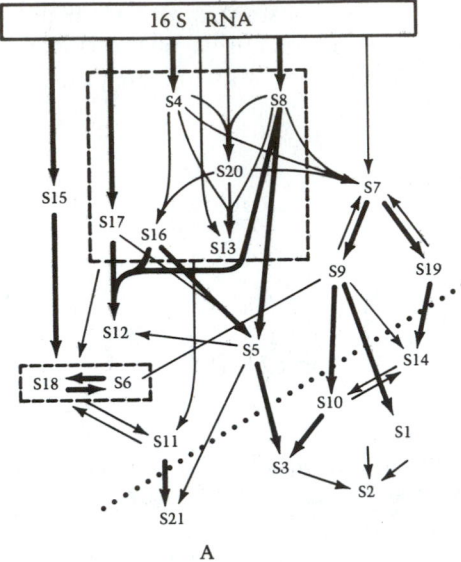

A

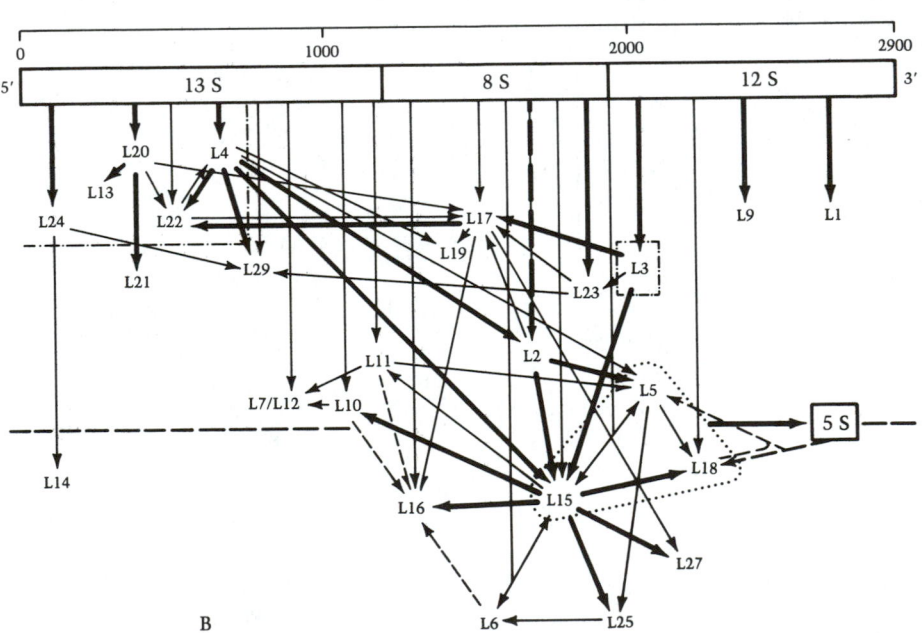

B

Figure 19.

Assembly maps of ribosomal subunits. Arrows indicate the facilitating effect on the binding of one protein by another. Thick arrows indicate major relationships, thin arrows weaker relationships. Arrows from RNA to a protein indicate binding of that protein directly to the RNA in the absence of other proteins. For other details consult the appropriate reference. A. The 30 S subunit. Proteins above the dotted line are those that form the RI_{30} particles. (Modified from Held, 1974.) The assembly reaction of protein S1 was added by Laughrea, 1978. B. The 50 S subunit. The horizontal broken line separates proteins of the $RI_{50}^*(1)$ particle from the ones added to give the $RI_{50}(2)$ particle. See text for more detail. (From Nierhaus, 1982.)

ASSEMBLY: THE CYTOSOL

If one could very carefully remove the envelope, the nucleoid, and the polysomes from a bacterial cell, what would remain is the cytosol. Mostly it is protein (approximately 80% by dry weight), but it also contains tRNA and a variable amount of storage polymer (glycogen in *E. coli*), plus small quantities of an extremely large number of metabolites, building blocks (sugars, amino acids, nucleotides, etc.), vitamins, cofactors, ATP, and inorganic ions. All but a small fraction of the fueling and biosynthetic reactions of cell metabolism occur here. Their products (building blocks and ATP) can be supplied directly to the macromolecule-producing machinery of the cell, because no membranes separate the cytosol from the polysomes (where proteins are made), the surface of the nucleoid (where RNA and DNA are made), or the inner surface of the cell membrane (where wall and membrane subunits are found).

This barrier-free design facilitates the high rate of metabolism and growth of bacterial cells and has implications for regulation of gene expression as well.

It is a matter of some debate whether the cytosol is highly organized or is a disordered solution: Is the cytosol a "bag of enzymes," or must it be assembled in some way to create structural order?

The answer is not known with accuracy, but some information is available. First, the cytosol is a concentrated solution; the total solute concentration is very high. (One limitation to in vitro studies of biochemical reactions and processes is that cell-free extracts rarely approach the protein concentration in the cytosol.) As a result it is quite possible that protein–protein interactions can provide a degree of molecular organization in vivo that is totally lost upon rupture of the cell and dilution of its contents. It seems likely that enzymes with cooperative or sequential functions (as, for example, the members of a biosynthetic pathway) are in physical contact in more or less loose clusters.

Direct as well as indirect evidence supports this view. Certain enzymes of the arginine biosynthetic pathway, for example, are mutually attracted and aggregate even in solutions more dilute than in vivo. Similar evidence exists for other pathways. The indirect evidence is more extensive: concentrations of most metabolic intermediates in biosynthetic pathways are vanishingly low, as though these molecules are handed batonlike from one enzyme to the next.

This situation is far from universal—and for good reason. The products formed by many enzymes, particularly products of the fueling reactions, must be supplied to many pathways that branch out from these central reactions. In these cases, the metabolites are found in readily measurable amounts. The rapidity of solute diffusion over very small distances compensates for what is likely to be an enzyme system with little structural order.

Whatever order exists in the cytosol is probably brought about by self-assembly of protein aggregates. Other potential sources of order could include miniature chemical gradients—as, for example, might be created by ATP generation at the cell membrane and ATP consumption in central regions of the cell. Nothing is known of such matters. In general, studies with radioactive tracers have provided no convincing evidence of diffusion barriers within the cytosol of *E. coli* and most other bacteria.

There are some interesting exceptions to this picture. The cytosol of certain bacterial species (of considerable ecological importance) contain organelles consisting of vesicles bounded by protein (rather than phospholipid) membranes. Many phototrophic and chemolithotrophic aquatic bacteria produce GAS VESICLES believed to regulate the buoyancy of the cell and thereby enable it to float at an appropriate depth in the water. Cells of one group of photosynthetic green bacteria enclose all their photosynthetic pigments in cigar-shaped bodies called CHLOROBIUM VESICLES; a large number of photosynthetic bacteria, nitrifying bacteria, and thiobacilli (sulfur oxidizers) contain POLYHEDRAL BODIES or CARBOXYSOMES; these vesicles contain an enzyme, carboxydismutase, that is important in CO_2 fixation, which presumably occurs within them. Neither the proteins that form these various vesicles nor the assembly process have been thoroughly studied. References may be found in Stanier (1976).

The remaining assembly reactions of the cytosol of *E. coli* and other bacteria are actually macromolecule-processing reactions. Transfer RNA, whether originating as a joint transcript with rRNA or separately, must be chemically modified by a moderately large number of reactions. These modifications, carried out by a battery of specific enzymes, occur throughout the synthesis and folding of the primary transcript. In general, procaryotic tRNA is somewhat less decorated than eucaryotic tRNA, but nevertheless a wide variety of specific methylations, thiolations, and hydrogenations of the four major nucleosides occur, along with the construction of elaborate substituents that hypermodify certain bases at particular positions. Because there is at least one tRNA-modifying enzyme for each kind of modified nucleoside at each specific position, there must be over 40 such enzymes in *E. coli*. Only a few have been studied intensively (for reviews, see Bjork, 1982; Soll, 1980).

The polypeptides of many proteins must be processed after their translation. In many cases these modifications are authentic assembly reactions—cleavage of the signal peptide from secreted proteins and addition of the diglyceride to the lipoprotein of the outer membrane are two examples we have already encountered—but in many others they are reactions that modulate the enzymatic activity of the protein rather than determine its assembly.

In the assembly category one would include the attachment of prosthetic groups and the binding of cofactors (organic and inorganic) to enzymes. In the modulation category are reactions that methylate, acetylate, adenylylate, or phosphorylate proteins. Usually these modifications can be reversed, and the cell can swiftly activate or inactivate enzymes by these devices. Table 3 lists a few examples. In bacteria there may be fewer proteins subject to these covalent modifications to regulate activity than in slower growing, multicellular organisms. Induction and repression of enzymes in bacteria are very fast and very effective devices in rapidly growing cells and, of course, they are more economical than building an unused protein.

The granular inclusions found in many bacteria during certain growth conditions (e.g., limiting nitrogen but ample carbon supply) consist of reserve material: either glycogen, poly-β-hydroxybutyrate, or both. Procaryotic cells do not store neutral fats. Glycogen is deposited fairly evenly throughout the cytoplasm, and its granules are detectable usually only by electron microscopy; no special assembly reaction is involved. The granules of poly-β-hy-

Table 3. Examples of protein modifications in *E. coli*

Type of modification	Proteins
Chiefly assembly reactions	
signal peptide cleavage	secreted proteins, periplasmic binding proteins
formation of -S-S-bonds	many enzymes
addition of lipid moiety	murein lipoprotein
addition of sugar moiety	membrane glycoprotein
attachment of prosthetic group	many enzymes
Chiefly modulation reactions	
phosphorylation	ribosomal protein S6, isocitrate dehydrogenase
methylation	chemotaxis signal transducers
acetylation	ribosomal protein L7
adenylylation	glutamine synthetase

droxybutyrate are larger and readily seen as refractile bodies with light microscopy. In many species these granules appear to be bounded by a protein membrane, about which little is known. Similarly, granules of polyphosphate (VOLUTIN GRANULES) and of inorganic sulfur, where they occur, are bounded by a protein, non-unit membrane. Nothing is known of the assembly of these protein, membrane-bounded granules. What is known of their structure has been comprehensively reviewed by Shively (1974).

SUMMARY

1. The bacterial cell synthesizes itself by as many as 1000–2000 chemical reactions. These can be usefully organized as (1) assembly, (2) polymerization, (3) biosynthetic, and (4) fueling reactions based on their primary function in growth. Assembly reactions involve the association of macromolecules to form cellular structures: envelope, appendages, nucleoid, polysomes, inclusions, and enzyme complexes. Polymerization reactions consist of directed, sequential linking of activated molecules, called building blocks (amino acids, nucleotides, sugars, and fatty acids), into large polymers. Biosynthetic reactions produce the building blocks needed for polymerization, as well as coenzymes and related metabolic factors; the various biosynthetic pathways begin with one or more of a small group of precursor metabolites. Fueling reactions produce these precursors, plus the ATP and reducing power needed for biosynthesis.
2. Assembly of cell structures occurs both by spontaneous aggregation (self-assembly) and by special, specific mechanisms. Some macromolecules are made at the sites of assembly and others must be transported to them.

3. Assembly of the cell membrane occurs by intercalation of protein and phospholipid in defined regions. Enzymes responsible for phospholipid synthesis are in the membrane. Proteins to be incorporated into the membrane may be made on membrane-attached ribosomes but are not, in normal cases, made with a signal sequence; intercalation by a membrane-trigger mechanism based on the membrane-induced alteration of the conformational state of the protein seems more likely as the normal mechanism. Lipid–lipid and lipid–protein interactions determine the specific composition of the two leaflets. Hydrophobicity is an important driving force in membrane assembly, but specific, directive forces must also be at work.

4. Proteins secreted into the periplasmic space of Gram-negative cells appear to be made initially with a signal peptide sequence that facilitates passage through the cell membrane.

5. Assembly of the peptidoglycan layer is accomplished by polymerization of the glycan chain after translocation of the precursors on a lipid carrier through the cell membrane. In Gram-positive cells, the thick multilayer peptidoglycan is in a dynamic state of synthesis at the membrane surface and sloughing off at the periphery.

6. Assembly of the outer membrane of Gram-negative cells occurs by independent assembly of each component (phospholipids, lipopolysaccharide, and proteins) rather than of multicomponent subunits. LPS subunits are synthesized in the cell membrane in two parallel processes: one produces the core polysaccharide on lipid A, the other the repeating polysaccharide side chain. A transfer enzyme unites the two parts after translocation across the cell membrane. Phospholipids are made in the cell membrane and interact with lipid A and neighboring lipids to self-assemble in the inner leaflet of the outer membrane. Outer membrane proteins, probably made on attached ribosomes, as translocated by some process like the bent loop model to the outer membrane, where LPS facilitates intercalation. Transport of proteins, LPS, and phospholipids to the outer membrane occurs at adhesive patches (called Bayer's junctions) where outer membrane and cell membrane are in intimate association. A special lipoprotein is important in anchoring the outer membrane to the peptidoglycan layer. Some of the major outer membrane proteins, called porins, self-aggregate to form diffusion pores.

7. Flagella are formed first by self-assembly of the proteins of the basal body and their appropriate intercalation into the envelope layers, followed by synthesis of flagellin and its transport through the center of the growing flagella to the tip of its filament.

8. Capsular material assembles outside the cell from subunits that are transported through the membrane(s) by a lipid carrier.

9. The nucleoid forms by processes that involve methylation, supercoiling, and folding of newly replicated DNA. Some of these are enzymatic reactions that are well understood, but fascinating puzzles about nucleoid assembly (and function) remain.

10. Polysomes form by the factor-promoted association of ribosomal subunits to special sequences on mRNA. The ribosomal particles themselves are the product of elaborate assembly reactions that involve the processing of 5 S, 16 S, and 23 S rRNA and the sequential attachment of over 50 different proteins to those molecules to generate the 30 S and 50 S ribosomal subunits. In vitro reconstitution of these structures from purified components has been achieved, but the factors that might guide this assembly in vivo are unknown.

11. The cytosol is a fluid structure but probably has a loose organization brought about by affinities among proteins with cooperative or sequential functions. Some polypeptides must be processed in a variety of ways to become functional proteins. Some bacterial species contain organelles within the cytosol, including vesicles of various sorts bounded by protein membranes; their assembly has not been thoroughly studied.

REFERENCES

Apirion, D. and P. Gegenheimer, 1980. Processing of bacterial RNA. FEBS Letters 125:1.

Bassford, P. J., T. J. Silhavy and J. R. Beckwith. 1979. Use of gene fusion to study secretion of maltose-binding protein into *Escherichia coli* periplasm. J. Bacteriol. 139:19.

Bayer, M. E. 1968. Areas of adhesion between wall and membrane of *Escherichia coli*. J. Gen. Microbiol. 53:345.

Bayer, M. E. 1975. Role of adhesion zones in bacterial cell-surface function and biogenesis. In *Membrane Biogenesis*, A. Tzagloff, ed. Plenum Press, New York, p. 393.

Bayer, M. E. and H. Thurow. 1977. Polysaccharide capsule of *Escherichia coli*: microscopic study of its size, structure, and sites of synthesis. J. Bacteriol. 130:911.

Begg, K. J. and W. D. Donachie. 1977. Growth of *Escherichia coli* cell surface. J. Bacteriol. 129:1524.

Beher, M. G. and C. A. Schnaitman. 1981. Regulation of the OmpA outer membrane protein of *Escherichia coli*. J. Bacteriol. 147:972.

Björk, G. R., A. S. Byström, T. G. Hägervall, K. J. Hjalmarsson, K. Kjellin-Stråby and P. H. R. Lindström. 1982. In *Transmethylation*, E. Ustin, R. T. Borchardt and C. R. Creveling, eds. Macmillan Publishers Ltd., London.

Buchanan, C. E. 1981. Topographical distribution of penicillin-binding proteins in the *Escherichia coli* membrane. J. Bacteriol. 145:1293.

Cole, R. M. and J. J. Hahn. 1962. Cell wall replication in *Streptococcus pyogenes*. Science 135:722.

Cronan, J. E., Jr. 1978. Molecular biology of bacterial membrane lipids. Ann. Rev. Biochem. 47:163.

Datta, D. B., C. Kramer and U. Henning. 1976. Diploidy for a structural gene specifying a major protein of the outer cell envelope membrane from *Escherichia coli* K12. J. Bacteriol. 128:834.

Davis, B. D. and P. -C. Tai. 1980. The mechanism of protein secretion across membranes. Nature 283:433.

Donohue-Rolfe, A. M. and M. Schaechter. 1980. Translocation of phospholipids from the inner to the outer membrane of *Escherichia coli*. Proc. Natl. Acad. Sci. USA 77:1867.

Foulds, J. and T. -J. Chai. 1978a. Chromosomal location of a gene (*nmpA*) involved in expression of a major outer membrane protein in *Escherichia coli*. J. Bacteriol. 136:501.

Foulds, J. and T. -J. Chai. 1978b. A new major outer membrane protein found in an *Escherichia coli tolF* mutant resistant to phage *tulb*. J. Bacteriol. 133:1478.

Frehel, C. and A. Ryter. 1979. Peptidoglycan turnover during growth of a *Bacillus megaterium* DAP-LYS-mutant. J. Bacteriol. 137:947.

Held, W. A. and M. Nomura. 1973. Rate-determining step in reconstitution of *Escherichia coli* 30S ribosomal subunits. Biochemistry 12:3273.

Inouye, M. and S. Halegoua. 1980a. Secretion and membrane localization of proteins in *Escherichia coli*. CRC Critical Reviews in Biochemistry.

Joseleau-Petit, D. and A. Képès. 1980. Phospholipid synthesis and turnover during the cell cycle in *E. coli*. In *Duplication of Bacteria*. EMBO Workshop.

Képès, A. and F. Autissier. 1972. Topology of membrane growth in bacteria. Biochim. Biophys. Acta. 265:443.

Modrich, P. 1979. Structures and mechanisms of DNA restriction and modification enzymes. Q. Rev. Biophys. 12:315.

Morris, D. R. and D. Lockshon. 1981. Polyamines, chromosome conformation, and macromolecular synthesis in prokaryotes: a hypothesis. In *Advances in Polyamine Research*, Vol. 3, C. M. Caldarera, ed. Raven Press, New York.

Osborn, M. J. and H. C. P. Wu. 1980. Proteins of the outer membrane of Gram-negative bacteria. Ann. Rev. Microbiol. 34:369.

Pierucci, O. 1980. Phospholipid and membrane protein synthesis during the division cycle of *Escherichia coli* B/r. In *Duplication of Bacteria*, EMBO Workshop.

Pooley, H. M. 1978. Localized insertion of new cell wall in *Bacillus subtilis*. Nature 274:264.

Shively, J. M. 1974. Inclusion bodies of prokaryotes. Ann. Rev. Microbiol. 28:167.

Soll, D., J. Abelson and R. R. Schimmel. 1980. *Transfer RNA: Biological Aspects*. Cold Spring Harbor Laboratory, Cold Spring Harbor, New York.

Spillman, S., F. Dohme and K. H. Nierhaus. 1977. Assembly in vitro of the 50S subunit from *Escherichia coli* ribosome: proteins essential for the first heat-dependent conformational change. J. Mol. Biol. 115:513.

Spratt, B. G. 1977. Properties of the penicillin-binding proteins of *Escherichia coli* K12. Eur. J. Biochem. 72:341.

Troy, F. A., II. 1979. The chemistry and biosynthesis of selected bacterial capsular polymers. Ann. Rev. Microbiol. 33:519.

Verwer, R. W. H. and N. Nanninga. 1980. Pattern of meso-DL-2,6-diaminopimelic acid incorporation during the division cycle of *Escherichia coli*. J. Bacteriol. 144:327.

Wickner, W. 1979. The assembly of proteins into biological membranes: the membrane trigger hypothesis. Ann. Rev. Biochem. 48:23.

Wickner, W. 1980. Assembly of proteins into membranes. Science 210:861.

Yamada, H. and S. Mizushima. 1978. Reconstitution of an ordered structure from major outer membrane constituents and the lipoprotein-bearing peptidoglycan sacculus of *Escherichia coli*. J. Bacteriol. 135:1024.

Yuan, R. and D. L. Hamilton. 1982. Restriction and modification of DNA by a complex protein. American Scientist 70:61-69.

Chapter Three

Chemical Synthesis of the Bacterial Cell: Polymerization, Biosynthesis, Fueling Reactions, and Transport

INTRODUCTION

The assembly processes described in Chapter 2 form cellular structures at prodigious rates. They utilize macromolecules that for the most part are chemically similar to those of eucaryotic cells but that must be supplied far faster (more than tenfold faster) than in any eucaryotic cell.

In this chapter we shall examine the polymerization, biosynthetic, and fueling reactions to learn how these rates are achieved. Our treatment of some of these topics will be somewhat unconventional because we wish to highlight certain special features of a metabolism geared to growth. Some biochemical details will be presented—but only the minimum number necessary to identify general principles of the chemistry of bacterial growth. The reader is directed to a modern text of general biochemistry [that by Stryer (1981) is particularly rich in information about *E. coli*] and to the texts on bacterial metabolism by Mandelstam and McQuillen (Mandelstam, 1973) and by Gottschalk (1979) for a detailed presentation of individual processes, particularly in *E. coli*. Other texts (e.g., Stanier, 1976) describe the metabolic variations evolved by specialized groups of bacteria, variations that permit them to grow in unusual environments. Several excellent treatments of the molecular biology of gene expression and macromolecular synthesis emphasize these processes in *E. coli* and its viruses (e.g., Watson, 1976).

To provide a framework for our discussion of the chemical synthesis of bacterial cells, let us consider a specific task—the formation of 1 g of (dry) bacterial mass of our reference population of *E. coli* B/r cells growing in glucose minimal medium (Chapter 1, Table 1). Assembly reactions (Chapter 2) require 550 mg of protein, 205 mg of RNA, 31 mg of DNA, 91 mg of phospholipids, 34 mg of lipopolysaccharide, 25 mg of peptidoglycan, and 25 mg of glycogen

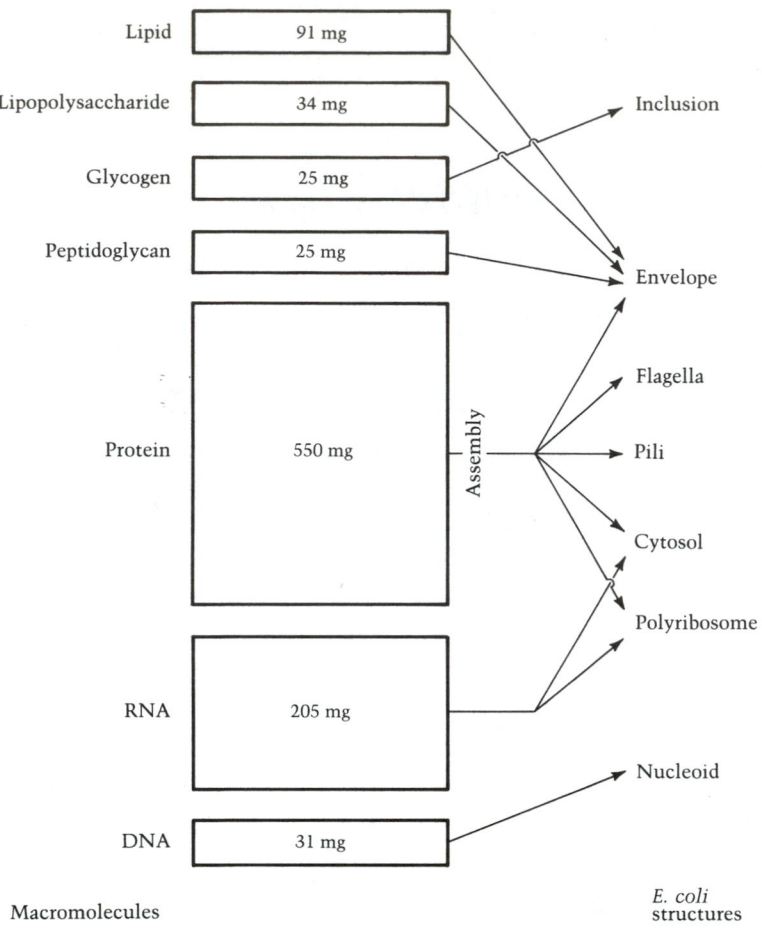

Figure 1. Macromolecules needed to produce 10^{12} cells (1 g, dry weight) of *E. coli* B/r under our reference conditions. The chemical composition of the cell described in Chapter 1, Table 1, was used to derive the amounts of each macromolecule needed for the assembly processes described in Chapter 2. The capsule has been omitted.

to produce the cellular structures of the approximately 1.05×10^{12} cells constituting this biomass (Figure 1). We shall examine how each of the macromolecules is made and shall identify the quantity of building blocks and energy (ATP) needed for these polymerization reactions. Next we shall examine how the required amounts of building blocks are made by the biosynthetic reactions. We shall then be able to calculate the requirements for intermediary metabolites, energy, and reducing power that must be generated by the fueling reactions.

POLYMERIZATION: DNA

The essence of a plausible mechanism by which DNA is replicated was included in the J. D. Watson and F. H. Crick paper (1953) that announced the double helix structure of the molecule. They stated that "each chain acts as a template for formation on itself of a new companion chain." This model suggested that at the replication point the double-stranded molecule separates and nucleotides pair with their complementary bases on the two exposed single strands and are polymerized. Such replication is semiconservative. The succeeding three decades of research have confirmed the correctness of the model but have revealed that the process biochemically is quite complex [cf. review by Ogawa (1980) and the book by Kornberg (1980)]. At least 13 proteins are present in a loose multienzyme complex, sometimes termed the REPLICATION APPARATUS, that catalyzes the reaction (Table 1). The

Table 1. Proteins involved in the discontinuous replication of *E. coli* DNA[a]

Protein	Gene	Function	Site of action in vivo (shown in Figure 3)
Protein i (X)		Prepriming	
Protein n (Z)		Prepriming	
Protein n' (Y)		DNA-dependent ATPase Prepriming	
Protein n"		Prepriming	
dnaC protein	*dnaC*	Formation of *dnaB-dnaC* Protein complex Prepriming	
dnaB protein	*dnaB*	DNA-dependent rNTPase Mobile promotor Prepriming Priming	2 4
Helix destablizing protein (DNA binding protein; single-strand binding protein)	*ssb*-1	Binding to single-stranded DNA	3
Primase	*dnaG*	Primer synthesis	4
DNA polymerase III holoenzyme (polymerase III)	*polC* (*dnaE*)	DNA elongation	2 5
γ (*DNAZ* protein)	*dnaZ*		
ρ (EF III)			
β (EF I)			
DNA polymerase I	*polA*	Primer degradation Gap filling	6
DNA ligase	*lig*	Joining of short chains	7
DNA gyrase	*gyrA* (*nalA*)	Supertwisting DNA-dependent ATPase	1
	gyrB (*cou*)	Relaxation of supercoils	
rep protein	*rep*	DNA-dependent-ATPase Strand separation	1

[a]Adapted from Ogawa (1980).

89

reaction proceeds with remarkable speed and accuracy. Almost 3000 nucleotides are polymerized per second (at 37°C), and only approximately 1 mistake (incorrect pairing) is made per 10^{10} nucleotide copies.

Strand separation, the first step in the process, requires the participation of several proteins and the utilization of ATP because energy is required to break the hydrogen bonds between the complementary bases and to unwind the strands. The cost of melting a base pair varies from approximately 1.2 kcal (for an A·T pair next to an A·T pair) to approximately 5.0 kcal (for a G·C pair next to a G·C pair); the energy available from the hydrolysis of a molecule of ATP is approximately 14 kcal. The energy cost of strand separation is reduced by the action of one of the proteins, the HELIX DESTABILIZING or HD-PROTEIN, which is present in the replication apparatus; HD-protein binds specifically and cooperatively to single-stranded DNA. A second set of proteins, including DNA-UNWINDING ENZYME I and REP PROTEIN, are collectively termed unwinding proteins. They actively separate the DNA strands while hydrolyzing ATP to ADP and P_i. Separation of strands wound in a helix, however, generates loops, termed supercoiled twists, in the single strands. The action of another protein, DNA GYRASE, prevents this, as we have already discussed in our consideration of assembly of the nucleoid. This enzyme periodically breaks a phosphodiester bond in one of the strands of the double helix, thereby allowing free rotation of the opposite strands. Later, the same enzyme reforms this bond. A schematic representation of the strand separation process is shown in Figure 2.

Once the strands are separated, 5'-deoxynucleoside triphosphates pair with their complementary bases exposed in the single-stranded region and, by the action of DNA POLYMERASES, the triphosphates are joined through their α-phosphate group by a phosphodiester bond to a terminal 3'-OH group on the growing strands; pyrophosphate (P_i-P_i) is released. Thus, in order to synthesize DNA, the DNA polymerases require (1) dATP, dGTP, dCTP, and dTTP as substrates; (2) a single-stranded region of DNA that serves as a TEMPLATE; and (3) nucleic acid with a 3'-OH end, which is termed the PRIMER of the reaction. However, the two strands of DNA in the double helix are antiparallel (Figure 2), that is, certain cuts across the double helix would reveal a free 3'-OH on one strand and a free 5'-OH on the other. Thus, the two strands of DNA at the replication fork are replicated in opposite directions: the one being replicated in the direction of movement of the fork is termed the leading strand; the other, the lagging strand (Figure 3). Replication of the leading strand is presumably largely continuous; replication of the lagging strand is necessarily discontinuous. Direct evidence for discontinuous synthesis was obtained by R. Okazaki, who exposed growing cells to [^{3}H] thymidine for brief periods (a technique termed PULSE LABELING) and observed that some newly synthesized DNA could be isolated as short, single-stranded fragments (1000 to 2000 nucleotides in length). These are termed OKAZAKI FRAGMENTS.

As stated earlier, DNA polymerases require a primer molecule; in other words, they are incapable of initiating synthesis of Okazaki fragments. This function is served by an RNA polymerase (encoded by the *dnaG* gene) that does not require a primer. After a short length of RNA is synthesized, which can serve as a primer for DNA polymerase, DNA replication proceeds.

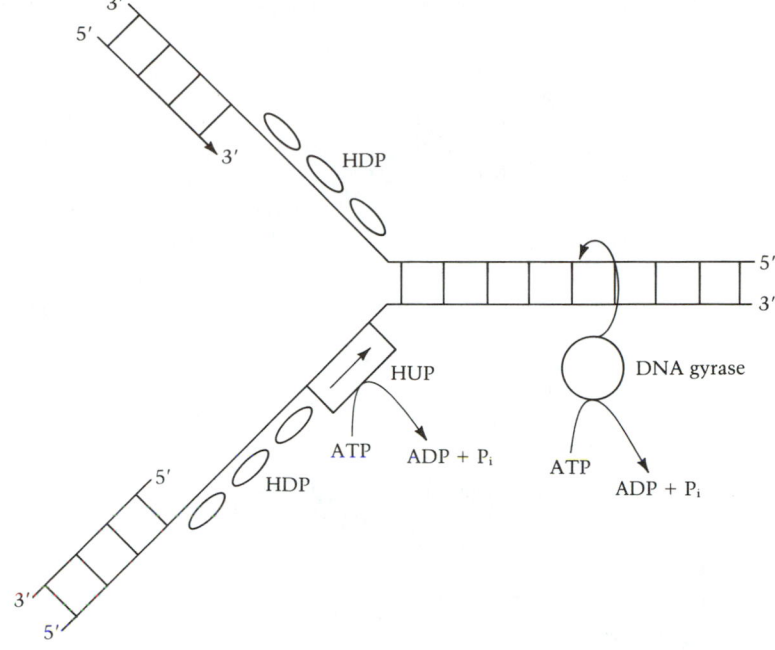

Figure 2. Schematic representation of the process of strand separation at the replication fork. The action of DNA gyrase in allowing free rotation within the double helix is illustrated by the circular arrow. A helix-unwinding protein (HUP) is shown at the point of separation; its direction of movement is indicated by the arrow. Molecules of helix-destabilizing proteins (HDP) are shown attached to the single-stranded regions. As indicated, the action of DNA gyrase and unwinding proteins involve hydrolysis of ATP. (After Alberts, 1977.)

One may ask why a DNA polymerase did not evolve that was primer-independent, thus obviating the need for an RNA primer on each Okazaki fragment. The answer seems to lie in the necessity for PROOFREADING. It has been estimated that the intrinsic mistake frequency in replication is not less than 1 incorrect base pairing per 10^5 nucleotides, a level of inaccuracy far too great to preserve the cell's required level of genetic stability. Proofreading, which reduces error frequency to approximately 10^{-10}, is accomplished partly by a built-in 3'-to-5' exonuclease activity of DNA polymerase III, which removes its own mismatches by moving backward; it does not catalyze replication in the forward direction unless there is a properly matched nucleotide pair behind it. Thus, such a self-correcting polymerase cannot start chains de novo.

The growing Okazaki fragment eventually arrives at the 5' end of the previously synthesized RNA primer molecule. The 5'-to-3' exonuclease activity of DNA polymerase I simultaneously hydrolyzes the primer, replacing it with DNA. Finally the 3'-OH-to-5'-OH gap, termed a NICK, is sealed by formation of two phosphodiester bonds, a reaction catalyzed by DNA ligase, which utilizes NAD as a phosphate donor.

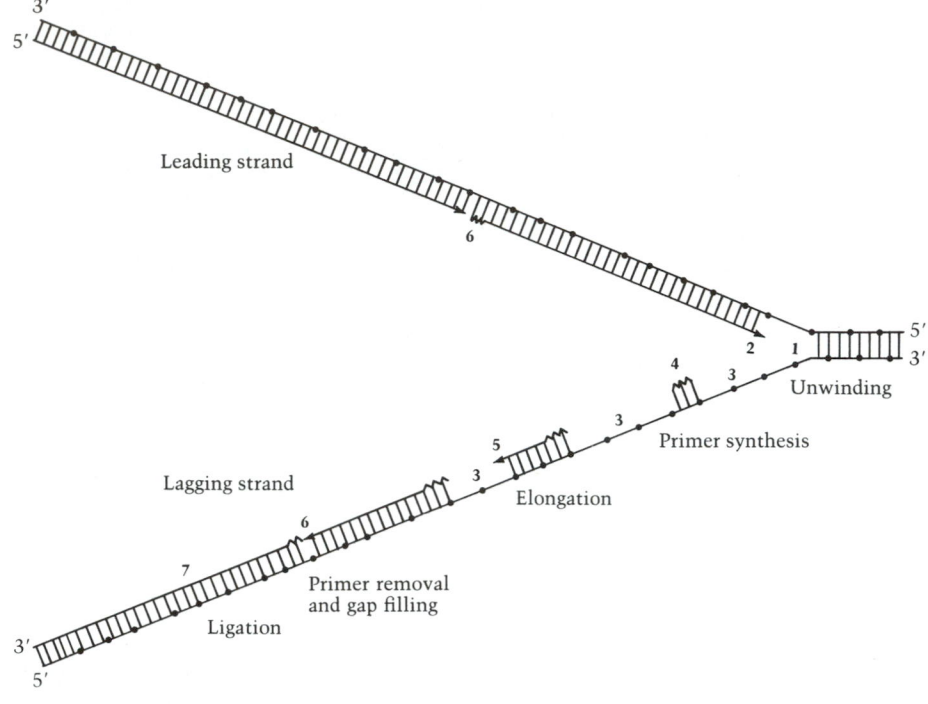

Figure 3. DNA synthesis at a replication fork. The two separated strands are replicated in opposite directions. In the leading strand, replication is continuous and proceeds in the direction (indicated by the arrow) of fork migration; in the lagging strand, replication occurs in the opposite direction and is discontinuous, creating short chains termed Okazaki fragments. Synthesis of an Okazaki fragment on the lagging strand begins with the polymerization of a short piece of RNA (wiggly line) that serves as a primer for the synthesis of DNA (straight line). When a previously synthesized primer is reached, it is degraded as DNA synthesis continues. The final nick is sealed by the action of DNA ligase. As the Okazaki fragment is synthesized, strand separation continues and exposes a length of single-stranded DNA where the next Okazaki fragment will be made. Closed circles in the parental strand represent the potential initiation sites for primer synthesis. Sites of action of replication proteins (Table 1) are indicated by the numbers. (From Ogawa, 1980.)

To this point we have discussed the movement of a preexisting replication fork. Replication of the bacterial chromosome begins at a specific site, termed *oriC*; two replication forks are formed and travel in opposite directions until they meet at the terminus. Then the two daughter chromosomes separate. Thus, replication of the bacterial chromosome is said to be BIDIRECTIONAL. The experimental basis for bidirectional replication is presented in Chapter 6, along with evidence that replication of the chromosome requires 40–45 minutes in *E. coli* under our reference conditions. The chromosome comprises approximately 3.8×10^6 base pairs, so DNA must polymerize at an overall rate of 1500 base pairs per second (3.8×10^6 base pairs per 2600 seconds), with each fork progressing at 750 base pairs per second.

Replication of the bacterial chromosome is regulated at the step of initiation: the frequency of initiation events (which varies with growth rate) rather than the rate of polymerization (which is essentially constant) determines the cell's complement of DNA (Chapter 6). The biochemical details of initiation are only incompletely understood, but sufficient information has accumulated to establish that they are complex. A number of proteins participate. One of these proteins must be consumed since protein synthesis is necessary at or shortly before the time of initiation. In addition, RNA synthesis is required, probably to form a primer. The components of the initiation system have been identified largely by analysis of heat-sensitive mutants (Chapter 4) that are unable to initiate chromosome replication at the restrictive temperature. Mutations in these strains define at least five distinct genes: *dnaA*, *dnaB*, *dnaC*, *dnaP*, and *dnaI*. The protein products of these genes probably form part of a multicomponent complex, possibly located in the membrane, that interacts with RNA polymerase and the *oriC* locus itself. Evidence for the existence of a complex and its interactions also comes from mutant analysis. Certain secondary (suppressor) mutations (*dasA–G*) that overcome the effects of mutations in *dnaA* have been shown to be located near *oriC*. One, *dasD*, maps in the structural gene *rpoB*, which encodes a subunit of RNA polymerase. Suppression (or correction of the original defect) is allele specific, that is, only certain *das* mutations are capable of suppressing a given mutation in *dnaA*. Thus, the *das* mutation does not substitute for *dnaA* function, rather it must alter a member of the complex in such a way that when it binds with the mutant *dnaA* protein in the initiation complex its function is restored.

Factors contributing to the speed of replication

The intrinsic speed of the bacterial replication apparatus (750 base pairs polymerized per second at each fork) is greater than that of eucaryotic cells. It is 10 times faster than the rate of function of RNA polymerase. The replication rate is all the more remarkable considering the great need for accuracy. It seems likely, therefore, that the proofreading function of DNA polymerases in bacteria evolved hand-in-hand with the development of rapid strand-separating and polymerizing activities of the replication apparatus.

The modification and assembly reactions (Chapter 2) occur as replication is proceeding, and they keep pace with it; within 20 minutes of completion of a new DNA molecule, it is segregated into a new daughter cell. Furthermore, as we shall see in Chapter 6, during rapid growth a new round of replication is initiated before the former round has been completed. Multiple replication forks provide an explanation of how the cell, in a rich medium, can double its DNA in less time than is required for one pair of replication forks to synthesize a new DNA molecule. A corollary is that there is no period in the cell cycle during which DNA synthesis cannot occur (i.e., no obligatory "G" period).

It is the inclusion of the entire genome in a single molecule (chromosome) that makes the procaryotic mode of DNA replication and cell division workable. Many of the streamlined features of DNA structure, synthesis, and function in bacteria are made possible by not having to segregate a set of multiple chromosomes accurately into daughter cells.

DNA repair

Owing to the special role of DNA as a repository of information gained throughout evolutionary history, a variety of cellular processes have developed to preserve its integrity. We have already discussed the proofreading mechanism, which minimizes errors introduced during replication. Proofreading by itself is insufficient because the completed molecule is susceptible to chemical damage, and therefore a set of cellular mechanisms, some of them redundant, have evolved to repair such damage.

Although DNA, as compared with RNA or protein, is a remarkably stable macromolecule, chemical damage to it occurs at a significant rate in most environments. A base may be altered chemically by hydroxylation, alkylation, or deamination; the two strands may be chemically cross-linked; large molecules may be attached to one of the strands; dimerization may occur between adjacent pyrimidines. PYRIMIDINE DIMERS, a product of exposure to ultraviolet (UV) radiation, are most frequently formed between molecules of thymine, but they also are formed between cytosines or between a thymine and a cytosine. A dimer is formed when the double bond between carbons 5 and 6 opens on each of two adjacent pyrimidines and new bonds are established between the two molecules to form a four-carbon ring (Figure 4). Once formed, the pyrimidines that constitute the dimer are pulled closer together, an action that distorts the backbone of the double helix and renders impossible the hydrogen bonding of the affected pyrimidines and the two bases on either side of them. A pyrimidine dimer interrupts both replication and transcription.

A variety of bacterial mechanisms can repair DNA that has a pyrimidine dimer defect. The existence of one, termed PHOTOREACTIVATION, became apparent when it was observed that the otherwise lethal effects on bacteria of irradiation by UV light could be largely prevented by subsequent exposure to short-wavelength, visible light or long-wavelength, UV light. The process is enzymatic. A protein binds to the dimer, absorbs light energy, and cleaves the bonds that join the adjacent pyrimidines, thereby directly reversing the reaction by which the dimer was formed (Figure 5B).

A second, more general mechanism (because it corrects a variety of defects

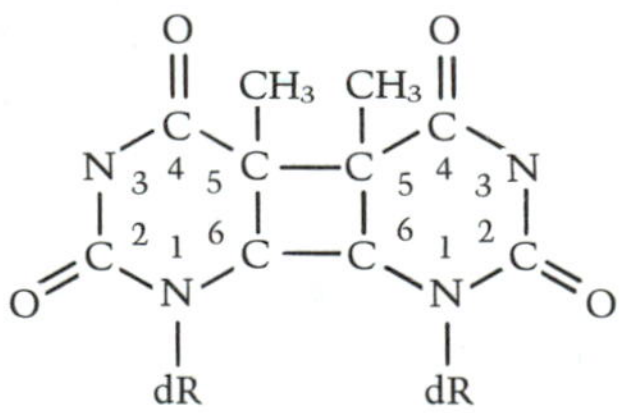

Figure 4.

Structure of a thymine dimer. Thymine dimers are the most frequent type of pyrimidine dimers to form in DNA during UV irradiation. The monomers of the dimer are attached to adjacent deoxyribose (dR) residues in one strand of the double helix.

in addition to pyrimidine dimers) is termed EXCISION REPAIR or, sometimes, DARK REPAIR (because it can occur in the complete absence of light). This mechanism involves the participation of at least five proteins: DNA ligase, DNA polymerase I, and those proteins encoded by the genes *uvrA*, *uvrB*, and *uvrC*. The protein products of *uvrA* and *uvrB* bind to the damaged region of DNA, and in the presence of the *uvrC* product they nick (introduce a single-strand break) the affected strand at the 5′ end of the damaged region (Figure 5A). Then, still in the presence of the *uvrC* product, DNA polymerase binds to the nick and adds nucleotides that are complementary to those on the intact strand. When the damaged region has been passed, the polymerase introduces a second nick, releasing a single-stranded fragment bearing the damaged region. The polymerase may then continue for a variable distance, polymerizing and releasing individual nucleotides, thereby moving the nick in the 5′-to-3′ direction. Finally the nick is sealed by the action of DNA ligase.

If the replication fork passes over the damaged region before it has been corrected by photoreactivation or excision repair, replication is interrupted over the damaged region because base pairing is not possible within it (Figure 5C). Replication reinitiates beyond the damage. The resulting lesion (a damaged region opposite a single-strand gap) cannot be repaired by excision because it lacks a complementary strand for DNA polymerase I to use as a template. However, it can be repaired by a process termed POSTREPLICATION REPAIR. By a recombinational event mediated by the *recA* gene product (see Chapter 4), the complementary strand to the damaged region, which now is present in the sister duplex, is inserted across from the damaged region. The products of the *recA*-mediated exchange are one strand with a single-strand gap and another intact duplex bearing the damaged region. The gap can be repaired by the combined activities of DNA polymerase I and DNA ligase; the damaged region can be repaired by excision repair.

DNA that is damaged because it contains chemically altered purine and pyrimidine bases can be repaired by a class of enzymes termed N-GLYCOSIDASES. One such N-glycosidase recognizes uracil in DNA and cleaves the N-glycosidic bond by which it is attached to the sugar–phosphate backbone of the double helix. Endonuclease II then attaches to the backbone at the 3′ position of the now-baseless deoxyribose and now cleaves it. The 3′-OH end thus formed creates a site at which exonuclease III can act to remove the unsubstituted deoxyribose and probably subsequent nucleotides as well. Finally the single-strand gap is refilled and ligated by the process previously described in excision repair.

It is interesting to speculate on the selective pressures that led to the emergence of the N-glycosidase mechanism of removing uracil from DNA. Because dUTP is an intermediate in the synthesis of dTTP and because it can serve as a substrate for DNA polymerase, some dUTP must constantly be incorporated into DNA (the amount must be quite small because the pool of dUTP is small). However, uracil has the same base-pairing properties as thymine, so uracil incorporation into DNA would be innocuous. However, uracil in DNA can also be derived from cytosine by deamination, a relatively frequent type of damage that is potentiated at low pH. Were these cytosine-derived uracils not removed, permanent damage (mutation, see Chapter 4)

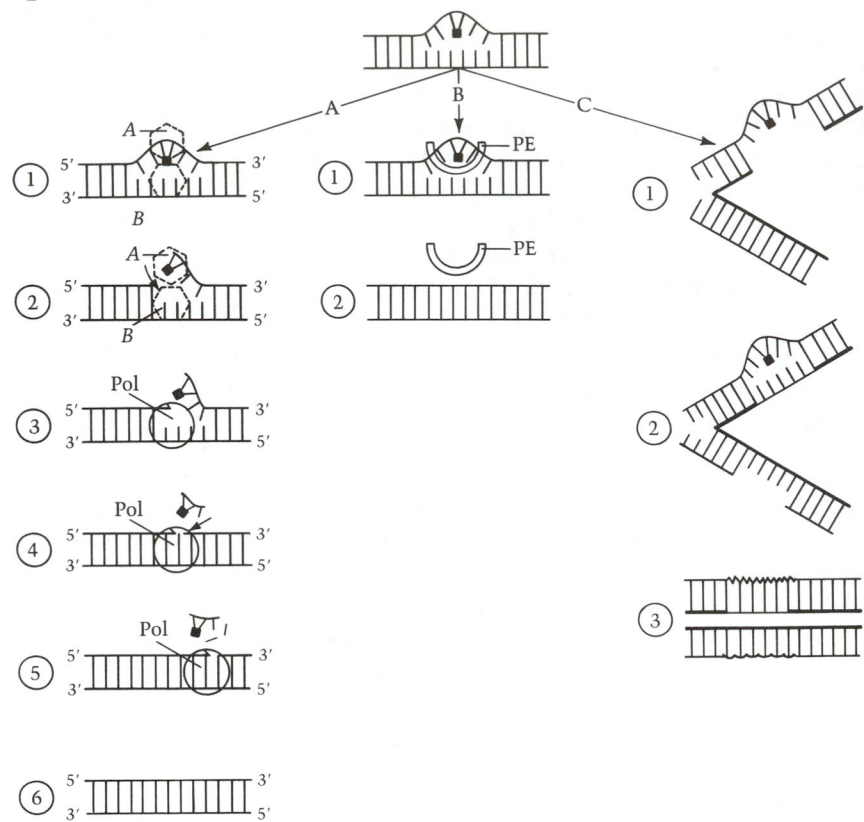

Figure 5. Mechanisms of repair of pyrimidine dimers. A. Excision repair. 1. The *uvrA* (*A*) and *uvrB* (*B*) proteins bind to the damaged region. 2. With the possible involvement of the *uvrC* protein (not shown), they nick the affected strand at the 5′ end of the damage (↓). 3. DNA polymerase I (Pol) then binds to the 3′ end of the nick and polymerizes in the 3′ direction (→) using the undamaged strand as template. 4. At the end of the damaged region (↓), polymerase I nicks the damaged strand again, releasing it. 5. The polymerase continues replication, releasing individual nucleotides (–) as it proceeds and thereby translating the nick in the 3′ direction. 6. The polymerase is released and the nick is sealed by the action of DNA ligase. (After P. Howard-Flanders, 1981.) B. Photoreactivation. 1. The photoreactivation enzyme (PE) binds to the dimer and, utilizing light energy, cleaves the four-membered ring. 2. The enzyme is released and normal hydrogen bonds reform in the duplex. (After P.C. Hanawalt, 1972.) C. Postreplication repair. 1. If a replication fork passes the pyrimidine dimer before it is repaired, the newly synthesized strand (heavy line) is not synthesized opposite the region of damage. 2. By a *recA*-mediated, single-strand cross-over event, the pyrimidine dimer region is converted to its original state and a single-strand gap is created in the sister duplex. 3. The dimer (〰) is repaired by photoreactivation or excision repair; the gap (〰) is repaired by the combined activities of DNA polymerase I and DNA ligase. (After P. Howard-Flanders, 1981.)

would result: by deamination of cytosine and subsequent replication, a cytosine–guanine pair would be converted to a thymine–adenine pair. Removal of the cytosine-derived uracil before it is replicated preserves the original base sequence. Successful operation of the *N*-glycosidase mechanism of repairing DNA damage caused by deamination of cytosine is obviously dependent on the fact that DNA (unlike RNA) contains thymine rather than uracil. Indeed, the best explanation of why thymine-containing rather than uracil-containing DNA is the modern product of natural selection is that the former composition permits *N*-glycosidase-mediated excision of cytosine-derived uracil.

Escherichia coli, and probably other bacteria as well, have evolved a mechanism termed the SOS SYSTEM, which comes into play only following large doses of UV light (cf. review by Little, 1982). In normal environments, *E. coli* cells contain only small quantities of the products of the *uvr* gene (approximately 10 to 20 molecules of *uvrA* gene product). Following large doses of UV, the excision repair system becomes saturated, large numbers of pyrimidine dimers remain, and, following replication, large numbers of single-stranded regions are produced in the chromosome. The RecA protein (product of the *recA* gene) binds to these regions and in so doing is activated to cleave another protein, the LexA protein, which is a repressor (see Chapter 7) of the several operons that constitute the SOS system. It is not completely clear that single-stranded DNA is the major inducing signal generated by UV damage; other possibilities include changes in nucleotide pools and changes in nucleoid structure or its degree of superhelicity. In any event, the expression of operons governed by the LexA–RecA regulatory system is greatly increased by UV light. A number of genes (nine or more) are encoded in these operons, but most importantly for repair, they include *recA*, *uvrA*, and *uvrB*. Note that the system has an autocatalytic component that decreases response time: *recA*, the trigger of the system, is itself under SOS control. As a consequence, the concentration in the cell of the products of *uvrA* and *uvrB* are rapidly increased, as is the cell's capacity to carry out both excision repair and postreplication repair. Ultraviolet light has long been known to be mutagenic, and now it appears that this is a consequence of the SOS response. Among the proteins synthesized in response to UV irradiation (under LexA repression) is, most probably, a DNA polymerase that is more tolerant to defects in the template because it observes base-pairing rules less strictly than the usual polymerase. This ERROR-PRONE polymerase fills the UV-caused gaps. It serves the purpose of increasing survival at the expense of an increased frequency of mutation. Another consequence of induction of the SOS system is inhibition of cell division brought about by high-level expression of the *sfiA* gene; presumably this inhibition aids the overall repair process, but in unknown ways. Repair of the DNA eventually leads to a drop in the level of the inducing signal, and LexA repression of the SOS operons is restored.

Requirements for producing the DNA of 1 g of cells

Let us return to the task of producing the 31 mg of DNA in our desired gram of *E. coli* cells; we now recognize the need for (1) appropriate amounts of dATP, dGTP, dTTP, and dCTP for polymerization per se, (2) some

ATP for the action of the unwinding proteins, (3) energy (as ~P equivalents) for the synthesis of the Okazaki fragments of the lagging strand and for ligating them together, (4) energy for the proofreading function of DNA polymerase III, (5) energy for the final adjustment of the torsional tension of each domain of the chromosome (Chapter 2), and (6) additional energy and methyl groups for the strain-specific modification of the polymerized DNA (Chapter 2). The first of these requirements—the provision of deoxyribonucleoside triphosphates as building blocks—is the only one that can be accurately specified. The other five requirements are of more or less uncertain magnitude, but collectively they are significant. Table 2 contains estimates of these needs and the basis for each estimate.

POLYMERIZATION: PROTEIN

Owing to their rapid growth rate and high protein content, bacteria have evolved as highly efficient synthesizers of protein: the cell discussed in Chapter 1 can add approximately 300,000 amino acids each second onto growing peptide chains. But the challenge of protein synthesis is more than a quantitative one; over 1000 different kinds of proteins are synthesized and all these must be produced in the correct proportions, within somewhat restricted tolerance ranges, if the bacterium is to be able to compete successfully in nature. Moreover, successful competition demands, as we will see in Chapter 7, that the proportions of the many types of proteins be modulated quantitatively and qualitatively in response to changes in the chemical or physical environment. As we shall discuss in Chapter 4, each type of protein is encoded by a separate gene (or in a few cases, duplicate copies of a gene), and the overwhelming majority of genes encode proteins. Only a few (22 for rRNA and 42 known for tRNA, see Appendix B) encode RNA as a final product. Thus, protein synthesis and gene expression are essentially a single topic.

With very few exceptions,[1] all bacterial proteins are synthesized according to the same general two-step scheme: (1) using one of the strands of DNA as template, a complementary molecule of RNA (one to several genes in length) is synthesized by a process termed TRANSCRIPTION because it involves the "rewriting" of the gene's informational content in the form of a different informational molecule; (2) transfer RNA (tRNA) molecules with amino acids attached (aminoacylated) bind to three sequential bases (a CODON) on the product of transcription [messenger RNA (mRNA) or simply a TRANSCRIPT]; and through the intervention of a ribosome, the amino acids become polymerized into a protein in the order indicated by the sequence of bases in the mRNA. This process is termed TRANSLATION, because the informational content of the mRNA is converted into an amino acid sequence.

---

[1] Certain polypeptide antibiotics and the short peptides contained in peptidoglycan are synthesized by a process in which ribosomes do not participate. Each amino acid is added to these peptides by a distinct enzyme-catalyzed reaction.

Table 2. Requirements for DNA production

1 g cells contains 31 mg DNA; 31 mg DNA = 100 μmol nucleotides; (1 μmol nucleotides = 309 μg nucleotides)

	Amount of building block required (μmol)	Amount of energy required (μmol ~P)
Building block[a]		
dATP	24.7	
dTTP	24.7	
dGTP	25.4	
dCTP	25.4	
Total nucleotides	100.2	
Energy for unwinding the helix[b]		100.0
Energy for discontinuous synthesis[c]		0.6
Energy for proofreading[d]		35.0
Energy for negative supercoiling		0.5
Energy for methylation		0.1
Total energy		136.2

[a]Based on A+T/G+C = 0.97.
[b]Based on 2 ATP → 2 ADP per nucleotide base pair separated.
[c]Based on synthesis and hydrolysis of a pentaribonucleotide from nucleoside triphosphates for every 1000 bases in the lagging strand (10 ~P) plus 2 ~P to regenerate NAD used for ligation.
[d]Based on in vitro evidence that indicates that to achieve the fidelity observed in vivo proofreading by DNA polymerase III contributes a factor of 10 to 200 to specificity (depending on the nature of the mispair) and costs 10–13% of the dATP and dTTP and 6% of the dGTP and dCTP being hydrolyzed to the respective nucleoside monophosphates (Fersht, 1982). For the quantity of DNA made for our gram of cells, 8.72 μmol of deoxyribonucleoside triphosphates would be hydrolyzed to monophosphates, or 8.72 × 2 ~P = 17.44 μmol ~P. We assume here that postreplication repair contributes an equal increment of accuracy, and at approximately equal cost, though this is not known.

Transcription

The transcription reaction is biochemically quite simple. It can be catalyzed by the single enzyme, DNA-dependent RNA polymerase (RNA polymerase, RNA-P). This enzyme utilizes any ribonucleoside triphosphate (ATP, CTP, GTP, or UTP) as substrate and one DNA strand as template. A single step of the reaction can be written:

$$\text{RNA} + \text{ATP} \xrightarrow{\text{RNA-P}} \text{RNA-AMP} + \text{PP}_i$$

In this step the products are an RNA molecule lengthened by a single adenyl group (C, G, or U residues are added by the same reaction) and one molecule of pyrophosphate (PP_i). The reaction is a nucleotidyl transfer from pyrophosphate to the 3'-OH of a molecule of RNA.

In spite of the chemical simplicity of the polymerization reaction, it presents an interesting topológical problem (Chapter 6). Also, its control is extremely complex and only now is an understanding of it beginning to emerge. The relevant questions of control concern where on the DNA template transcription is initiated; how frequently initiation occurs at a given site; where on the template transcription is terminated; and what the probability of termination at these sites is.

All control of transcription is effected through probability of initiation or termination (a special type of termination that is quite important in modulating gene expression is termed ATTENUATION and is discussed in Chapter 7). Once initiation has occurred, polymerization occurs at a constant rate regardless of template. As our use of the term *site* implies, much of the control on initiation and termination of transcription is exerted by specific sequences of bases in DNA, but accessory proteins and even the molecular composition of RNA-P itself also play an important role in certain cases.

Initiation of transcription

Initiation of transcription occurs at sites on the DNA termed PROMOTERS. Ribonucleic acid polymerase recognizes these and binds to them, causing the strands of the double helix to separate (melt). Then RNA-P begins to synthesize RNA, using one DNA strand (the sense strand) as a template. As the polymerization reaction proceeds, RNA-P moves down the sense strand.

Ribonucleic acid polymerase from *E. coli*, by far the most thoroughly studied, is a large molecule ($\cong$500,000 MW) composed of five subunits: one copy each of the two large proteins, β (about 145,000 MW) and β' (about 155,000 MW), two copies of a small protein, α ($\cong$40,000 MW), and one copy of a less tightly associated protein, σ (85,000–95,000 MW). Thus, the complete enzyme, or HOLOENZYME, can be represented as $\beta\beta'\alpha_2\sigma$; without the loosely associated sigma (σ), it is termed the CORE ENZYME. Sigma participates only in recognition of the promoter (more properly it decreases affinity for non-promoter sequences) and in initiation of transcription. Soon after chain elongation has begun, σ dissociates from the core enzyme.

Ribonucleic acid polymerase binds to the promoter over a relatively extended region from approximately 20 base pairs downstream (direction of transcription) to approximately 45 base pairs upstream from the point at which transcription is initiated (Figure 6). On the basis of the frequency of binding sites between RNA-P and the promoter, two sets of sequences within the promoter appear to be particularly important to promoter function. Not surprisingly, the sequence of bases within these regions is highly conserved among the various known promoters on the *E. coli* chromosome, an observation that first led to a recognition of their importance. One of these sequences (TATAAT), termed the PRIBNOW BOX, lies approximately 10 base pairs upstream (-10) from the point of initiation of transcription. The other (TTGACA) lies, as its name (-35 REGION) indicates, approximately 35 base pairs upstream. Though highly conserved, these *precise* CONSENSUS SEQUENCES

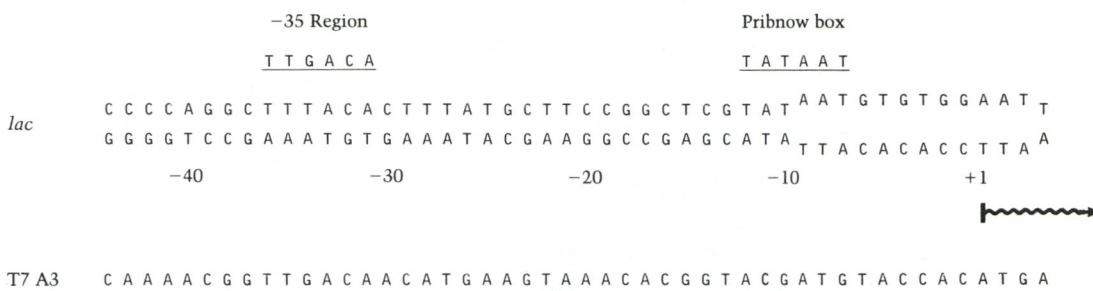

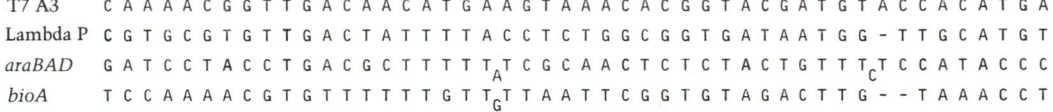

Figure 6. Base sequences of certain promoters in *E. coli* and its phages. The two upper rows (*lac*) show the sequence of bases in a mutant lactose promoter (*lacUV5*) that is transcribed at high frequency in the absence of accessory proteins. The numbering follows the convention of assigning to base pairs increasing negative numbers from the point of initiation of transcription in the direction opposite to transcription. The point of initiation of transcription (+1) and direction of transcription is indicated by the end and point, respectively, of the wiggly arrow. The separated bases are those that are unwound when RNA-P binds to the promoter. Above the *lac* promoter are shown the most probable sequences of bases in the nontemplate strand of known promoters of the two highly conserved regions: the -35 region and the Pribnow Box. The four lower rows show the sequence of bases (in the same register as *lac*) in the nontemplate strand of four other promoters: the A3 promoter of the coliphage T7 (T7 A3); the P promoter of the coliphage λ (Lambda P); the promoter of the arabinose operon (*araBAD*); and the promoter of the biotin A gene (*bioA*). Extra bases between the highly conserved regions are shown below the lines; lesser numbers of bases are indicated by dashes (–). (From Siebenlist, 1980.)

are found in few *E. coli* promoters. The rules relating promoter structure to function are far from being understood, as illustrated by the fact that single base changes in the non-conserved region can markedly alter a promoter's function (Chapter 7, Figure 4).

Of course, different promoters initiate transcription at different frequencies. If the frequency is high, the promoter is said to be a STRONG PROMOTER—if it is low, a WEAK PROMOTER. Most probably the entire promoter sequence contributes to its strength, with the highly conserved Pribnow box and −35 region being the most important portions. Marked, but lesser, conservation is seen in the surrounding regions; the distances between the −35 region, the Pribnow box, and the site of initiation vary only by a few base pairs at most. It is not yet possible to predict a promoter's strength from its base sequence.

Termination of transcription

At the end of a unit of transcription (a gene or an operon) lies a DNA sequence, a TERMINATOR, which signals RNA-P to stop synthesizing RNA. The completed transcript and RNA-P are released from the DNA. An accessory protein, rho, participates in the termination reaction. As in promoters, the precise sequence of terminators varies. Some are strong (they cause termination with high frequency) and others are weak. Rho seems to increase the strength of all terminators, raising it to nearly 100%. Thus, intrinsically strong terminators are sometimes termed *rho-independent*, and intrinsically weak ones, *rho-dependent*. The action of rho involves the hydrolysis of ATP. In some way the energy derived from this process facilitates the termination reaction.

The sequences of a number of terminators from *E. coli* have been determined. All the strong terminators share three features (Figure 7): (1) a set of A·T base pairs, preceded by (2) a set of G·C base pairs, preceded (not contiguously) by (3) an inverted repeat of approximately one-half of the G·C and A·T

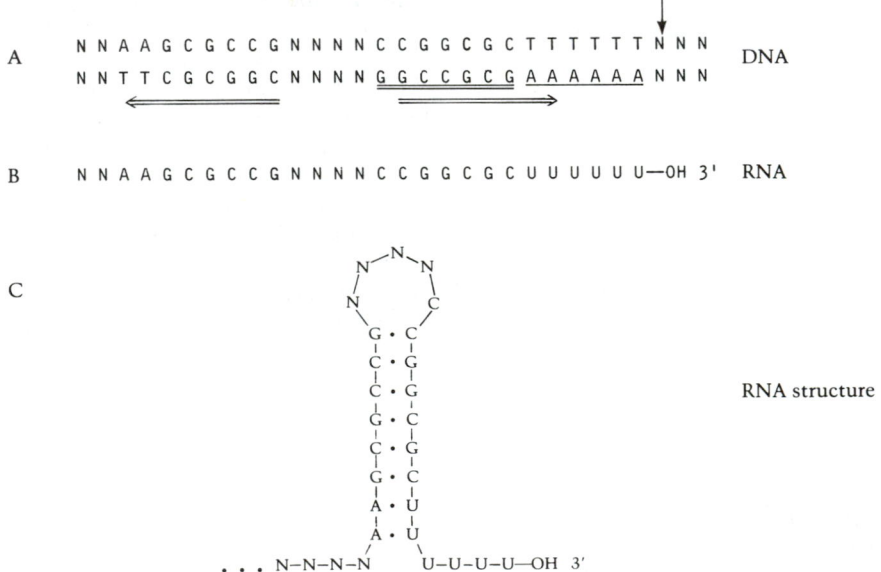

Figure 7. A. A hypothetical base sequence of a strong terminator. A set of A·T pairs (single underscore) is preceded by a set of G·C pairs (double underscore) and by an inverted repeat of some of these (horizontal arrows). Termination occurs at the site of the vertical arrow. Noncritical nucleotides are designated N. B. The terminal portion of the transcript formed from the gene in A. C. The stem-and-loop structure that can form in the transcript as a consequence of hydrogen bonding (·) between regions encoded by the inverted repeat. This hypothetical terminator is derived from a number of terminator sequences in *E. coli* and its phages. (After D. Pribnow, 1979.)

sets. Termination occurs within the A·T set. As a consequence of the inverted repeat, the terminated molecule can form a "stem-and-loop" structure, which probably acts to push the core enzyme off the template.

The role of RNA-P in recognition of promoters

The observations that σ functions only in the promoter-recognition step of transcription and that it is only loosely bound to the core protein suggests that σ-type proteins might play an important role in regulation of gene expression. If a bacterium had the capacity to produce several types of sigma proteins with differing affinities for various promoters, the proportion of these σ's at any particular time would certainly modulate the relative expression of genes in that bacterium. Additional σ proteins in *E. coli* have been looked for but not found, although certain small proteins (e.g., ω) of unknown function are sometimes associated with RNA-P in this organism; and there is preliminary evidence that the *E. coli* heat-shock response (Chapter 5) may involve a substitute σ protein. But the situation is quite different in the case of the aerobic spore-former *Bacillus subtilis*. This organism, which undergoes morphogenesis late in the growth cycle when endospores develop within many vegetative cells, produces a number of σ proteins that confer promoter specificity on the core enzyme. At least four σ proteins (σ^{28}, σ^{29}, σ^{37}, and σ^{55}), designated by their molecular weight ($\times 10^{-3}$), have been demonstrated. Additional forms have been reported. Thus, *B. subtilis* can produce at least four different RNA-P holoenzymes. The sporulating cell contains a complex mixture of RNA-P holoenzymes; the relative proportions of the forms are dependent on the stage of sporulation. Even vegetative cells contain more than a single form.

Although the existence in *B. subtilis* of multiple forms of RNA-P, the modulation of their relative proportions during development, and the different promoter specificities of the various forms are now established facts, the general significance of multiple forms of RNA-P to expression of procaryotic genes is by no means clear. Possibly, multiple forms of RNA-P will be found only in procaryotes that undergo morphogenesis. Possibly, multiple forms will prove to be more widespread among procaryotes.

The general molecular structure of RNA-P from all eubacteria (all procaryotes other than archaebacteria) studied so far is similar to RNA-P from *E. coli*, that is, they have the basic $\beta\beta'\alpha_2\sigma$ structure. However, none of the archaebacteria have this type of RNA-P. *Sulfolobus acidocaldarius*, the most thoroughly studied archaebacterium in this respect, has an RNA-P composed of two large subunits and eight different small ones. The subunit structure of its RNA-P resembles RNA-P from yeasts more than those from eubacteria.

Translation

As mentioned earlier, use of the terms *transcription* and *translation* to describe the steps of protein synthesis draws a useful analogy between this set of biochemical reactions and the processing of language. Translation from one language to another is dependent on someone's knowing both languages. Translation of the informational content of mRNA into a sequence of amino acids in a protein is dependent on a molecule that interacts with mRNA and with amino acids. In this sense tRNA is the key component of the translation system: one portion of the molecule, the ANTICODON, pairs (according to base-pairing rules) with codons on mRNA; another portion carries a molecule of the amino acid that corresponds to that codon. One may imagine that tRNA must have been the first component of the translation system to evolve. The other components of the system—ribosomes, various accessory proteins, and aminoacyl-tRNA synthetases—evolved to increase the rate and accuracy of the process.

Amino acid attachment to tRNA, catalyzed by the aminoacyl-tRNA synthetases, takes place in two stages: first, the amino acid is activated by reacting with ATP to form an enzyme-bound molecule of aminoacyl-adenylate; then by a transfer reaction, aminoacyl-tRNA is formed.

(1) ATP + amino acid → aminoacyl-AMP + pyrophosphate
(2) aminoacyl-AMP → aminoacyl-tRNA + AMP

As in the case of DNA synthesis, protein synthesis is far more accurate than some of its individual steps. The fidelity actually observed derives from the existence of a number of points at which proofreading corrects errors that would result in an incorrect polypeptide sequence (Hopfield, 1974). The aminoacylation of tRNA by synthetases seems to include at least two proofreading reactions. As a result, the frequency of mischarging isoleucine tRNA, for example, with the highly similar amino acid, valine, is reduced from 1 per 100 to 1 per 10,000. From an equimolar mixture of these two amino acids, the isoleucine enzyme will incorrectly form valyl-AMP once for every 100 times it forms isoleucyl-AMP, but two additional discrimination steps, one of which involves the tRNA for isoleucine (Baldwin, 1967), generates the desired degree of fidelity. The energy cost of proofreading in this instance has been estimated at 0.1 ATP per isoleucyl-tRNA eventually passed on to the ribosome (Freter, 1980). For more dissimilar amino acids, the cost of accuracy will be lower. On the ribosome there must be additional proofreading steps of some sort because codon–anticodon interactions studied in model systems lead to expectations of errors in the frequency range of 10^{-2} (Hopfield, 1974), whereas the measured error in vivo is only 10^{-4} (Gallant, 1980).

It is convenient to divide the mRNA-directed polymerization of amino acids into two phases: a microcycle in which one amino acid residue is added to a growing peptide chain and a macrocyle composed of many microcycles along with initiation and termination of translation.

Translation: the microcycle

A microcycle (Figure 8) begins with the 70 S ribosome bearing a partially completed peptide chain attached at a certain point to an mRNA. A variety of studies has established that a 70 S ribosome has two sites—with differing properties—that bind tRNA. One of these, designated the A SITE,

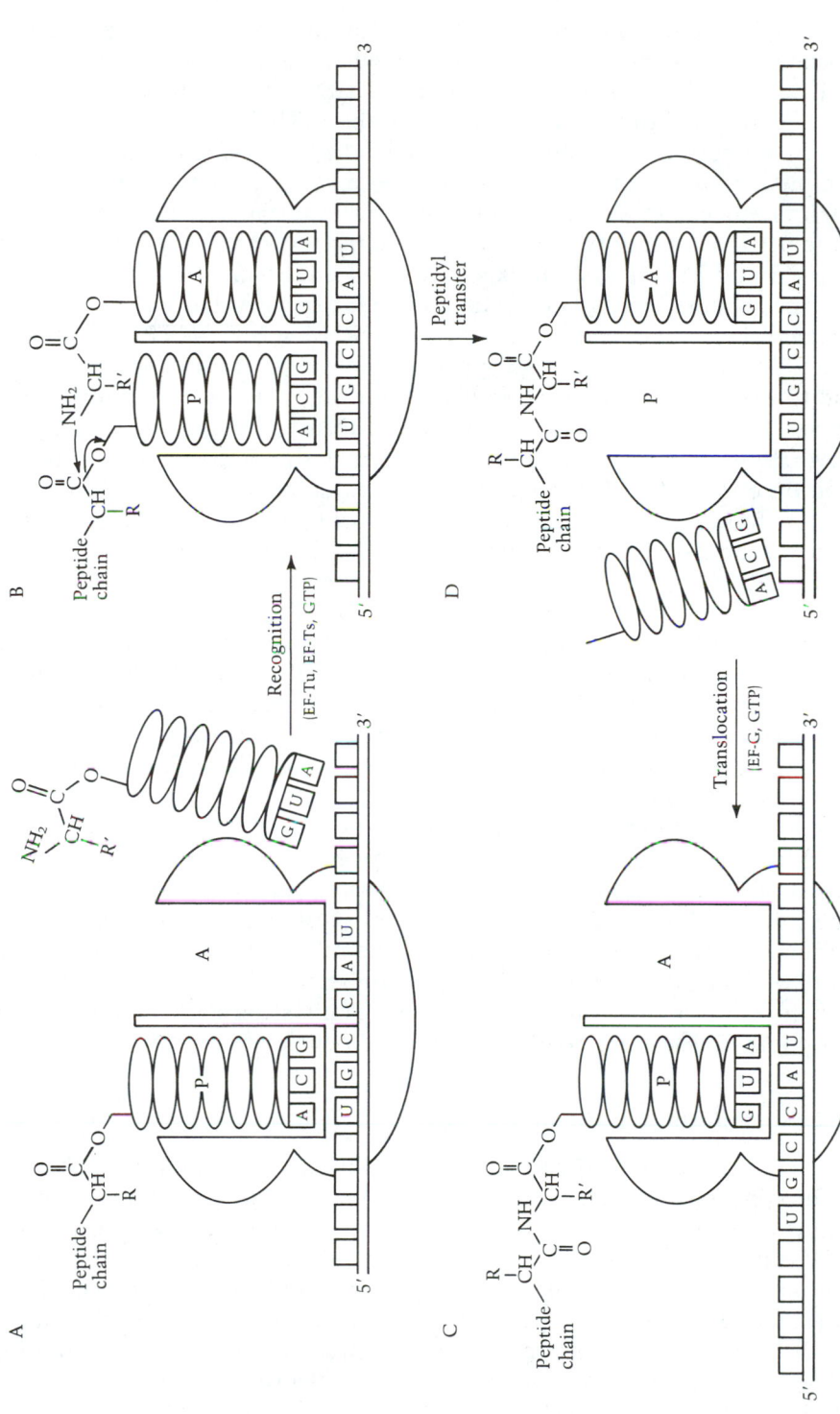

Figure 8.

The microcycle of protein synthesis. A. A tRNA with a peptide attached is located in the P site, with the anticodon (ACG) of the last-added amino acid paired with its codon (UGC) on the mRNA. The codon (CAU) of the next amino acid to be added is aligned with the empty A site. By the recognition reaction, the next aminoacyl-tRNA enters the A site (B). A peptide bond is formed by peptidyl transfer, and the uncharged tRNA is released (C). The peptide-bearing tRNA is transferred to the P site by translocation, and the ribosome advances by one codon in the 3' direction down the mRNA molecule (D).

accepts the molecule of aminoacyl-tRNA. At this time the other site, the P SITE, is occupied by a molecule of tRNA to which is attached a partially completed peptide chain. When a molecule of tRNA is in either site, its anticodon is positioned to pair with a codon of the mRNA.

In the first reaction of the microcycle—RECOGNITION—an aminoacylated tRNA complementary to the exposed anticodon enters the A site. This reaction, which is subject to specific inhibition by the tetracycline antibiotics, requires the participation of two accessory proteins [ELONGATION FACTORS Tu and Ts (EF-Tu and EF-Ts)] and an expenditure of energy in the form of hydrolysis of the γ phosphate of GTP. The molecular species that enters the A site is actually the ternary complex, aminoacyl-tRNA–EF-Tu–GTP; after hydrolysis of GTP, EF-Tu–GDP and phosphate are released. The two states (free and bound to EF-Tu–GTP) of aminoacyl-tRNA in the A site may provide two opportunities for rejecting a wrong aminoacyl-tRNA.) The tRNAs in the A and P sites are so positioned in the ribosome that the amino group of the amino acid on one tRNA lies next to the terminal acyl group of the peptide on the other. Breakage of the acyl bond and formation of a peptide bond (PEPTIDE TRANSFER) results in transfer of the peptide (now one amino acid longer) to the tRNA in the A site—a reaction catalyzed by the 50 S moiety of the ribosome and requiring no expenditure of energy. The uncharged tRNA is then released from the P site in preparation for TRANSLOCATION, a process by which the peptide-bearing tRNA is transferred to the P site and the ribosome is moved one codon in the 3′ direction down the mRNA. Translocation requires the participation of a third accessory protein—ELONGATION FACTOR G (EF-G)—and hydrolysis of a second γ phosphate of GTP. Translocation completes the microcycle.

A number of microcycles, corresponding to the number of amino acids in the protein, results in the synthesis of a complete protein. But additional factors are involved in protein synthesis. Translation must begin at the proper site on the mRNA. (Most mRNA molecules are significantly longer than the proteins they encode. An extreme example is the mRNA for the A protein of the R17 and MS2 bacteriophage genomes; this messenger has 129 nucleotides in front of the point where translation begins.) Translation also must end at the proper point. These steps constitute the macrocycle.

Translation: the macrocycle

Synthesis of all bacterial proteins begins with a formyl-substituted methionine residue (encoded by the codons AUG or GUG). However, these codons are not sufficient to designate the start point because they also encode the internal methionine and valine residues that are present in most proteins. Two Australian biochemists, Shine and Dalgarno, solved this dilemma (Shine, 1974). They sequenced the 3′ terminus of 16 S rRNA and found that it contains a sequence that is complementary to the base sequences that precede the start codon by approximately 10 nucleotides on most mRNAs (Figure 9). It is now generally believed that this complementarity helps designate the start codon; hydrogen bonding between the terminus of the 16 S

araB	U U U U U U G G A U G <u>G G A G</u> U G A A A C G A **U G** G C G A U U
trpE	G A A C A A A A U U A <u>G A G</u> A A U A A C A A **U G** C A A A G A
trpA	G A A A G C A C G <u>A G G</u> G G G A A A U C U G A **U G** G A A C G C
lacZ	A A U U U C A C A C <u>A G G</u> A A A C A G C U **A U G** A C C A U G
lacI	A G U C A A U U C A G G <u>G G U G</u> G <u>U</u> G A A U G **U G** A A A C C A
galT	T A T C C C G A T T <u>A A G G A</u> A C G A C C **A T G** A C G C A A

16 S rRNA 3′ end   HO—A U U C C U C C A C U A G - - - - - -

Figure 9. Sequences of the start regions of six mRNAs and the 3′ terminus of 16 S rRNA from *E. coli*. Start codons are indicated in boldface. Regions complementary to 16 S tRNA are underscored. Possible sites of G·U pairing are indicated by a dot below the base. (From J. Steitz, 1979.)

rRNA and mRNA properly positions the ribosome at the start codon. Of 74 initiator regions of *E. coli* and coliphage mRNA that had been sequenced by 1980, all contained a segment complementary to the 3′ end of 16 S rRNA. The complementarity varies from 3 to 9 bases, which are centered approximately 10 bases upstream from the initiator codon; the intervening distance shows considerable variation. Nevertheless, there are indications that this complementarity may be *necessary* but not *sufficient* for translation initiation. Some sequences that contain an initiator codon appropriately spaced from a region of complementarity are not sites of translation initiation; also, there are striking effects on initiation frequency of very small changes in the distance between the Shine-Dalgarno sequence and the initiator codon. There is still more to learn about the structure and function of initiator sites (cf. review by Steitz, 1980).

After a messenger has been read, the 70 S ribosome dissociates into its components, a 50 S and 30 S subunit (possibly catalyzed by an accessory protein, initiation factor 3); the 70 S ribosome then reassembles at the start codon. The 30 S subunit and formylmethionyl-tRNA form an INITIATION COMPLEX at the start codon; one molecule of GTP is hydrolyzed and three accessory proteins (initiation factors 1, 2, and 3) participate. Then the 50 S ribosome joins the complex to form a 70 S ribosome. Translation terminates when the ribosome encounters one or more nonsense (stop) codons (UAA, UAG, or UGA). Three additional factors, termination factors 1, 2, and 3, participate in this process.

Factors contributing to the speed of protein synthesis

Although protein synthesis is fundamentally similar in all living cells (use of the same genetic code and the same kinds of machinery), there are some significant differences between procaryotic and eucaryotic protein synthesis. These differences can best be interpreted in light of the tenfold or higher rate of protein synthesis in bacteria.

To sustain the rate of protein synthesis needed for doubling times that may be as short as 20 minutes, the bacterial cytosol must be rich in the machinery of protein synthesis. The PROTEIN-SYNTHESIZING SYSTEM (PSS)—which consists of ribosomes, tRNA, mRNA, the enzymes that make and modify these molecules, aminoacyl-tRNA synthetases, and the protein factors required for the initiation, elongation, and termination reactions—constitutes over one-half the total mass of rapidly growing cells. (How the quantity of this system is adjusted will be discussed in Chapters 6 and 8.) Natural selection has resulted in leaner, more streamlined PSS components than those found in eucaryotic cells. Bacteria use only 3 initiation factors, compared to 10 used by eucaryotic cells (Grunberg-Manago, 1980). Bacterial ribosomes are smaller and have only one-half the protein content of eucaryotic ribosomes (Wool, 1980). Bacterial mRNA is in many cases POLYCISTRONIC (containing the coding sequence of two or more contiguous genes), and initiation of translation can occur at internal sites. Eucaryotic mRNA is functionally monocistronic, with a single site for translation initiation. Eucaryotic mRNA must be transported from its site of synthesis in the nucleus to the ribosomes in the cytoplasm. As a result, it must be protected against degradation during transport; this protection is achieved by CAPPING (attachment of a 7-methylguanosine triphosphate) at its 5′ terminus, by METHYLATION at internal residues, and by POLYADENYLATION at the 3′ terminus. The central feature of the procaryotic body plan—a DNA genome that is directly accessible to the cytosol—enables translation and transcription of genes to go on simultaneously and thereby eliminates the need for all of the devices involved in protecting and transporting mRNA.

The structural streamlining of the PSS in bacteria has several functional consequences, one of which is a difference in the initiation reaction (cf. reviews by Grunberg-Manago, 1980; Steitz, 1980). The use of base sequence complementarity between mRNA and 16 S rRNA to help establish correct initiation of protein synthesis at the appropriate AUG (or GUG) codon turns out to be exclusively a bacterial device. Eucaryotic mRNA has been found in most instances not to contain a Shine-Dalgarno sequence at its 5′ terminus. It appears that cells of higher organisms can afford an initiation process (presumably slower) in which the 40 S ribosomal subunit containing a methionyl-tRNA–GTP–initiation factor 2 complex binds near the 5′ end of an mRNA and then scans the message for an AUG initiation codon, after which the numerous other initiation factors come into play. The eucaryotic initiation process presumably provides more opportunity for regulation of protein synthesis by restricting translation initiation of some mRNAs under certain conditions—a device not widely used to control bacterial gene expression.

Requirements for producing the protein of 1 g of cells

The production of 550 mg of protein requires (1) appropriate amounts of each of the 20 amino acids found in *E. coli* protein, (2) sufficient energy (as ∼P equivalents) for the activation of each amino acid molecule and its incorporation into protein, (3) energy for mRNA synthesis, (4) energy for proofreading at the synthetase level and ribosome level, and (5) energy for assembly and modification reactions. The amino acid building block requirement is obtained by an amino acid analysis of total *E. coli* protein. The high-energy phosphate requirement per peptide bond consists of one ATP → AMP (i.e., 2 ~P) for activation and formation of aminoacyl-tRNA, one GTP → GDP (i.e., 1 ~P) for the EF-Tu-mediated transfer of the aminoacyl-tRNA to the ribosomal A site, and one GTP → GDP (i.e., 1 ~P) for the EF-G-mediated translocation of peptidyl-tRNA to the P site; a total of 4~P are therefore needed per aminoacyl residue incorporated. There are two independent ways to calculate the energy requirement for mRNA synthesis. One, given in Chapter 6, is based on structural information about polysomes; the other is based on measurements of the rates of incorporation of nucleotides into RNA and of amino acids into protein by appropriate pulse labeling of growing cells. Both methods yield an estimate of an average of 25–35 protein copies made from a single mRNA transcript. If we take 30 copies per transcript as our estimate and recall that each amino acid incorporated must be specified by a codon of three bases in mRNA, we arrive at a cost of 0.1 nucleotide (3/30) polymerized into mRNA per amino acid incorporated into protein; because each nucleotide polymerized into mRNA costs 2~P, the ~P cost per amino acid residue is 0.2 (0.1 × 2). The energy for proofreading is very uncertain, but some estimates indicate it might be 0.1 ~P per amino acid incorporated. Finally, the cost for assembly and modification is also unknown. We shall estimate it by assuming that 10% of the cell's protein molecules are made with a signal sequence of 20 aminoacyl residues, which are cleaved off and hydrolyzed during translocation through the membrane. Altogether, synthesis of that quantity of oligopeptide is equivalent to approximately 0.6% of the total protein of the cell (10% of the cell's protein made with a signal peptide averaging 6% of the length of an average protein [20 is approximately 6% of 350 amino acids]). The cost of assembly has therefore been estimated at an additional 0.6% of the cost of making protein. The values are summarized in Table 3.

POLYMERIZATION: STABLE RNA

The polymerization of DNA from deoxyribonucleoside triphosphates in vivo requires a dozen or so proteins. The polymerization of protein from aminoacyl-tRNAs (exclusive of regulatory and processing reactions) involves more than 60 proteins. In striking contrast, the single oligomeric enzyme RNA polymerase seems capable of correctly making RNA. For many transcripts, special factors such as rho or the NusA protein are needed to assure proper termination, and perhaps most transcription is modulated by regulatory proteins that affect initiation. Nevertheless, these proteins all appear to be auxiliary factors not required in principle for accurate polymerization of RNA from a DNA template.

Eucaryotic cells use different DNA-dependent RNA polymerases to make tRNA, rRNA, and mRNA. *E. coli,* and presumably other bacteria, utilize the same RNA polymerase for all three functions.

We have already described the subunit makeup of this enzyme while

Table 3. Requirements for protein production

1 g cells contains 550 mg protein; 550 mg protein = 5080 μmol amino acids; (1 μmol amino acids = 108 μg amino acids)

	Amount of building block required (μmol)	Amount of energy required (μmol ~P)
Building blocks[a]		
Alanine	488	
Arginine	281	
Aspartate, asparagine	458	
Cysteine	87	
Glutamate, glutamine	500	
Glycine	582	
Histidine	90	
Isoleucine	276	
Leucine	428	
Lysine	326	
Methionine	146	
Phenylalanine	176	
Proline	210	
Serine	205	
Threonine	241	
Tryptophan	54	
Tyrosine	131	
Valine	402	
Total amino acids	5081	
Energy for activation and incorporation[b]		20,324
Energy for mRNA synthesis[c]		1,016
Energy for proofreading[d]		508
Energy for assembly and modification[e]		122
Total energy		21,970

[a]Based on amino acid analysis of total *E. coli* B/r protein by Teresa Phillips; except the cysteine and tryptophan values are from Roberts (1955), and no distinction is made between glutamate and glutamine and between aspartate and asparagine. The data have been corrected to exclude peptidoglycan amino acids.

[b]We have not included the very small additional cost of 1 ~P per polypeptide as a result of 1 GTP → GDP during formation of the 70 S mRNA–fMet-tRNA initiation complex. Also ignored is the ~P cost of N-formylation of methionine on the initiator tRNA. Energy is derived from summing 2 ~P for ATP → AMP and 2 ~P for 2 GTP → 2 GDP (= 4 ~P) and multiplying times 5081 μmol amino acids.

[c]An estimate of 30 translations per mRNA transcript has been made based on the life history of polysomes (Chapter 6) and independently from the information that 6.04×10^5 nucleotides are polymerized into mRNA per second by one genome-equivalent of cell mass while 6.46×10^6 amino acids are simultaneously incorporated into protein (Dennis, 1974). Because the coding ratio is 3 bases to 1 amino acid, the mRNA sequences must on average be translated $[(6.46 \times 10^6)/(6.04 \times 10^5)] \times 3 = 32$ times. Each amino acid is incorporated at a cost of 0.1 nucleotide polymerized into mRNA, the cost of which is 0.1×2 (~P per nucleotide polymerized) = 0.2 ~P per amino acid residue. Cost = 0.2 ~P × 5081 residues.

[d]The estimate of 0.1 ~P per amino acid for proofreading is extremely soft—direct measurements of excess ~P dissipation for proofreading *in vivo* is not technically feasible. We have chosen this value because it has been estimated as the cost of reducing the valine/

discussing mRNA synthesis, and we have seen how it recognizes promoter sites and initiates transcription. The same general sort of initiation signals function in stable RNA formation as function in the formation of mRNA. Two sequences (one approximately 10 and another 35 base pairs upstream from the transcription initiation nucleotide) constitute the promoter and are involved in initiation of stable RNA synthesis, as in mRNA synthesis. As noted in Chapter 2, five of the seven rRNA operons have been sequenced and found to contain two tandem promoters each. The significance of this finding is yet to be learned.

Transfer RNA genes are of two sorts: those included in the seven rRNA operons, and the remainder, which are scattered throughout the genetic map (Appendix B). The 14 tRNA genes included in the *rrnA–rrnG* operons (see Chapter 2, Table 2) code for 6 of the major tRNA species in the cell ($tRNA_1^{Ile}$, $tRNA_{1B}^{Ala}$, $tRNA_2^{Glu}$, $tRNA_1^{Asp}$, $tRNA^{Trp}$, $tRNA^{Thr}$). The other tRNAs are produced from transcripts that contain one or more tRNA sequences. In all cases the direct transcript is larger than the mature tRNA molecules and must be processed by nucleases to yield the one or more tRNA products.

All stable RNA molecules in the cell, therefore, are made from precursor RNA molecules that must be processed by nucleases and then extensively modified to produce the mature product.

Factors contributing to the speed of stable RNA synthesis

Although the significance of tandem promoters is yet to be elucidated, other aspects of the genetic organization of stable RNA sequences have clear implications. The sevenfold redundancy of rRNA genes (eightfold for 5 S rRNA) enables synthesis of the large number of rRNA molecules needed during rapid growth. Also, the clustering of these seven operons in the half of the genome proximal to *oriC* results in a desirable gene dosage effect at fast growth rates, because genes near the replication origin are present in more copies per cell, on the average, then genes near the replication terminus.

Requirements for producing the RNA of 1 g of cells

Production of the 197 mg of stable RNA (167 mg of rRNA, 30 mg of tRNA) in 1 g of reference *E. coli* cells requires (1) the appropriate amounts of ATP, GTP, UTP, and CTP as building blocks judged from the final products, (2) an amount of energy (as ~P) equivalent to the segments of primary transcript that are removed and hydrolyzed, and (3) the energy, methyl groups, and assorted minor constituents involved in the modification of these RNA species. Because the cost of mRNA turnover has been charged to protein synthesis (Table 3), the cellular content of mRNA can be treated here as

---

isoleucine error (at tRNA charging) by 100-fold, and this error reduction is what is achieved for most tRNA-codon matches on the ribosome. Cost = 0.1 ~P × 5081 residues.

[e]No estimate has been made for modification reactions; the cost of assembly is estimated as the cost of making the signal sequences of translocated proteins as described in the text. Cost = 0.6% × 20,324 µmol ~P.

Table 4. Requirements for RNA production

1 g cells contains 197 mg stable RNA and 8.3 mg mRNA; (197 mg stable RNA = 167 mg rRNA + 30 mg tRNA)

	Amount of building block required (μmol)	Amount of energy required (μmol ~P)
Building blocks[a]		
ATP	165	
GTP	203	
CTP	126	
UTP	136	
Total nucleotides	630	
Energy for discarded segments of primary transcripts[b]		242
Energy for modification[c]		14
Total energy		256

[a]The average nucleotide residue was taken to have a molecular weight of 325. The A:G:C:U ratios were those of Roberts (1955), corrected for DNA. The 630 μmol of building blocks are sufficient for the pool of mRNA as well as the stable RNA species.
[b]Approximately 20% of the primary transcript of *rrn* genes is discarded (Apirion, 1981) and a similar value is reasonable for tRNA gene transcripts. The energy cost is therefore 0.02 × 605 μmol of nucleotides in stable RNA, or 121 μmol times 2 ~P per nucleotide polymerized = 242 μmol ~P.
[c]Approximately 15% of the bases in tRNA are modified (Altman, 1978). We have estimated a cost of 1 ~P for each of the 92 × .15 = 14 μmol of modified tRNA bases, but this is undoubtedly an underestimate, given the great variety of complex decorations involved.

though it were stable in calculating the building block requirement. These requirements are estimated in Table 4.

POLYMERIZATION: PEPTIDOGLYCAN

The biological challenge in peptidoglycan synthesis is not to produce a macromolecule with remarkable fidelity (as in the case of protein and nucleic acid synthesis) but to construct a single saclike molecule to surround the cell totally. By its very nature, the peptidoglycan sac must be polymerized outside the cell proper, that is, exterior to the cytoplasmic membrane.

An overview of peptidoglycan synthesis is presented in Figure 10, which is arranged to emphasize two special features of the process. First, there are three cellular locations involved. The building blocks are made in the cytosol; the subunit is constructed at the inner surface of the cell membrane and translocated to the outer surface by a lipid carrier; polymerization and assem-

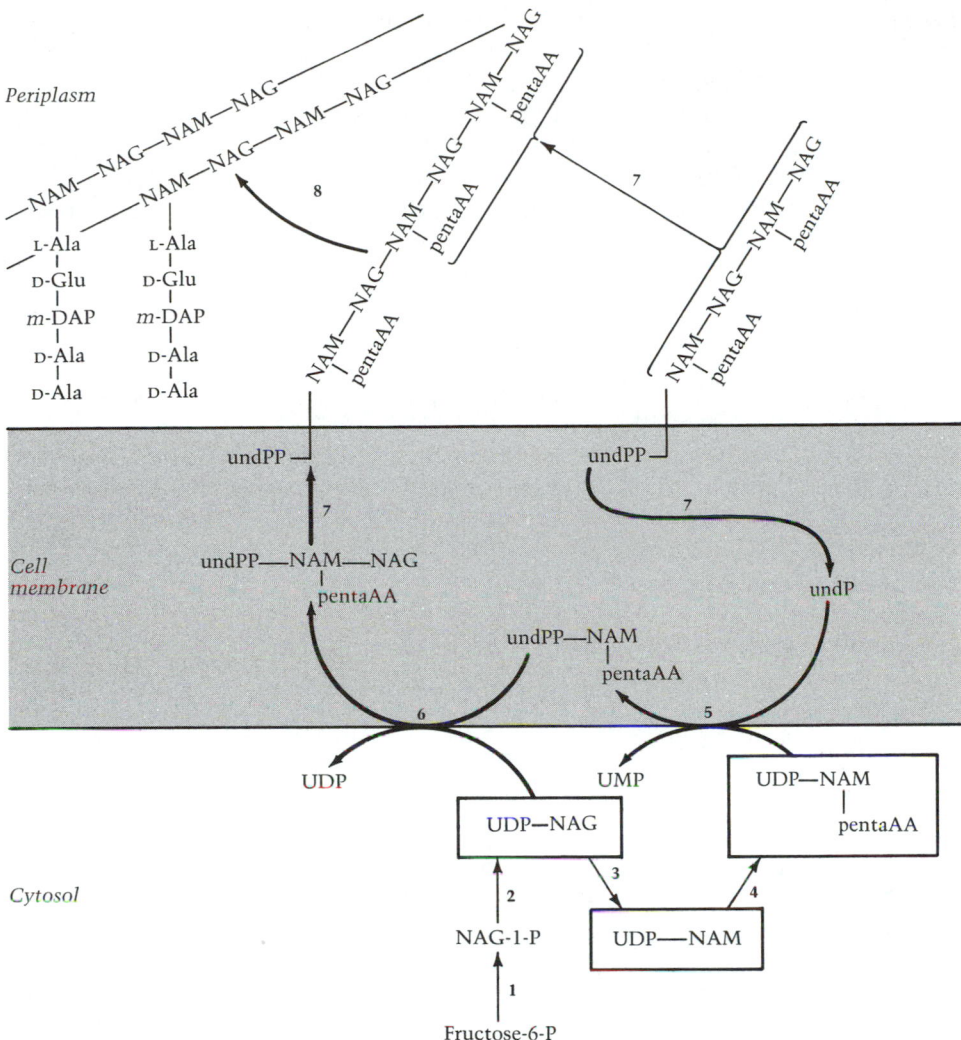

Figure 10. Polymerization of peptidoglycan. NAG, *N*-Acetylglucosamine; NAM, *N*-Acetylmuramic acid; und, undP, undPP, undecaprenol and its phosphate and pyrophosphate derivatives; penta-AA, the pentapeptide L-alanine-D-glutamate-*m*-diaminopimelate-D-alanine-D-alanine; UMP, uridine monophosphate; UDP, uridine diphosphate. The structures of key compounds are given in Chapter 2, Figure 11. See the text for a discussion of the steps labeled with boldface numbers.

bly occur in the periplasmic space. Second, all the energy needed for polymerization seems to be used in the cytosolic reactions in preparing the activated subunits—no ~P need be used in the periplasmic steps.

In the scheme presented in Figure 10, the first three steps (labeled **1, 2, 3**) can be considered to be biosynthetic rather than polymerizing. They utilize three of the key intermediary metabolites (fructose-6-P, acetyl CoA, and phosphoenolpyruvate) plus ~P and reducing power to make two sugar derivatives, UDP-*N*-acetylglucosamine (UDP-NAG), and UDP-*N*-acetylmuramic acid (UDP-NAM), which are activated building blocks for the glycan chains. Step **4** consists of five reactions that sequentially add D-glutamic acid, *m*-diaminopimelic acid, and two D-alanines, in that order, to the lactyl residue of UDP-NAM. In analogy with amino acid activation for protein synthesis, these

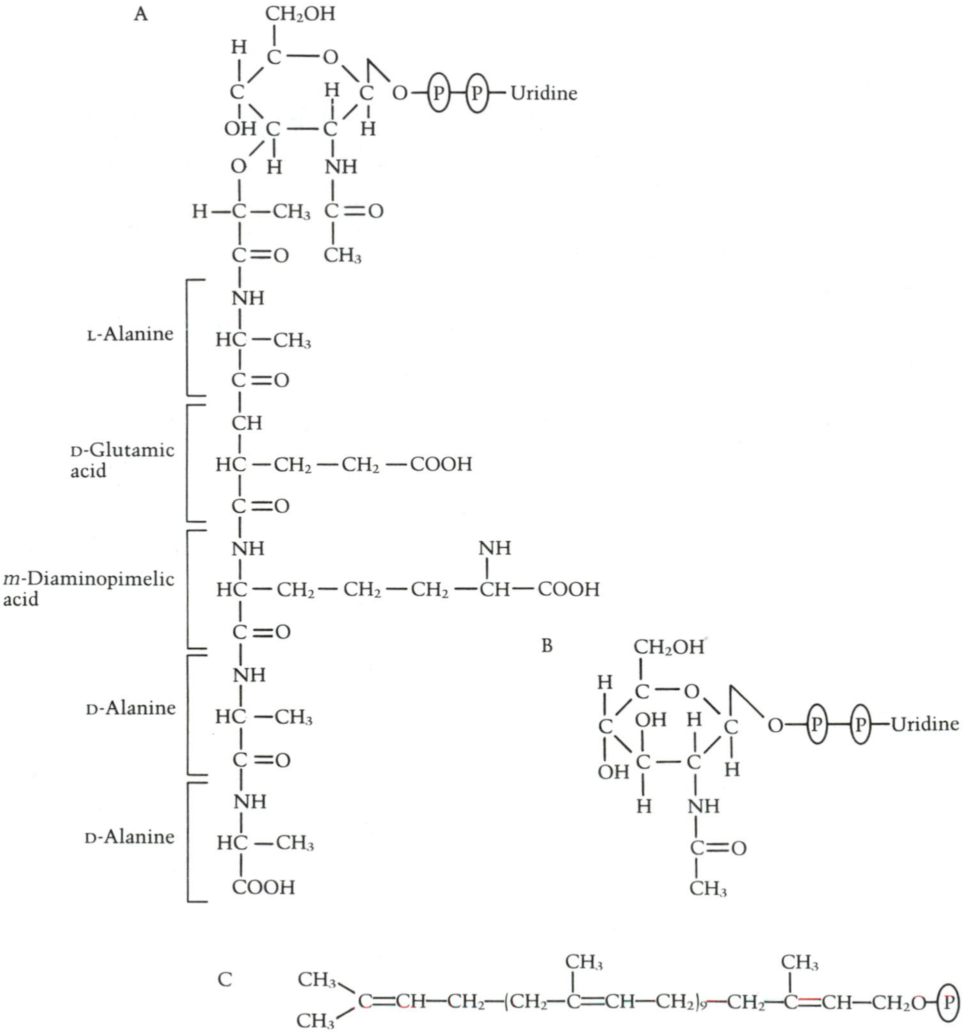

Figure 11. Structures of the subunits of peptidoglycan and their lipid carrier. A. Uridine diphosphate-*N*-acetylmuramic acid pentapeptide (UDP-NAM-pentaAA). B. Uridine diphosphate-*N*-acetylglucosamine (UDP-NAG). C. Undecaprenol phosphate (UndP; also called bactoprenol phosphate).

reactions can be considered part of polymerization. Formation of the penta-peptide on UDP-NAM, however, uses different enzymes than the synthetases that attach amino acids to tRNA; ATP is hydrolyzed to ADP (rather than AMP); and no tRNA is involved.

Steps **1** through **4** are cytosolic, and from this point on there is no known further input of energy. The next three steps occur in the cytoplasmic mem-brane. In steps **5** and **6** two sugar derivatives are attached to the lipid carrier undecaprenol phosphate (undP) to form the carrier-bound disaccharide subunit of peptidoglycan. [The structures of UDP-NAG, UDP-NAM-pentapeptide (UDP-NAM-pentaAA), and undP are shown in Figure 11]. At the outer surface of the membrane, the lipid-bound disaccharide subunit receives a growing glycan chain from an adjacent lipid carrier, enabling the latter, after hydrolysis of a phosphate, to return for a new subunit (step **7**).

In this way chains of perhaps 20–80 subunits are polymerized with their free ends projecting into the periplasm at points where they can become linked by interchain peptide bonds between the fourth residue (D-alanine) of one pentapeptide and the third residue (*m*-diaminopimelic acid) of another. This bond forms at the expense of the energy of hydrolyzing away the fifth residue (D-alanine), thereby obviating the need for a supply of ATP in the periplasm (step **8**).

In Table 5 we have summarized the building blocks and energy required to produce the peptidoglycan of our reference mass of cells.

Table 5. Requirements for peptidoglycan production

1 g cells contains 25 mg peptidoglycan; 25 mg peptidoglycan = 27 μmol disaccharide subunits; (1 μmol disaccharide = 904 μg disaccharide)

	Amount of building block required (μmol)	Amount of energy required (μmol ~P
Building blocks[a]		
UDP-NAG	27.6	
UDP-NAM	27.6	
L-Alanine	27.6	
D-Alanine	27.6	
m-Diaminopimelic acid	27.6	
D-Glutamic acid	27.6	
Energy for forming the pentapeptide[b]		138.0

[a]We consider the sugar building blocks to be their UDP derivatives because NAM (*N*-acetylmuramic acid) is made from the UDP derivative of NAG (*N*-acetylglucosamine). The amino acid building blocks will differ somewhat in different bacterial species (Chapter 1), and in particular there will in many instances be an oligopeptide bridge between adjacent chains; in *Staphylococcus aureus*, for example, it is a pentaglycine bridge. In such cases, the peptide is added between steps **6** and **7**, Figure 10.
[b]These ~P's represent the 5ATP → ADP hydrolyses that accompany construc-tion of the pentapeptide in step **4**.

POLYMERIZATION: PHOSPHOLIPIDS

The lipids of *E. coli* are all polar lipids and are located exclusively in the two membranes; there are no neutral lipids. In this section we shall describe the formation of phosphatidylethanolamine and phosphatidylglycerol, which together constitute nearly 95% of *E. coli*'s phospholipids. Though not macromolecules in a strict sense of the word (their average molecular weight is under 1000), their insolubility and the close assemblage of phospholipids with each other and with proteins to form membranes has led them to be considered with macromolecules.

The branched pathway shown in Figure 12 produces the two major *E. coli* phospholipids from sugar, amino acid, and fatty acid building blocks. All five steps of each pathway take place at the cell membrane; the soluble building blocks in the cytosol are utilized by enzymes bound tightly to the inner surface of the membrane (cf. Cronan, 1978). Reactions **1** and **2**, catalyzed by glycerol-*P*-acyltransferase, transfer fatty acyl residues from the ACYL CARRIER PROTEIN (ACP) on which they were synthesized to a glycerol-3-phosphate molecule. By means currently unknown, only saturated fatty acids are incorporated in reaction **1**, and only unsaturated fatty acids in reaction **2**. Reaction **3** utilizes CTP and releases pyrophosphate, the only direct input of ~P that is necessary because of the activated form of the fatty acyl building blocks. The CDP-diglyceride formed in reaction **3** is then used to produce each of the phospholipid species. There are interesting (and unanswered) questions concerning the regulation of the proportion of each species produced, the selection of different fatty acids, and the coordination of phospholipid synthesis with the other chemical processes producing the envelope.

Table 6 summarizes the requirements of phospholipid synthesis if all the membrane lipids (other than lipid A) of 1 g of cells were to be made by the phosphatidylethanolamine pathway.

Table 6. Requirements for phospholipid production

1 g cells contains 91 mg phospholipid other than lipid A; 91 mg phospholipid = 129 μmol phosphatidylethanolamine (PPE); (1 μmol PPE = 750 μg PPE)

	Amount of building block required (μmol)	Amount of energy required (μmol ~P)
Building blocks[a]		
Glycerol-3-phosphate	129	
Serine	129	
Fatty acyl-ACP ($C_{16:0}$, $C_{16:1}$, $C_{18:1}$)	258	
Energy for synthesis of phosphatidylethanolamine[b]		258

[a]Phosphatidylethanolamine, the major (75%) phospholipid in the cell membrane is formed from glycerol-3-phosphate, serine, and fatty acids in the ratio of 1:1:2 by the reactions shown in Figure 12. The next most prevalent phospholipid, phosphatidylglycerol, is formed from glycerol-3-phosphate and fatty

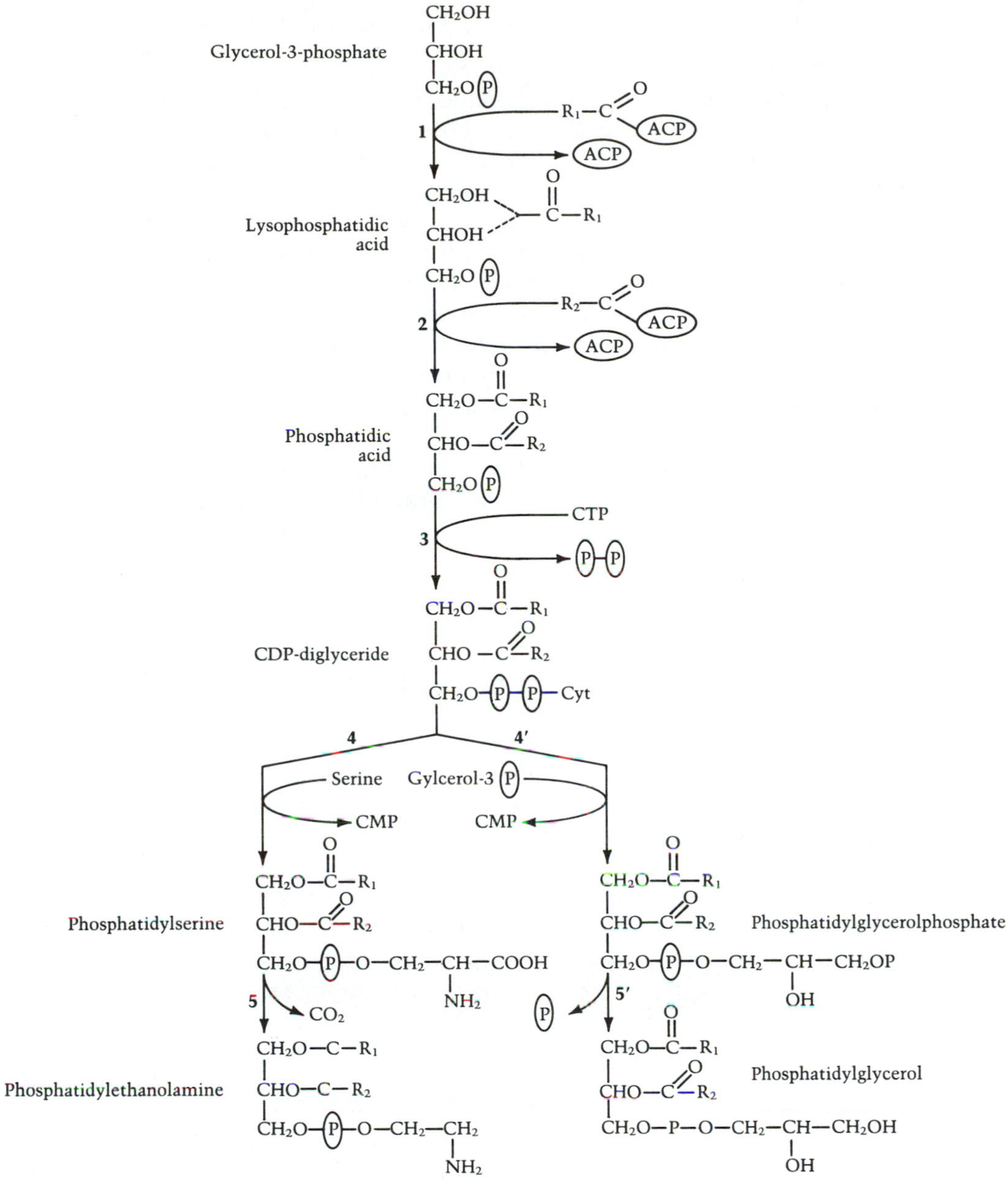

Figure 12. Synthesis of two major phospholipids in *E. coli*.

acids in equal molar ratios (Figure 12). The difference (1 serine replaced by 1 glycerol-3-phosphate) is small, and for simplicity the total biosynthetic demand here and in Table 10 will be calculated as though 100% were phosphatidylethanolamine. Also, the three major fatty acids are combined as a single requirement.

[b]Two ~P's are released in reaction (3) of Figure 12. The fatty acids are provided as already activated building blocks—conjugated to acyl carrier protein (ACP). The cost of activation is therefore included later as a biosynthetic cost.

POLYMERIZATION: LIPOPOLYSACCHARIDE

The outer leaflet of the outer membrane in *E. coli* and other Gram-negative cells contains the unique phospholipid, LPS (lipopolysaccharide). Formation of this complex molecule occurs on the cytoplasmic membrane (Figure 13). Soluble building blocks (nucleoside sugars and fatty acyl-ACP) synthesized in the cytosol are brought to the cytoplasmic membrane, where polymerization occurs in stepwise fashion and each addition is catalyzed by a different enzyme in the membrane. Lipid A is first constructed by the addition of KDO (3-deoxy-D-mannooctulosonic acid), fatty acids, and other substituents (e.g., ethanolamine) to a glucosamine disaccharide (Figure 13A). Lipid A then serves both as a primer and as a lipid carrier for the stepwise addition of sugars from nucleoside diphosphate derivatives to form the CORE POLYSACCHARIDE, the structure of which is similar but not identical among different genera.

Meanwhile, on the other lipid carrier, undecaprenol phosphate—encountered in our discussion of peptidoglycan formation—polysaccharide chains are being polymerized from nucleoside sugars to form the REPEATING SIDE CHAINS. Short oligosaccharides (2–5 sugar residues) are assembled on the carrier by the successive action of separate transferase enzymes. These repeat units are then linked together in long chains. As shown in Figure 13, a growing polysaccharide chain is transferred from its carrier to the distal sugar of a recently completed unit on another carrier molecule. The final synthetic step is the transfer of a suitably long polysaccharide repeat chain to the distal sugar of a core polysaccharide already completed on a lipid A moiety. Several LPS molecules may become joined by phosphodiester bonds linking their lipid A disaccharides. Finally, the completed LPS is translocated to the outer leaflet of the outer membrane, presumably through Bayer's zones of adhesion (Chapter 2).

Our reference cell makes a truncated version of the core polysaccharide, and no repeating chain at all. The B/r strain of *E. coli* has apparently sometime in its history suffered mutational loss of several of its transferase enzyme activities. This is fairly common, particularly among strains cultivated under laboratory conditions for extensive periods; the mutants presumably grow faster. In nature there must be selective pressures (e.g., adhesion to surface and escape from phagocytosis) that maintain the "smooth" cellular surface conferred by the LPS repeating chains. These, incidentally, are antigenic and constitute the O-ANTIGENS of the enteric group of Gram-negative bacteria including *E. coli*. In Table 7 we have summarized the requirements for making the abbreviated LPS structure of 1 g of strain B/r cells.

POLYMERIZATION: POLYSACCHARIDE

Glycogen

As discussed in Chapter 1, bacteria under certain conditions store carbon sources and other substances by producing large amounts of storage polymers (e.g., poly-β-hydroxybutyrate, glycogen, polymetaphosphate).

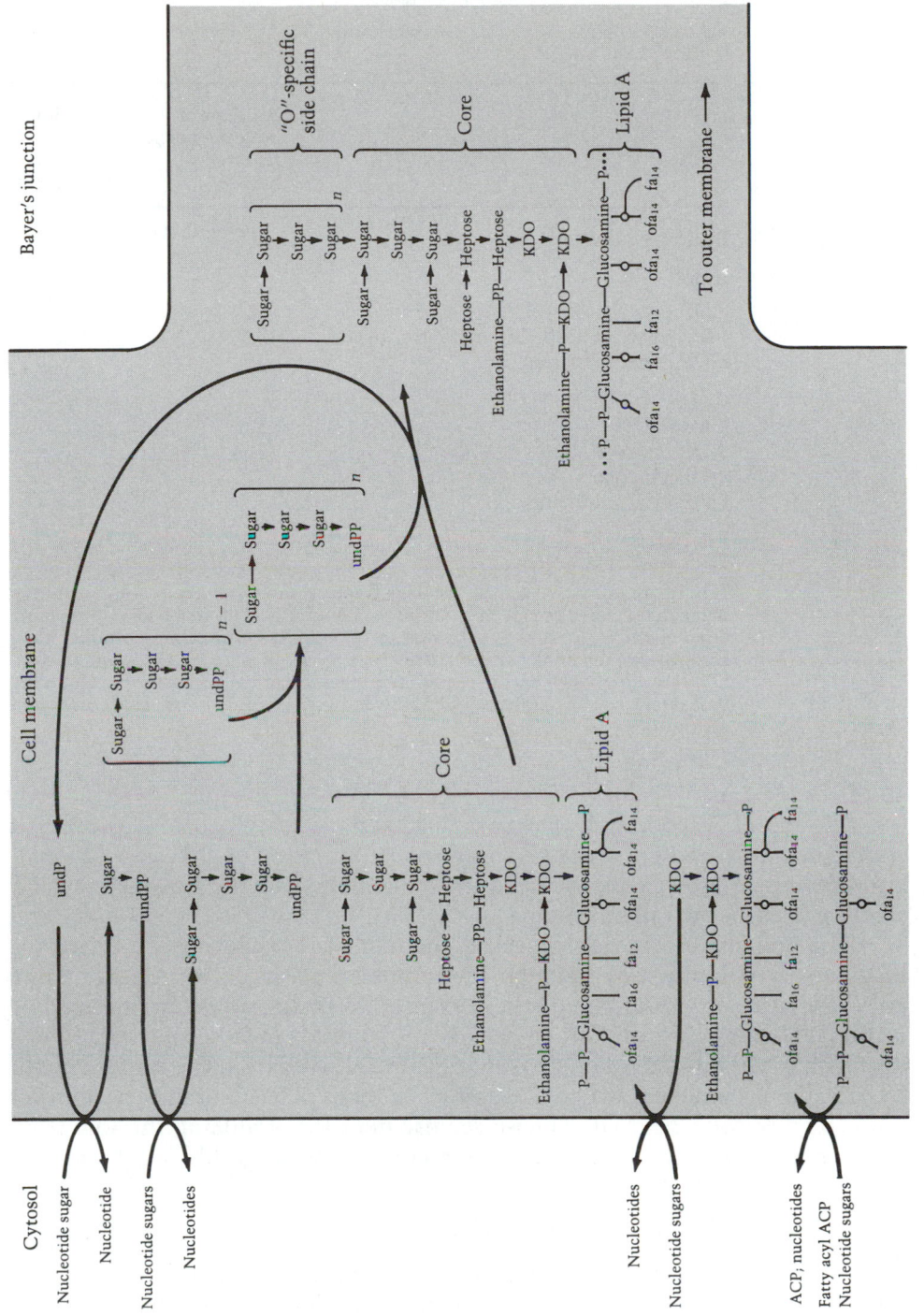

Figure 13. Polymerization of LPS. Abbreviations are in Chapter 1, Figure 9.

Table 7. Requirements for LPS production

1 g cells contains 34 mg lipopolysaccharide; 34 mg LPS = 7.8 μmol LPS unit; (1 μmol LPS = 4350 μg LPS)

	Amount of building block required (μmol)	Amount of energy required (μmol ∼P)
Building blocks[a]		
Lipid A		
TDP-glucosamine	15.7	
Fatty acyl-ACP (β-OH-myristic, others)	47.0	
CDP-ethanolamine	7.8	
Core		
CMP-KDO	23.5	
ADP-heptose	23.5	
UDP-glucose	15.7	
CDP-ethanolamine	15.7	
Energy[b]		0

[a]These are not based on the precise structure shown in Figure 13, but on the truncated version found in some *E. coli* B strains. The mechanism of adding ethanolamine residues is assumed to be via CDP-ethanolamine rather than through serine and decarboxylation.
[b]The seemingly low energy requirement is misleading, because all the building blocks are used as activated derivatives with an energy cost charged to biosynthesis.

Growth of *E. coli*, and many other bacteria, with glucose as the major source of carbon and energy leads to some accumulation of glycogen—a small amount unless growth is restricted (as by depletion of the nitrogen source) while glucose is still plentiful.

The building block for glycogen is the nucleoside diphosphate derivative of glucose, ADP-glucose. Through the action of glycogen synthetase, large polysaccharide chains of glucose, joined in α-1,4 linkages, are synthesized. A primer at least four residues in length is required. In a second reaction—transglucosylation—small oligosaccharides are cleaved from the end of a chain by a branching enzyme, which then attaches them in α-1,6 linkage at another point (Figure 14). The ADP-glucose requirement for producing the low level of glycogen present in our reference biomass is given in Table 8.

Capsular polysaccharide

In strains of *E. coli* that produce a polysaccharide capsule (B/r does not), the mode of polymerization and translocation is similar to that described for the repeating oligosaccharide chains of LPS. That is, polymerization of sugar building blocks, presented as nucleoside diphosphate sugars at the inner surface of the cell membrane, occurs on the same lipid carrier

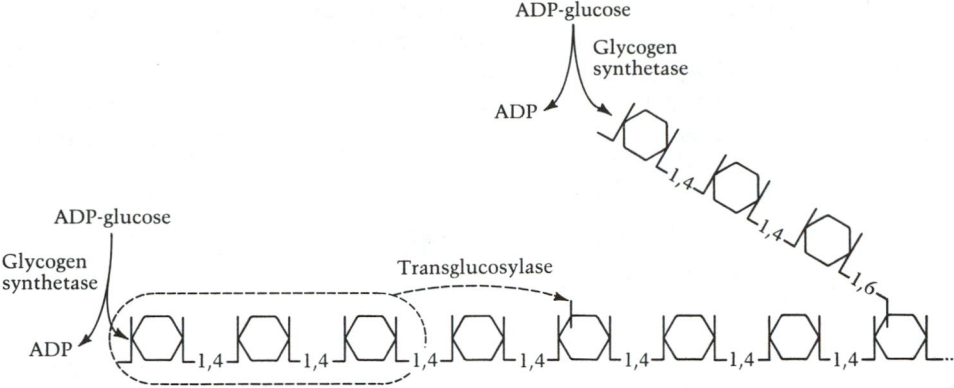

Figure 14. Glycogen synthesis in *E. coli*. A small portion of a growing glycogen molecule is shown to illustrate the addition of glucose residues from ADP-glucose to the nonreducing ends of chains, catalyzed by glycogen synthetase, and the formation of a branch by transfer of an oligosaccharide to an internal glucose residue, catalyzed by transglucosylase.

(undecaprenol phosphate; Figure 11C) that functions in peptidoglycan (Figure 10) and LPS (Figure 13) synthesis. Subsequent assembly of oligosaccharide units and translocation have been described (Chapter 2).

POLYMERIZATION: THE ENERGY COSTS

In Tables 2–8 we have tallied the requirements of building blocks and energy for producing the macromolecules of 1 g of cells. These values will be useful as we progress with our examination of the chemistry of growth. We shall be concerned with the synthesis of ribonucleoside and deoxyribonucleoside triphosphates, amino acids, fatty acyl-ACP, and sugar nucleoside diphosphates in the next phase of metabolism. Before proceeding, however, it will be interesting to take one look at the energy debit we have generated.

Table 8. Requirements for glycogen production

1 g cells contains 25 mg glycogen; 25 mg glycogen = 154 μmol glucosyl residues; (1 μmol residues = 162 μg residues)

	Amount of building block required (μmol)	Amount of energy required (μmol ~P)
Building block		
ADP-glucose	154	
Energy for polymerization		None

Table 9 summarizes the high-energy phosphates needed to polymerize the building blocks of each class of macromolecules. (The figures are simply lifted from Tables 2–8). The astounding fact is that of 22,800 μmol of ~P required for all polymerizations, 22,000 μmol are required for protein synthesis. True, this value is somewhat biased. The building blocks for macromol-

Table 9. Energy requirements for polymerization

1 g cells contains 961 mg macromolecules

Macromolecule	Amount of energy required[a] (μmol ~P)
DNA (made from activated building blocks)	150
RNA (made from activated building blocks)	256
Protein (made from free amino acids)	21,970[b]
Peptidoglycan (made in part from activated building blocks)	138
Phospholipid (made in part from activated building blocks)	258
LPS (made from activated building blocks)	0
Polysaccharide (glycogen; made from activated building blocks)	0
Total energy	22,772

[a]Data compiled from Tables 2–8.
[b]If already activated building blocks (aminoacyl-tRNA) are used, the cost is reduced to 10,000 μmol.

ecules other than protein are wholly or partially in an activated form, whereas the protein building blocks (amino acids) are not. This bias is readily corrected. Let us remove the energy cost (2 ~P) of activating each amino acid from our protein cost. This correction reduces by 10,200 μmol (2 × 5081 μmol aa.) the ~P cost of protein synthesis, but leaves us still with a cost of 11,800 μmol to polymerize protein, and 12,600 μmol for all polymerization. *Protein synthesis, therefore, requires approximately 95% or more of all the energy used for polymerization reactions, no matter what convention is used for defining a building block.*

BIOSYNTHESIS

Elucidation of how a living cell produces the building blocks needed for macromolecular synthesis was the first triumph of bacterial mutant methodology. The isolation and analysis of bacterial auxotrophic mutants, supplemented by two other important approaches—in vivo studies with isotope-labeled metabolites and in vitro studies of enzymatic activities—led to the discovery of the several hundred biosynthetic reactions that produce amino acids, nucleotides, sugars, nucleotide-sugar derivatives, fatty acids, coenzymes, and enzyme prosthetic groups.

This work, which reached peak intensity during the 1950s and 1960s, was of general significance in biology because, as it turned out, there is a basic similarity of biosynthetic routes in all living cells. To a first approximation, the same biochemical text can serve as a reference on biosynthesis for the botanist, microbiologist, and zoologist. There are, of course, many exceptions to this rule; some involve the biosynthesis of compounds unique to a given organism and others are instructive indications of special biochemical constraints imposed on some organisms by their peculiar ecological niche.

All 75–100 building blocks, coenzymes, and prosthetic groups are synthesized from 12 PRECURSOR METABOLITES by sequential enzymatic reactions that employ energy, reducing power, and sources of nitrogen, sulfur, and single carbon units (Figure 15, Table 10). The several hundred biosynthetic reactions are organized into functional units called BIOSYNTHETIC PATHWAYS. The reactions of a pathway are generally regulated en bloc by controls that operate on enzyme synthesis as well as on the enzymatic activity of the pathway. As we shall see, there are many other characteristics of pathways, which make them readily recognizable.

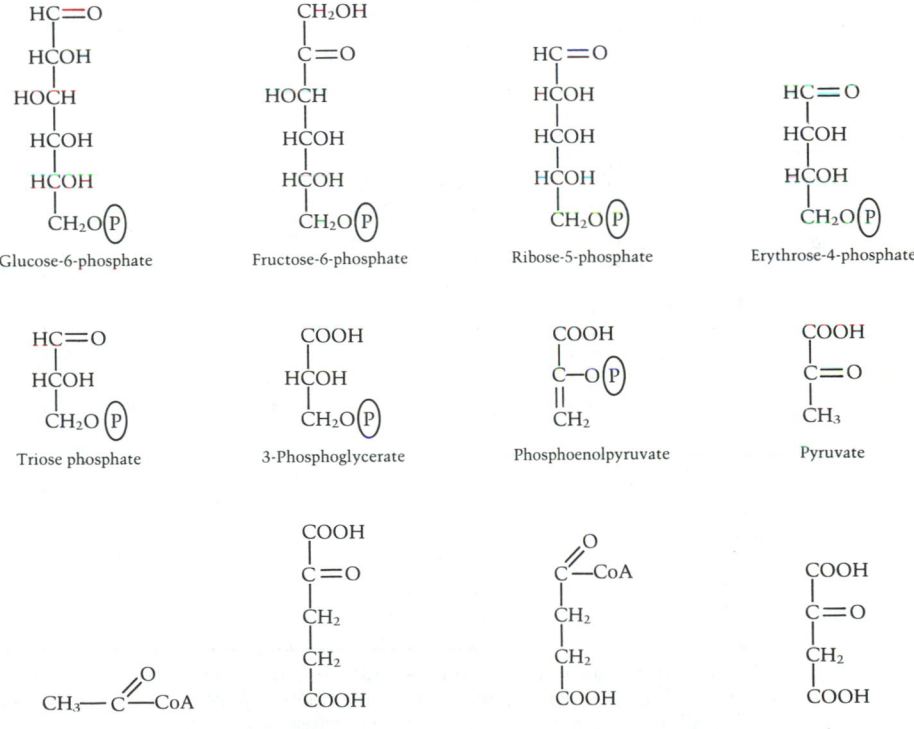

Figure 15. Structures of the 12 precursor metabolites.

Table 10. Building blocks needed to produce 1 g of *E. coli* protoplasm

Building block	Amount present in *E. coli* B/r (μmol/g dried cells)	Metabolites[a]	Cost of making 1 μmol of each of these building blocks (μmol/μmol)					
			ATP	NADH	NADPH	1-C	NH₃	S
Protein amino acids								
Alanine	488	1 pyr	0	0	1	0	1	0
Arginine	281	1 αkg	7	−1	4	0	4	0
Asparagine	229	1 oaa	3	0	1	0	2	0
Aspartate	229	1 oaa	0	0	1	0	1	0
Cysteine	87	1 pga	4	−1	5	0	1	1
Glutamate	250	1 αkg	0	0	1	0	1	0
Glutamine	250	1 αkg	1	0	1	0	2	0
Glycine	582	1 pga	0	−1	1	−1	1	0
Histidine	90	1 penP	6	−3	1	1	3	0
Isoleucine	276	1 oaa, 1 pyr	2	0	5	0	1	0
Leucine	428	2 pyr, 1 acCoA	0	−1	2	0	1	0
Lysine	326	1 oaa, 1 pyr	2	0	4	0	2	0
Methionine	146	1 oaa	7	0	8	1	1	1
Phenylalanine	176	1 eryP, 2 pep	1	0	2	0	1	0
Proline	210	1 αkg	1	0	3	0	1	0
Serine	205	1 pga	0	−1	1	0	1	0
Threonine	241	1 oaa	2	0	3	0	1	0
Tryptophan	54	1 penP, 1 eryP, 1 pep	5	−2	3	0	2	0
Tyrosine	131	1 eryP, 2 pep	1	−1	2	0	1	0
Valine	402	2 pyr	0	0	2	0	1	0
RNA nucleotides								
ATP	165	1 penP, 1 pga	11	−3	1	1	5	0
GTP	203	1 penP, 1 pga	13	−3	0	1	5	0
CTP	126	1 penP, 1 oaa	9	0	1	C	3	0
UTP	136	1 penP, 1 oaa	7	0	1	0	2	0
DNA nucleotides								
dATP	24.7	1 penP, 1 pga	11	−3	2	1	5	0
dGTP	25.4	1 penP, 1 pga	13	−3	1	1	5	0
dCTP	25.4	1 penP, 1 oaa	9	0	2	0	3	0
dTTP	24.7	1 penP, 1 oaa	10.5	0	3	1	2	0
Lipid components								
Glycerol phosphate	129	1 triosP	0	0	1	0	0	0
Serine	129	1 pga	0	−1	1	0	1	0
C₁₆:₀ fatty acid (43%)		8 acCoA	7	0	14	0	0	0
C₁₆:₁ fatty acid (33%)		8 acCoA	7	0	13	0	0	0
C₁₈:₁ fatty acid (24%)		9 acCoA	8	0	15	0	0	0
Average fatty acid	258	8.2 acCoA	7.2	0	14	0	0	0

[a]acCoA, Acetyl CoA; eryP, erythrose-4-phosphate; fruP, fructose-6-phosphate; gluP, glucose-6-phosphate; αkg, α-oxoglutarate; oaa, oxalacetate; penP, ribose-5-phosphate; pep, phosphoenolpyruvate; pga, 3-phosphoglycerate; pyr, pyruvate; triosP, triose phosphate.

Table 10. (Continued)

Building block	Amount present in *E. coli* B/r (μmol/g dried cells)	Metabolites[a]	Cost of making 1 μmol of each of these building blocks (μmol/μmol)					
			ATP	NADH	NADPH	1-C	NH₃	S
LPS components								
UDP-glucose	15.7	1 gluP	1	0	0	0	0	0
(CDP) ethanolamine	23.5	1 pga	3	−1	1	0	1	0
OH-myristic acid	23.5	7 acCoA	6	0	11	0	0	0
C₁₄:₀ fatty acid	23.5	7 acCoA	6	0	12	0	0	0
(CMP) KDO	23.5	1 penP, 1 pep	2	0	0	0	0	0
(NDP) heptose	23.5	1.5 gluP	1	0	−4	0	0	0
(TDP) glucosamine	15.7	1 fruP	2	0	0	0	1	0
Peptidoglycan monomers								
UDP-*N*-acetylglucosamine	27.6	1 fruP, 1 acCoA	3	0	0	0	1	0
UDP-*N*-acetylmuramic acid	27.6	1 fruP, 1 pep, 1 acCoA	4	0	1	0	1	0
Alanine	55.2	1 pyr	0	0	1	0	1	0
Diaminopimelate	27.6	1 oaa, 1 pyr	2	0	3	0	2	0
Glutamate	27.6	1 αkg	0	0	1	0	1	0
Glycogen monomers								
Glucose	154	1 gluP	1	0	0	0	0	0
1-Carbon requirement								
Serine	48.5	1 pga	0	−1	1	0	0	0
Polyamines								
Ornithine equivalents	59.3	1 αkg	2	0	3	0	2	0

Other (small) molecules (less than 3% of cell dry weight)

Coenzymes: NAD, NADP, CoA, CoQ, bactoprenoid, tetrahydrofolate, cyanocobalamin, pyridoxal phosphate

Prosthetic groups: FMN, FAD, biotin, cytochromes, lipoic acid, thiamine pyrophosphate

Pool of unpolymerized monomers: average approximately 1% of amount in macromolecules

Biosynthetic pathways

A simple pathway, consisting of three sequential enzymatic reactions, might be represented as:

$$\text{Precursor Metabolite (PM)} \xrightarrow[\text{(E}_1\text{)}]{\text{Enzyme 1}} A \xrightarrow[\text{(E}_2\text{)}]{\text{Enzyme 2}} B \xrightarrow[\text{(E}_3\text{)}]{\text{Enzyme 3}} \text{Building Block (BB)}$$

where A and B are intermediate products in the pathway. Recognizing that these reactions are controlled as a physiological unit during cell growth, we might more simply represent the pathway as:

$$PM \xrightarrow{3E} BB$$

Some pathways produce a building block that, in turn, is converted by a second pathway into another building block:

$$PM \xrightarrow{xE} BB_1 \xrightarrow{yE} BB_2$$

where x and y represent the number of enzymes in the two pathways. In some cases the pathway is branched:

$$PM \xrightarrow{xE} BB_1 \left[\begin{array}{l} \xrightarrow{yE} BB_2 \\ \xrightarrow{zE} BB_3 \end{array} \right.$$

Most of the 12 precursor metabolites actually serve as the starting point for several pathways:

$$PM \left[\begin{array}{l} \xrightarrow{xE} BB_1 \\ \xrightarrow{yE} BB_2 \\ \xrightarrow{zE} BB_3 \end{array} \right.$$

Finally, in many cases more than one precursor molecule is involved in the biosynthesis of a building block:

$$\left. \begin{array}{l} PM_a \\ PM_b \end{array} \right] \xrightarrow{xE} BB$$

Branching and interlocking of this sort is common among biosynthetic pathways. Building blocks that are produced from some common precursor are referred to as a FAMILY. The *aspartate family* consists of seven amino acids (aspartate, asparagine, threonine, isoleucine, methionine, diaminopimelate, and lysine) that are synthesized from the common precursor metabolite oxalacetate, through aspartate (Figure 16).

A fascinating amount of biochemical detail is known about the biosynthetic reactions. These details are important for many purposes, including an understanding of the elegant biochemical tricks that have evolved to handle various problems in organic synthesis. Any of several biochemistry texts (e.g.,

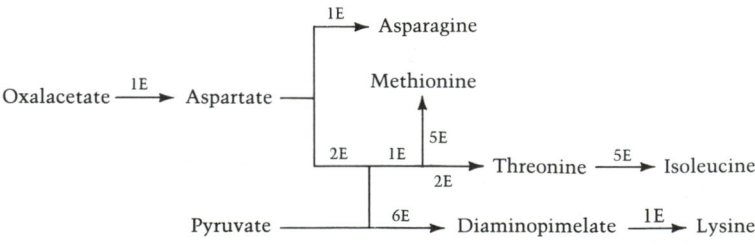

Figure 16. Pathways of biosynthesis of the aspartate family of amino acids in *E. coli.* The numbers over the arrows indicate the number of enzymes involved in each segment of the pathway.

Mandelstam, 1973; Stryer, 1981) can be consulted for this detail. We shall concentrate on the role of biosynthetic pathways in cell growth and on the resources required for their operation. To learn what resources are needed to produce all the building blocks and coenzymes for 1 g of *E. coli* cells we are less concerned about what occurs at each enzymatic step than about the overall cost of biosynthesis. We shall find it useful, therefore, to deal with each pathway as a UNIT FUNCTION, noting its components (number of enzymes) and its metabolic cost (consumption of energy, reducing power, NH_3, sulfate, and one-carbon units). We have constructed unit functions for most of the major pathways—those producing the building blocks for protein, RNA, DNA, phospholipids, LPS, peptidoglycan, and glycogen (Figure 17).

Requirements for biosynthesis

The amount of each building block needed for polymerization (Tables 2–8) is listed in Table 10, together with the metabolic cost of making 1 μmol of each, as compiled from the biosynthetic charts of Figure 17. (Table 10 includes the cost of producing ornithine for the formation of polyamines (Chapter 2) but not the costs of producing the coenzymes, prosthetic groups, and related substances. Biosynthesis of these important small molecules is vital to cell growth. Inspection of their biosynthetic pathways is worthwhile because one can gain some appreciation of the large number of enzymes necessary for this task and for the consequent dedication of a significant part of the genome for coding for these enzymes (Appendix B). The metabolic cost (carbon compounds, energy, etc.) of producing these substances, however, is quite small relative to the cost of macromolecular synthesis.

From the information in Table 10 one can calculate the molar demand for precursor metabolites, energy, reducing power, and nitrogen and sulfur for our biosynthetic task (Table 11).

Assimilation of nitrogen and sulfur

None of the precursor metabolites contain nitrogen or sulfur. These elements enter the cell as constituents of a variety of organic molecules or as inorganic compounds or ions, but they always enter cellular metabolism in inorganic form: nitrogen as ammonia (NH_3) and sulfur as hydrogen sulfide (H_2S). The entry occurs in certain biosynthetic reactions (Figure 17).

Of the three elements nitrogen, sulfur, and phosphorus, nitrogen is quantitatively the most significant, constituting 14% of the dry weight of our reference bacterial cell. Like nitrogenous compounds in the bacterial cell itself, most utilizable organic sources of nitrogen contain nitrogen in the -3 oxidation state as amino groups, imino groups, or heterocyclic nitrogen atoms. These compounds are metabolized by pathways of various lengths yielding NH_3 without the necessity of reducing the nitrogen atom. As will be discussed later, these pathways are usually under the repressive control of the NH_3 ion: when NH_3 is present in adequate concentration, the enzymes catalyzing the catabolism of these ammonia-yielding reactions are not synthesized.

Bacteria as a group also utilize inorganic sources of nitrogen: most sig-

Table 11. Requirements for biosynthesis of building blocks

Precursor metabolites	Amount required[a] (μmol/g cells)	Precursor metabolites	Amount required[a] (μmol/g cells)
Glucose-6-phosphate	205	Succinyl CoA	—[b]
Fructose-6-phosphate	70.9	Oxalacetate	1,786.7
Ribose-5-phosphate	897.7	~P	18,485
Erythrose-4-phosphate	361	NADH$_2$	(−3,547)
Triose phosphate	129	NADPH$_2$	18,225
3-Phosphoglycerate	1,496	1-C (accounted for as serine)	—[c]
Phosphoenolpyruvate	519.1	NH$_3$	10,180
Pyruvate	2,832.8	S	233
Acetyl CoA	3,747.8		
α-Oxoglutarate	1,078.9		

[a]Calculated from the information in Table 10 by multiplying the amount of each building block (first column of figures) by the molar requirement of precursor metabolites or other components to produce them.
[b]Succinyl CoA is used as a cofactor in the synthesis of building blocks, that is, succinate is later released from the biosynthetic intermediates. Therefore, this results in the net expenditure of 1 ~P in the synthesis of lysine and methionine. Succinyl CoA is a precursor metabolite, however, in the synthesis of heme and hemelike compounds.
[c]The requirement for 1-C fragments in the synthesis of methionine, purines, and thymine has been arbitrarily accounted for by synthesizing the requisite number of serine molecules.

nificant of these are NH$_3$ itself, nitrate ion (NO$_3^-$), and dinitrogen gas (N$_2$). Probably all bacteria can utilize NH$_3$; ability to utilize NO$_3^-$ and N$_2$ is more restricted; those organisms that can assimilate NO$_3^-$ reduce it to nitrite ion (NO$_2^-$) by ASSIMILATORY NITRATE REDUCTASE, a molybdenum-containing enzyme and then, by NITRITE REDUCTASE, to NH$_3$. Both of these enzymes are repressed if NH$_3$ is present. In general, it can be deduced that a bacterium able to utilize NO$_3^-$ as a source of nitrogen for biosynthesis must possess both assimilatory nitrate and assimilatory nitrite reductases; but this is not always the case. Certain bacteria, including *E. coli*, reduce NO$_3^-$ to NO$_2^-$ for a totally different function. Nitrate, rather than O$_2$, serves as the terminal electron acceptor in a respiratory process in which ATP is generated anaerobically and NO$_2^-$ is produced. The enzyme catalyzing this reduction is bound to the membrane and is termed DISSIMILATORY NITRATE REDUCTASE; it is synthesized only in the absence of O$_2$. *Escherichia coli* also produces a nitrite reductase, which reduces NO$_2^-$ to NH$_4^+$ without generating ATP. The function of this nitrite reduction is thought to be detoxification, that is, removal of the toxic NO$_2^-$ that would otherwise accumulate during anaerobic respiration of NO$_3^-$. Thus, anaerobically *E. coli* can utilize NO$_3^-$ as a total source of nitrogen; but aerobically it requires a reduced nitrogen source in order to grow.

Many natural environments contain inadequate amounts of biologically available nitrogen to support optimal growth of organisms, but the earth's atmosphere contains approximately 80% dinitrogen gas (N_2). Certain bacteria, and only bacteria, have the capacity to reduce N_2 to NH_3, a process termed NITROGEN FIXATION. The process is catalyzed by a single enzyme complex, NITROGENASE, which is composed of two iron–sulfur proteins: AZOFERREDOXIN

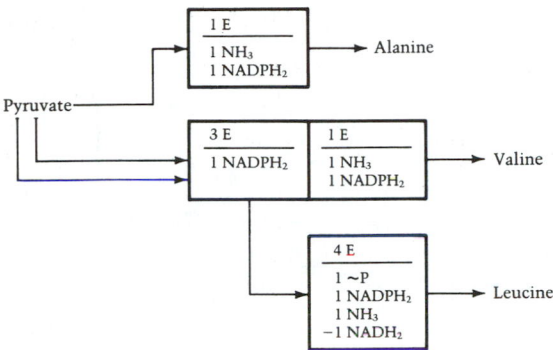

Figure 17. Major bacterial biosynthetic pathways. Each pathway is depicted as a unit function. Individual reactions and pathway intermediates are not shown. Instead, the conversion of each precursor metabolite into a building block is summarized as a single "transfer function," and the number of enzymes (E) and the metabolic costs involved are indicated. Each box represents an entire pathway or pathway branch. This figure and Tables 10 and 13 are derived from the presentations of metabolic economics by Fraenkel (1962) and Umbarger (1977). Another helpful treatment of the same subject is by Stouthammer (1973). In our charts and tables, metabolic costs are calculated and expressed by a set of conventions closely patterned after the cost accounting system introduced by Umbarger (1977) for teaching the subject of metabolic pathways. Costs are tallied for each individual reaction of a pathway, and a running summary is kept. Positive values indicate *net utilization* in the unit function, negative values, *net production*. All costs are converted to precursor metabolites, energy ~P, reducing power ($NADH_2$; $NADPH_2$), ammonia, and sulfur. Thus, the use of glutamine as amino donor is charged as 1 ~P and 1 NH_3 (the cost of regenerating glutamine from the glutamate produced in the reaction). Similarly, transamination using glutamate as an amino donor is charged as 1 $NADPH_2$ and 1 NH_3, and acetylation using acetyl CoA in a pathway in which acetate is later released is charged as 1 ~P. By this convention succinyl CoA appears only in the one pathway (heme biosynthesis) in which it serves as a precursor metabolite and not where it serves as a cofactor equivalent to 1 ~P. In addition to standard texts (Stryer, 1981; Mandelstam, 1973; Gottschalk, 1979), the review of amino acid biosynthesis by Umbarger (1979) is recommended to the reader.

Figure 17. (Continues)

Figure 17. (Continued)

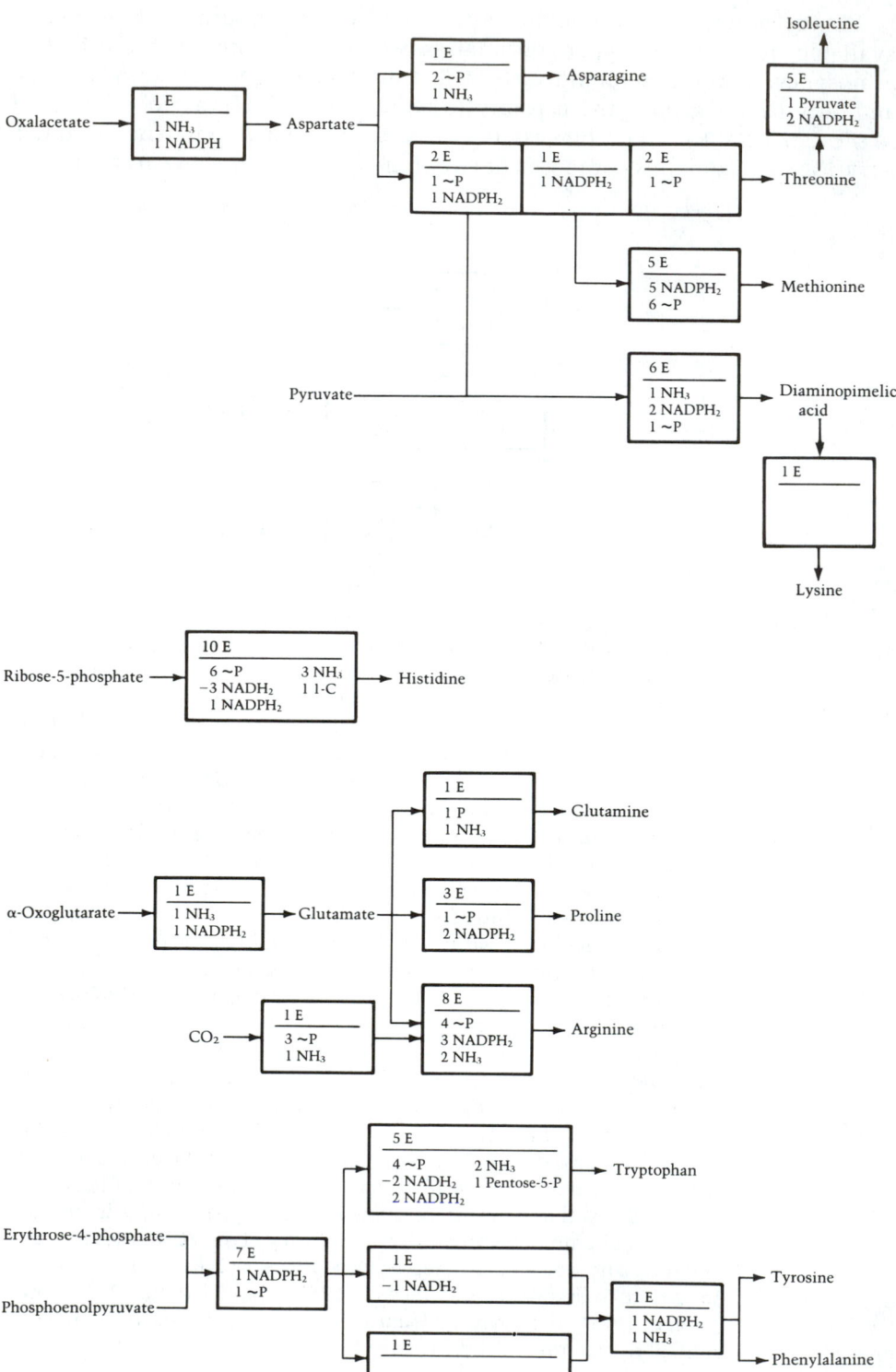

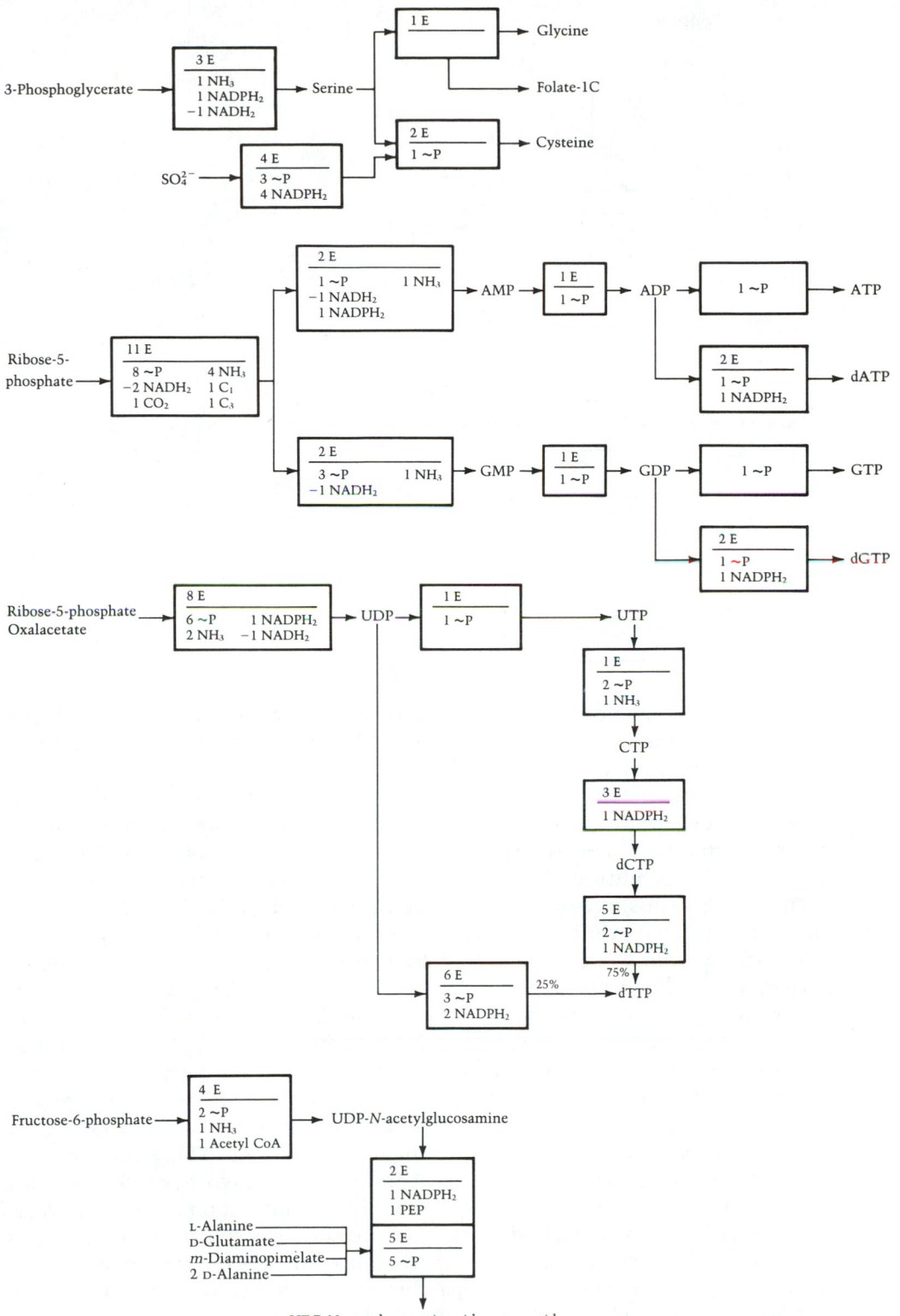

Figure 17. (Continues)

131

Figure 17. (Continued)

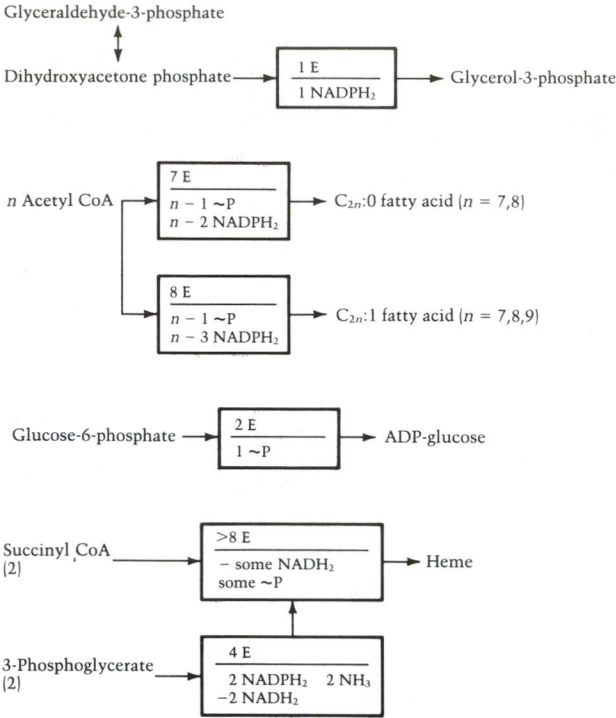

and MOLYBDOFERREDOXIN (which contains two molybdenum atoms in addition to the iron-sulfur centers). Although the reduction of N_2 to NH_3 is exothermic, N_2 is an extremely unreactive molecule. Reduction to NH_3 requires considerable activation energy, which is supplied in the form of ATP. In vitro, 6–15 moles of ATP are hydrolyzed for each mole of N_2 fixed. In vivo, the requirement is probably closer to 6 moles of ATP; it is an extremely costly reaction in terms of expenditure of ATP. Nitrogenase is also remarkable in its sensitivity to O_2. Thus, nitrogen-fixing bacteria have all evolved mechanisms to protect nitrogenase from O_2. Certain nitrogen fixers (e.g., *Clostridium pasteurianum*) are strict anaerobes and therefore only grow in O_2-free environments. Others (e.g., *Rhodospirillum rubrum*) are facultative anaerobes and fix nitrogen only when the environment is anaerobic. Certain strict aerobes that fix nitrogen (e.g., *Azotobacter vinelandii*), maintain an anaerobic intracellular environment by possessing an extremely active respiratory capacity that utilizes all available oxygen at the cell surface. Other aerobes have evolved structures that exclude O_2. In the case of the blue-green bacteria, nitrogen fixation occurs in specialized cells, termed HETEROCYSTS. These cells are impervious to oxygen and, lacking the appropriate photosynthetic component, they do not produce it. The rhizobia (e.g., *Rhizobium japonicum*) form symbiotic relationships with plant roots. They invade the roots and cause nodules to develop that become packed with rhizobia; in this environment, pO_2 is maintained at a lower level compatable with nitrogenase activity, by the action of an O_2-binding protein, LEGHEMAGLOBIN.

Although there is considerable diversity in the way bacteria derive ammonium ion (directly from the environment; from organic compounds; from NO_3^-; from N_2) there are only two pathways by which ammonium ion is assimilated into organic constituents of the cell: (1) directly into glutamate via a reaction catalyzed by L-GLUTAMATE DEHYDROGENASE (GDH) and by a cycle of reactions catalyzed by GLUTAMINE SYNTHETASE (GS) and GLUTAMATE SYNTHASE (sometimes termed glutamine: α-oxoglutarate aminotransferase, or GOGAT). The glutamate dehydrogenase pathway is an $NADPH_2$-specific reductive amination of the precursor metabolite, α-keto- (or oxo-) glutarate, to produce glutamate (Figure 17), the α-amino group of glutamate coming directly from NH_4^+, and by subsequent reactions being transferred to many other nitrogenous constituents of the cell. In the GS–GOGAT system, the α-amino group is derived from the amide group of glutamine. Thus, in the GS–GOGAT system, the primary assimilation of NH_4^+ occurs in the reaction catalyzed by glutamine synthetase. Glutamine synthetase and GDH differ in two important ways: the affinity of GS for NH_4^+ is considerably higher than that of GDH; the GS-catalyzed reaction requires ATP and the GDH reaction does not. Thus, at the expense of ATP, the GS system can assimilate ammonia when its concentration is low (< 1 mM). Some bacteria have only one of these systems for assimilating ammonia; others (e.g., *E. coli* and *S. typhimurium*) have both. As might be expected, when both pathways are present, glutamine synthetase is repressed if ammonia is present in high concentrations. Indeed, recent experiments have shown that GS is under the same system of represssive NITROGEN CONTROL as those ammonia-yielding catabolic pathways mentioned earlier. The activity of GS is also subject to elaborate control, being inhibited in a cumulative way by each of the eight nitrogenous compounds derived from glutamine. Its activity is further modulated by POSTTRANSCRIPTIONAL MODIFICATION, a rather rare control mechanism in bateria. A specific enzyme modifies GS by adenylylating (adding an adenyl group) it and thus converting it to a less active form; a second enzyme deadenylylates GS. Both of these enzymes are under nitrogen control.

The major reaction by which sulfur is assimilated is catalyzed by O-ACETYLSERINE SULFHYDRYLASE. In this reaction H_2S reacts with O-acetyl-L-serine to produce L-cysteine, acetate, and water; L-cysteine is the source, directly or indirectly, of sulfur for most other sulfur-containing compounds in the cell. Exogenous sources of sulfur must, therefore, be converted to H_2S in order to be assimilated. Some bacteria, including methanogens, require H_2S as a sulfur source, but most bacteria can reduce oxidized sulfur compounds. The most common source of sulfur in laboratory media (but probably not in nature) is sulfate. Many bacteria, including *E. coli*, have active transport systems that concentrate sulfate within the cell. Reduction of sulfate requires that it be activated. Three high-energy phosphate bonds are utilized for each sulfate molecule reduced. Sulfate reacts with ATP, yielding adenosine-5′-phosphosulfate and pyrophosphate (P_i-P_i). A second molecule of ATP reacts with adenosine-5′-phosphosulfate, yielding ADP and adenosine-3′-phosphate-5′-phosphosulfate. In this form, sulfate is reduced to free sulfite, with thioredoxin serving as the electron donor. The cofactor $NADPH_2$ serves as the reductant for sulfite reductase, which reduces sulfite to H_2S, the form in which it is assimilated. As mentioned earlier, sulfate may not be the major source of sulfur for microbial growth in nature. Certainly, this is the case in aerobic soil, where only a small percentage of the total sulfur occurs as sulfate, and little if any occurs as elemental sulfur or sulfide. The bulk of sulfur is found in organic compounds: as organic sulfates [including ester sulfates ($R—CH_2—O—SO_3^-$), sulfonates ($R—CH_2—SO_3^-$), sulfamates ($R—NH—SO_3^-$), oxime sulfates ($R—C(R)\!\!=\!\!N—OSO_3^-$)], amino acids, and other C-bonded sulfur compounds. With the exception of amino acids, which are probably assimilated directly, the organic sulfur compounds are metabolized to sulfate. Although sulfate-yielding metabolism has not been studied in great detail, regulation of the process seems to differ from NH_3-yielding metabolism. The latter is under repressive control of NH_3, whereas enzymes of sulfate-yielding metabolism (for example, sulfohydrolases) are induced by organic sulfates.

Nutritional diversity

Escherichia coli, biosynthetically, is the complete biochemist. Most strains, including our reference strain B/r, possess intact pathways for producing all of its organic constituents from whatever single source of carbon and energy is provided in the medium. Not all bacteria share this property. Many lack one or more pathways; and, in fact, the spectrum of biosynthetic competency extends from cells that can produce all their building blocks and other small molecules from inorganic carbon (CO_2) to cells that lack virtually all of the pathways shown in Figure 17 along with the biosynthetic pathway producing coenzymes and related molecules. *Because all cells require the same building blocks, there is a reciprocal relationship between biosynthetic capability and nutritional requirements.* As might be expected, bacteria that grow in environments that are relatively rich in organic material have tended to dispense with redundant biosynthetic pathways.

Growth in rich media

The energy and reducing power required for biosynthesis is a heavy cost to the cell (Table 11) and, depending on the carbon/energy source available for growth, may severely limit the growth rate. It is not surprising, therefore, to find a significant selective pressure favoring auxotrophy (substitution of growth factor requirements in place of biosynthetic capability). As we shall describe more fully later in this chapter, bacteria (including *E. coli*) that genetically have a full complement of biosynthetic pathways have evolved mechanisms (repression and end product inhibition) by which exogenously supplied building blocks cause the near cessation of their own endogenous biosynthesis. These cells can therefore take full advantage of nutrients supplied to them, and it is customary to observe an increase in growth rate when amino acids or other substances are added to a minimal medium. Our reference *E. coli* B/r cells, for example, double their growth rate when a large number of nutritional supplements are added to glucose minimal medium. The increase is even greater upon supplementation of minimal media containing poorer substrates than glucose.

How great a savings can be brought about by nutritional provision of

Table 12. Cost of biosynthesis in rich medium[a]

Cellular component	Energy cost (μmol ~P/g cells)		Reducing power cost (μmol NADPH/g cells)	
Protein	0	(7,287)	0	(11,523)
RNA	1,890	(6,540)	0	(427)
DNA	300	(1,090)	0	(200)
Lipid	129	(2,578)	0	(5,270)
LPS	125	(470)	0	(564)
Peptidoglycan	55	(248)	0	(193)
Glycogen	154	(154)	0	(0)
1-Carbon	0	(0)	0	(48)
Polyamines	118	(118)	0	(0)
Total	2,771	(18,485)	0	(18,225)

[a]The medium is assumed to contain the following utilizable nutrients (in addition to glucose and inorganic salts): 21 amino acids, ribonucleosides and deoxyribonucleosides, glycerol phosphate, fatty acids, ornithine, glucosamine, ethanolamine, heptose, and KDO. Each nucleoside triphosphate is assumed to be made by consecutive reactions with ATP that consume 3 ~P per NTP produced. Formation of sugar nucleotide derivatives are assumed to occur by direct reaction with the appropriate nucleoside triphosphate. For each macromolecule, the number listed is the cost to produce the respective building blocks from the substances in the medium. The numbers in parentheses are the normal cost of biosynthesis from the precursor metabolites.

building blocks and other small organic compounds needed by the cell? In principle, if every compound listed in the first column of Table 10 could be supplied pre-formed to the cells, the total energy (>18,000 μmol of ~P per gram of cells) and reducing power (>18,000 μmol of H per gram of cells) used in biosynthesis would be saved. In practice the savings approaches but does not reach this limit. The difference is brought about largely by the failure of some building blocks (particularly the nucleoside triphosphates and conjugated nucleotides) to penetrate the cell envelope. Amino acids, nucleosides, sugars, aminosugars, and fatty acids, on the other hand, enter many bacterial cells fairly readily, as do precursors of coenzymes and prosthetic groups. In Table 12 the cost of producing the standard building blocks from compounds that can be supplied in the medium is calculated for each class of macromolecules. This cost is only 15% (2771 compared to 18,458 μmol ~P/g cells) of the energy cost of de novo biosynthesis; or, viewed the other way, nutritional supplementation can save 85% of the energy required in biosynthesis and all of the need for reducing power (18,000 μmol supplied as NADPH$_2$) can be spared.

This analysis deals only with part of the consequences of supplementation. If biosynthetic pathways are shut off, the demand for precursor metabolites is diminished. In some cases the reduced demand will be highly significant because during growth on certain substrates the formation of precursor metabolites is energetically costly (e.g., during growth on acetate, formation of each micromole of glucose-6-phosphate requires 8 μmol of high-energy phosphate). Because this cost varies strikingly with the nature of the substrate, we must examine the fueling reactions before completing our discussion of nutritional supplementation. Furthermore, there is some cost in the entry of nutrient into the cell, and these costs for cells growing on a single substrate must be compared to those for cells growing on a rich medium before we can know the net effects of supplementation.

FUELING REACTIONS

Up to this point the chemical processes of growth have had one marvelous simplicity: virtually all biosynthetic pathways, polymerization reactions, and assembly processes are fundamentally similar for different bacteria. We have encountered differences in biosynthesis—but these are trivial because they concern simply the presence or absence of different pathways in a given species of bacteria. Even the coordination of biosynthetic pathways can be easily understood because they involve adjustments in flow through essentially unidirectional pathways with a few alternative routes to a given building block.

It is in the fueling reactions that one encounters the extraordinary diversity and versatility of bacteria. Two features of natural environments have been extremely influential in the evolution of fueling reactions: (1) the enormous diversity of potentially usable sources of carbon, energy, and reducing power, and (2) the great ranges of pH and of oxidation–reduction potential and the variable availability of electron acceptors. These environmental factors have favored the evolution of bacteria with a collective ability to utilize virtually every carbon compound on this planet (coal, diamonds, and more newly synthesized polycarbon plastics are the major exceptions). Highly aerobic and strictly anaerobic environments at a variety of pH values have been successfully colonized by different bacterial species. But it is not solely the diversity of microbial environments that imposes complexity on the fueling area of metabolism. Much of the complexity is inherent in the physiological role of the fueling reactions—no matter what the nature of the carbon source and the environmental circumstance, the fueling reactions must provide 13,400 μmol of precursor metabolites in precise proportions, plus 41,200 μmol of ~P and 18,500 μmol of NADPH for biosynthesis and polymerization (Tables 9 and 11). *Because different substrates vary in their potential yield of these three components, the cell must be able to generate them in the appropriate proportion by varying its metabolic pathways.*

The fueling pathways

The precursor metabolites are derivable from each other by a large number of reactions that constitute the CENTRAL PATHWAYS. In *E. coli*, these pathways include the EMBDEN-MEYERHOF-PARNAS PATHWAY (EMP), which leads to the overall conversion of glucose-6-phosphate to pyruvate; the TRICARBOXYLIC ACID CYCLE (TCA), which can oxidize acetyl CoA to CO_2 and the PENTOSE PHOSPHATE CYCLE, which can oxidize glucose-6-phosphate to CO_2 and water, but which also supplies $NADPH_2$ and C_4 and C_5 compounds. The central pathways also include various reactions that connect these major cycles (such as pyruvate dehydrogenase and pyruvate-formate lyase, which produces acetyl CoA from pyruvate), several reactions that enable an overall reversal of carbon flow through the pathways (such as phosphoenolpyruvate synthase, which forms phosphoenolpyruvate from pyruvate), and finally, several other reactions that enable a bypass of certain portions of a pathway (such as the glyoxylate shunt). The connecting, reversing, and bypassing reactions are termed ANAPLEUROTIC REACTIONS, meaning that they replenish intermediates in the central pathways that are drained off for biosynthesis. Other bacteria have additional or alternative central pathways, such as the ENTNER-DOUDOROFF route from glucose-6-phosphate to pyruvate used by many pseudomonads. *E. coli* uses enzymes of the Entner-Doudoroff pathway to metabolize gluconate, not glucose. The central pathways of *E. coli* are outlined in Figure 18.

The operations of the central pathways reduce large quantities of NAD to $NADH_2$. This reduction must be coupled to pathways that reoxidize the coenzyme. FERMENTATION PATHWAYS accomplish this anaerobically; electron transport via the RESPIRATORY CHAIN accomplishes it aerobically (and with certain terminal acceptors other than oxygen). These pathways will be discussed later when we consider ATP generation.

There is enormous flexibility in the operation of the central pathways in order to serve all three fueling functions (producing ATP, reducing power, and precursor metabolites). The central pathways can function cyclically and can lead to the complete degradation of glucose to CO_2 and water, producing large amounts of ATP and $NADH_2$. They can also function unidirectionally to replace precursor metabolites drained off for biosynthesis. (This dual role has led to their being called AMPHIBOLIC.) How the central pathways operate depends on such factors as the nature of the substrate available for growth, what electron acceptors are present, and what building blocks are already supplied in the medium. For example, anaerobically the reactions of the TCA pathway function solely to produce precursor metabolites. Aerobically it additionally oxidizes acetyl CoA to CO_2 and, coupled to the respiratory chain, generates most of the ATP for growth. Many adjustments are necessary for these two modes of operation. Similar major adjustments occur throughout the central pathways.

Almost any carbon source can be used by some bacterial species or another. Growth on substances that are not intermediates in the central pathways is brought about by PERIPHERAL PATHWAYS, sometimes consisting of very many steps, that convert the particular substance to a central pathways compound. Bacterial species are so diverse in their ability to utilize different substrates by peripheral pathways that they, as well as fermentation pathways, are very useful for taxonomic purposes. We cannot present even the metabolic versatility of *E. coli*, much less that of a large number of other important and well-studied bacteria. In the following sections, however, we hope to emphasize the *fact* if not the details of fueling diversity.

A useful approach to the complex world of the fueling reactions is first

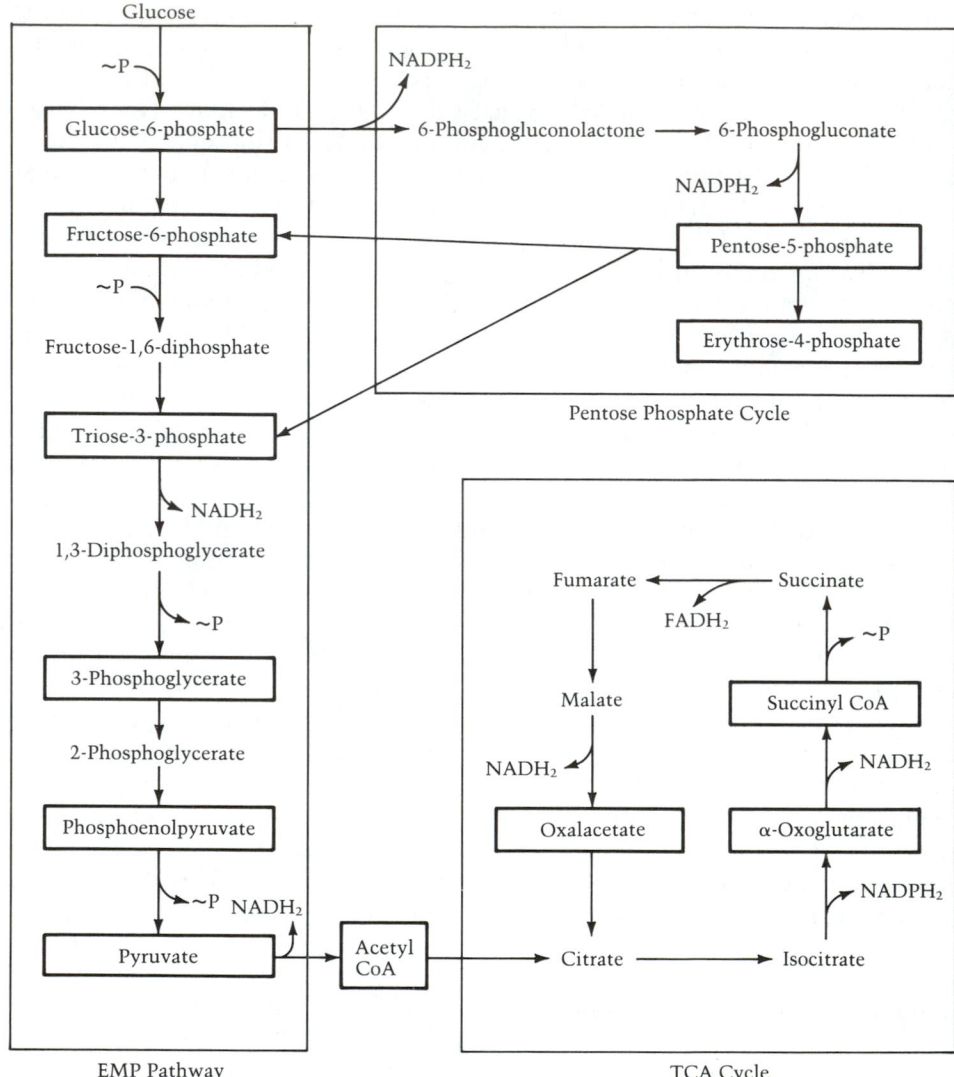

Figure 18. Central pathways of fueling reactions in *E. coli*. The three central pathways (EMP pathway, Embden-Meyerhof-Parnas pathway; TCA cycle, tricarboxylic acid cycle; and the pentose phosphate cycle) are shown to outline their individual reactions, not to indicate how these reactions operate during growth on any one substrate. Anapleurotic reactions and peripheral pathways (including fermentation and respiration) are not shown. The reductive branch of the pentose phosphate pathway is simplified to a single overall step from pentose-5-phosphate to the C_6 and C_3 compounds of the EMP pathway. The 12 precursor metabolites are boxed.

to use them solely to generate the precursor metabolites needed for growth. Then one can examine the balance sheet to learn of any unmet needs for energy and reducing power.

Generation of precursor metabolites

Figures 19–21 depict the formation of each of the 12 precursor metabolites under three different conditions: aerobic growth on glucose, anaerobic growth on glucose, and aerobic growth on malate.

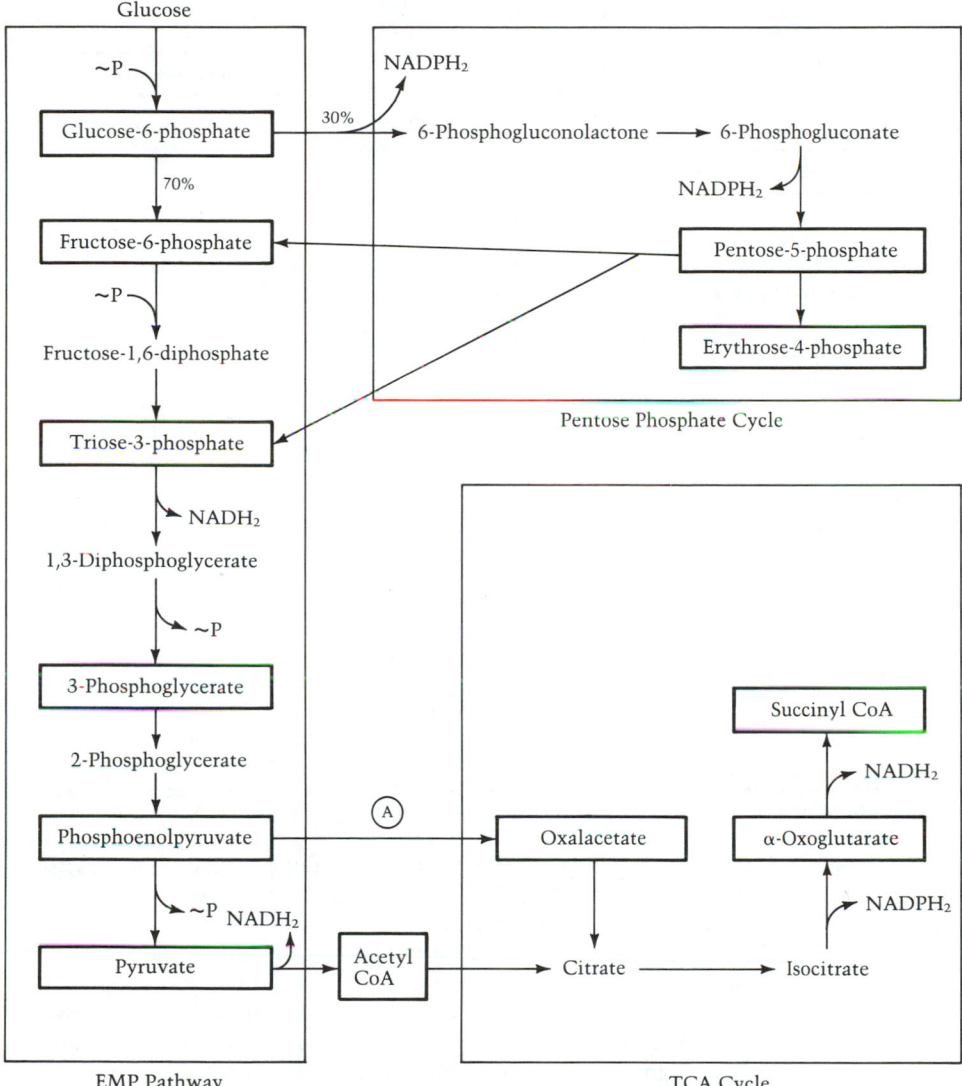

Figure 19. Formation of the precursor metabolites during aerobic growth of *E. coli* on glucose. The major anapleurotic reaction (A) forms oxalacetate by carboxylation of phosphoenolpyruvate and is catalyzed by phosphoenolpyruvate carboxylase. As shown, components of the TCA cycle function almost exclusively to provide the three TCA precursor metabolites for biosynthesis and not as an energy-generating cycle during unrestricted aerobic growth on glucose. The 12 precursor metabolites are boxed; other conventions are as in Figure 18.

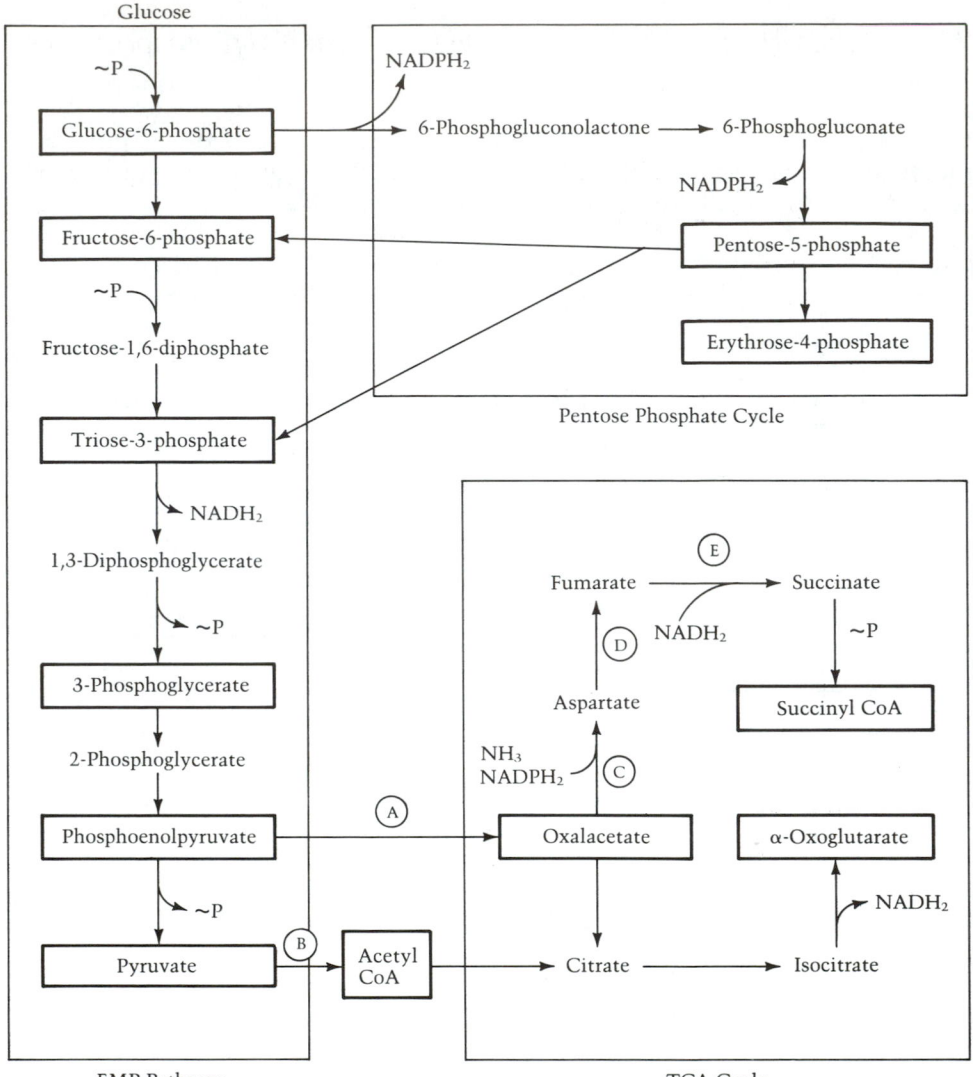

Glucose

~P

Glucose-6-phosphate → 6-Phosphogluconolactone → 6-Phosphogluconate

NADPH₂

NADPH₂

Fructose-6-phosphate ← Pentose-5-phosphate

~P

Fructose-1,6-diphosphate

Erythrose-4-phosphate

Triose-3-phosphate

Pentose Phosphate Cycle

NADH₂

1,3-Diphosphoglycerate

~P

3-Phosphoglycerate

2-Phosphoglycerate

Phosphoenolpyruvate

~P

Pyruvate

Acetyl CoA

EMP Pathway

E
Fumarate → Succinate

D   NADH₂   ~P

Aspartate   Succinyl CoA

NH₃
NADPH₂   C

Oxalacetate   α-Oxoglutarate

NADH₂

Citrate → Isocitrate

TCA Cycle

Figure 20. Formation of the precursor metabolites during anaerobic growth of *E. coli* on glucose. There is evidence that the pentose phosphate cycle functions anaerobically, though a lesser proportion of the glucose is handled through it anaerobically than aerobically. The major anapleurotic reactions are indicated by circled letters; precursor metabolites are boxed. Reaction (A) is the major replenishing reaction for the TCA cycle (as in aerobic growth depicted in Figure 19) and is catalyzed by the same enzyme. Reaction B is catalyzed by pyruvate-formate lyase instead of by the aerobic pyruvate dehydrogenase; as a result no NADH₂ is produced. Reaction C, interestingly, is a bypass of a portion of the cycle. Apparently, oxalacetate undergoes reduction and amination by transamination with glutamate to yield aspartate (one NADPH₂ and one NH₃ are used to regenerate glutamate from α-oxoglutarate). A new enzyme, aspartase, then produces fumarate by deamination (reaction D). Reaction E is catalyzed by an enzyme (fumarate reductase) that is different from the succinate dehydrogenase that functions in the opposite direction. Reactions C, D, and E serve to produce succinate, from which succinyl CoA is formed as a precursor for heme and a cofactor in amino acid biosynthesis; in

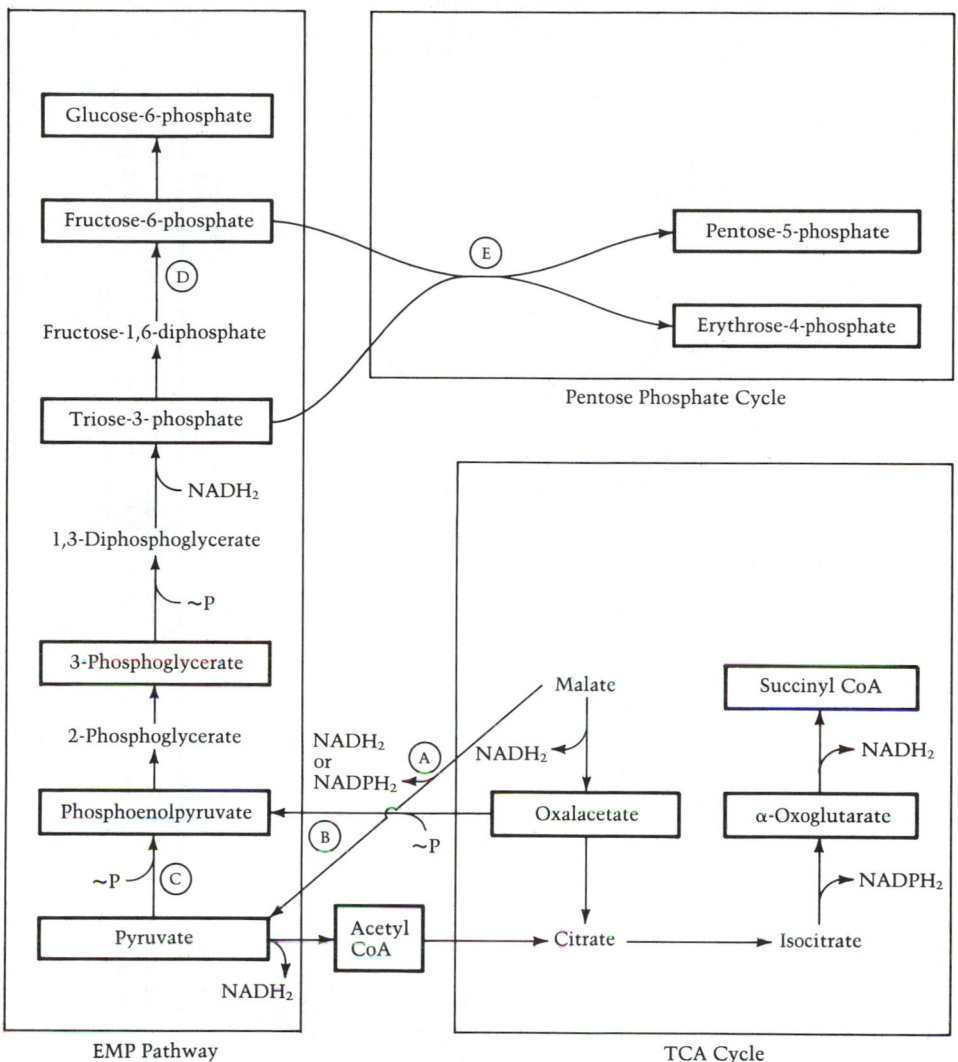

Figure 21. Formation of the precursor metabolites during aerobic growth of *E. coli* on malate. Two routes lead from malate to the EMP pathway; reaction A, catalyzed by either of two malate enzymes, produces pyruvate and either $NADH_2$ or $NADPH_2$ directly; reaction B, catalyzed by phosphoenolpyruvate carboxykinase, produces phosphoenolpyruvate after oxidation of malate to oxalacetate and $NADH_2$. The pyruvate formed by reaction A serves as a source of acetyl CoA and, by a reaction that uses 2 ~P from ATP, serves also as a source of phosphoenolpyruvate. Reaction C is catalyzed by phosphoenolpyruvate synthetase, which is one of the two special enzymes needed to reverse the flow of carbon in the EMP pathway to glucogenesis; the other is fructose-1,6-diphosphatase, which catalyzes reaction D. Pentose and erythrose phosphate compounds are formed by the anaerobic branch of the pentose phosphate cycle (reactions E).

essence it is formed by a reductive arm of the TCA cycle. The oxidative arm leads from oxalacetate to α-oxoglutarate. There is no need for oxidation of this precursor, and the enzyme α-oxoglutarate dehydrogenase is not formed anaerobically.

Table 13. Amounts and costs of the 12 key metabolites needed to produce monomers for 1 g of *E. coli* protoplasm from two substrates

Metabolite	Amount needed to produce *E. coli* B/r (μmol/g dried cells)	Cost of making 1 μmol of each of the metabolites (μmol/μmol)					Cost of making 1 μmol of each of the metabolites (μmol/μmol)				
		Glucose	~P	NADH$_2$	NADPH$_2$	CO$_2$	Malate	~P	NADH$_2$	NADPH$_2$	CO$_2$
Glucose-6-phosphate	205	1	1	0	0	0	2	4	0	0	-2
Fructose-6-phosphate	70.9	1	1	0	0	0	2	4	0	0	-2
Pentose-5-phosphate	897.7	1	1	0	-2	-1	1.7	3.3	0	0	-1.7
Erythrose-4-phosphate	361	1	1	0	-4	-2	1.3	2.6	0	0	-1.3
Triose phosphate	129	0.5[a]	1	0	0	0	1	2	0	0	-1
3-Phosphoglyceric acid	1,496	0.5	0	-1	0	0	1	1	-1	0	-1
Phosphoenolpyruvate	519.1	0.5	0	-1	0	0	1[e]	1	-1	0	-1
Pyruvate	2,832.8	0.5	-1	-1	0	0	1[f]	0	-1	0	-1
Acetyl CoA	3,747.8	0.5	-1	-2	0	-1	1	0	-2	0	-2
α-Oxoglutarate	1,078.9	1[c]	-1	-3	-1	-1	2	0	-3	-1	-3
Succinyl CoA	—[b]	1[c]	-1	-4	-1	2	2	0	-4	-1	-4
Oxalacetate	1,786.7	0.5[d]	0	-1	0	1	1	0	-1	0	0

[a]The values shown are those for the glycolytic pathway. By the pentose cycle, the cost is 0.6 glucose, 1.0 ~P, −0.6 CO$_2$, and −1.2 NADPH$_2$.
[b]Succinyl CoA is the precursor metabolite for tetrapyrroles (e.g., heme). It functions as a cofactor in the synthesis of major building blocks.
[c]The values shown are those for the TCA cycle functioning under aerobic conditions. Anaerobically the cost is 0.5 glucose, 1.0 ~P, 1 NADH$_2$, and 1 CO$_2$.
[d]Formed by carboxylation of PEP.
[e]The PEP values assume formation from oxalacetate.
[f]The pyruvate values assume formation by malic enzyme.

Features to note in these pathways include (1) the special anapleurotic reactions needed to supplement the EMP, TCA, and pentose phosphate cycles in each situation; (2) the new enzymes needed to reverse the flow of carbon in parts of these pathways; and (3) the different yield of $NADH_2$ and $\sim$P under the different conditions.

The latter feature is the subject of Table 13. Examination of the biochemical details of the operation of the central pathways during aerobic growth on glucose and malate enable us to assign molar costs (or yields) of $\sim$P, $NADH_2$, and $NADPH_2$ in generating each precursor metabolite. From these values and from the value for the molar demand for each precursor metabolite, the total cost (yield) of the fueling reactions can be calculated (Table 14).

The information in Table 14 is notable in two respects. Aerobic growth on these two substrates involves very different proportions of metabolites, $\sim$P, and H. And neither of these substrates have yielded sufficient ATP to fill our needs (41,200 μmol of $\sim$P). In terms of energy, it should be clear why malate is called a poor substrate and glucose a good substrate for growth. Growth on a poor substrate, one that places a heavy energy burden on the cell, usually is slower than on a good substrate, and this observation will be dealt with further in Chapters 5 and 8.

Generation of ATP

Not only does the cell need energy for biosynthesis and polymerization (and in some instances, to form the precursor metabolites), but also it consumes ATP in the active transport of some solutes into the cell, in the retention of metabolite pools within the cell, in the maintenance of proper osmotic pressure, and in motility. The magnitude of these additional energy requirements can only be guessed. The fueling reactions generate all the ATP required by the cell. They do so by two chemical mechanisms: SUBSTRATE LEVEL PHOSPHORYLATION and ELECTRON TRANSPORT.

In substrate level phosphorylation, ATP is formed from ADP and a phosphorylated intermediate of a catabolic pathway. The pathways of aerobic glucose catabolism (EMP and TCA pathways) contain three reactions that yield substrate level phosphorylations: those catalyzed by 3-phosphoglycerate kinase and pyruvate kinase in the Embden-Meyerhof-Parnas pathway and by succinate thiokinase in the tricarboxylic acid cycle. These reactions partici-

Table 14. Total cost of fueling reactions operating to produce precursor metabolites for 1 g of *E. coli* from two substrates

Substrate	Carbon source (μmol)	$\sim$P (μmol)	$NADH_2$ (μmol)	$NADPH_2$ (μmol)
Glucose	7,869	(−5,593)	(−16,965)	(−4,319)
Malate	15,109	7,152	(−16,965)	(−1,079)

pate in the pathway after glucose has been cleaved to trioses; from glucose, therefore, six molecules of ATP are generated by substrate level phosphorylation. In addition, 24 hydrogen atoms are released (from glucose and water) and contribute to the generation of approximately 24 additional molecules of ATP by electron transport.

A theory to explain the process by which electron transport could be coupled to the generation of ATP was proposed by Peter Mitchell in 1961. The succeeding 20 years' research have established that his CHEMIOSMOTIC THEORY is almost certainly correct. An electron transport chain is located in the cell membrane; the oxidation–reduction potentials of the compounds that constitute this chain are poised such that each succeeding member is capable of being reduced by the reduced form of the preceding component. Thus, reducing power (as H atoms or electrons) flows through the chain of carrier molecules. The overall operation of the chain consists of the transfer of hydrogen molecules from an organic compound to oxygen, but it is not necessary at each step to transfer a complete hydrogen atom (H). When at one step reduction occurs by transfer of an electron from a hydrogen carrier, a proton (H^+) is released; reduction of a hydrogen carrier by an electron carrier consumes a proton. In the chemiosmotic system, hydrogen carriers always alternate with molecules that carry electrons, and therefore the flow of reducing power through the chain of carrier molecules can serve as a proton pump—protons being taken from solution on one side of the membrane (in the process of reduction of a hydrogen carrier) by an electron carrier and being released on the other side in the process of reduction of an electron carrier by a hydrogen carrier. The membrane itself is impermeable to H^+ and OH^- ions. In procaryotes, the flow of protons is to the exterior of the cell, and it creates a difference in hydrogen ion concentration (pH) and in electrical potential (membrane potential); the sum of these is termed PROTONMOTIVE FORCE. A recent proposal for the structural arrangement of the components of the carrier system that functions in *E. coli* during oxidative metabolism (the respiratory chain) is summarized in Figure 22.

Protonmotive force drives a variety of energy-linked processes including the entrance against a concentration gradient of certain substrates into the cell; the turning of flagella; the reverse transport of electrons through the respiratory chain, enabling reduction of NAD^+ when the supply of NADH is inadequate; and the generation of ATP from ADP via a membrane-bound structure termed an ATPASE (because hydrolysis of ATP to ADP and P is one of the reactions it catalyzes). These membrane processes are shown schematically in Figure 23. The action of membrane-bound ATPase is reversible: it can generate membrane potential by hydrolyzing ATP or it can generate ATP from ADP by utilizing membrane potential. Thus, the protonmotive force represents a store of free energy.

Membrane-bound ATPase is, in fact, a small cellular organelle. Electron micrographs of the cell membrane reveal conspicuous knobs that protrude to the interior of the cell and that are composed of six different protein subunits. This structure, termed F_1, can be removed readily from the membrane; in soluble form, it retains ATPase activity. In situ it is attached to another set

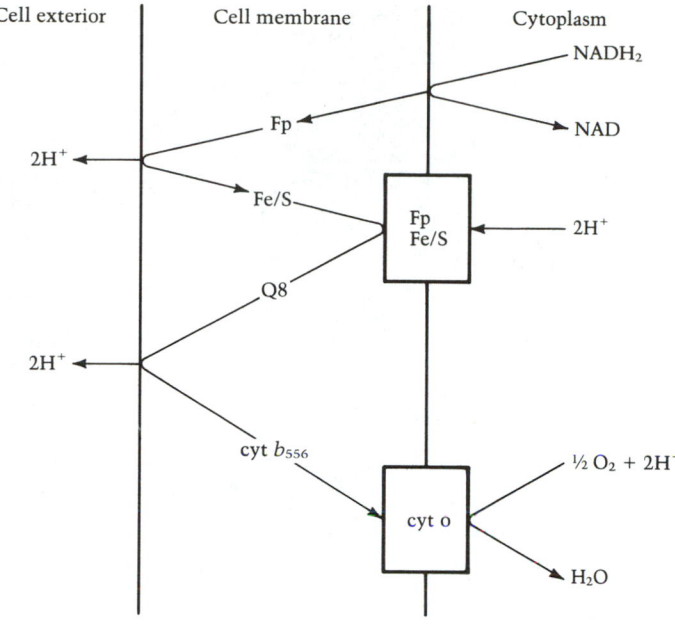

Figure 22. Proposed functional organization in *E. coli* of the hydrogen carriers (Fp, flavoprotein; Q8, ubiquinone) and electron carriers (Fe/S, iron-sulfur protein; cyt b_{556}, cytochrome b_{556}) that function during certain conditions of aerobic respiration. Cyt *o* (cytochrome *o*) transfers electrons to O_2, thereby reducing it to water. Components of the chain and the number of protons extruded vary with growth conditions. (After Haddock, 1977.)

of proteins, termed F_0, that are embedded in and pass through the membrane, being released only when the membrane is destroyed with detergents. Evidence indicates that one molecule of ATP is produced from ADP for each pair of protons passing through the F_1–F_0 complex.

The mechanism of ATP synthesis in the F_1–F_0 complex remains unresolved. Mitchell proposed that a phosphate ion binds to an active site in F_1 and that the two protons passing through the F_0 channel activate the phosphate ion by removing an oxygen atom from it to form a molecule of water. The unattached phosphorus bond then binds directly to ADP to form ATP. Others have proposed that the passage of protons changes the conformation of one of the proteins of the complex and that this energy renders thermodynamically feasible the synthesis of ATP from ADP.

Generation of ATP during aerobic growth

In the pattern of aerobic respiration that we have discussed, the flow of reducing power from $NADH_2$ to O_2 causes four protons to be expelled from the cell and two molecules of ATP to be synthesized. In such

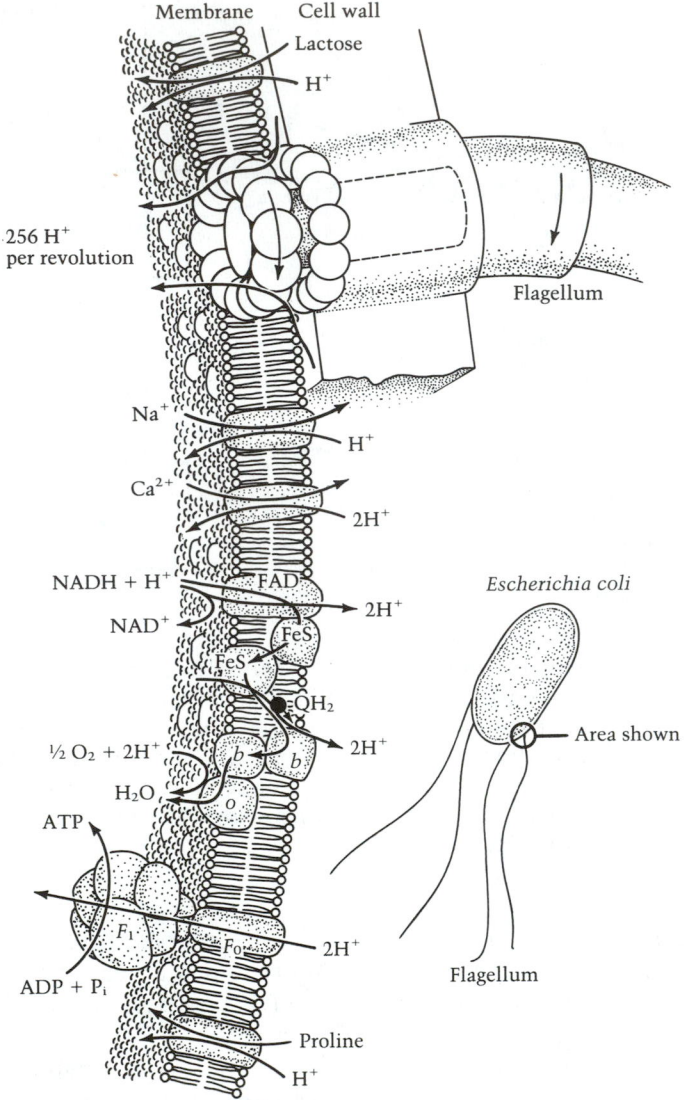

Figure 23. Activities of the cytoplasmic membrane involving proton transfer. (From Hinkle, 1978.)

a system, the H^+:O ratio (protons expelled to oxygen atoms consumed) is 4:1, and because the H^+:P ratio (protons reentering the cell to molecules of ATP synthesized) is 2:1, the overall P:O ratio is 2:1. This value of P:O is not universal for aerobic respiration chains. In eucaryotic mitochondria, the P:O ratio is closer to 3:1. Even in *E. coli*, the P:O ratio varies with environmental conditions. As might be expected, oxidation of certain substrates yields less free energy, thus reducing the P:O ratio. Such is the case for L-α-glycerophosphate, D-lactate, and succinate. Electrons from these substrates enter the redox chain at the Fp–Fe/S site (Figure 22), a situation that results in the expulsion of only two protons and the generation of a single molecule of ATP. However, profound changes in the redox chain occur with variations in growth phase, carbon source, and bacterial strain. Nine different cytochromes have been identified in *E. coli*, including two *c*-type cytochromes (c_{550} and c_{548}), five *b*-type cytochromes (b_{556}, b_{558}, b_{562}, $b_{556}NO_3$, and o), cytochrome *a*, and cytochrome *d*. Not all of these participate in redox chains (for example, cytochrome c_{550} is found in the periplasm), but presumably many do under various conditions. Also, *E. coli* synthesizes menaquinone-8 in addition to ubiquinone-8 (Figure 22). One significant variation in the redox chain occurs in response to changes in oxygen tension. Under conditions of high aeration, the redox chain shown in Figure 22 is thought to function. However, at low oxygen tension, only a single pair of protons is expelled because the first proton-expelling loop is short-circuited. The final electron transport step is mediated by cytochrome b_{558} rather than cytochrome b_{556}, and reduction of O_2 is catalyzed by cytochrome *d* rather than by cytochrome o. The physiological advantage of this and other modifications brought about by changes in the environment remains obscure, but these changes must reflect important adaptive capabilities of *E. coli* and other bacteria that generate ATP from the chemical energy of organic compounds.

Another class of bacteria, termed CHEMOLITHOTROPHS, are able to oxidize inorganic compounds to generate ATP. There are five classes of chemolithotrophic bacteria (Table 15): the HYDROGEN BACTERIA, the SULFUR BACTERIA, the IRON BACTERIA, the AMMONIA OXIDIZERS, and the NITRATE OXIDIZERS. These bacteria generate all of their ATP by transport of electrons from the reduced form of the inorganic compound to oxygen. In this respect ATP generation by chemolithotrophs is similar to that by chemoorganotrophs, but

Table 15. Redox potentials of reactions in the metabolism of chemolithotrophs

	Half-reaction of substrate (oxidation)	E'_0(volts)
Class of chemolithotroph		
Hydrogen bacteria	$H_2 \rightarrow 2H^+ + 2e^-$	−0.41
Sulfur bacteria	$H_2S \rightarrow S + 2H^+ + 2e^-$	−0.25
	$S + 3H_2O \rightarrow SO_3^{2-} + 6H^+ + 4e^-$	+0.005
	$SO_3^{2-} + H_2O \rightarrow SO_4^{2-} + 2H^+ + 2e^-$	−0.28
Iron bacteria	$Fe^{2+} \rightarrow Fe^{3+} + e^-$	+0.79
Ammonia oxidizers	$NH_4^+ + 2H_2O \rightarrow NO_2^- + 8H^+ + 6e^-$	+0.44
Nitrate oxidizers	$NO_2^- + H_2O \rightarrow NO_3^- + 2H^+ + 2e^-$	+0.35
Oxidation of NADH	$NADH + H^+ \rightarrow NAD^+ + 2e^- + 2H^+$	−0.32
Reduction of O_2	$O_2 + 4H^+ + 4e^- \rightarrow 2H_2O$	+0.86

there is an important distinction: the redox potential of the oxidation reactions catalyzed by all chemolithotrophs (with the exception of the hydrogen bacteria) is more positive than the redox potential (E'_0) for the oxidation of $NADH_2$. Thus, oxidation of these substrates cannot be coupled directly to reduce NAD to $NADH_2$—such reduction must occur by reverse electron transport (i.e., transport in a thermodynamically unfavorable direction) from the inorganic substrate to NAD, the flow of electrons being driven by the proton-motive force generated by the flow of reducing power from the inorganic substrate to O_2. Thus, bacteria are able to utilize oxidative reactions with a relatively positive E'_0 as an energy source.

The redox potential of the oxidation of hydrogen gas (H_2), like that of most dehydrogenations of organic compounds, is significantly more negative than the potential of $NADH_2$ oxidation. Hydrogen bacteria and chemoorganotrophs couple substrate oxidation directly to reduction of NAD.

Generation of ATP during anaerobic growth

Chemotrophs (organisms that utilize chemical energy for growth) generate ATP by one or both of two modes of metabolism: ANAEROBIC RESPIRATION, in which the terminal electron acceptor in an electron transport chain is a compound other than O_2; and FERMENTATION, in which ATP is generated by substrate level phosphorylation as a consequence of oxidizing one organic compound and reducing another.

The capacity for anaerobic growth is variable among bacteria. Some bacteria are incapable of growth in the absence of O_2; these are called STRICT AEROBES. Others, the ANAEROBES, grow only in the absence of oxygen; some of these, the STRICT ANAEROBES, are rapidly killed by oxygen.

The FACULTATIVE ANAEROBES can grow in the presence or absence of oxygen; generally, they alter their manner of generating ATP, depending on whether or not oxygen is present. However, the lactic acid bacteria are exceptional in this respect. They can grow in the presence or absence of oxygen, but under both conditions they generate ATP by a fundamentally anaerobic process—fermentation.

A variety of compounds can serve as terminal electron acceptors in anaerobic respiratory chains. These compounds include fumarate, nitrate, and sulfate, which are reduced to succinate, nitrite, and sulfide, respectively. Anaerobic respiration mediated by certain bacteria brings about the reduction of nitrite to dinitrogen gas; this process is termed DENITRIFICATION because it depletes fixed nitrogen in terrestrial and aquatic environments. In such anaerobic respiratory chains, organic compounds are usually the primary electron donors; but certain chemolithotrophs, including hydrogen bacteria and a sulfur bacterium, can generate ATP by anaerobic respiration in which the primary electron donor is inorganic and the terminal electron acceptor is NO_3^-.

The basic mechanism of ATP generation by anaerobic respiration is the same as by aerobic respiration, but the alternate electron acceptors are less powerful oxidizing agents than O_2, and, as a consequence, lesser amounts of

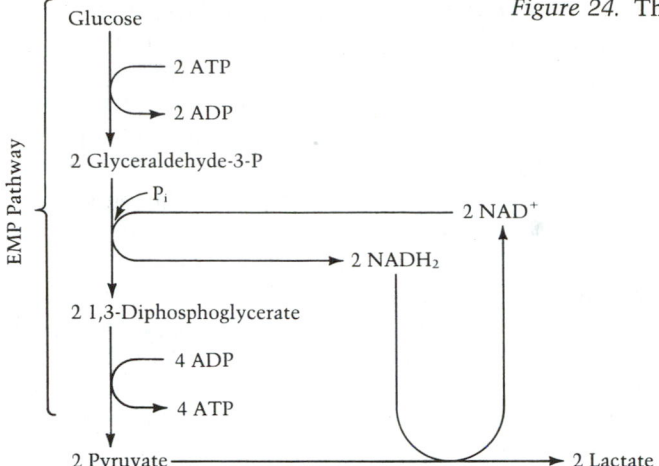

Figure 24. The homolactic fermentation.

ATP are generated per electron (or hydrogen atom) pair transported through the redox chain.

The other principal mode of generation of ATP is fermentation. The principles of fermentation are illustrated by a consideration of the HOMOLACTIC FERMENTATION, which is a biochemically simple process (in relative terms) that is mediated by certain lactic acid bacteria (Figure 24). By means of the EMP pathway, one molecule of glucose is cleaved and oxidized to yield two molecules of pyruvic acid. In this process, a net yield of two molecules of ATP is realized by substrate level phosphorylation and two molecules of NAD^+ are reduced. The two molecules of NAD^+ are regenerated when pyruvate is reduced to lactate in a reaction catalyzed by lactate dehydrogenase. Although the substrates and products in the myriad bacterial fermentations differ widely, all share certain features with the homolactic fermentation:

1. Almost all ATP is generated by substrate level phosphorylation.
2. Oxidative and reductive reactions occur in the fermentative pathway, but a strict oxidation–reduction balance is maintained, that is, the average oxidation state of the products is the same as that of the substrate (see Table 16). Of course, some substrate carbon is assimilated into cell material, but this amount is relatively small (see point 5, below). Moreover, the oxidation state of cell carbon is approximately that of the usual substrates of fermentation.
3. In order for both oxidation and reduction of the substrate to occur, substrates of fermentation are usually at an intermediate state of oxidation. Hence, the substrates of fermentation are usually sugars.
4. Most (but not all) pathways of bacterial fermentation involve pyruvate as a metabolic intermediate; the diversity of end products produced in various bacterial fermentations depends largely on the reactions by which pyruvate is metabolized.

5. Because the yield of ATP from fermentations is relatively low (2 moles per 1 mole of glucose in the homolactic fermentation), large quantities of substrate are utilized when growth occurs as a consequence of fermentative metabolism; as a consequence, most of the carbon from the metabolized substrate can be recovered in fermentation products (Table 16).

Escherichia coli is a facultative anaerobe that mediates a relatively complex fermentation known as the MIXED ACID FERMENTATION (Table 16). In contrast to the homolactic fermentation in which only a single product is produced, seven products result from the mixed acid fermentation. With the single exception of succinate, which is made from phosphoenolpyruvate, all of the other products are made from pyruvate (Figure 25). The ATP yields of the mixed acid fermentation and the homolactic fermentation are similar, with the exception that in the mixed acid fermentation one additional molecule of ATP is synthesized by substrate level phosphorylation for each molecule of acetate produced. The production of succinate involves a step that is not fermentative in the strict sense. Succinate is formed from fumarate, which

Table 16. Parameters of the mixed acid fermentation of glucose by *E. coli*

Product	Moles formed/ 100 moles glucose fermented	Moles carbon	O/R value	O/R sum (O/R value × moles) +	−
Formate	2.4	2.4	+1	2.4	
Acetate	36.5	73.0	0		
Lactate	79.5	238.5	0	10.7	
Succinate	10.7	42.8	+1		
Ethanol	49.8	99.6	−2		99.6
2,3-Butyleneglycol	0.3	1.2	−3		0.9
CO_2	88.0	88.0	+2	166.0	
H_2	75.0	—	−1		75.0
Total		545.5		179.1	175.5

$$\text{carbon recovery} = \frac{545.5}{100 \times 6} \times 100 = 91\%$$

O/R balance = 179.1/175.5 = 1.02[a]

[a]A number of methods can be used to calculate the oxidation–reduction balance of a fermentation. In the method used here, the O/R value of each product and of the substrate is calculated and multiplied by the number of moles of that product or substrate to give the O/R sum. When totaled algebraically, the total O/R of the products should equal that of the substrate used. In this fermentation, the substrate, glucose, has an O/R value of 0. Thus, the total of the positive O/R products approximately equals the negative O/R products. O/R value is calculated by assigning arbitrarily an O/R value of 0 to compounds with the empirical formulas of a carbohydrate $(CH_2O)_x$ and expressing each 2H in excess as a −1 value and each 2H shortage as +1. Thus, formate (CHOOH) falls two hydrogens short of being at the oxidation state of carbohydrate and is assigned an O/R value of +1. (Data from Wood, 1961.)

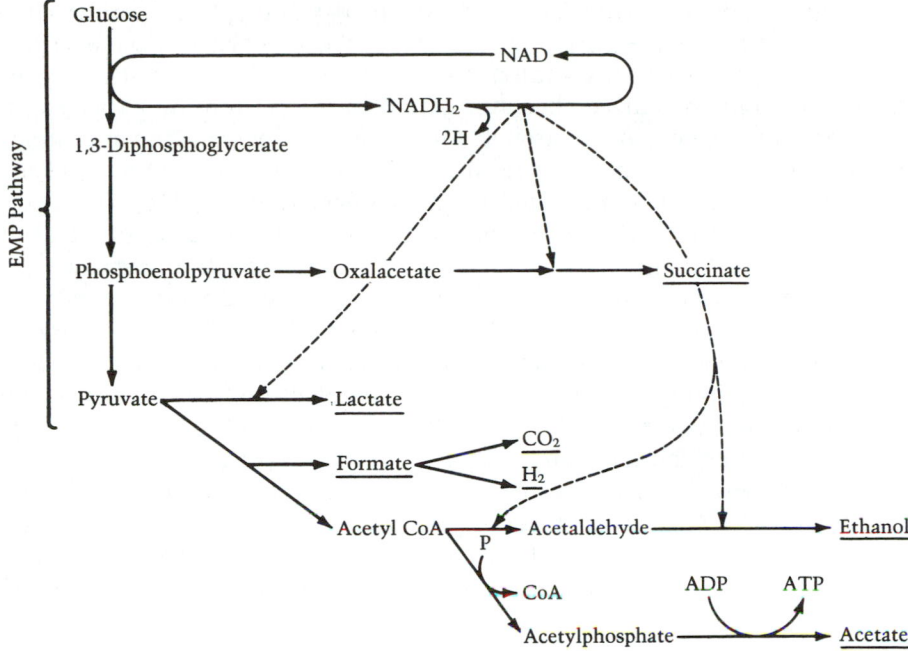

Figure 25. The mixed acid fermentation of *E. coli*. The seven principal fermentation end products are underlined. (An eighth product, 2,3-butanediol, is not prominent in the metabolism of *E. coli*, but is produced via α-acetolactate and acetoin by other species of enterobacteria.) The dashed lines show the four reactions that generate NAD from the NADH₂ formed during glycolysis.

serves as the terminal electron acceptor in an electron transport chain, that is, the formation of succinate is a product of anaerobic respiration.

If nitrate is present in the medium, *E. coli* is capable of an anaerobic respiration in which this terminal electron acceptor is reduced to nitrite.

Generation of ATP by photosynthesis

The process by which light energy is utilized to drive the synthesis of ATP from ADP is termed PHOTOSYNTHESIS. Organisms that have this capability—termed PHOTOTROPHS—include the higher plants, the algae, and certain bacteria. The type of photosynthesis carried out by higher plants, algae, and one group of bacteria (the cyanobacteria or blue-green bacteria) is termed OXYGENIC PHOTOSYNTHESIS because it is always accompanied by the evolution of molecular oxygen. The other bacterial phototrophs (the Rhodospirillaceae, or purple nonsulfur bacteria; the Chromatiaceae, or purple sulfur bacteria; and the Chlorobiaceae, or green bacteria) carry out ANOXYGENIC PHOTOSYNTHESIS; this process does not produce oxygen nor does it occur in the presence of oxygen.

A major portion of the ATP generated by oxygenic photosynthesis and all of the ATP generated by anoxygenic photosynthesis is synthesized by a process termed CYCLIC PHOTOPHOSPHORYLATION, a process that in many respects is analogous to the oxidative phosphorylation in respiration. By passage of electrons and hydrogen atoms through a redox chain located in a membrane, a protonmotive force is created and drives a reaction catalyzed by ATPase; in this reaction ATP is synthesized by phosphorylation of ADP. The two processes differ with respect to primary electron donors and terminal electron acceptors of the chain. As we have seen, in respiration the primary electron donor is an organic or inorganic substrate and the terminal electron acceptor is an oxidizing agent—O_2, or NO_3^-, or SO_4^{2-}. In cyclic phosphorylation, chlorophyll, in two different states, is both the electron donor and the electron acceptor. Light energy causes an ejection of an electron from chlorophyll with a negative E'_0; the electron flows through the redox chain to the chlorophyll molecule lacking the electron and hence having a positive E'_0.

The process of cyclic phosphorylation in bacteria is represented schematically in Figure 26. Photosynthesizing bacteria have an increased area of membrane, which accommodates the photosynthetic apparatus; the membrane invaginates into the cytosol, and the patterns thus created are characteristic

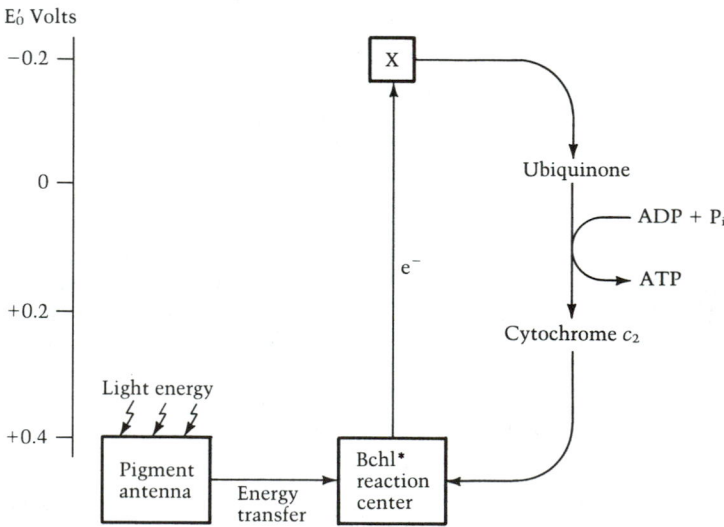

Figure 26. Cyclic photophosphorylation. Light energy is harvested by the pigment antenna and transferred to the reaction center, where an electron (e^-) is ejected from bacteriochlorophyll (Bchl*) and transferred first to a primary acceptor (X) and then to a redox chain (ubiquinone and cytochrome). Passage of reducing power through the chain back to the oxidized bacteriochlorophyll generates ATP through the intermediary of a protonmotive force. The approximate E' value of the various redox reactions is indicated by their relative positions on the scale at the left.

of various types of photosynthetic bacteria. Some species produce rounded vesicles; some produce regular layers of membrane that run parallel to the wall; others produce stacks of membranes that protrude into the cytosol at right angles to the wall; still others produce elaborate tubular structures that run through the cell (Stanier, 1976). The photosynthetic apparatus can be divided into three functional components: the PIGMENT ANTENNA, the REACTION CENTER, and the ELECTRON TRANSPORT CHAIN or REDOX CHAIN.

Light is absorbed by the pigment antenna, which is composed of carotenoids, bacteriochlorophyll, and proteins, the light absorption properties of which determine the wavelengths of light that support phototrophic growth. This action spectrum varies among phototrophs, but, in general, two major regions of light absorption occur—one between 400 and 500 nm and another between 700 and 1000 nm. Light energy is transferred by the pigment antenna to the reaction center, which is composed of bacteriochlorophyll, bacteriopheophytin (Mg-free bacteriophylin), a carotenoid, ubiquinone, and proteins. The interaction of these molecules causes the bacteriochlorophyll in the reaction center to be particularly reactive, so that light energy can eject an electron from it with an energy content sufficient to reduce an unknown acceptor molecule (probably a ferredoxin) with a negative E'_0. This compound (designated X in Figure 26) donates electrons to an electron chain, which is composed of ubiquinone and cytochrome c_2 and through which the reducing power flows back to the oxidized molecule of bacteriochlorophyll (the molecule from which an electron was ejected).

By cyclic photophosphorylation, ATP is generated but no reducing power is stored in the form of $NADH_2$ or $NADPH_2$ for subsequent use in biosyntheses. In the oxygenic phototrophs, NADP is reduced to $NADPH_2$, by electrons derived from water in a light-driven process, termed NONCYCLIC PHOTOPHOSPHORYLATION (Figure 27). Two types of reaction centers operate in this process: reaction center II (RCII) and reaction center I (RCI). The oxidized chlorophyll molecule in RCII (a molecule of chlorophyll from which an electron has been ejected) is at a sufficiently positive E'_0 value to oxidize water by removal of electrons (thus producing molecular oxygen) and transferring it to an electron carrier (Q) with an E'_0 value near 0. This electron flows through a redox chain (thus generating ATP) to reaction center I, where light energy ejects the electron to a carrier molecule (Z) with a sufficiently negative value of E'_0 to reduce a molecule of ferredoxin, which in turn can reduce NADP. Thus, in sequential light-driven steps, reducing power from water is rendered capable of reducing NADP.

Anoxygenic phototrophs lacking RCII cannot utilize water as a source of reducing power. Some of these organisms (purple sulfur bacteria and green bacteria) derive reducing power from reduced sulfur sources; and others (purple nonsulfur bacteria) derive it from organic compounds or H_2. Because the redox potential of sulfur oxidation is not sufficiently negative to reduce NAD or NADP, the phototrophic sulfur bacteria, like the chemolithotrophic ones, reduce pyridine nucleotides by reverse electron transport. The purple nonsulfur bacteria, like the chemoorganotrophs, can couple the reduction of pyridine nucleotides directly to the oxidation of organic substrates.

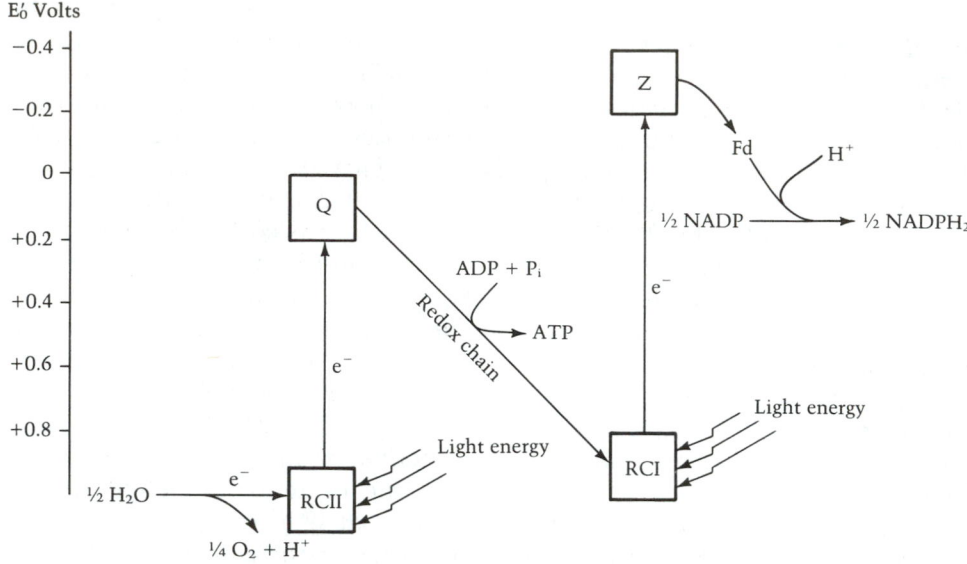

Figure 27. Noncyclic photophosphorylation. Light energy at reaction center II (RCII) causes ejection of an electron from chlorophyll; the resulting oxidized molecule of chlorophyll is an oxident powerful enough to oxidize water, a process that produces O_2. The ejected electron (e^-) is transferred to an acceptor (Q), which donates electrons that flow through a redox chain to chlorophyll in reaction center I (RCI) that was oxidized by light-driven ejection of electrons. These electrons are transferred to an acceptor (Z) that transfers them through ferredoxin (Fd) to reduce $NADP^+$.

Assimilation of phosphorus

Assimilation of phosphorus differs from assimilation of nitrogen or sulfur in one important respect: phosphorus is neither oxidized nor reduced in the assimilation process. Organic phosphates and phosphate ion constitute the only sources of phosphorus for the cell. Phosphate ion enters *E. coli* by a specific transport system. With few exceptions, organic phosphate compounds do not permeate the cell membrane, rather they are hydrolyzed outside the cell, or in the periplasm, to phosphate ion. In *E. coli*, most hydrolysis of organophosphates in the periplasm is catalyzed by a relatively nonspecific enzyme, ALKALINE PHOSPHATASE, the synthesis of which is repressed when phosphate ion is present in the medium. Also present in the periplasm of *E. coli* is a 5'-nucleotidase that splits phosphate ion from nucleotides. Once in the cell, phosphate is incorporated into the cellular components by several central pathway reactions; in aerobes, most phosphate is first incorporated into ATP by oxidative phosphorylation. ATP is the major phosphate donor in cellular phosphorylations.

TRANSPORT

Our account of the chemical processes by which a cell is made from substances provided in the environment is almost complete. There remains the matter of how these nutrients enter the cell rapidly enough. That bacterial cells maintain internal substrate concentrations that are quite different from those in the medium can be established by observing the growth curve of a batch culture in a minimal medium (Chapter 5): growth rate usually remains constant until the concentration of the limiting nutrient in the medium has been reduced to an extremely low value. Obviously, mechanisms exist to concentrate nutrients within the cell and, thereby, to maintain nutrient concentrations that saturate the internal enzyme systems. Membranes, both the outer membrane and the cytoplasmic or cell membrane, by their chemical nature, constitute barriers to the passage of nutrients. The outer membrane of Gram-negative cells contains protein pores that in a relatively nonspecific way allow the passage of most small molecules; still, it is a significant barrier to a variety of toxic agents (see the section on the outer membrane in Chapter 1). It is the cell membrane that maintains the major distinction between the intracellular and external environments. The cell membrane contains proteins that allow the passive passage of certain small molecules, actively concentrate others within the cell, and actively pump still others out of the cell.

The processes by which molecules (or ions) cross the cell membrane may be classified as follows:

1. DIFFUSION. The cell membrane is not a barrier to all molecules. Some molecules (for example, water) simply diffuse or flow through it.
2. FACILITATED DIFFUSION. Stereospecific transmembrane proteins (CARRIERS) that allow a specific compound(s) to diffuse through the membrane are said to participate in facilitated diffusion. Although common among the transport systems of eucaryotes, facilitated diffusion is rarely encountered in procaryotes. In *E. coli*, only glycerol is known to enter by facilitated diffusion. Interestingly, glycerol also enters a number of other bacteria, including *Salmonella typhimurium* and species of *Pseudomonas*, *Klebsiella*, *Shigella*, *Bacillus*, and *Nocardia*, by facilitated diffusion. Indeed, in every bacterium in which it was studied, glycerol was found to enter by facilitated diffusion.

    Facilitated diffusion, as the name implies, does not concentrate the substrate within the cell, it merely allows it to diffuse in, down a concentration gradient. In this respect, it resembles simple diffusion, but in another respect it differs fundamentally because the substrate enters only at the sites in the membrane where specific carrier proteins are located. Because the number of specific carrier sites is limited, the rate of substrate entry exhibits saturation kinetics typical of enzyme-catalyzed chemical reactions.
3. ACTIVE TRANSPORT. Active transport resembles facilitated diffusion in that stereospecific, membrane-located carrier proteins (sometimes termed PER-

MEASES) mediate the process, and saturation kinetics are seen; but it differs from facilitated diffusion in that a concentration gradient is established. By active transport, the concentration of a substrate within the cell can be maintained at a level many fold higher than its concentration in the medium. Obviously, creation of a concentration gradient requires the expenditure of energy. The variety of ways by which metabolic energy is coupled to active transport is discussed later in this section.

4. GROUP TRANSLOCATION. The mechanism by which a substrate is chemically altered to an impermeable derivative as it crosses the cell membrane is termed group translocation. Although technically this is not active transport (because a concentration gradient is not established), the same metabolic function is accomplished: the concentration of the substrate moiety of the derivative within the cell is higher than the concentration of the unsubstituted substrate in the medium. Entry of a substrate by group translocation rather than by active transport usually constitutes a net saving of metabolic energy. Although the transport-related derivatization reaction expends one or more high-energy phosphate bonds, the derivative is the same compound as the product of the first intracellular metabolic reaction. Had the substrate entered by active transport, intracellular derivatization would require the same expenditure of energy, and the total expenditure of energy would be the sum of the energy used in the derivatization reaction and the energy used in active transport of the unmodified substrate. In view of the energy savings associated with group translocation mechanisms, it is not surprising that they are most frequently encountered in strict and facultative anaerobes.

Binding proteins

If Gram-negative bacteria are suspended in a buffered 20% sucrose solution containing ethylenediamine tetraacetic acid (EDTA), centrifuged, and rapidly resuspended in 0.5 mM MgCl$_2$ at 0°C, the proteins normally located in the periplasm leak into the medium. This sudden change in osmotic strength of the external environment—or OSMOTIC SHOCK, as it is called—apparently damages the outer membrane, which is also the outer barrier of the periplasm. A number of enzymes are released along with a group of proteins, termed BINDING PROTEINS. These bind tightly (with binding constants in the range of 10^{-7}) and specifically to certain nutrients. The first binding protein, one that interacts specifically with SO$_4^{2-}$, was discovered by A. Pardee in 1966. Since that time a number of others, that bind to other ions, amino acids, sugars, and various other compounds, have been isolated. Although binding proteins are clearly not carriers, because they have a periplasmic rather than a membrane-associated location, they do play a vital role in certain transport systems. Those systems in which binding proteins participate are said to be SHOCK-SENSITIVE because osmotically shocked cells lack or have greatly diminished activities of the binding proteins. Transport systems that remain in osmotically shocked cells are termed SHOCK-INSENSITIVE. The role of binding proteins in shock-sensitive transport systems is twofold. By reversibly binding small molecules, the binding proteins sequester the small molecules within the periplasm, making them available to the carrier proteins for transport into the cell; by interacting with the carrier proteins, the binding proteins stimulate transport activity.

Shock-sensitive and shock-insensitive active transport systems differ with respect to the form of metabolic energy that drives them. Shock-sensitive active transport systems are usually coupled to the hydrolysis of ATP or some other source of a high-energy phosphate bond; shock-insensitive active transport systems are usually directly coupled to protonmotive force.

Secondary active transport

The processes of respiration and photosynthesis, as has been previously discussed, extrude protons from the cell, creating a pH gradient and a membrane potential (collectively termed protonmotive force). Proton extrusion can be considered as a metabolically linked active transport system and is sometimes termed PRIMARY ACTIVE TRANSPORT. As was discussed earlier, hydrolysis of ATP by the membrane-associated ATPase drives protons out of the cell; this, too, is primary active transport. Protonmotive force, which is a source of energy that can be used to generate ATP via the membrane-bound ATPase, can be used to move molecules across the cell membrane. Such movement of a molecule across the cell membrane at the expense of a previously established gradient of another molecular species is called SECONDARY ACTIVE TRANSPORT.

There are three types of secondary active transport systems: SYMPORT, ANTIPORT, and UNIPORT (Figure 28).

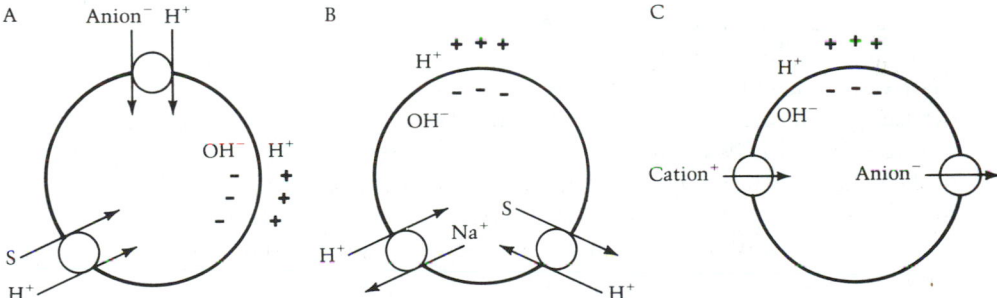

Figure 28. Secondary active transport systems. A. Symport reactions. On the right is shown a pH gradient created by primary active transport; the pH gradient drives (on the left) an electrogenic symport of an uncharged solute (S) with a proton and (at the top) an electroneutral symport of an anion with a proton. B. Antiport reactions. The pH gradient (at the top) drives (lower left) the electroneutral exchange of a cation for a proton and (lower right) the electrogenic exchange of an uncharged solute (S) for a proton. C. Uniport reactions. The pH gradient (at the top) drives a cation into the cell (left) or an anion out (right). (After Rosen, 1978.)

Symport is the transporting of two substrates simultaneously by a single carrier: if one of these flows down its (previously created) concentration gradient, the other flows with it. If the driving force is an ionic gradient (for example, a proton gradient; Figure 28A), either an oppositely charged ion or a neutral molecule can be brought into the cell. In the former case, only the concentration gradient is diminished; in the latter, both the concentration gradient and the membrane potential are diminished. In this example, the product of primary active transport (a pH gradient) drives the symport system, but by other secondary active transport systems (see later) this primary gradient can create other ionic gradients to serve as driving forces for other symports.

Antiport is the simultaneous transport, mediated by a common carrier, of two materials in opposite directions (Figure 28B). These processes also are concentration driven, with or without an electrostatic component.

Active uniport (Figure 28C) is a flow of ions driven directly by an electrostatic gradient. (Facilitated diffusion can be considered as passive uniport of a neutral compound.)

Active transport systems linked to phosphate bond energy

As stated earlier, shock-sensitive active transport systems seem largely to be driven by phosphate bond energy, whereas shock-resistant active transport systems are driven by charged membranes, that is, are examples of secondary active transport. The latter are more completely understood because they can be studied in MEMBRANE VESICLES. These laboratory artifacts are closed vesicles of the cell membrane; they lack cytoplasmic contents and are unassociated with periplasmic constituents. Their use offers distinct advantages to studies on active transport because they lack the intracellular enzymes that would otherwise metabolize the transported substrate, thereby complicating measurements. They also lack substrate level mechanisms of phosphorylation, thereby facilitating the identification of the source of energy that drives the particular transport system under study. Membrane vesicles function with remarkably high efficiency for shock-insensitive secondary active transport systems using primary gradients produced by the metabolism of exogenous substrates such as D-lactate. But lacking periplasmic proteins, including binding proteins, their shock-sensitive transport systems are inactive.

Preparation of membrane vesicles depends on an intrinsic property of membranes: fragments of them fuse spontaneously at their edges to create closed vesicles. Preparation of vesicles involves conversion of bacteria to osmotically fragile spheroplasts or protoplasts (from Gram-negative and Gram-positive bacteria, respectively) by treatment with EDTA and lysozyme or with penicillin; centrifugation and resuspension in an osmotically protective phosphate buffer containing 20% sucrose, $MgCl_2$, and nucleases; rapid dilution (300- to 500-fold) in phosphate buffer; and low-speed centrifugation to remove intact cells. If the spheroplasts or protoplasts are lysed in a solution of a particular material, the vesicles that will be formed will be PRELOADED within it. The technique of preloading vastly increases the types of experiments in which vesicles can be used.

But as stated earlier, the powerful technique of using membrane vesicles is not available for studies on shock-sensitive transport systems. More difficult and equivocal studies with intact cells have to be employed. But these studies have shown, in most cases, that phosphate bond energy rather than membrane potential drives this class of transport systems.

It is instructive to consider the experiments of Berger (1973); these experiments established the direct involvement of phosphate bond energy in shock-sensitive systems. Using *E. coli* cells that were depleted of endogenous energy sources by starving them of a carbon source in the presence of the uncoupler 2,4-dinitrophenol, Berger studied the dependency on exogenous energy supplies of a shock-sensitive system (glutamine transport) and a shock-insensitive system (proline transport). In wild-type cells, both systems were energized by either exogenous D-lactate or glucose. Metabolism of either substrate establishes a protonmotive force by primary active transport; metabolism of lactate generates high-energy phosphate bonds by reentry of protons through membrane-bound ATPase; metabolism of glucose generates them by this mechanism and by substrate level phosphorylation as well. In mutant cells that lack functional membrane-bound ATPase, a protonmotive force is generated by metabolism of either D-lactate or glucose; but only the metabolism of glucose generates high-energy phosphate bonds.

Using starved cells of ATPase-less mutants, Berger showed that proline transport was energized by either glucose or D-lactate, whereas glutamine transport was energized only by glucose. Thus, the shock-sensitive glutamine transport system was not activated by protonmotive force and, therefore, is probably dependent on high-energy phosphate bonds. These conclusions have subsequently been confirmed by a variety of other experiments, and the generalization has developed that most shock-sensitive transport systems are so dependent.

Group translocation

The most well established of the group translocations is the PHOSPHOTRANSFERASE SYSTEM (PTS) by which certain sugars are brought into the cell (Figure 29). The system is somewhat complex, involving the participation of at least four separate proteins that function within the cell as phosphocarriers of the high-energy phosphate group from phosphoenolpyruvate to the incoming sugar. The last member of the chain, Enzyme II (EII), which is an integral part of the membrane, also serves as the carrier protein that brings the sugar across the membrane. The penultimate one, Enzyme III (EIII), is a peripheral membrane protein. The first two members of the system, Enzyme I (EI) and the low-molecular-weight, histidine-containing protein (HPr) are soluble and nonspecific in their action. That is, the same two proteins function in all PTSs in the cell, whereas a different EII and EIII are required

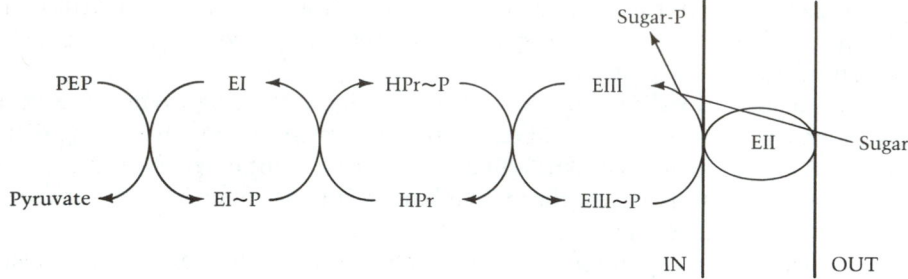

Figure 29. Schematic representation of the phosphotransferase system (PTS). A high-energy phosphate bond from phosphoenolpyruvate (PEP) is transferred through a chain of carriers—Enzyme I (EI), a small histidine-containing protein (HPr), Enzyme II (EII), and Enzyme III (EIII)—to the incoming sugar. Enzyme II also serves as the carrier protein of this transport system. (After Dills, 1980.)

for each sugar transported. The genes encoding EI and HPr (*ptsI* and *ptsH*, respectively) form an operon (Chapter 5), the expression of which is stimulated when cells are exposed to a PTS-transported sugar.

As mentioned earlier, PTS transport is energy conserving; and owing to the low ATP yields of fermentations as compared to respirations, one might expect to find PTSs predominately associated with fermentative organisms. As a generalization, this is true. The strict aerobes *Azotobacter, Micrococcus, Mycobacterium,* and *Nocardia* do not possess PTSs, although *Arthrobacter pyridinolis* does. The facultative anaerobes known to possess PTSs include *Escherichia, Salmonella, Staphylococcus, Photobacterium,* and *Vibrio;* the strict anaerobes include *Clostridium* and *Fusobacterium.*

In addition to PTSs, other less well established examples of group transfer systems have been proposed. A coenzyme A transferase system mediated by acyl:CoA-synthetase is presumed to play the same role in the uptake of fatty acids as PTSs do in the uptake of certain sugars. Similarly, phosphoribosyl-transferases, which catalyze the class of reactions:

$$\text{purine or pyrimidine base} + \text{PRPP} \xrightarrow{\text{phosphoribosyltransferase}} \text{nucleoside monophosphate} + \text{PP}_i$$

are thought to constitute a group translocation mechanism for the uptake of adenine, guanine, hypoxanthine, xanthine, and uracil.

Uptake of specific substrates

Of the various transport mechanisms, almost all (including facilitated diffusion, shock-sensitive active transport, secondary active transport, and group translocation) participate in the uptake of one or another sugars or sugar alcohols by *E. coli* (Figure 30). Interestingly, there is no pattern among bacteria for the uptake of a particular sugar.

Amino acids are actively transported by shock-sensitive or shock-insensitive systems. In *E. coli*, there are 14 distinct transport systems known for amino acids: glycine-alanine; threonine-serine; leucine-isoleucine-valine; phenylalanine-tyrosine-tryptophan; methionine; proline; lysine-ornithine-arginine; cystine; asparagine; glutamine; aspartate; glutamate; histidine; and probably cysteine (Table 17). Several patterns emerge. Some of the systems transport a group of amino acids with similar structures. Often these have subsystems specific for only one of the amino acids. This apparent redundancy serves a purpose. One of the systems is high affinity (K_t in the submicromolar range), low flow; and the other is low affinity (K_t 10 times or more greater), high flow. Each has obvious advantages in particular environments. Many of

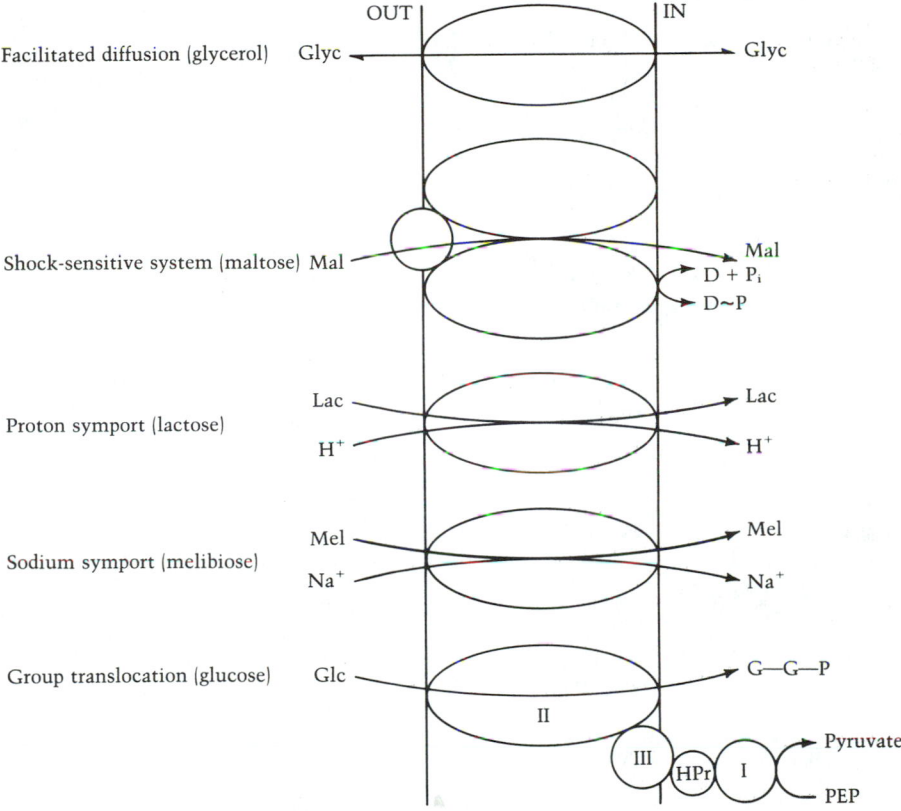

Figure 30. Representative carrier-mediated mechanisms for the uptake of sugars and sugar alcohols by *E. coli*. Glycerol (Glyc) enters by facilitated diffusion. Maltose (Mal) enters by a shock-sensitive system dependent on a binding protein (small circle) and a high-energy phosphate donor (D~P). Lactose (Lac) enters by a proton symport of secondary active transport; melibiose (Mel) enters by a sodium ion symport. Glucose (Glc) enters by PTS through the mediation of Enzyme I (I), HPr, Enzyme II (II), and Enzyme III (III); the intracellular product of the process is glucose-6-phosphate (G-6-P). (After Dills, 1980.)

Table 17. Transport systems for amino acids in E. coli[a]

Transport system	Substrate(s)	$K_t(\mu M)$[b]	Competitive inhibitor(s)	$K_i(\mu M)$	Encoding gene (map position in units)
1. Glycine-alanine	Glycine	2.8–4.8	D-Alanine	—	cycA(94)
	Alanine	27	D-Serine	—	
1a. Alanine	Alanine	2	Leucine	68	dagA(94)
2. Threonin-serine	Threonine	2	Serine	1.4	
2a. Threonine	Threonine	0.39	Alanine	8.2	
			Leucine	0.55	
			Isoleucine	0.37	
			Valine	0.78	
3. Leucine-isoleucine-valine I					brnQ(8)
					brnR(8)
Leucine-isoleucine-valine II					brnS(1)
					brnR(8)
4. Phenylalanine-tyrosine-tryptophan	Phenylalanine	0.47			aroP(2)
	Tyrosine	0.57			
	Tryptophan	0.40			
4a. Phenylalanine	Phenylalanine	2.0			
4b. Tyrosine	Tyrosine	2.2			
5. Methionine I	Methionine	0.075	L-Ethionine	23.0	metD(5)
Methionine II	Methionine	40.0	Selenomethionine		
6. Proline	Proline	0.44	L-3,4-Dehydroproline	2.6	
			L-Azetidine-2-carboxylate	2.4	
7. Lysine-arginine-ornithine	Lysine	0.5	D,L-Canavanine		argP(62)
7a. Lysine	Lysine	10			
8. Cystine I	Cystine	0.02	Selenocystine		
Cystine II	Cystine	0.3	Diaminopimelic acid	10	
9. Asparagine I	Asparagine	3.5	Aspartate	50	
			5-Diazo-4-oxonor-valine	4.6	
			β-Hydroxyl amyl-L-aspartate	10	
Asparagine II	Asparagine	80			
10. Glutamine I	Glutamine	0.15	α-Glutamyl hydrazine	75	
			α-Glutamyl-hydroxamate		
Glutamine II	Glutamine	2	Glutamate	640	
11. Aspartate I	Aspartate	3.7	Glutamine	840	
			D-Aspartate	200	
Aspartate II	Aspartate	39	Succinate		
			Fumarate		
			D,L-Malate		
			Oxalacetate		
12. Glutamate	Glutamate with 40 mM NACl	0.7	Aspartate	60	
	Glutamate without NaCl	10	Aspartate	5	
			D-Glutamate	245	

13. Histidine Not studied in E. coli, extensively studied in Salmonella typhimurium

14. Cysteine Not studied

these transport systems have been identified and studied by use of specific competitive inhibitors of them (Table 17).

In addition, a variety of transport systems occur for cofactors, various ions, and metabolic intermediates (see Appendix B).

In the preceding discussion, we have emphasized the role of transport systems in bringing substrates into the cell. They play other roles as well. One of them is to keep metabolites in the cell. Endogenously synthesized metabolites, such as amino acids for which transport mechanisms exist to capture exogenous supplies, constantly leak out and are pumped back. Other transport systems have the function of pumping certain materials out of the cell. The transposon Tn10 (see Chapter 5), which confers resistance to the antibiotic tetracycline, does so by encoding a transport system that actively pumps the antibiotic out of the cell.

COORDINATION

Biosynthesis of a bacterial cell requires the participation of over a thousand separate assembly, polymerization, biosynthetic, and fueling reactions. The relative rates of all these reactions must be coordinated if orderly growth is to occur. Elaborate autocontrol devices are employed for this purpose. A detailed analysis of some of these mechanisms is found in Chapter 7, but an indication of their precision and comprehensiveness can be illustrated here by a few simple observations on *E. coli*.

1. A growing culture, even a rapidly growing one, spills very few intermediates of biosynthesis into the medium, that is, the relative rates of formation of building blocks (amino acids, nucleotides, etc.) just match their rates of polymerization into macromolecules.

2. The macromolecular composition of a cell varies with the growth rate supported by different growth media in a manner that appears to maximize growth rate. In cells growing at high rates, which is a condition that requires a rapid rate of synthesis of protein, ribosomes constitute a greater portion of cellular mass than they do in slowly growing cells. At low rates of growth, the higher ribosome content would be largely unused and their synthesis would constitute a useless metabolic burden.

3. When certain building blocks (e.g., amino acids) become available in the growth medium, endogenous biosynthesis of them stops immediately, as does synthesis of the enzymes that catalyze the biosynthesis.

4. Enzymes in certain catabolic pathways are synthesized only if the substrate of the pathway is present.

---

[a]After Anraku (1978).

[b]Concentration of substrate at which the rate of transport is half-maximal.

[c]The multiple systems for transport of branched-chain amino acids are probably the most complex and certainly the most thoroughly studied systems in *E. coli*. One that transports all three amino acids has an extremely high affinity for the three substrates ($K_t \cong 0.01$ μM). Another has the same substrate specificity but lower affinity ($K_t \cong 2$ μM). There are three different low-affinity systems ($K_t \cong 10$–100 μM), each specific for one branched-chain amino acid. Some of these are shock-sensitive; others are shock-resistant.

5. If substrates for two distinct catabolic pathways are simultaneously present in the medium, only those enzymes that catalyze the substrate supporting the faster growth rate are synthesized. When it has been exhausted, enzymes of the other catabolic pathway are synthesized.
6. Growth is possible on a variety of substrates that differ in their molar yield of precursor metabolites, energy, and $NADPH_2$.

The mechanisms by which these (and many other) metabolic controls are effected can be divided into two broad classes: control of gene expression, that is, determination of the kinds and amounts of macromolecules that are made; and control of enzyme activity. The former class of mechanisms is discussed in Chapter 7. Enzyme activity is modulated in a variety of ways, but the most important metabolic coordinations are inhibition or activation of allosteric proteins by their specific effectors and variation in rate of reactions in response to concentration of substrate. Closely related to this latter mechanism is the shift in equilibrium of certain cellular reactions that occurs in response to changes in concentration of one of the products of that reaction.

The systems that regulate the pool levels of building blocks (e.g., amino acids and nucleotides) rely principally on FEEDBACK INHIBITION by the building block of an ALLOSTERIC PROTEIN catalyzing its biosynthesis. The phenomenon of feedback inhibition, discovered in 1958 simultaneously by H. E. Umbarger and by R. A. Yates and A. B. Pardee, depends on the properties of allosteric enzymes, the activities of which can be modified by small-molecule effectors that are structurally unrelated to the substrates or products of the reactions. The effector binds to a specific site on the enzyme—a site distinct from the catalytic site—thereby inducing or favoring a conformational change in the enzyme that alters enzyme activity either by changing K_m or V_{max}. An example of the sensitivity of one allosteric enzyme (aspartate transcarbamoylase) to one of its feedback effectors (CTP) is shown in Figure 31. In the presence of CTP, the affinity of the enzyme for one of its substrates (aspartate) is markedly decreased (K_m is increased). Thus, enzyme activity decreases with an increase in concentration of the end product (CTP) of the pathway (pyrimidine biosynthesis) in which the enzyme participates. Feedback inhibition serves to maintain constant internal concentrations of building blocks in the face of changing demands for them and availability of them from the medium.

Although bacteria are highly variable with respect to the types of allosteric inhibition that occur in various biosynthetic pathways, certain generalizations can be drawn: (1) only the first enzyme of a particular pathway is feedback inhibited (inhibition of a later enzyme in the pathway could cause wasteful or even damaging accumulation of intermediates of the pathway); (2) end products of a biosynthetic pathway (building blocks) act as feedback inhibitors. This second generalization introduces another property of the 12 precursor metabolites that are essential for biosynthesis: *these 12 compounds enter biosynthetic pathways via reactions catalyzed by feedback inhibited enzymes.*

The simple rule that the end product of a biosynthetic pathway is the feedback inhibitor of the first reaction becomes somewhat complicated in the case of pathways that lead to the synthesis of more than one building block,

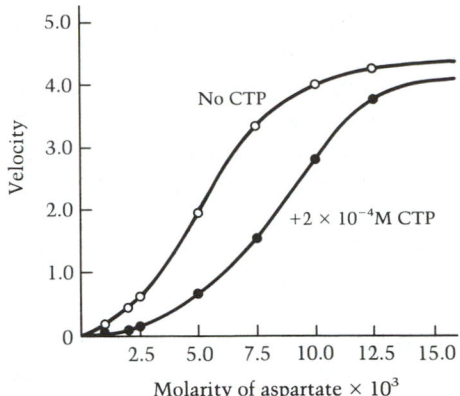

Figure 31.

Allosteric effects on aspartate transcarbamoylase. The figure demonstrates the sigmoidal relationship between substrate (aspartate) concentration and reaction velocity, and the effect of the allosteric inhibitor, CTP, on this activity. The topological distinctness of the allosteric (regulatory) and catalytic sites of this enzyme is easily demonstrated because they are found on two separate subunits of the enzyme. (After Gerhart, 1962.)

either by sequential synthesis through a linear pathway (e.g., threonine and isoleucine) or through branches from a common stem (e.g., tryptophan, phenylalanine, and tyrosine). In such cases, feedback inhibition of the first reaction by any single end product could starve the cell for the others. A number of patterns of feedback inhibition have evolved to avoid this metabolic problem.

1. ISOFUNCTIONAL ENZYMES. In one type of feedback inhibition, there are two or more species of the enzyme catalyzing the first reaction on the common stem. Each form is sensitive to feedback inhibition by one of the end products of the pathway.
2. CUMULATIVE FEEDBACK INHIBITION. In this type of feedback inhibition, there is a single species of enzyme that has distinct allosteric binding sites for each end product of the pathway; the inhibitory consequence of their being bound is cumulative. (A kinetic variation on this pattern in which singly bound effectors have no effect is sometimes termed CONCERTED FEEDBACK INHIBITION.)
3. SEQUENTIAL FEEDBACK INHIBITION. This is a type of inhibition in which each end product inhibits the first reaction of its own branch and the intermediate at the branch inhibits the first reaction of the common stem.
4. More complex pathways in which intermediates synthesized by one branch are fed into another are sometimes regulated by a combination of allosteric inhibition and activation; the end product of one pathway is the inhibitor and the intermediate product at the joining point on the other is an activator. This pattern, which is found to regulate carbamoylphosphate synthase in *E. coli*, has been termed INHIBITION PLUS ACTIVATION.

These various patterns of feedback inhibition are highly variable among bacteria: examples of each pattern can be found to regulate different pathways in a single bacterium, and the same pathway (e.g., biosynthesis of aromatic amino acids) in different bacteria has been found to be regulated in many ways.

Allosteric inhibition and activation by intermediates play an important role in regulating the flow through fueling pathways as well. However, the generalization that the end product inhibits the first reaction does not apply: such controls are internal to the pathway. For example, phosphoenolpyruvate, an intermediate of glycolysis, inhibits phosphofructokinase, an enzyme catalyzing a reaction three steps back in the same pathway; and α-oxoglutarate, an intermediate of the tricarboxylic acid cycle, inhibits citrate synthase, which catalyzes the reaction two steps back. In addition to the formation of the 12 essential precursors of biosynthesis, the important products of catabolism are ATP and reduced pyridine nucleotides (both $NADH_2$ and $NADPH_2$). Fluctuations in the steady state of these compounds also modulate the rate of catabolism. Because ATP synthesis and utilization involves a cyclic flow through ADP and/or AMP, it is not surprising that all three adenylates play regulatory roles in catabolism as well as biosynthesis. Some enzymes are regulated primarily by the concentration of ATP, ADP, or AMP; others respond to the ratios of [ATP]:[ADP] or [ATP]:[AMP]. An important unifying concept of these diverse regulatory patterns was made by D. Atkinson (1968) with his introduction of the idea of ENERGY CHARGE, which he defined as

$$\text{energy charge} = \frac{[\text{ATP}] + \frac{1}{2}[\text{ADP}]}{[\text{ATP}] + [\text{ADP}] + [\text{AMP}]}$$

Energy charge reflects the relative number of high-energy phosphate bonds (anhydride-bound phosphate groups) in the adenylate pool. ADP, with a single anhydride-bound phosphate, is assigned one-half the charge value of ATP, which has two. The activity of ADENYLATE KINASE provides a physiological reason, in addition to the mathematical one, for assigning to ADP one-half the charge value of ATP. This enzyme, the activity of which is quite high in bacteria, catalyzes the reaction:

$$\text{AMP} + \text{ATP} \rightleftharpoons 2\text{ADP}$$

thereby establishing an equilibrium between ADP and ATP with a stoichiometry of 2 to 1. Energy charge is calculated from the actual intracellular concentrations of the three adenylates, but it reflects their relative concentrations. Energy charge of a cell can vary mathematically from 0 to 1; but, in fact, the energy charge of bacteria under normal conditions will not lie outside the range 0.87 to 0.95, and it is invariant with growth rate. The energy charge of cells deprived of a substrate for long periods decreases slowly; when it reaches a value of 0.5, the cells are dead (see also Chapter 8).

Regulation of enzyme activity cannot, of course, be effected by an abstract concept like energy charge. Rather, it is effected by the absolute concentration of one, two, or all three of the adenine nucleotides. However, using the energy charge concept, certain important generalizations can be drawn.

Regulated enzymes in ATP-replenishing pathways respond quite differently to energy charge than do regulated enzymes of ATP-utilizing pathways (Figure 32). The former are inhibited at high values of energy charge; the latter are stimulated. At the level of energy charge in growing cells, both curves of Figure 32 are steep. Hence, energy charge is sensitively and precisely poised.

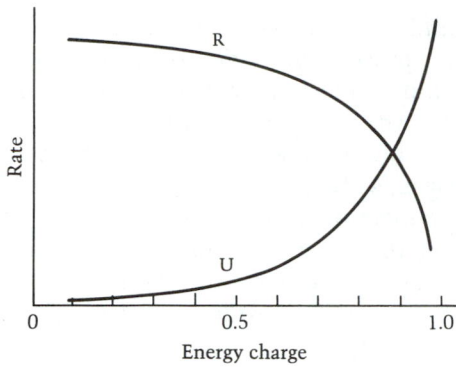

Figure 32.

Generalized response to energy charge of the rate of reactions catalyzed by regulated enzymes in ATP-regenerating (R) and ATP-utilizing (U) pathways. Examples of R-type enzymes are phosphofructokinase, pyruvate dehydrogenase, citrate synthase, and isocitrate dehydrogenase; of U-type are phosphoribosyl-pyrophosphate (PRPP) synthetase, aspartate kinase, and phosphoribosyl-ATP pyrophosphorylase. (After Atkinson, 1968.)

Any decrease in it stimulates ATP formation and inhibits its utilization. An increase in energy charge has the reverse consequence. The critical nature of this regulation or homeostasis is further emphasized by the realization that the turnover time for ATP in a growing bacterium may be on the order of 0.01 second. The precise setting of intracellular levels of ATP and the speed with which the controlling mechanisms respond to environmental changes can be illustrated by a simple experiment. If a culture of *E. coli* growing under forced aeration on glycerol (a nonfermentable substrate) as the total source of carbon is suddenly deprived of air and sparged with nitrogen, growth and presumably regeneration of ATP stops abruptly, but the intracellular concentration of ATP is unchanged, as judged by precise analytical methods (see also Chapter 8).

It has been proposed that under conditions that rapidly stimulate the regeneration of ATP, an additional factor, the operation of FUTILE CYCLES, is activated to dampen the tendency to overproduce ATP. A futile cycle is the operation of two or more reactions that have the net effect of hydrolyzing ATP without changing the concentration of any other component of the system. An example is the concerted action of glutamine synthase and glutaminase:

$$\text{glutamate} + NH_3 + ATP \xrightarrow{\text{glutamine synthase}} \text{glutamine} + ADP + P_i$$

$$\text{glutamine} + H_2O \xrightarrow{\text{glutaminase}} \text{glutamate} + NH_3$$

$$\text{net reaction: } ATP + H_2O \longrightarrow ADP + P_i$$

A number of such futile cycles can be written from reactions known to exist in procaryotes. D. W. Tempest has suggested that "the spillage of ATP-associated energy may not, in reality, be the primary purpose of the futile cycle, but is the price that must be paid for maintaining the cells in a state where they can respond rapidly to a sudden change in the supply of a limiting nutrient (in this case NH_3)." (In homeothermic animals and in the organs of certain plants, futile cycles play the important role of generating heat.)

We have noted earlier the existence in bacterial and eucaryotic cells of two types of pyridine nucleotides (NAD and NADP) that serve as reservoirs and carriers of hydrogen atoms or reducing power. We have also noted that NAD participates principally in catabolic oxidoreduction reactions and NADP in biosynthetic reactions. The necessity for the existence of two species of pyridine nucleotides in cells undergoing aerobic metabolism seems apparent as a consequence of three facts: (1) dehydrogenase reactions in which pyridine nucleotides participate have equilibrium constants near 1; (2) in catabolic reactions, oxidized pyridine nucleotides are the reactant; and (3) in biosynthetic reactions, reduced pyridine nucleotides are the reactant. Thus, for catabolic reactions to proceed, the species of pyridine nucleotide must be largely oxidized; for biosynthetic reactions, it must be largely reduced. Indeed, within the cell, most NAD is in the oxidized form and most NADP in the reduced form. As suggested by D. E. Atkinson (1968) and extended by K. Andersen and K. von Meyenburg (Andersen, 1977) useful parameters of the oxidoreduction state of pyridine nucleotides are

$$\text{catabolic reduction charge (CRC)} = \frac{NADH_2}{NADH_2 + NAD}$$

$$\text{anabolic reduction charge (ARC)} = \frac{NADPH_2}{NADPH_2 + NAD}$$

CRC in growing cells is maintained at a low level of 0.03 to 0.07 and ARC at the higher level of 0.4 to 0.5.

The setting of these charges undoubtedly plays a vital role in the coordination of catabolism and biosynthesis. The way that they are set is considerably less clear. However, two facts must be considered in the development of a consistent hypothesis. (1) As a striking exception to the generalization that catabolic reactions require the participation of NAD, two reactions of the pentose phosphate pathway require NADP. These oxidative fueling reactions can proceed in spite of the unfavorable oxidation–reduction state of NADP because they are followed by a decarboxylation reaction, and the rapid loss of the volatile product, CO_2, shifts the reaction equilibrium toward product formation. Two other fueling reactions, catalyzed by the enzymes isocitrate dehydrogenase and malic enzyme, also preferentially utilize NADP; the former is immediately followed by a decarboxylation reaction, and the latter itself involves a decarboxylation. (2) Procaryotes and certain other cells synthesize an enzyme termed TRANSHYDROGENASE that catalyzes the reaction:

$$NADP + NADH_2 \rightleftharpoons NADPH_2 + NAD$$

The equilibrium of this reaction is shifted by the hydrolysis of ATP.

It is tempting to speculate that the aforementioned four fueling reactions are the major sources of $NADPH_2$ that drive biosynthesis and that transhydrogenase is essential to make adjustments in the settings of CRC and ARC. However, other unknown factors must be important also, because mutant strains of *E. coli* that lack the hexose monophosphate shunt or transhydrogenase, or both, grow well.

SUMMARY

1. The metabolism of bacteria is geared toward rapid growth. Maintenance processes exist and are important for survival between periods of growth, but the chemical processes of bacterial cells have been selected to produce new cells rapidly and with high efficiency under a wide variety of environmental conditions. Structural and functional streamlining during evolution has led to metabolic rates that are 10 to 100 times faster than those of eucaryotic cells.

2. Polymerization of DNA involves 12 or more proteins acting at a small number of replication sites (growing points), where DNA is synthesized from activated precursors at a nearly constant rate of 750 base pairs per second at 37°C. The frequency of initiation (and therefore the number of growing points), rather than the chain elongation rate, varies with cell growth rate. Proofreading reduces the error frequency in DNA synthesis from 10^{-5} to a level of 10^{-10}, at a cost of energy. Additional energy is required for strand unwinding, for introducing torsional tension (negative superhelicity), and for methylation of DNA. Rapid growth is aided by the inclusion of the entire genome in one molecule, by the simultaneous operation of multiple growing points in DNA, and by an inherently rapid rate of polymerization permitted by subsequent active correction of errors.

3. Proteins are made at the direction of structural genes in all cells by the two processes transcription and translation; in bacteria, these processes are closely coupled. Transcription (mRNA synthesis) is biochemically simple, but its regulation is complex. A single enzyme (in some bacteria, the enzyme has alternate subunits), RNA polymerase, is responsible for all mRNA synthesis from activated building blocks. Regulation occurs by variations in the frequency of initiation and termination of mRNA synthesis, as determined by the interaction of various regulatory proteins at promoter and terminator sequences of the DNA. Translation of mRNA involves many proteins plus transfer RNA and ribosomes in a process that, as a result of sequential proofreading steps, is accurate to within 1 error per 10^4 aminoacyl residues. The protein synthesizing system (PSS) constitutes over one-half the *E. coli* cell during rapid growth. Compared to eucaryotic cells, components of the bacterial PSS are streamlined for efficient and rapid operation: (1) initiation of translation requires fewer proteins, and ribosomes are smaller and simpler; (ii) bacterial mRNA is mostly polycistronic; and (iii) translation of mRNA occurs as it is being synthesized (RNA polymerase and ribosomes work at equivalent speeds—55 nucleotides and 17 amino acids polymerized per second, respectively) and therefore mRNA need not be processed and transported between cellular compartments. Polymerization of approximately 5000 μmol of amino acids to form 1 g of reference *E. coli* cells requires approximately 22,000 μmol of ~P—more than 95% of the energy needs for polymerization of total cellular macromolecules.

4. Stable RNA is polymerized from activated building blocks by the same polymerase that makes mRNA. All stable RNA molecules are made from

precursor molecules that must be processed by nucleases and then extensively modified to produce the mature product. There is sevenfold redundancy of rRNA genes, and they are clustered in the half of the chromosome proximal to the origin of replication; these factors help generate the high rate of RNA synthesis needed to produce cells rich in PSS.

5. The polymerization of peptidoglycan, phospholipid, LPS, and capsular polysaccharide occurs from activated building blocks that are polymerized or assembled within or on the exterior surface of the cytoplasmic membrane. The quantity of material and the energy involved are small compared to those for protein synthesis.

6. The building blocks needed for macromolecular polymerization are made by biosynthetic pathways that form a network leading from 12 precursor metabolites to the many amino acids, nucleotides, sugars, aminosugars, and other molecules. These pathways are similar in all species, but species differ in their complement of these pathways. All cells require essentially the same building blocks; those that cannot be produced by a cell must be obtained pre-formed from the environment. In addition to precursor metabolites, large quantities of $NADPH_2$, $\sim P$, N, and lesser quantities of S are used.

7. Fueling reactions consist of central and peripheral pathways that provide the precursor metabolites $NADPH_2$ and $\sim P$ needed for biosynthesis. The central pathways (EMP, TCA, and pentose phosphate pathways in *E. coli*) can function cyclically to degrade glucose and other substrates to CO_2 and water, producing large quantities of $NADH_2$ and ATP. They can also function unidirectionally to produce the 12 precursor metabolites that are consumed in biosynthesis. This dual, or amphibolic, role is made possible by anapleurotic reactions that replenish intermediates drawn off for biosynthesis. Some anapleurotic reactions connect the major cycles of the central pathways, some permit a reversal of flow and some bypass portions of a pathway. As a result growth is possible with any of several compounds serving as the major substrate. Fermentation and respiration pathways are coupled to the central pathways and regenerate NAD from the large quantities of $NADH_2$ produced by many fueling reactions. The respiratory pathway supplements ATP formation in the central pathways with large amounts formed by oxidative phosphorylation. Because different substrates differ in their potential yield of $NADH_2$ and $\sim P$ relative to precursor metabolites, effective regulation is necessary to provide flexibility in operation of the fueling reactions. Peripheral pathways convert compounds that may be present in the environment into one or another intermediate in the central pathways. Bacterial species differ markedly in the variety of compounds they can utilize as growth substrates and in their products of fermentation.

8. Generation of the large quantities of $\sim P$ needed for cell growth occurs by three separate modes among different bacteria: substrate phosphorylation in fermentation, phosphorylation coupled to respiration, and photosynthesis. Many species can switch modes depending on the availability of oxygen or other electron acceptors.

9. Transport of substrates and nutrients into bacterial cells occurs chiefly by one or another form of active transport against a concentration gradient, or by group translocation, rather than by simple or facilitated diffusion. Transport systems also help in retaining endogenously made metabolites.

10. Coordination of the thousand or more separate assembly, polymerization, biosynthetic, and fueling reactions must be precise and comprehensive to enable rapid growth under a variety of conditions. Two types of metabolic control operate in the bacterial cell: control of gene expression and control of enzyme activity.

PROBLEMS

Unless specified otherwise, these problems refer to a culture of the reference strain of *E. coli* growing aerobically in glucose minimal medium at 37°C.

1. Considering only the carbon requirements for producing precursor metabolites, what fraction of metabolized glucose must be shunted through the pentose pathway?

2. Approximately 18,000 μmol of $NADPH_2$ are needed for biosynthesis of the building blocks for 1 g of cells (Table 11). Only 3,600 μmol of $NADPH_2$ are produced incidentally in the formation of precursor metabolites (Table 14). How much additional glucose would have to be shunted through the pentose pathway to provide all the additional $NADPH_2$?

3. If the additional pentose molecules produced in generating $NADPH_2$ in Problem 2 were returned to the EMP pathway (by transketolase/transaldolase 3 pentose-5-phosphate molecules form 2 fructose-6-phosphate molecules plus 1 triose phosphate), what would be the new total glucose requirement to make the precursor metabolites?

4. Polymerization and biosynthesis require 23,000 and 18,500 μmol of ~P, respectively, to produce 1 g of cells (Tables 9, 11). Only 5,600 μmol of ~P are produced incidentally in forming the precursor metabolites from glucose (Table 14). (a) If the ~P debit were made up exclusively by combined operation of the EMP pathway and the TCA cycle, with all $NADH_2$ being used to generate ~P by oxidative phosphorylation, how much additional glucose would have to be consumed? (b) This strategy is not used by *E. coli*; instead, large quantities of glucose carbon are excreted as acetate, and much ATP seems to be generated through substrate level phosphorylation. Assume that this glucose is converted to acetate, generating 6 μmol of ~P per μmol of glucose, and that the $NADH_2$ produced is respired to generate additional ~P. How much glucose would be needed for the ~P debit?

5. The values in Table 12 ignore the effect of supplementation on the need for generating precursor metabolites. Calculate the *total* savings (in mol of ~P per g of cells) by having all 20 amino acids provided ready made. Assume that no amino acid biosynthetic reactions are needed (Table 10)

and no precursor metabolites need be produced for these amino acids (Table 13).

6. Up to now our calculations have ignored the cost of transport. Repeat the calculations of Problem 5 incorporating the additional assumptions that the entry of 1 μmole of any amino acid costs 1 μmole of ~P, and the entry of 1 μmol of NH_3 costs 0.5 μmol of ~P. (These are semi-reasonable, but arbitrary, assumptions; the precise costs of entry are unknown.) Ignore any cost for the uptake of phosphate. Glucose uptake has already been paid for. (How?)

7. Growth on malate (Figure 21; Tables 13, 14) presents *E. coli* with a greater challenge for ~P production. Calculate how much malate would have to be consumed (beyond that needed for the precursor metabolites) to generate the ~P needed to produce 1 g of cells. Assume the malate is converted to pyruvate by malic enzyme and the pyruvate oxidized to acetyl CoA for complete oxidation in the TCA cycle. Ignore all costs for active transport.

8. Repeat the calculation of Problem 7 incorporating the additional assumption that entry of 1 μmol of malate costs 1 μmol of ~P, and the entry of 1 μmol of NH_3 costs 0.5 μmol of ~P (see Problem 6).

9. Would you expect the addition of a full supplement of amino acids to be of different benefit to cells growing on malate than on glucose? Test your guess by calculating the benefit (in μmol of ~P per g of cells) including the estimates of entry costs given in Problem 8.

10. Expand Tables 13 and 15 by adding entries for cells growing (a) anaerobically on glucose and (b) aerobically on acetate.

REFERENCES

Andersen, K. B. and K. von Meyenburg. 1977. Charges of nicotinamide adenine nucleotides and adenylate energy charge as regulatory parameters of the metabolism in *Escherichia coli*. J. Biol. Chem. 252:4151.

Atkinson, D. E. 1968. The energy charge of the adenylate pool as a regulatory parameter. Interaction with feedback modifiers. Biochem. 7:430.

Baldwin, A. N. and P. Berg. 1967. Transfer ribonucleic acid-induced hydrolysis of valyladenylate bound to isoleucyl ribonucleic acid synthetase. J. Biol. Chem. 241:839.

Berger, E. A. 1973. Ph.D. thesis, Cornell University. Described in Rosen, B. P. and E. R. Kashket, 1978. Energetics of active transport. In *Bacterial Transport*, B. P. Rosen, ed. Marcel Dekker, New York.

Cronan, J. D., Jr. 1978. Molecular biology of bacterial membrane lipids. Ann. Rev. Biochem. 47:163.

Fraenkel, D. G. 1962. The control of ribonucleic acid synthesis in *Aerobacter aerogenes*. Ph.D. Thesis. Harvard University, Cambridge, Mass.

Freter, R. R. and M. A. Savageau. 1980. Proofreading systems of multiple stages for improved accuracy of biological discrimination. J. Theoret. Biol. 85:99.

Gallant, J. and D. Foley. 1980. On the causes and prevention of mistranslation, p. 615. In *Ribosomes: Structure, Function, and Genetics*, G. Chambliss, G. R. Craven, J. Davies, K. Davis, L. Kahan and M. Nomura, eds. University Park Press, Baltimore.

Grunberg-Manago, M. 1980. Initiation of protein synthesis seen in 1979, p. 445. In *Ribosomes: Structure, Function, and Genetics*, G. Chambliss, G. R. Craven, J. Davies, K. Davis, L. Kahan, and M. Nomura, eds. University Park Press, Baltimore.

Gottschalk, G. 1979. *Bacterial Metabolism*. Springer-Verlag, New York.

Hopfield, J. J. 1974. Kinetic proofreading: a new mechanism for reducing errors in biosynthetic processes requiring high specificity. Proc. Natl. Acad. Sci. USA 71:4135.

Kornberg, A. 1980. *DNA Replication*. W. H. Freeman and Co., San Francisco.

Little, J. W. and D. W. Mount. 1982. The SOS regulatory system of *Escherichia coli*. Cell 29:11.

Mandelstam, J. and K. McQuillen. 1973. Biochemistry of Bacterial Growth, 2nd Edition. John Wiley & Sons, New York.

Mitchell, P. 1961. Coupling of phosphorylation to electron and hydrogen transfer by a chemiosmotic type of mechanism. Nature 191:144.

Ogawa, T. and T. Okazaki. 1980. Discontinuous DNA replication. Ann. Rev. Biochem. 49:421.

Pardee, A. B. 1966. Purification and properties of a sulfate-binding protein from *Salmonella typhimurium*. J. Biol. Chem. 241:5886.

Shine, J. and L. Dalgarno. 1974. The 3'-terminal sequence of *Escherichia coli* 16S ribosomal RNA: complementary to nonsense triplets and ribosome binding sites. Proc. Natl. Acad. Sci. USA 71:1342.

Stanier, R. Y., E. A. Adelberg, and J. Ingraham. 1976. *The Microbial World*, 4th Edition. Prentice-Hall, Englewood Cliffs, NJ.

Steitz, J. A. 1980. RNA-RNA interactions during polypeptide chain initiation, p. 479. In *Ribosomes: Structure, Function, and Genetics*, G. Chambliss, G. R. Craven, J. Davies, K. Davis, L. Kahan, and M. Nomura eds. University Park Press, Baltimore.

Stouthammer, A. H. 1973. A theoretical study of the amount of ATP required for synthesis of microbial cell material. Antonie van Leeuwenhoek 39:545.

Stryer, L. 1981. *Biochemistry*, 2nd Edition. W. H. Freeman and Co., San Francisco.

Tempest, D. W. and O. M. Neijssel. 1980. Growth yield values in relation to respiration. In *Diversity of Bacterial Respiratory Systems*, Vol. 1, C. J. Knowles, ed. Chemical Rubber Company Press, Boca Raton, FL.

Umbarger, H. E. 1956. Evidence for a negative-feedback mechanism in the biosynthesis of isoleucine. Science 123:848.

Umbarger, H. E. 1977. A one-semester project for the immersion of graduate students in metabolic pathways. Biochem. Education 5:67.

Umbarger, H. E. 1978. Amino acid biosynthesis and its regulation. Ann. Rev. Biochem. 47:533.

Watson, J. D. 1976. *Molecular Biology of the Gene*, 3rd Edition. W. A. Benjamin, Menlo Park, CA.

Watson, J. D. and F. H. C. Crick. 1953. Molecular structure of nucleic acid. A structure for deoxyribose nucleic acid. Nature 171:737.

Wool, I. G. 1980. The structure and function of eukaryotic ribosomes, p. 797. In *Ribosomes: Structure, Function, and Genetics*, G. Chambliss, G. R. Craven, J. Davies, K. Davis, L. Kahan and M. Nomura, eds. University Park Press, Baltimore.

Yates, R. A. and A. B. Pardee, 1956. Control of pyrimidine biosynthesis in *Escherichia coli* by a feedback mechanism. J. Biol. Chem. 221:757.

Chapter Four

The Bacterial Genome

We have previously emphasized growth rate as a primary factor in natural selection of bacteria and in the consequent adaptability and economy of their metabolism. The bacterial cell also is remarkable with respect to genetic, as well as physiological, adaptability and economy. The bacterial genome, which encodes all the structures and functions discussed in the previous chapters, can change rapidly through mutation and through genetic exchange among cells. Because the genome is haploid, recessive mutations are rapidly expressed. Furthermore, the small size of the genome and the simplicity of the chromosome permit efficiency and economy in replication and expression of genetic information. In this chapter genetic mechanisms will be discussed along with their utility in answering questions about bacterial growth and physiology.

GENETIC ORGANIZATION OF THE BACTERIAL GENOME

The sole function of the bacterial genome is to serve as a repository of the information that is encoded in the sequence of bases in its DNA. In this respect the genomes of bacteria and all other cellular organisms are identical. But in many other aspects they differ markedly, particularly with respect to the molecular architecture of the bacterial chromosome and the way genes are arranged on it. Most notably, the bacterial chromosome is naked DNA unassociated with HISTONES, which are the basic proteins that give eucaryotic chromosomes their typical staining pattern, contribute to their form, and probably participate in the modulation of gene expression. Also, the arrangement of genes on the chromosomes differs between bacteria and other cellular organisms in two important respects. Bacterial genes encoding related

175

functions are often linked in transcriptional units termed OPERONS, and all known bacterial genes are colinear with their protein products. They are not known to contain INTRONS, which are noncoding, intervening sequences that must be excised; nor are scattered fragments of bacterial genes joined after transcription. Splicing of mRNA is unknown in bacteria.

In addition to the single chromosome, the bacterial genome (and the eucaryotic genome as well) often includes smaller circular molecules of DNA termed PLASMIDS. They vary in size from a few kilobase (kb) pairs to several hundred and in number of different kinds within a single cell from 0 to as many as 10 or more. To date, no plasmids are known to encode indispensible cellular functions: if a plasmid is lost, the cell may be able to metabolize a smaller number of carbon sources, to tolerate a smaller number of toxic materials, or to synthesize fewer toxins, but it remains viable. Plasmids, therefore, constitute an accessory chromosome that may or may not be present in a particular cell, even if it is in other cells of the same strain. Plasmids and their properties are discussed later in this chapter.

The arrangements of genes on a number of bacterial chromosomes are now known in detail, but, as is the case for so many bacterial properties, more is known about the chromosome of *E. coli* than that of any other species. It serves as the paradigm to which others are compared.

The latest gene map of the *E. coli* chromosome (Bachmann, 1980) shows the location of over 900 genes. Of these, approximately 260 are known to be, or (on the basis of their relative location) are highly likely to be, located in a total of approximately 75 operons. Clustering of genes into operons is highly characteristic of *E. coli* and of many other bacteria as well.

Another aspect of gene arrangement becomes apparent from inspection of the genetic map of *E. coli*. In the process of gene expression, only one of the strands of DNA (the SENSE STRAND) is transcribed into message. Because transcription always proceeds from 5' to 3', a direction of transcription on the chromosome determines which DNA strand is transcribed. The latest map of *E. coli* shows the direction of transcription of 50 individual genes or operons: 27 are transcribed in the clockwise direction and 23 in the counterclockwise direction. Therefore, with approximately equal frequency, either strand of DNA serves as the sense strand.

The approximately 900 mapped genes of *E. coli* are cataloged in Appendix B according to the physiological role of the function they encode. A glance at this table reveals that almost all of the genes encoding certain aspects of cellular function have already been identified. Sufficient genes have been identified to encode almost all of the enzymes known to be required to synthesize amino acids and other essential small molecules and to catabolize most known carbon and nitrogen sources; and almost all the genes to encode the component parts of ribosomes also have been recognized. In these cases, one can be assured of the near-complete listing of genes because the biochemical reactions of the processes and the macromolecular structure of the organelles are known. Thus, the required number of proteins can be estimated. But no such guidelines exist to predict the number of genes required for more complex aspects of metabolism. Hence, the total number of genes of the chromosome of *E. coli* is not known, nor can it be estimated accurately from our current knowledge of biochemistry. On the basis of total coding capacity, one might expect that it contains approximately 3500 genes because the molecular weight of the chromosome is approximately 2.5×10^9, which corresponds to 3.8×10^3 kilobase pairs or 3500 genes that are each 1.1 kilobase long. If these assumptions are correct, only one-quarter of the genes on the chromosome have been identified. However, the assumption that all of the DNA of the chromosome encodes something is not established. Indeed, the unequal distribution of known genes on the genetic map raises the suspicion that certain regions of the chromosome may be genetically nonfunctional. The answer to this question awaits a detailed analysis of the number of proteins encoded by various short regions of the chromosome.

Genes and operons on the chromosome of *E. coli* are not grouped by their function in metabolism: genes governing biosynthesis and catabolism are found in all quadrants. However, there is a distinct tendency for genes with similar metabolic functions to be clustered in four groups that are separated by 90°. This observation has led to the hypothesis that the chromosome of *E. coli* might have evolved by two successive duplications of a smaller chromosome (Riley, 1978). Genes located near the point at which replication of the chromosome starts (*oriC*) are present in a rapidly growing cell in greater number than those located near the terminus of replication. Thus, location of a gene on the chromosome could affect its level of expression in *E. coli* by a factor as large as 2. But other regulatory mechanisms (Chapter 6) exert much larger effects. There is at present no rationale to explain the relative location of genes on the chromosome, but it does appear that gene location is a highly conserved character. This conclusion comes from a comparison of the genetic maps of various enteric bacteria. The linkage maps of *E. coli* strains K12 (Figure 1), *E. coli* strain B, *E. coli* strain C, *Salmonella typhimurium*, *S. abony*, *Shigella dysenteriae*, *Citrobacter freundii*, *Klebsiella pneumoniae*, and *Enterobacter aerogenes* are quite similar. The detailed maps of *E. coli* strains K12 and *S. typhimurium* are virtually identical except for a region of approximately 10% of the chromosome in which the gene order in *S. typhimurium* is the inverse of that in *E. coli*. The similarity of gene order between these species is remarkable in view of the fact that *they have diverged to the point where recombinants between them are infrequent*. The frequency of recombinational integration into the chromosome of a fragment of DNA introduced by an *E. coli* × *E. coli* cross is usually 0.3 to 0.8; by an *E. coli* × *S. typhimurium* cross, it is approximately 10^{-8}. In spite of the remarkable conservation of gene order within the enteric group, gene order on the chromosomes of more distantly related bacteria varies widely. The arrangement of genes on the chromosomes of other bacteria (Figure 1) in which it is known in any detail (*Bacillus subtilus*, *Pseudomonas aeruginosa*, *Acinetobacter calcoaceticus*, *Streptomyces coelicolor*, *Rhizobium japonicum*) is quite different from enteric bacteria and from each other.

The relative arrangement of the genes that encode the enzymes of tryp-

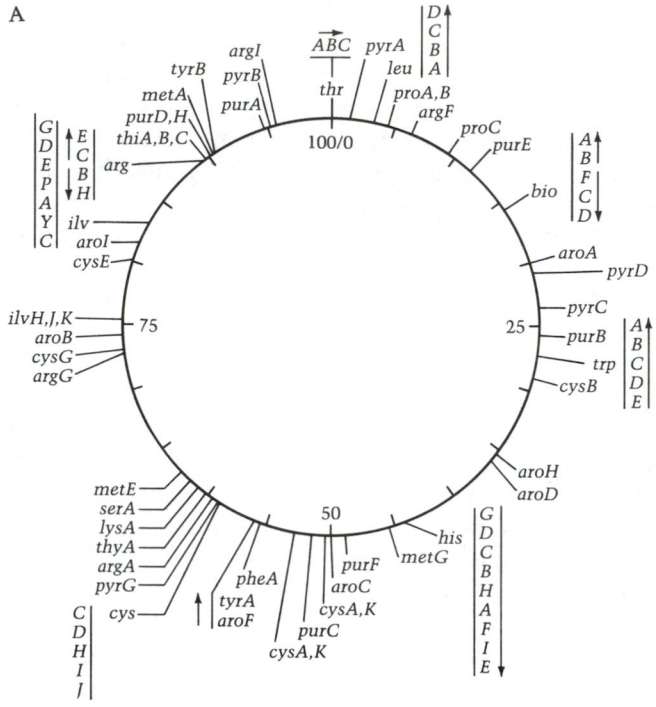

Figure 1. Maps of selected genes on the chromosomes of *Escherichia coli*, *Bacillus subtilis*, and *Pseudomonas aeruginosa*. Genes encoding enzymes in biosynthetic pathways are shown. Gene designations of biosynthetic pathways are *arg*, arginine; *aro*, aromatic amino acids; *bio*, biotin; *cys*, cysteine; *gly*, glycine; *gua*, guanine; *his*, histidine; *ilv*, isoleucine-valine; *leu*, leucine; *lys*, lysine; *met*, methionine; *phe*, phenylalanine; *pro*, proline; *pur*, purine; *pyr*, pyrimidine; *ser*, serine; *thi*, thiamin; *thr*, threonine; *thy*, thymine; *trp*, tryptophan; *tyr*, tyrosine. A. The map of *E. coli* is divided into 100 units, which correspond to the number of minutes required for a chromosome from an Hfr to enter an F⁻ cell at 37°C. The map of *Salmonella typhimurium* is similarly divided and the gene arrangement is almost identical. Genes arranged in operons are shown joined by a horizontal or vertical bar. The arrows indicate direction of transcription; 93 of a total of 900 mapped genes are shown. B. The map of *B. subtilis* is divided into 360°; 55 of 360 mapped genes are shown. (After Henner, 1980.) C. The *P. aeruginosa* chromosome is divided into minutes on the basis of the time taken for the chromosome to enter a recipient cell by FP2-mediated conjugation. FP2 is inserted in donor strains at 0 minutes. Late markers have been located by a conjugation system mediated by R68.45 plasmid. Although the chromosome has been demonstrated genetically to be circular, the total length of the chromosome in minutes is not yet established. The location of 41 of 95 mapped genes is shown. (After Royle, 1981.)

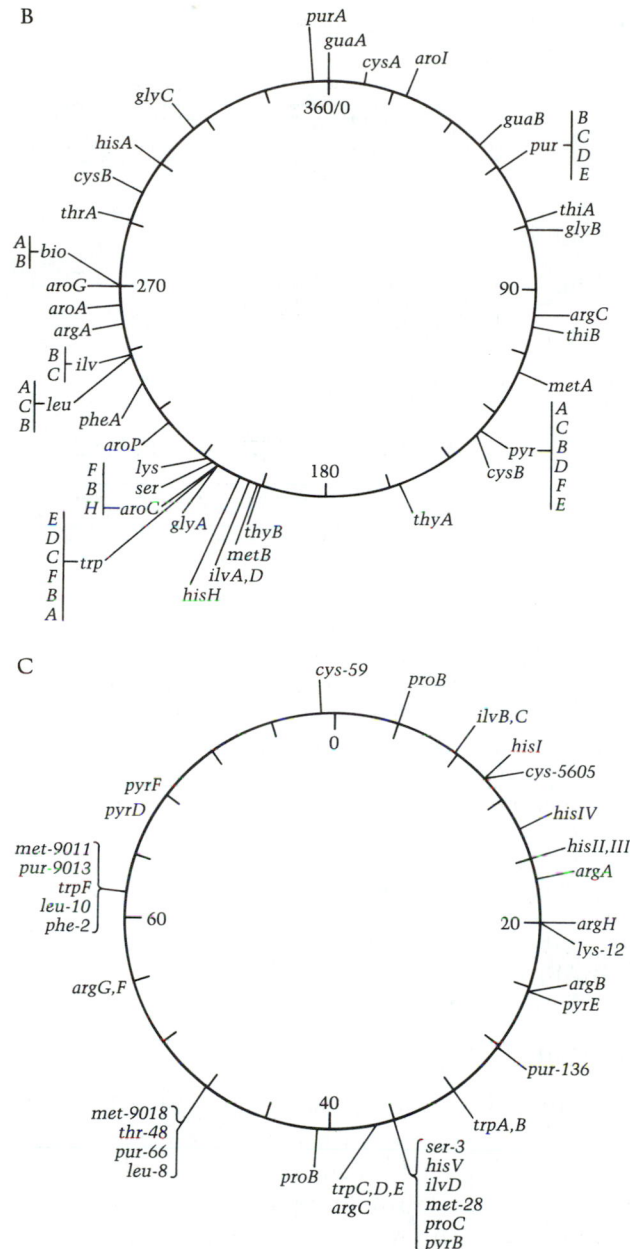

B

purA
guaA
cysA   aroI
360/0
guaB — B / C / D / E — pur
glyC
hisA
cysB
thrA
thiA
glyB
A/B — bio
aroG — 270
aroA
argA
90
argC
thiB
B/C — ilv
A/C/B — leu
pheA
metA
A/C/B/D/F/E — pyr
cysB
aroP
F/B/H — lys / ser — aroC
180
E/D/C/F/B/A — trp
glyA   thyB
metB
ilvA,D
hisH
thyA

C

cys-59   proB
ilvB,C
0
hisI
cys-5605
pyrF
pyrD
hisIV
hisII,III
argA
met-9011 / pur-9013 / trpF / leu-10 / phe-2
60
20
argH
lys-12
argB
pyrE
argG,F
pur-136
met-9018 / thr-48 / pur-66 / leu-8
40
trpA,B
ser-3 / hisV / ilvD / met-28 / proC / pyrB
proB   trpC,D,E
argC

tophan biosynthesis illustrates the principle of conservation and variation of gene arrangement (Figure 2). These genes are arranged very similarly in all members of the enteric group that have been examined. Such an arrangement has not been found in any other bacterium, nor do any two of these species

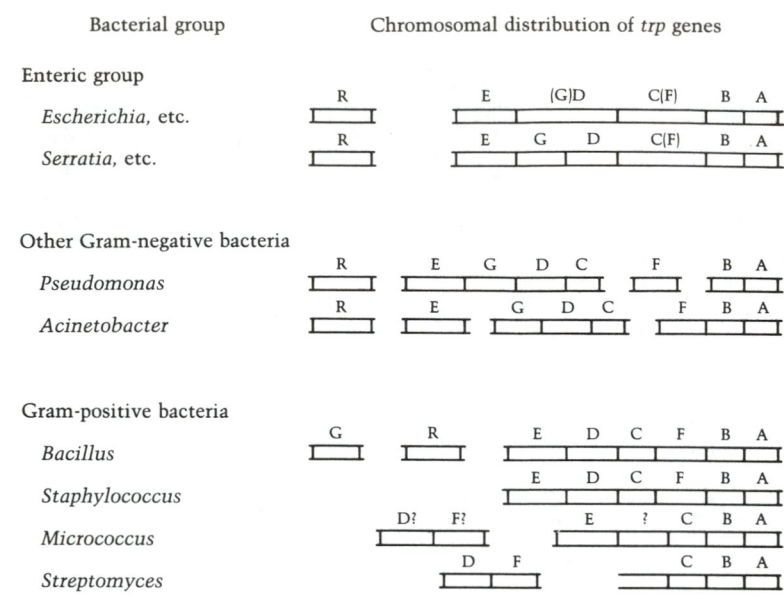

Figure 2. Arrangement on the chromosome of various bacteria of the genes (*trp*) encoding tryptophan biosynthesis. Continuous double lines indicate contiguous regions of DNA. Capital letters represent the genes (*trpA* through *trpG*) encoding enzymes that catalyze the various biosynthetic steps and the gene *trpR* that regulates their expression. Gene termini are indicated by vertical lines. Letters in parentheses indicate that the function is also catalyzed by the protein product of the gene designated by the letter without parentheses. (After Crawford, 1975.)

have the same gene order. Still, certain patterns of clustering can be discerned (e.g., the conserved order E-C-B-A).

MUTANT METHODOLOGY

Advances in our understanding of the process of microbial growth accelerated markedly when it was realized that the MUTANT METHODOLOGY of Beadle (1941) could be used to analyze any phenotype in bacteria. In principle, this methodology is remarkably simple: by comparing the properties of a genetically altered cell line (a MUTANT STRAIN) with those of the cell line from which it was derived (the PARENT STRAIN), the function of the affected gene product can be deduced. In practice, mutant methodology is useful in answering a variety of fundamental questions: (1) How many biochemical reactions constitute a particular pathway? (2) Are there redundant mechanisms to accomplish a certain biological function? (3) What is the biological significance of the activities of a particular enzyme or structural macromolecule?

Certain questions can be answered unequivocally *only* by using the mutant technique, as illustrated by the following example. Assume that we have studied the properties of a certain enzyme in vitro and found that it is allosterically inhibited by the end product of the pathway in which the enzyme participates. A reasonable conclusion from such an observation might be that this allosteric inhibition functions to maintain constant intracellular levels of the end product under various conditions of growth. But such a conclusion is based on a number of unproved assumptions that can be lumped into the following question: Is the inhibition that was observed in vitro physiologically significant in vivo? Proof of the conclusion can be attained by isolating a mutant strain that produces an enzyme with altered sensitivity to the end product in vitro and showing that its intracellular concentration is correspondingly altered. Often the mutant technique is the only way to evaluate the physiological significance of measurements made in vitro.

Mutations are alterations in the sequence of the bases in DNA; frequently they alter the cell's phenotype. Thus, mutations can be classified either by the change in DNA or the change in phenotype. The molecular classification of mutations will be considered later in this chapter along with the treatments that cause them (MUTAGENESIS). On the basis of their phenotypes (Table 1), mutants that have lost the ability to synthesize an essential building block— for example, an amino acid, a purine, a pyrimidine, or a vitamin—are termed AUXOTROPHS (Gr. *auxein*, to increase + *trophe*, nourishment). Wild-type strains, which have this ability, are termed PROTOTROPHS. CARBON SOURCE MUTANTS and NITROGEN SOURCE MUTANTS are those that have genetically lost the ability to catabolize a carbon source or a nitrogen source, respectively. CRYPTIC MUTANTS (Gr. *kryptikis*, secret) are unable to transport a particular metabolite into the cell—their internal enzymatic machinery becomes "hidden" by the mutation.

Certain mutations are CONDITIONALLY EXPRESSED, yielding a wild-type phenotype under one set of conditions and a mutant phenotype under another. Temperature-conditional mutants, termed TEMPERATURE-SENSITIVE (ts), produce gene products that are inactive in one temperature range but active in another. Those active at low temperature but inactive at high temperature are termed HEAT-SENSITIVE; those with the opposite phenotype are termed COLD-SENSITIVE. Most temperature-sensitive mutants produce a gene product that is intrinsically unstable at the RESTRICTIVE TEMPERATURE and sufficiently active at the PERMISSIVE TEMPERATURE to express a near wild type phenotype; others are sensitive to the restrictive temperature only when they are being synthesized—probably because secondary, tertiary, or quaternary structures are not correctly formed. Such mutants are termed TEMPERATURE-SENSITIVE SYNTHESIS (tss) MUTANTS.

Other classes of conditionally expressed mutants include OSMOTIC REMEDIAL MUTANTS, which produce gene products the proper functioning of which depends on the osmotic strength of the medium, and STREPTOMYCIN REMEDIAL MUTANTS, the proper function of which depends on translational errors induced by the presence of a streptomycin or of similarly acting ami-

Table 1. Phenotypes of bacterial mutants

Class	Mutational Defect
Auxotrophic	Defect in a biosynthetic pathway—growth of the mutant is dependent on the presence in the medium of the end product of the pathway (amino acid, purine, pyrimidine, vitamin, etc.)
Carbon source	Defect in a metabolic pathway of a compound that serves as a carbon or energy source—the mutant cannot grow if carbon sources prior to the block are the only ones present
Nitrogen source	Defect in a catabolic pathway that yields ammonia—the mutant cannot grow if nitrogen sources prior to the block are the only ones present
Cryptic mutants	Mutants that have lost a particular function but retain the intracellular enzymatic activities to catalyze the reactions of the function; usually such mutants have lost a permease
Conditionally expressed	
Temperature-sensitive	Gene product cannot function or be synthesized correctly (temperature-sensitive synthesis) at the restrictive temperature; functions or is synthesized at near-normal levels at permissive temperature
Heat-sensitive	High temperature is restrictive, low temperature is permissive (usually 42° and 30°C, respectively, in the case of *E. coli*)
Cold-sensitive	Low temperature (20°C) is restrictive, high (37°C) is permissive
Osmotic remedial	Permissive and restrictive conditions are set by the osmotic strength of the medium
Streptomycin remedial[a]	Permissive medium contains streptomycin, restrictive does not

[a]Other aminoglycosides including kanamycin and neomycin also suppress certain missense mutations.

noglycoside antibiotics. Suppressible mutants of bacteriophage are also conditionally expressed: the phage gene product is expressed in a bacterial strain that carries the proper suppressor gene but not in the wild type.

    Conditionally expressed mutants (sometimes termed CONDITIONAL LETHAL MUTANTS, a confusing term because they can occur in genes whose loss is not normally lethal) can occur as a consequence of mutations in almost all genes (including those that encode tRNAs), but their great utility has been in studying those cellular functions that are essential to bacterial growth under all cultural conditions, including the various steps in macromolecular synthesis, modification, or assembly into supramolecular structures.

DESIGNATION OF MUTATIONS AND MUTANT STRAINS

Before proceeding to discuss the isolation of mutant strains, it is useful to consider the convention (Demerec, 1966) for designating mutations and mutant strains. Three sets of designations have been adopted: one to designate the mutation, another to designate the most obvious phenotype, and a third to designate the bacterial strain that carries the given mutation and possibly others as well.

Mutations are designated by a lower case, italicized, three-letter code followed by an italicized capital letter and an italicized number, for example, *lacZ57* (Figure 3). The three-letter code designates a set of genes with similar metabolic function, that is, *lac* for those genes that encode the catabolism of lactose. The capital letter designates a particular gene in that set, that is, *lacZ* encodes β-galactosidase and *lacY* encodes galactoside permease. The numbers, which are assigned sequentially, indicate a particular mutation, that is, *lacZ57* is mutation number 57 in the gene encoding β-galactosidase, which participates in the catabolism of lactose. If only a single gene is involved in a given cellular function, the capital letter is not used; a hyphen is. Thus, *cod-7* is used to designate a particular mutation in the single gene that encodes cytosine deaminase. A particular phenotype is designated by three-letter codes in roman type, the first letter of which is capitalized. Superscripts are frequently added. Thus, Lac⁻ might be used to designate the phenotype of a strain carrying the mutation *lacZ57*.

Strains should be designated by two capital letters followed by a number. Each laboratory uses its own set of letters and numbers new strains sequentially as they are constructed. Strain AA99 might carry *lacZ57* and thereby express a Lac⁻ phenotype.

ISOLATION OF MUTANT STRAINS

Procedures and schemes employed to isolate mutant strains are almost as numerous and varied as the types of mutants that have been

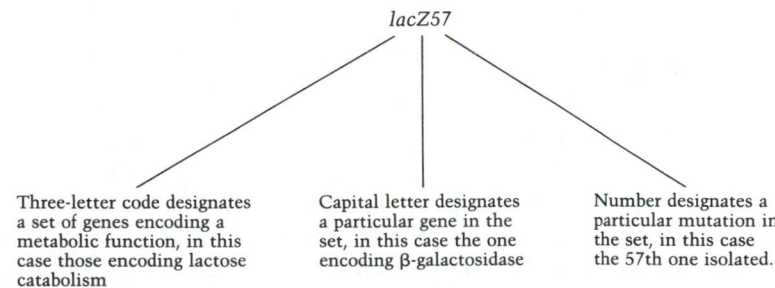

Figure 3. System of designating bacterial mutations. A mutation in β-galactosidase is used as an example.

isolated. At one extreme, mutant strains that are resistant to an inhibitory or lethal agent can be isolated simply by spreading a large population of bacteria on plates that contain the agent and picking the colonies that subsequently develop. At the other extreme, elaborate schemes must be employed to isolate the desired mutants. In all cases, some or all of the following steps are employed.

1. *Mutagenesis*: A culture of the parent strain is treated with a chemical, physical, or biological agent that generates mutations and thereby increases the frequency of all types of mutants, including the desired one, in the treated culture.
2. *Phenotypic expression*: The culture is grown for an interval long enough for the phenotype of the mutant allele to be expressed.
3. *Enrichment*: The culture is subjected to procedures that increase the abundance of the desired mutant in the culture.
4. *Detection*: A test is employed to identify clones of the desired mutant.

Although mutagenesis is the usual first step in a regimen of mutant isolation, it is convenient to postpone a discussion of it until other aspects of the procedure have been considered.

Phenotypic expression

If the mutant allele is dominant, the period required for phenotypic expression is brief. As soon as sufficient product encoded by the mutant allele has been synthesized, the phenotype of the cell becomes mutant. The time required for this to happen can be as short as a few minutes. However, if the mutant allele is recessive, and this is the more common possibility, a considerably longer period of growth must occur before the cell's phenotype is mutant. Although haploid, bacterial cells may be multinucleate. Both nuclear segregation and dilution of parental-type gene products are prerequisite to expression of the mutant phenotype. The number of nucleoids that a bacterial cell contains varies with species and conditions of growth (see Chapter 6), and a rapidly growing cell of *E. coli* often contains as many as four nucleoids. For such a cell, a minimum of two cell divisions must take place subsequent to the mutational event before a genetically homogeneous mutant is produced. But even then the cell would remain phenotypically parental because it would contain significant amounts of parental gene product (such as an enzyme) in the cytosol. Depending on the level of expression of the gene in question and the amount of product required to express the phenotype, a variable number of successive divisions must occur to complete the process of phenotypic expression. In the particular case of a mutation conferring resistance to the coliphage T4, an extensive period of phenotypic expression is required. Usually such a mutation inactivates a gene encoding phage receptors on the cell surface (the OmpC protein of the outer membrane) and thereby renders the cell resistant to infection by that phage. Up to 10^5 OmpC protein molecules are present on the cell surface, but one is presumably sufficient to allow phage attachment. As a consequence, approximately 11 cell divisions occur before a mutagenized culture of *E. coli* yields cells maximally resistant to phage T4.

Enrichment

Enrichment of the proportion of mutant cells in a population can be effected by using one or more of three distinct schemes: DIRECT SELECTION involves the establishment of growth conditions that favor partially or completely the growth of the mutant; COUNTERSELECTION involves the establishment of conditions that kill parental cells; PHYSICAL SEPARATION exploits an altered physical property of mutant cells that allows them to be separated from parental cells (Table 2).

Table 2. Examples of procedures that enrich the proportion of mutants in a population

Type	Affected Gene	Procedure
Direct selection		
Phage resistance	*tonB*	Plating with phage T1
Antibiotic resistance	*rpsL*	Plating with streptomycin
Chemical resistance	*chlC*	Plating with chlorate ion
Revertant of auxotroph	*argI*	Plating *argI⁻* strain on medium lacking arginine
Constitutive mutant	*lacI*	Pulse-feeding lactose
Fitter mutant	*lac*	Duplication selected in lactose-limited chemostat
Counterselection		
Penicillin		
Auxotroph	*argI*	Growth of culture without arginine and with penicillin
Base analogue		
Auxotroph	*argI*	Growth of culture without arginine and with 8-azaguanine
Radioactive suicide		
DNA synthesis	*dnaB*	Growth of culture with [^{3}H]thymidine and subsequent storage
Thymineless death		
Auxotroph	*argI*	Transfer of *thy⁻* culture to medium lacking thymine and arginine
Physical selection		
Nonmotile mutant	*flaA*	Picking from point of inoculation of plate with soft agar and low nutrient concentration
Divisionless		Filtering culture through paper

Direct selection can be used to isolate mutants that are resistant to a particular chemical, physical, or biological agent. If the agent is lethal to the parental cells, prior phenotypic expression is essential; then such resistant mutants can be isolated merely by spreading the culture on plates that contain the agent or that have been exposed to it. Mutants resistant to antibiotics, toxic chemicals, bacteriophage, bacteriocins, radiation, or extremes of temperature can be isolated in this way. Mutants that have gained a function that the parent lacks can be directly selected by requiring use of that function for growth. For example, a revertant of an auxotrophic strain can be selected by growing the culture in a medium lacking the nutrient for which the parent is auxotrophic, or a revertant of a carbon source mutant can be selected by growth in a medium in which that carbon source is the only one.

Less obvious direct selection procedures have also been devised. For example, CONSTITUTIVE MUTANTS (those that produce a particular enzyme regardless of the presence of an inducer or repressor; see Chapter 7) for utilization of a substrate can be isolated by a technique of PULSE FEEDING: small amounts of a growth-essential substrate are intermittently added to the culture. The constitutive cells in the culture are immediately able to utilize the added substrate and, thereby, increase their proportion of the population.

Of course, direct selection constantly operates to enrich for FITTER MUTANTS in any particular environment. Environments in which a particular factor is growth limiting can be used to isolate particular fitter mutants. For example, growth of a culture in a lactose-limited culture enriches for mutants that produce increased levels of the enzymes (galactoside permease and β-galactosidase) that catabolize lactose (and, incidentally, for mutants that adhere to the walls of the growth vessel).

Most counterselection (or indirect selection) techniques are based on the use of agents that selectively kill growing cells. Penicillin was the first such agent to be used (Davis, 1948). By inhibiting the formation of cross linkage in peptidoglycan, growing cells produce weakened walls that are unable to withstand the cell's internal osmotic pressure. Growing cells burst when exposed to penicillin, but nongrowing cells are unaffected by it. Thus, effective enrichment of mutants occurs if penicillin is added to a culture that supports growth of the parental cells but not the desired mutant cells. A number of other agents are useful as a consequence of their ability to kill growing cells selectively. Certain analogues of natural purines and pyrimidines (for example, 8-azaguanine) are incorporated into nucleic acids with lethal consequences, and high concentrations of glycine inhibit peptidoglycan synthesis by some growing cells.

Other agents that are by themselves not markedly lethal to growing cells become incorporated into macromolecules and thereby render cells grown in their presence sensitive to killing by subsequent treatments. Cells grown in the presence of the thymine analogue 5-bromouracil are particularly sensitive to subsequent irradiation by ultraviolet light (Bonhoeffer, 1965). Cells that have been grown in the presence of a radioactive substrate die rapidly during subsequent storage because of internal radiation and the change in elemental composition that occurs when the radioactive isotope decays. Counterselection based on this principle is termed RADIOACTIVE SUICIDE.

A phenomenon known as THYMINELESS DEATH can also be used effectively to counterselect. For reasons that are still not completely understood, thymine auxotrophs die if held in a medium that contains all nutrients necessary for growth except thymine. Thus, the prevalence of an amino acid auxotroph, for example, can be increased in a population of thymine auxotrophs if the suspending medium lacks thymine and the required amino acid.

Methods of enrichment that employ physical separation of mutant and parent can enrich effectively for certain types of mutants. For example, mutants that are unable to divide under certain conditions can be separated from their parents by filtration because they are longer. Mutants that overproduce nucleic acid can be separated from their parents by centrifugation because they are denser. Nonmotile or nontactic mutants (see Chapter 1) can be enriched following incubation on soft agar plates (containing low concentrations of agar) by picking from the point of inoculation because they are unable to swim through the agar from this nutrient-depleted region to the periphery of the zone of growth where nutrients are more abundant.

Detection of mutant strains

Even after the fraction of mutant cells in the population has been enriched, it usually is necessary to apply a test to individual clones to identify the desired mutant ones. If the desired mutant is an auxotrophic, carbon-source, or nitrogen-source mutant, it can be identified simply by picking candidate clones and determining if their growth phenotype is the expected one. For example, an arginine auxotroph will not grow on media lacking arginine but will grow on one containing it; a lactose mutant will not grow on a medium containing that compound as the only source of metabolizable carbon, but will grow if another is available.

A number of procedures have been developed to simplify the screening of clones. REPLICA PLATING (Lederberg, 1952) can be used to screen in a single operation all colonies on a plate. The plate containing the colonies to be tested is pressed face-down on a block covered with filter paper or velveteen cloth, part of each colony being thereby transferred to its surface. The covered block is then used to inoculate other sterile plates containing media that either will or will not support growth of the desired mutant. Because the pattern of inoculation is the same on all the replicate plates, they can be examined in register after incubation and desired mutant clones can be easily identified.

Another useful technique for identifying mutants with medium-dependent growth phenotypes is LIMITED ENRICHMENT. The culture to be screened is plated on a medium containing only small amounts of the required nutrient; mutant colonies can be identified as being smaller than wild type.

Fermentable carbon source mutants of enteric bacteria can easily be detected by plating enriched cultures on certain identification media that stain colonies differentially, depending on whether or not the fermentable carbon source present in the medium can be metabolized. Most of these distinctions depend on the fact that those enteric bacteria (including *E. coli*) that carry out a mixed acid fermentation produce sufficient acid to decrease significantly the pH of the medium around the colony. The medium is designed to support equally the growth of clones that can or cannot ferment the carbon source, but only those that ferment it become acidified. MacConkey medium contains the acid–base indicator neutral red, which causes fermenting colonies to stain red. Eosin–methylene blue medium (EMB) contains eosin and methylene blue, which when acidified react to form a dark blue pigment. Thus, if one wanted to detect a clone unable to ferment lactose it could be done by plating the enriched culture on an identification plate and selecting by the color reaction a colony that does not acidify the medium. The use of these and related indicators has been greatly increased by the development of certain new in vivo genetic procedures, including operon fusions by which the β-galactosidase gene is linked to a promoter of interest (Appendix D). A visit to a bacterial genetics laboratory is a colorful experience.

COLONY AUTORADIOGRAPHY is a generally applicable technique that can be used to identify mutant clones with altered capacity to take up or utilize nutrients. The enriched population is spread on cellulose nitrate filters placed on the surface of an agar medium that contains a radiolabeled nutrient. After colonies have developed, the filter is removed, dried, and used to expose X-ray film. On development, the intensity of exposure in the region of a colony is an index of the amount of the nutrient incorporated into the cells of the colony. On this basis, the desired mutant colony can be identified, and the dried colony (which contains viable cells) serves as the source of the clone (Beck, 1974).

Often a mutant clone can be identified only by measuring the activity of an enzyme or the amount of a cellular component. Although laborious, such procedures are feasible. Many important types of mutants have been isolated by screening clones by biochemical measurements. Indeed, many mutants have been isolated for which enrichment was not possible and biochemical analysis was required for detection. This "brute force" approach depends on the efficacy of mutagenic procedures. Following efficient mutagenesis, one can be virtually assured of isolating any particular loss-of-function mutant by screening 10,000 clones. The first mutant strain lacking DNA polymerase I was isolated by such a procedure (DeLucia, 1969).

Two-dimensional gel chromatography can also be used to examine mutagenized clones for defects in specific proteins: all nonsense, frameshift, deletion, and insertion mutants together with ⅓ of missense mutants are detectable by this procedure.

MUTATIONS

A MUTATION can be defined as a change in the base sequence of DNA. Mutations can be conveniently divided into two broad classes: MICROLESIONS, in which a single base pair has been altered; and MACROLESIONS, in which more extensive changes have occurred. Macrolesions include DELETIONS, DUPLICATIONS, INVERSIONS, TRANSLOCATIONS, and INSERTIONS (Table 3).

Table 3. Macrolesions of DNA

Type	Molecular change[a]	
Deletion	abcdefghi	abc-ghi[b]
Duplication	abcdefghi	abcdef-defghi
Inversions	abcdefghi	abc-fed-ghi
Translocations (insertions)	abcdefghi	uvw-def-xyz

[a]Letters are meant to indicate genes on double-stranded DNA.

[b]- indicates improper junction.

Deletions

Deletions of segments of DNA occur at a significant rate during growth. Deletions constitute approximately 12% of spontaneously occurring mutations. Because deletions completely remove part of a gene or all of one or more genes, cells in which such mutations have occurred remain viable only if the deletion does not extend into an essential gene. As a consequence, the location and extent of deletions is limited. Although most treatments with mutagens is not particularly effective in increasing the number of deletions in a population, a variety of simple schemes can be employed to isolate strains with certain deletions. All such procedures are based on the probability of a deletion extending beyond the limit of a single gene and thereby causing more than the loss of a single metabolic function. By selecting for the simultaneous loss of function of two closely linked genes, one selects for deletion mutations that cover both of them. B. Ames used this property to evaluate the efficacy of chemical mutagens as inducers of deletions (Figure 4).

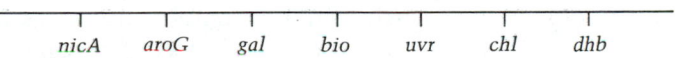

nicA   aroG   gal   bio   uvr   chl   dhb

Figure 4. Region of the *Salmonella typhimurium* chromosome in which deletions can be easily scored. Anaerobic growth in the presence of chlorate ion is possible only if the strain is mutant in the *chl* gene, because the normal product of this gene, nitrate reductase, which is formed only anaerobically, reduces chlorate ion to the highly toxic product, chlorite ion. Anaerobic or aerobic growth in the presence of 2-deoxygalactose (2DG) is possible only if the strain is mutant in the *gal* region, because the normal products of this gene cluster can reduce 2DG to the highly toxic product 2-deoxyglucose phosphate. Thus, by plating cells on media containing chlorate and 2DG and incubating anaerobically, the only clones that can develop are those that have simultaneously lost the functions of *gal* and *chl*. The most probable way for this to happen is as a consequence of a deletion mutation that extends at least from *gal* to *chl*. All such mutations will also have lost *bio* and *uvr* function. Some deletions selected in this way extend into the *dhb* and *nicA* genes. (After Alper, 1975.)

In an analogous way, direct selection for loss of function of a particular gene enriches for a population of deletion mutants, the lesions in some of which might extend into neighboring genes. For example, *tonB*, which encodes the receptor protein for bacteriophage T1, is close to the tryptophan operon on the chromosome of *E. coli*; deletions extending into the tryptophan operon can be readily isolated by selecting a population of cells that are resistant to phage T1 and screening clones of it for tryptophan auxotrophy.

Deletion mutants are particularly valuable in certain studies on bacterial growth because no product is formed from the deleted genes. They are useful in certain genetic studies because by using a set of them as recipients in genetic crosses the relative location of a set of point mutations in donor strains can be quickly determined (see Appendix C).

The obvious consequence of a deletion mutation is a complete loss of a portion of DNA. Because it cannot be regained by subsequent mutation, deletions are characterized as mutations that do not revert spontaneously or by mutagenic treatment. But lack of a detectable reversion rate is not limited to deletion mutations. Genetic proof that a mutation is a deletion can be obtained by showing that crosses between the presumed deletion strain and strains carrying point mutations in affected genes do not yield wild-type recombinants.

A less obvious consequence of a deletion is the formation of an IMPROPER JUNCTION (see Table 3), that is, the fusion of two previously separate regions of DNA. Sometimes such fusions join one gene to the promoter of another, thus facilitating studies on that promoter's function if the fused gene product is more easily assayed than the original one.

Duplications

The frequency of formation of duplications is remarkably high. Values in the range of 10^{-4} have been obtained for specific *E. coli* genes. Thus, a significant fraction of the cells in a culture would be expected to carry duplications somewhere on the genome. Duplications are highly unstable, being lost by homologous recombination between duplicate segments (Figure 5). The mechanism by which duplications are formed is considerably less obvious than the one by which they are lost. Probably it involves recombination between daughter chromosomes, by a process that would appear to involve ILLEGITIMATE RECOMBINATION because recombination occurs between nonhomologous regions. However, in strains that carry a mutation in *recA* and are thereby unable to undergo homologous recombination, the frequency of formation of duplications is decreased approximately 1000-fold. Thus, most duplications must form between preexisting redundant regions of the chromosome. This certainly occurs in the case of the genes that encode ribosomal RNA. Four of the seven sets of these genes are clustered in the region of the *E. coli* chromosome between 83 and 90 units (Figure 6). In media that support growth at a high rate (i.e., growth conditions that require large numbers of ribosomes), cells accumulate duplications that extend between homologous regions of these gene clusters. They are lost when such cultures are transferred

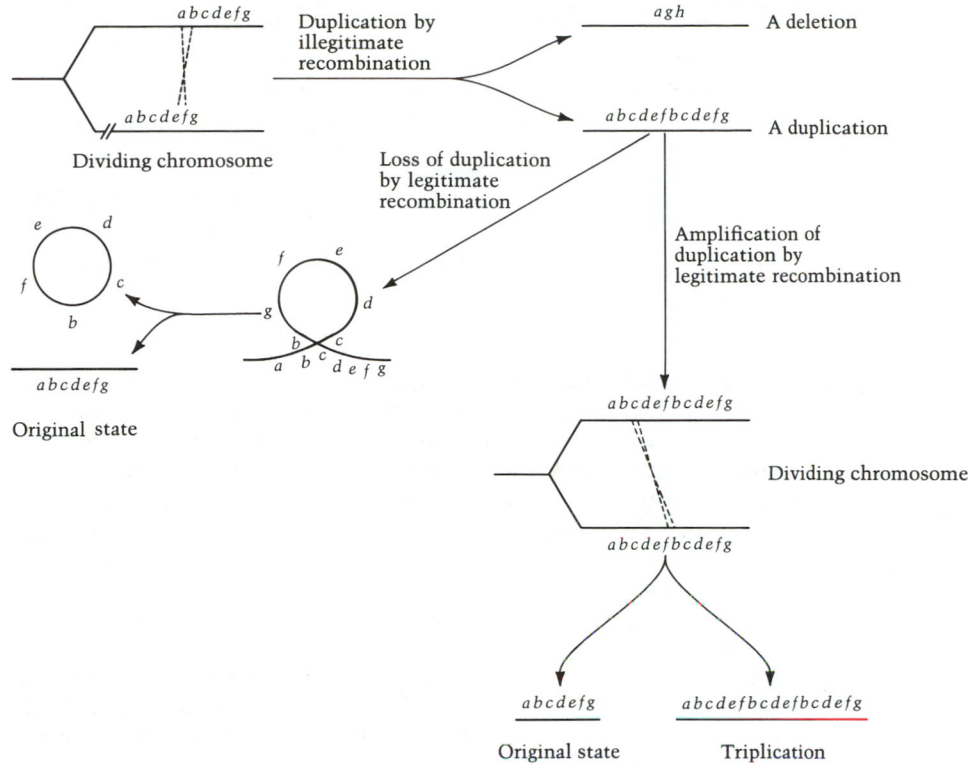

Figure 5. Probable mechanisms for the generation, loss, and amplification of duplication mutations. (After Anderson, 1977.)

to poorer media in which optimal growth does not require large numbers of ribosomes.

Once a duplication is formed, a large region of homology is created, a condition that can lead to further amplification of these genes (Figure 5) or to loss of the duplication, as indicated earlier.

There are a number of interesting physiological, genetic, and evolutionary consequences of duplications. Duplications are the only means by which the amount of the cell's resident DNA can be increased. Such mutations must, therefore, play essential roles in evolution. Indeed, as stated earlier, the arrangement of genes on bacterial chromosomes provides some evidence for duplication in the relatively recent evolution of bacterial chromosomes.

The obvious physiological consequence of a duplication is the synthesis of greater amounts of the products of the duplicated genes (a GENE DOSE EFFECT), and it is on this basis that duplication-carrying clones can be isolated. Clones with duplications of the *lac* operon accumulate in a lactose-limited chemostat, sometimes to be followed by further amplification (Horiuchi, 1963). Strains with duplications of the *lac* operon also predominate in cultures grown on sodium lactobionic acid, which is a carbon source metabolized inefficiently by the gene products of the lactose operon. Selection for rapidly

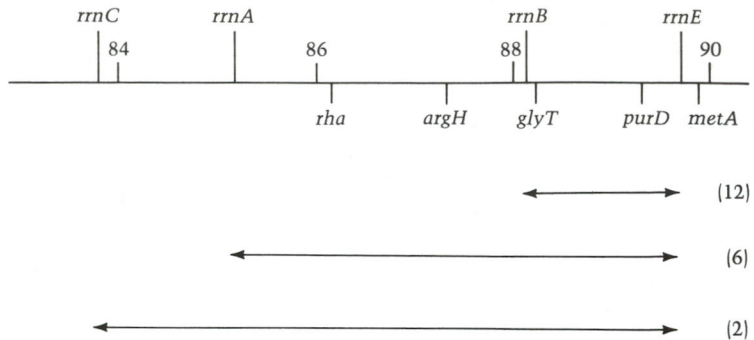

Figure 6. Extent of duplications in the region of the chromosome of *E. coli* that contains genes encoding ribosomal RNA. A set of 25 duplications of *glyT* were selected and their extent was determined by genetic and physical techniques. The majority were shown to terminate within operons encoding ribosomal RNA (*rrnA*, *rrnB*, *rrnC*, and *rrnE*), indicating that these duplications were formed by homologous recombination between these redundant genes. The top line represents a region of the chromosome of *E. coli*: numbers refer to chromosome units. In addition to those encoding ribosomal RNA, certain other genes are shown: *rha* encodes an enzyme that participates in the catabolism of the sugar rhamnose; *argH* encodes an enzyme of the arginine biosynthetic pathway; *glyT* encodes a glycine tRNA; *purD* and *metA* encode enzymes of the purine and methionine biosynthetic pathways, respectively. Double-headed arrows indicate the extent of the duplicated regions; numbers in parentheses to the right indicate the number of representatives (out of the total of 25) in that particular class. (After Hill, 1977.)

growing variants from strains with a defective glycyl-tRNA synthetase has produced strains in which the defective gene is duplicated (Folk, 1971). Similarly, duplications of genes are selected frequently whenever conditions are arranged that render the amount of a specific gene product growth-rate limiting.

Possibly the duplications most revealing of evolutionary processes are those that occur when selection pressures are imposed that require a single gene to encode two different products, that is, when duplication and mutation are simultaneously selected. Owing to the frequency of formation of duplications, such double events can occur with detectable frequency. For example, glycine-tRNA reads the codons GGA and GGG; a certain mutation in *E. coli* can be suppressed if the glycine-tRNA-encoding gene is mutationally altered to read the arginine codons AGA and AGG. But such a strain would grow extremely poorly because its capacity to read the glycine codons would become inadequate. Instead, the selection yields strains that carry duplicate glycine-tRNA genes; one copy makes a tRNA that recognizes GGA and GGG, the other makes one that recognizes AGA and AGG (Hill, 1969).

Duplications do not cause any loss of function if they extend beyond a gene or an operon, so there is no upper limit to their extent, but they do create

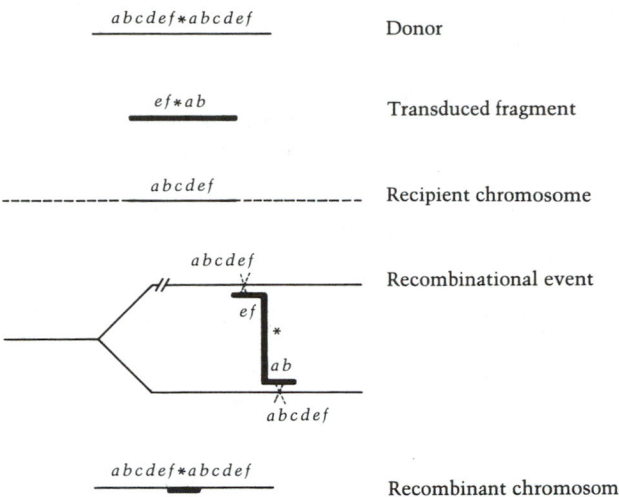

Figure 7. Mechanism of transduction of duplications larger than the phage genome. *indicates improper junction; see text for details.

an improper junction that can result in gene fusion. The existence of a duplication-generated improper junction creates the unusual situation in which a large duplication can be transferred from one cell to another by physically transferring only the small piece of DNA that includes the improper junction (Figure 7). This apparent anomaly occurs as a consequence of crossovers between the legs of a replicating chromosome at regions on either side of the improper junction.

Inversions

Inversions must occur spontaneously in bacteria because the chromosomes of *E. coli* and *S. typhimurium* differ significantly only by the reversal of gene sequence in one region of the chromosome. But there are few systematic studies on spontaneous formation of inversions. However, certain genetic elements (see section in this chapter on insertion sequences and transposons) can generate inversions as well as deletions. These elements translocate to different regions of the genome.

Translocations (Insertions)

Translocation, which is the movement of a fragment of DNA from one region of the genome to another, occurs rarely in bacteria. But the special genetic elements termed insertion sequences and transposons have the special property of replicative translocation, whereby a copy appears at a new location on the chromosome and the original copy remains intact. This event is termed an INSERTION MUTATION. The properties of these mutagenic elements and their special utility in strain construction is discussed later in this chapter.

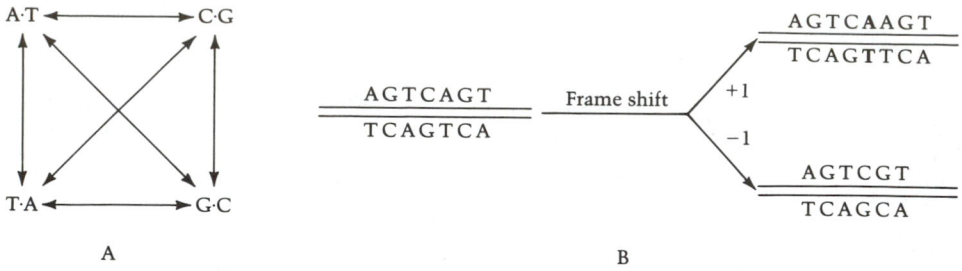

Figure 8. Classes of microlesion mutations. A. Base pair mutations. Horizontal and vertical lines indicate transversions; diagonal lines indicate transitions. B. Frameshift mutations illustrated by possible changes in a length of double-stranded DNA.

Microlesions

Those base pair changes in which one purine is substituted for another purine, and consequently one pyrimidine for another pyrimidine, are termed TRANSITION MUTATIONS; substitution of a purine for a pyrimidine and vice versa are termed TRANSVERSION MUTATIONS (Figure 8). Additions or losses of one or two base pairs are termed FRAMESHIFT MUTATIONS.

The possible consequences of a base pair substitution are several and depend on the specific circumstances. There is no effect whatsoever if the gene product is a protein and the substitution produces an alternate codon for the amino acid encoded by the wild-type allele. For example, as stated earlier, the AGA and AGG codons both encode arginine (Figure 9). A transition mutation resulting in change from one of these codons to another would have no effect on the protein product. Base substitutions that cause a different amino acid to be inserted into the protein product (mutations termed MISSENSE MUTATIONS) might or might not have a significant effect on the activity of the protein, depending on the extent to which the affected amino acid changes the properties of the protein. As is discussed in Chapter 5, most missense mutations do not detectably affect the catalytic activity of an encoded enzyme; temperature lability of the protein is more frequently affected, as is susceptibility to proteolytic attack.

The class of base substitution mutations termed NONSENSE MUTATIONS almost always lead to production of a nonfunctional gene product because generation of a nonsense codon within a gene causes translation to stop at the point of the nonsense codon, thereby producing a truncated protein. Such proteins are usually nonfunctional.

Frameshift mutations

Frameshift mutations change the register and hence the coding properties of the portion of the gene transcriptionally distal to the site of the mutation. This miscoded region is termed GIBBERISH. Hence, frameshift mu-

U U U ⎫ Phe	U C U ⎫	U A U ⎫ Tyr	U G U ⎫ Cys
U U C ⎭	U C C ⎬ Ser	U A C ⎭	U G C ⎭
	U C A ⎪	U A A ⎫ Term	U G A Term
U U A ⎫	U C G ⎭	U A G ⎭	U G G Trp
U U G ⎪			
C U U ⎬ Leu	C C U ⎫	C A U ⎫ His	C G U ⎫
C U C ⎪	C C C ⎬ Pro	C A C ⎭	C G C ⎪ Arg
C U A ⎪	C C A ⎪	C A A ⎫ Gln	C G A ⎬
C U G ⎭	C C G ⎭	C A G ⎭	C G G ⎭
A U U ⎫	A C U ⎫	A A U ⎫ Asn	A G U ⎫ Ser
A U C ⎬ Ile	A C C ⎬ Thr	A A C ⎭	A G C ⎭
A U A ⎭	A C A ⎪	A A A ⎫ Lys	A G A ⎫ Arg
A U G Met	A C G ⎭	A A G ⎭	A G G ⎭
G U U ⎫	G C U ⎫	G A U ⎫ Asp	G G U ⎫
G U C ⎬ Val	G C C ⎬ Ala	G A C ⎭	G G C ⎪ Gly
G U A ⎪	G C A ⎪	G A A ⎫ Glu	G G A ⎬
G U G ⎭	G C G ⎭	G A G ⎭	G G G ⎭

Figure 9. The genetic code. The possible triplet codons of mRNA are listed with the amino acids they encode. A, C, G, and U represent the bases adenine, cytosine, guanine, and uracil, respectively. Amino acids are indicated by their standard abbreviations: Ala, alanine; Arg, arginine; Asn, asparagine; Asp, aspartic acid; Cys, cysteine; gln, glutamine; Glu, glutamic acid; Gly, glycine; His, histidine; Ile, isoleucine; Leu, leucine; Lys, lysine; Met, methionine; Phe, phenylalanine; Pro, proline; Ser, serine; Thr, threonine; Trp, tryptophan; Tyr, tyrosine; and Val, valine. Nonsense codons, which cause termination of translation, are indicated by "Term."

tations usually render the gene product completely inactive. Nonsense codons are frequent among the new codons generated by this mutational shift in reading frame. Thus, truncated proteins are a common product of frameshift mutations.

MUTAGENESIS

Mutations occur spontaneously because the process of replicating the chromosome is not completely free of errors. For this reason, the rate of formation of spontaneous mutations, termed MUTATION RATE (a), is calculated as the number formed per cell doubling, according to the formula:

$$a = \frac{m}{\text{cell generations}} = \frac{m \ln 2}{n - n_0}$$

where m is the number of mutations formed as the number of cells increases from n_0 to n.

Environmental conditions, including the presence of low levels of certain agents that induce mutations (MUTAGENS), can be evaluated by their effect on mutation rate, but mutation rate is not a good index of efficacy of most mutagenic treatments used in the laboratory. Many mutagens do not depend on chromosomal replication to be fully active. The frequency of mutant cells in the surviving population is a better index of the effect of a mutagenic treatment.

The three principal mutagenic treatments used in the laboratory are treatment with chemicals, irradiation, and insertion of DNA in the form of transposons or insertion sequences (see later) into the gene.

Mutagenic chemicals can be divided into three broad classes (Figure 10): (1) BASE ANALOGUES, which are sufficiently close structural analogues of one of the four DNA bases to become incorporated in their stead into DNA; their

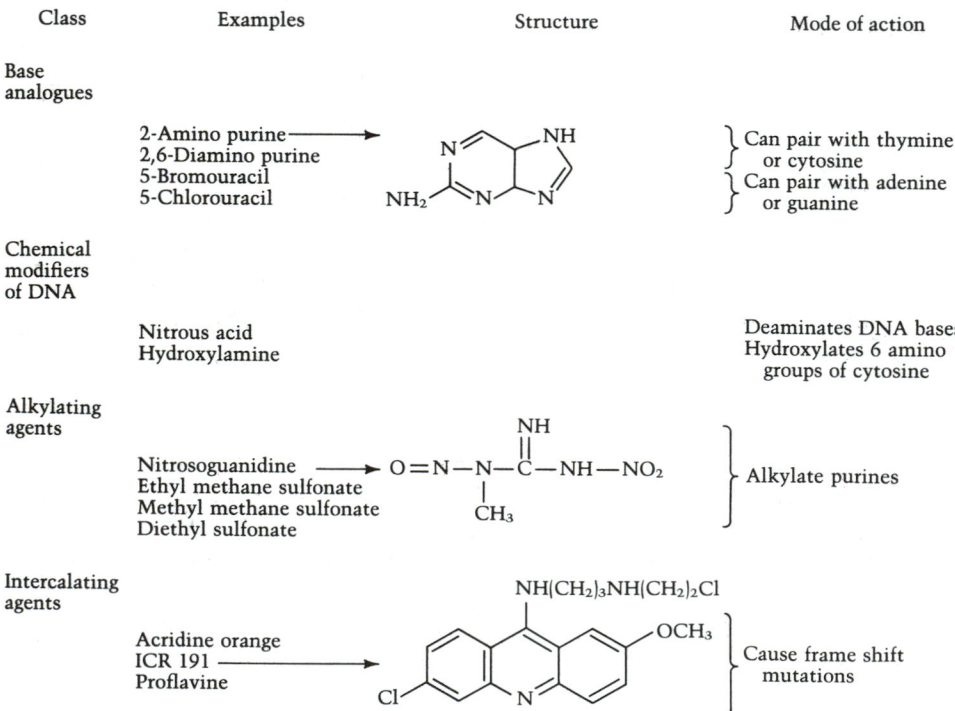

Class	Examples	Structure	Mode of action
Base analogues	2-Amino purine⟶ 2,6-Diamino purine 5-Bromouracil 5-Chlorouracil		} Can pair with thymine or cytosine } Can pair with adenine or guanine
Chemical modifiers of DNA	Nitrous acid Hydroxylamine		Deaminates DNA bases Hydroxylates 6 amino groups of cytosine
Alkylating agents	Nitrosoguanidine ⟶ Ethyl methane sulfonate Methyl methane sulfonate Diethyl sulfonate	$O=N-N-\overset{\overset{\displaystyle NH}{\|}}{C}-NH-NO_2$ $\underset{\displaystyle CH_3}{\mid}$	} Alkylate purines
Intercalating agents	Acridine orange ICR 191 ⟶ Proflavine	$NH(CH_2)_3NH(CH_2)_2Cl$ OCH_3 Cl N	} Cause frame shift mutations

Figure 10. Classes and examples of mutagenic chemicals. Nitrosoguanidine is the trivial name for *N*-methyl-*N'*-nitro-*N*-nitrosoguanidine.

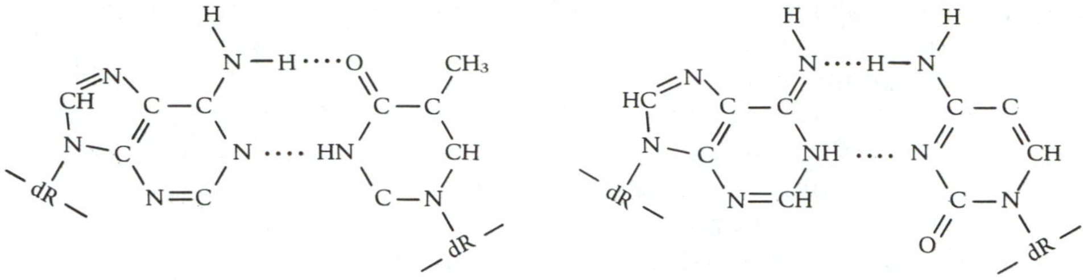

Adenine (amino form)   Thymine (keto form)   Adenine (imino form)   Cytosine (keto-amino form)

Figure 11. Pairing of DNA bases in different tautomeric states. In the normal keto-amino form, thymine pairs with adenine and cytosine with guanine. If a base undergoes a tautomeric shift, its pairing rules are reversed: thymine would pair with guanine and cytosine with adenine. Adenine in its amino form is shown paired with thymine and in its imino form with cytosine. Hydrogen bonds are shown as dotted lines; —dR— indicates deoxyribose in the backbone of the DNA strand.

differing base-pairing properties cause mutations when DNA containing them is replicated; (2) DNA REACTING MUTAGENS, which react directly with DNA, thereby changing its coding properties; and (3) INTERCALATING AGENTS, which are flat molecules that can intercalate between base pairs in the central stack of the DNA double helix, thereby distorting the structure and causing subsequent replication errors.

Base analogues incorporated into DNA cause mutations by an amplification of the same process by which most spontaneous mutations occur. The base-pairing properties [i.e., adenine (A) pairs with thymine (T) and guanine (G) with cytosine (C)] are a consequence of the hydrogen bonds that form between these pairs when the bases are in their keto-amino state (Figure 11). In the tautomeric, enol-imino state, opposite base pairing rules apply, that is, A·C and G·T. When a tautomeric shift occurs at the time of incorporating a base into DNA or replicating an incorporated one, a transition mutation is a possible consequence. The subsequent change in DNA that occurs as a consequence of mispairing during incorporation is the opposite of mispairing during replication. Therefore, a tautomeric shift in any base can cause a transition in both directions, i.e., A·T to G·C and G·C to A·T. With effective base analogue mutagens, the equilibrium of the tautomeric shift is altered toward the enol-imino form, thereby increasing the rate of formation of transition mutations.

DNA-reacting mutagens depend for their activity on a variety of different chemical modifications of DNA, and therefore they cause a variety of different types of mutations to occur. Nitrous acid converts amino groups on the bases to hydroxyl groups, thereby converting adenine, guanine, and cytosine to hypoxanthine, xanthine, and uracil, respectively. Certain of these products pair differently from the reactants: hypoxanthine pairs preferentially with cytosine and uracil with adenine. Thus, replication of nitrous acid-treated DNA generates all types of transition mutations: $A \cdot T \rightleftharpoons G \cdot C$.

Hydroxylamine is more specific in its action. It reacts significantly only with cytosine, converting it to 6-hydroxylaminouracil, which pairs with adenine, a process that causes $G \cdot C$ to $A \cdot T$ transitions on replication. Owing to its specific action, hydroxylamine is a useful reagent in determining the nature of mutations. If a mutation can be caused to revert to wild type by treatment with hydroxylamine, it must have been caused by an $A \cdot T$ to $G \cdot C$ transition.

Curiously, hydroxylamine can generate nonsense mutations but cannot cause them to revert, because DNA encoding the three nonsense codons contains only two $G \cdot C$ pairs. Conversion of either of them to $A \cdot T$ creates another nonsense codon (Figure 12).

The most powerful DNA-reacting mutagens (measured by frequency of causing mutations) are the ALKYLATING AGENTS, which add methyl or ethyl groups to nitrogen atoms in the heterocyclic rings of DNA bases. Although their mode of action is not fully understood, they are capable of causing a variety of mutations including transitions, transversions, and -1 (but not $+1$) frameshifts. Most powerful of the known alkylating agents is nitrosoguanidine. Treatment of a culture with nitrosoguanidine under standard conditions (Adelberg, 1965) causes at least one mutation in each surviving cell. Although nitrosoguanidine is similar to other alkylating agents in the types of mutations it causes, it differs from the others in the fact that it acts preferentially at the point of replication of the chromosome. Owing to this specificity, it is necessary to treat exponentially growing cultures, in which the replication points are spread over the chromosome, to assure random mutagenesis. Multiple mutations are produced in clusters at the growing point. The power of the mutagen causes these clusters to contain many mutations. It has been estimated that a strain of *E. coli* mutagenized by nitrosoguanidine and selected to carry a mutation in a particular gene will also have undergone 50 other base changes in a 100-gene region surrounding the selected mutation (Guerola, 1971). As a consequence, nitrosoguanidine-generated mutations have diminished value for the mutant technique; one cannot be assured that all the altered phenotypic properties of the mutant strain are a result of the selected mutation.

Intercalating agents, by distorting the stacked base pair column at the center of the DNA double helix, generate frameshift mutations on subsequent replication of the molecule. One or more base pairs can be added $(+1, +2)$ or deleted from the molecule. The most useful of the intercalating agents for mutagenizing bacteria is one synthesized at the Institute for Cancer Research (ICR) in Fox Chase, Pennsylvania, and designated ICR 191. Proflavine holds a special place among mutagens because its ability to intercalate and cause frameshifts was exploited by F. Crick to deduce that the genetic code was triplicate and comma-less. He showed that combinations of $+$ or $-$ frameshifts that added to 0 or to ± 3 restored the correct reading frame of the gene, thereby encoding a partially active (PSEUDOWILD) gene product (Crick, 1961).

Various types of radiation are mutagenic. X rays cause chromosomal breaks and consequently various macrolesions. Ultraviolet (UV) light is a

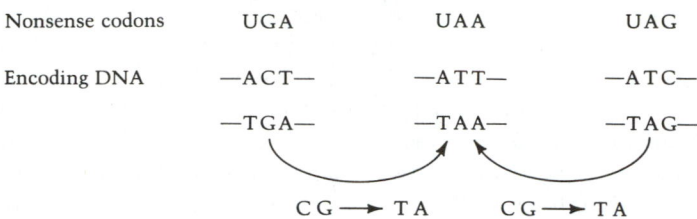

Figure 12. Effect of hydroxylamine on nonsense codons. Hydroxylamine treatment of DNA converts some cytosine residues to 6-hydroxymethyl uracil, which has coding properties like thymine. Thus, hydroxylamine causes G·C ⇌ A·T transitions. As shown, the DNA encoding the three nonsense codons contains only two G·C pairs. Conversion of either of these to an A·T generates another nonsense codon (UAA).

commonly used mutagen. It catalyzes the joining of adjacent pyrimidines, principally thymine, at positions 4 and 5. The cell's repair system (see Chapter 3) is then activated to excise these PYRIMIDINE DIMERS. The process of repair of these single-strand gaps, which are first extended by nucleases, is less accurate than normal chromosomal replication. This ERROR-PRONE REPAIR brings about some frameshifts and transitions, but mostly transversions. Ultraviolet mutagenesis is accompanied by a high level of killing, which is commonly monitored as an index of the dose required for effective mutagenesis. Commonly, a dose is used that kills 99.9% of the culture. However, the correlation between killing and mutations does not hold for all mutagenic treatments. Nitrosoguanidine and other alkylating agents can produce high levels of mutagenesis with only slight killing.

Insertion sequences and transposons

For some time it has been known that certain mutant alleles, termed MUTATOR GENES, increase the rate of spontaneous mutation in other parts of the genome. Although the molecular mode of action of most mutator genes remains unexplained, one of them has been shown to encode a defect in DNA polymerase (Speyer, 1965), and it is assumed that other mutator genes also affect some aspect of the replication process, making it more susceptible to error.

Recently a new class of endogenous mutator elements has been discovered; they are mutagenic because they have the capacity to translocate in a special way to new locations in the genome. These elements, termed INSERTION SEQUENCES (IS) and TRANSPOSONS (Tn), translocate by duplication: one copy of the element appears at a new site in the genome and the other copy remains at the original site. Insertion at the new site is mutagenic because the element usually interrupts the continuity of a gene and destroys its function. Most transposable elements carry strong transcriptional stop signals; therefore, the mutations they cause are highly POLAR—if they are inserted in a gene within an operon, the expression of all genes distal to the one inactivated by insertion of the element is greatly reduced. IS sequences and transposons are distinguished by size and by the amount of information they encode. The former are small (800 base pairs to approximately 2 kilobase pairs) and encode only the ability to transpose. The latter are more than 2 kilobase pairs long and encode a variety of enzymes in addition to transposition (Table 4). The structures of all known transposons have certain common features. At their ends are short IS-like regions that in any particular transposon are almost identical. These REPEATS, as they are called, can lie in the same direction (DIRECT REPEATS) with respect to each other, or, as is more commonly the case, in the opposite direction (INVERTED REPEATS). A variety of enzyme activities are encoded in the central region (the CORE) between the repeated ends. Most of these have the capacity to inactivate a specific antibiotic, by acetylating, phosphorylating, or adenylylating it. Some encode membrane proteins that catalyze the active efflux of antibiotics from the cell or encode enzymes that catalyze the catabolism of a substrate, for example, lactose.

The process of transposition of IS sequences and transposons does not involve homologous recombination; it occurs unimpeded in strains mutated in *recA*. However, there is specificity with respect to the sites into which transpositions occur. In certain cases (e.g., Tn5 and Tn10), the apparent specificity is low because the variety of insertions that have been isolated suggest that they must occur in most, if not all, genes. But more detailed analysis has shown that insertion occurs with greatest frequency at certain sites within a gene. On the other hand, the insertion of other transposons is highly specific; Tn7 inserts with detectable frequency only at a single site on the chromosome of *E. coli*, and at a limited number of sites on the chromosomes of other bacteria.

Table 4. Properties of certain insertion sequences and transposons

Insertion sequences		Transposons			
Designation	Size (base pairs)	Designation	Size (kilobase pairs)	Core encodes	Arrangement of repeats
IS1	768				
IS2	1327				
IS3	1400				
		Tn1	5.0	Ampicillin resistance	Inverted
		Tn5	5.7	Kanamycin resistance	Inverted
		Tn7	14.0	Streptomycin–trimethoprim resistance	Inverted
		Tn9	2.6	Chloramphenicol resistance	Direct
		Tn10	9.3	Tetracycline resistance	Inverted

Tn5, Tn10, and presumably other transposons inhibit their own translocation through the mediation of a soluble product that they encode (Bukhari, 1977). As a consequence, translocation frequency is highest when a transposon is freshly introduced into a cell, either by being a part of the genome of an infecting phage particle or carried within an entering plasmid (Bukhari, 1977). In addition to causing mutations, the introduction of transposons in the bacterial genome creates regions of homology that are useful in a number of genetic manipulations (Kleckner, 1975). These will be discussed later in this chapter.

MECHANISMS OF GENETIC EXCHANGE AMONG BACTERIA

Bacterial genes can be transferred from one cell to another by three different mechanisms. The process by which certain bacteria are able to take soluble DNA from their surroundings is called TRANSFORMATION. The process by which certain bacterial viruses occasionally transfer bacterial genes from one cell to another is termed TRANSDUCTION. The process of direct transfer of bacterial genes from one cell to another by cell-to-cell contact is termed CONJUGATION.

In spite of the fundamental differences among these three mechanisms, they share certain important common features. With rare exceptions only a portion of the genome of the DONOR cell is transferred to the RECIPIENT cell. In this sense, genetic exchange is always polarized. Because there is not a complete mixing of genomes, the product of genetic exchange is not a true zygote. The term MEROZYGOTE is applied to emphasize this partial mixing of genomes. Because the transferred DNA (EXOGENOTE) is only a fragment of the donor genome, it usually lacks those features that allow replication; it is not a REPLICON, which is a DNA molecule capable of self-replication. Thus, unless the exogenote becomes incorporated by two crossovers into a replicon (the chromosome or a plasmid) of the recipient cell's genome (ENDOGENOTE), it will not be spread among the progeny cells. Only a single cell of the clone that develops will contain the unincorporated exogenote.

Transformation

Transformation holds a special place in the history of molecular genetics and in the biology of procaryotes. It was the first of the three processes to be discovered—by A. Griffith (1928). The finding that the active agent of transformation (the TRANSFORMING PRINCIPLE) was DNA—by S. Avery (1944)—was the first direct evidence that genetic information was encoded in that particular macromolecule. Of the three known mechanisms of genetic exchange, transformation is the only one that appears to have evolved solely for the purpose of genetic exchange. The others—transduction and conjugation—cause genetic exchange as a consequence of errors in phage growth and plasmid transfer, respectively.

A wide variety of bacteria undergo transformation (Table 5), and the molecular details of the process are somewhat variable among them. But certain generalizations can be drawn about NATURAL TRANSFORMATION of Gram-positive bacteria, natural transformation of Gram-negative bacteria, and ARTIFICIAL TRANSFORMATION, which can be induced to occur in certain bacteria by treating them with calcium ions and temperature shocks to damage their envelopes (Smith, 1981).

Transformation of Gram-positive bacteria / Griffith discovered transformation in the Gram-positive bacterium *Streptococcus pneumoniae* when he observed that mice injected simultaneously with live cells of an avirulent strain and dead cells of a virulent strain died and that tissues of the dead mice contained bacterial cells that had genetic characteristics of both injected strains. Later it was established that bits of DNA from the dead cells entered the other strain, recombining with the endogenote. *Streptococcus* has remained a subject for studies on transformation. These organisms and *Bacillus* spp, which have also been intensively studied, are the paradigms of Gram-positive transformation (Figure 13).

Cells of a transformable strain of bacteria are not always transformable. When they are, they are said to be COMPETENT. A freshly inoculated culture of *Streptococcus* contains very few competent cells; however, during the later part of the exponential phase of growth, the majority of cells in the culture rapidly—at a rate much higher than the growth rate—become competent. The process of becoming competent appears to be infectious; and, in a sense, it is. Each cell produces a small amount of a low-molecular-weight protein (~10,000), termed a COMPETENCE FACTOR, that induces cells in the population to synthesize 8 to 10 new proteins that establish the competent state.

The process of release of DNA from donor cells has been little studied. It might occur only by the occasional lysis of a cell, or DNA might be extruded, possibly in response to a chemical signal from competent cells.

Table 5. Bacteria known to be capable of natural transformation

---

Natural transformation
　Gram-positive bacteria
　　Streptococcus pneumoniae, S. sanguis, Bacillus subtilis,
　　B. cereus, B. stearothermophilus

　Gram-negative bacteria
　　Neisseria gonorrheae, Acinetobacter calcoaceticus,
　　Moraxella osloensis, M. urethalis,
　　Psychrobacter sp., Azotobacter agilis, Haemophilus influenzae,
　　H. parainfluenzae, Pseudomonas stutzeri

Artificial transformation
　　Escherichia coli, Salmonella typhimurium, Pseudomonas aeruginosa

---

A

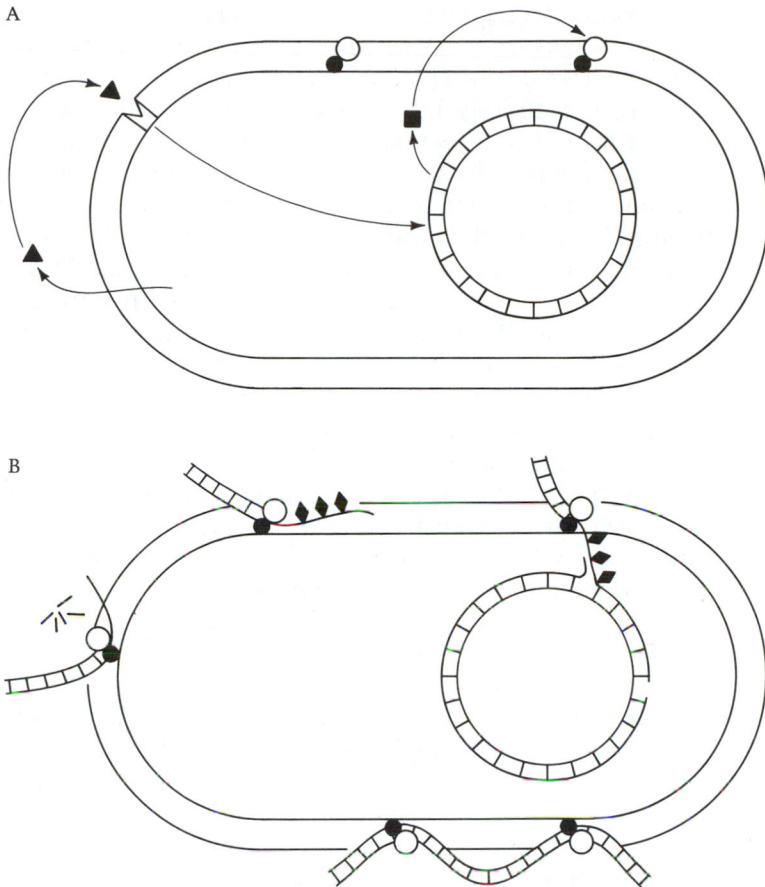

B

Figure 13. Certain features of transformation in Gram-positive bacteria. A. Development of competence. Competence factor (▲) interacts with a cell-surface receptor (M) and causes certain competence-specific proteins to be expressed. One of these is an autolysin (■), which exposes a membrane-associated DNA-binding protein (●) and a nuclease (○). B. Transformation. A long fragment of dsDNA is bound at several sites to the cell surface. It is progressively nicked and cut. One strand of the cut fragment is degraded by the nuclease, a process that releases nucleotides (−) into the medium. The other strand becomes associated with a competence-specific protein (♦). In this form it enters the cell and recombines by a process that involves replacement of one strand of the endogenote by the donor. (After Smith, 1981.)

Whatever the mechanism of its release, bits of double-stranded (ds) DNA become reversibly bound to specific proteins on the surface of competent cells. In the case of *S. pneumoniae*, there are 30 to 80 sites. Single-stranded (ss) DNA is not bound; it is completely unable to transform. When still on the surface of the cell—as demonstrated by its sensitivity to DNase—the dsDNA becomes irreversibly bound. Associated with the binding process, single-strand breaks or NICKS are introduced at 6- to 8-kb intervals.

Before the transforming DNA enters the cell, double-strand breaks occur in the DNA, presumably by cleavage opposite the nick introduced during binding. Then one of the strands of DNA is completely hydrolyzed by an envelope-bound exonuclease, and the resulting single strand becomes associated with a single, small (15.5 kilodaltons in the case of *S. sanguis*) polypeptide that coats it. DNA in this form (termed an ECLIPSE COMPLEX) enters the cell. Still in a single-stranded form, it becomes incorporated into the endogenote. If the transforming DNA differs at specific alleles from the endogenote, the immediate product of incorporation is a HETERODUPLEX because the two strands are not identical. Homogenotes are produced either as a consequence of replication of the heterogenote, a process leading to progeny that are genetically altered, or by a process sometimes termed CORRECTION in which the progeny are not genetically altered. Correction is brought about by DNA repair mechanisms. The imperfectly paired heteroduplex region necessarily forms single-stranded loops. Even if these regions are quite short, the cell's DNA repair mechanism is activated: the one strand in the mismatched regions is cut, the surrounding regions of that strand are hydrolyzed, and the resulting single-stranded region is refilled by polymerase action using the single strand as a template. If correction occurs, the recipient cell is not genetically altered because by an unknown mechanism the donor strand is selectively recognized. The level of correction and hence the efficiency of transformation varies with the markers transformed. Some are transformed at a high efficiency (1.0) and others at quite low efficiency (as low as approximately 0.05).

Because the process of transformation of Gram-positive bacteria involves the intermediary production of a single-stranded DNA intermediate and because single-stranded DNA cannot itself cause transformation (because it cannot be bound to surface receptors on the cell), DNA that has just entered a recipient cannot be recovered in a form that is capable of transforming other cells. The entering DNA is said to enter an ECLIPSE PERIOD, apparently disappearing as judged by a transformation assay and not reappearing until it integrates into the endogenote and again becomes double stranded.

Transformation of Gram-negative bacteria / With certain minor exceptions, transformation of other Gram-positive bacteria that can undergo this mechanism of genetic exchange is similar to the process just described. But transformation of Gram-negative bacteria, if one can generalize from the few well-studied examples, appears to be different in a number of important aspects. *Haemophilus influenzae* and *H. parainfluenzae* are by far the best studied of the transformable Gram-negative bacteria. The generalization to other Gram-negative bacteria of results obtained from studies on *Haemophilus* is based on fewer experiments, but the general pattern of transformation of naturally transformable Gram-negative bacteria appears to be similar to that in *Haemophilus*.

Soluble competence factors are unknown in Gram-negative bacteria, but cultures can be induced to greater competence by certain growth conditions. Cultures of *Haemophilus* become almost 100% competent when transferred to non-growth-supporting media that permit continued protein synthesis. *Acinetobacter* cells become competent as cultures enter the stationary phase of growth.

Like Gram-positive bacteria, competent Gram-negative bacteria absorb only dsDNA. But Gram-negative bacteria (as has been established for *Haemophilus* and *Neisseria*) only absorb DNA from closely related strains. In the case of *Haemophilus*, this remarkable specificity of absorption has been related to specific DNA sequences that its DNA contains. These RECOGNITION SITES contain a common 11-base pair sequence, 5'-AAGTGCGGTCA-3', most of which must be essential for uptake. There are approximately 600 such uptake sites on the chromosome of *Haemophilus*, or approximately one site per 4000 base pairs; even small fragments of DNA have a high probability of containing at least one site and thereby of binding to one of the four to eight receptors on the cell surface. The basis of specificity of DNA recognition by *Neisseria* is not known.

The dsDNA is not degraded to ssDNA as part of the process of entry into the cell. Thus, transforming DNA does not pass through an eclipse period as it does in Gram-positive bacteria. However, even in Gram-negative bacteria, only one strand of the transforming DNA is incorporated into the endogenote; the other strand and the displaced strand of the endogenote are degraded, probably simultaneously. Because no eclipse period occurs, there are no free, single-stranded intermediates in the process.

Artificial transformation / The mechanisms of transformation that we have discussed so far are elaborate in terms of the number of proteins—and in the case of *Haemophilus*, the DNA base sequence—that are required to mediate the process. One can hardly doubt that the process evolved in response to the selective advantages to be realized from the exchange of genetic material in nature. But many bacteria, including *E. coli*, have not evolved natural mechanisms of transformation. The process of transformation allows the investigator to treat soluble DNA chemically *in vitro* and then reintroduce it into a cell in order to evaluate the biological consequences of the treatment. *E. coli*, being so well studied in other respects, would be the ideal bacterium with which to do such studies. Efforts to introduce soluble DNA into *E. coli* by artificial means were successful when it was discovered that treatment of cells with Ca^{2+} in the cold accomplished this (Mandel, 1970). Further elaboration of the technique (Table 6) has made it possible to transform approximately 20% of the cells that survive the treatment, or approximately 4% of the cells in the original population. But artificial transformation of *E. coli* is effective only if the DNA used is capable of self-replication without being integrated into the endogenote (i.e., if it is a REPLICON). Thus, plasmids and viral genomes (transformation of viral genomes is termed TRANSFECTION) transform at high efficiency, but bits of chromosomal DNA transform at low efficiency. Apparently the linear bits of chromosomal DNA are hydrolyzed by intracellular nucleases before they can be integrated, because nuclease-deficient strains of *E. coli* ($recBC^-$, $sbcB^-$) can be artificially transformed by chromosomal DNA at significant frequencies.

Calcium treatment renders a number of other bacteria, including both

Table 6. A procedure for artificially transforming *E. coli*

1. Grow culture to OD_{590} 0.85.
2. Chill rapidly to 0°C and centrifuge.
3. Wash once in 0.5 volumes of 10 mM NaCl.
4. Suspend in same volume of 30 mM $CaCl_2$ and hold at 0°C for 20 minutes.
5. Centrifuge and resuspend in 0.1 original volume of 30 mM $CaCl_2$.
6. To a 0.2-ml sample of "competent" cells, add 0.1 ml DNA solution in 30 mM $CaCl_2$, and hold at 0°C for 60 minutes.
7. Heat mixture at 42°C for 2 minutes and again chill rapidly to 0°C.
8. Incubate appropriately.

After Cohen, 1972.

Gram-positive and Gram-negative species, susceptible to transformation of replicons; it appears to be almost universally applicable among bacteria. Although this technique has already had great impact on biotechnology—it is a cornerstone of the set of procedures termed GENETIC ENGINEERING or RECOMBINANT DNA TECHNOLOGY (Old, 1980)—the basis of its efficiency and the molecular details of the process have received little attention. However, it is clear that DNA enters and remains in double-stranded form.

Other artificial means have been developed for introducing replicons into bacteria. A freeze-thaw technique has been used in *Agrobacterium tumefaciens*; protoplast formation followed by polyethylene glycol treatment has been used in *B. subtilis*.

Transduction

As has been discussed, natural transformation systems are somewhat elaborate mechanisms that have evolved in certain bacteria to mediate genetic exchange between related strains of bacteria. In contrast, TRANSDUCTION, which accomplishes the same biological function, appears to be the consequence of errors that sometimes occur in the development of certain types of bacteriophage. During phage development, an occasional phage particle becomes filled with chromosomal DNA or a mixture of chromosomal and phage DNA rather than completely with phage DNA, as is normally the case. Such an aberrant phage, termed a TRANSDUCING PARTICLE, can attach to another bacterium and introduce bacterial, rather than just phage, DNA into it, thereby transferring bacterial DNA from one cell to another.

Two sorts of errors in phage development can yield transducing particles: one of these leads to GENERALIZED TRANSDUCING PARTICLES that might contain DNA from any region of the bacterial chromosome; the other leads to SPECIALIZED TRANSDUCING PARTICLES that contain phage DNA with DNA from one specific region of the chromosome covalently bound to it.

Generalized transduction / A number of phages that infect a wide variety of bacteria are known to mediate generalized transduction. Of these, the *Salmonella typhimurium* phage P22, in which the phenomenon was discovered in 1952 (Zinder), is probably the most thoroughly studied. The general scheme of formation of transducing particles by this phage, which is a consequence of the particular way it packages its DNA into the viral capsid, probably applies to all generalized transducing phages.

In infected cells, ds P22 DNA is synthesized in long pieces (CONCATEMERS) of DNA that consists of approximately 10 phage genomes tandemly joined (Figure 14). Certain sites in the genome (which tend to be clustered) are

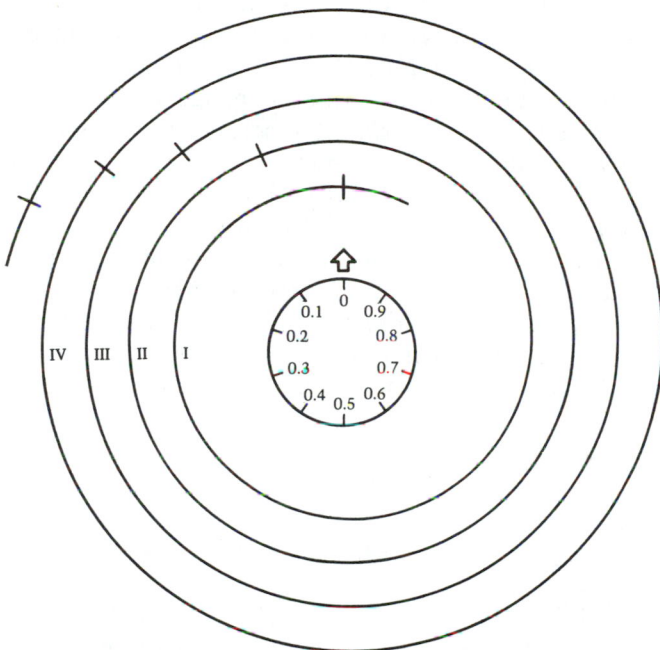

Figure 14. Mechanism of packaging of DNA by phages that mediate generalized transduction. The inner circle represents the genetic map of the generalized transducing phage P22, numbered according to the physical map coordinates. The outer spiral lines represent a concatemer of phage DNA aligned appropriately with the genetic map. The arrow indicates the *pac* site at which a phage-encoded nuclease cleaves the concatemer. Subsequent cleavages, indicated by the cross lines, are made at distances appropriate to produce a phage-headful amount of DNA. The headful amount of DNA exceeds the amount of DNA in the complete phage genome by approximately 2%. Thus, each sequentially packaged fragment of the concatemer (I through IV) has different ends. Generalized transducing particles are formed when certain sites on the host chromosome bearing some resemblance to the *pac* site are cleaved by the phage-encoded endonuclease and sequential headfuls of host DNA are encapsulated. (After Jackson, 1982.)

susceptible to cleavage by a nuclease. Then, by a so-called HEADFUL MECHA-NISM, successive cuts at intervals approximately 2% longer than a complete phage genome are made along the concatemer to yield a set of molecules that are packed into capsids to form mature phage particles (VIRIONS). Each contains a complete phage genome; each has a different starting and ending point within the genome; and each contains a 2%-long region at their ends that are identical. They are said to be TERMINALLY REDUNDANT. By such a packaging mechanism, phage particles all contain complete phage genomes, although they contain no specific nuclease-sensitive sites that delineate the genome. There is a low level of specificity for the primary cleavage of the concatemer, but subsequent cleavages are made solely on the basis of their distance from the primary cleavage.

If, as sometimes happens, the primary cut is made in the bacterial chromosome rather than in the concatemeric phage DNA, subsequent cuts in the chromosome by the headful mechanism yield fragments of chromosomal DNA that, when incorporated into the phage capsid, become generalized transducing particles. Presumably all generalized transducing phages package their DNA by a headful mechanism from concatemeric DNA.

The headful mechanism of formation of transducing particles fits with some long-known facts of generalized transduction. Not all chromosomal genes are transduced at the same frequency, presumably because they are at different distances from the sites at which primary cleavages occur. Deletion mutations near, but outside, the region between two closely linked markers affect the frequency with which those two markers appear in the same transducing particle (are COTRANSDUCED), presumably because the deletion changes

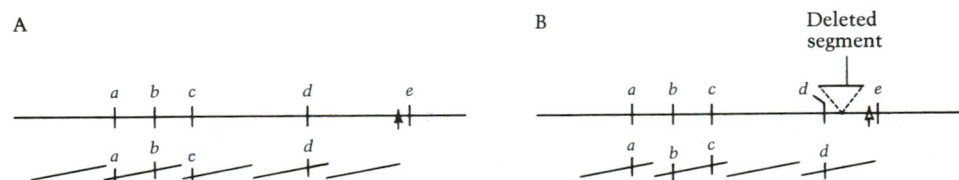

Figure 15. Model explaining effect of outside deletions on frequency of cotransduction. A. Representation of a region of the bacterial chromosome containing a site (↑) susceptible to cleavage by the nuclease encoded on the genome of a generalized transducing phage such as P22. The letters indicate various genetic markers. The sloping lines below represent fragments of chromosomal DNA that would be packaged into transducing particles by the headful mechanism. Note that packaging from this site (packaging from other sites also occurs) produces transducing particles that permit cotransduction of *a* and *b* but none that permit cotransduction of *b* and *c*. B. A deletion has occurred between markers *d* and *e* (outside the region *a-b-c* in which cotransduction was examined). This changes the relative position of the site of the primary nucleolytic attack (↑) and the markers of interest. Note that now transducing particles would be produced that permit cotransduction of *b* and *c* but none permitting cotransduction of *a* and *b*. (After Chelala, 1974.)

the register of cuts between the two markers and a nearby probable site for a primary cut (Figure 15).

Some years ago Schmieger (1972) isolated mutants of phage P22 (HT mutants) that were capable of transduction at a greater rate than wild type. These mutants have the properties to be expected if the nuclease that causes the primary cut was altered such that its site specificity was reduced. Wild-type phages incorporate only approximately 1% of chromosomal DNA into phage capsids; HT mutants incorporate up to 50% of it. Differences among the frequencies with which various markers are transduced by wild-type phage almost disappear in HT-mediated transductions (Table 7). Effects of outside deletions on transduction frequencies are not completely eliminated but they vary with each particular HT mutant as though each alters nuclease specificity uniquely.

Once injected into the recipient cell, DNA from transducing particles remains double-stranded and becomes incorporated into the endogenote by replacing both strands of the recipient chromosome. However, not all pieces of donor DNA are incorporated; at least 90% fail to recombine. Because they are not replicons, they are not replicated; however, because they persist and are double-stranded, their genes are expressed. Curiously, if incorporation does not occur immediately after injection, it almost never does. Cells that contain unincorporated donor DNA are termed ABORTIVE TRANSDUCTANTS. Under conditions of growth that require expression of a donated allele, colonies arising from abortive transductants can be easily distinguished from those arising from COMPLETE TRANSDUCTANTS (those in which the donated DNA fragment is recombined into the endogenote) because the former are still quite tiny at a time when the latter are full sized. They are approximately 10 times as abundant as complete transductants. Colonies arising from abortive transductants grow slowly because only a single cell can synthesize the growth-depen-

Table 7. Transductional frequencies of various mutations of *Salmonella typhimurium* mediated by parental-type phage P22 (H5) and a HT mutant (HT 12/4)[a]

Phage	Transduction frequency of:[b]			
	hisB22	*trp-8*	*arg*	*cysE8*
H5	1.1×10^{-6}	3.9×10^{-7}	3.5×10^{-9}	1.0×10^{-9}
HT12/4	9.5×10^{-5}	1.6×10^{-4}	1.6×10^{-4}	1.0×10^{-4}
Phage	*thy*	*thr*	*leu*	*ara*
H5	1.6×10^{-7}	5.1×10^{-8}	3.3×10^{-10}	
HT12/4	6.7×10^{-5}	5.7×10^{-5}	3.6×10^{-4}	3.4×10^{-4}

[a]From Schmieger, 1972.
[b]Calculated as the ratio of transductants to plaque-forming units.

dent gene product. Indeed, although the abortive transductant colony may contain many thousand cells, if it is picked and spread on another selective plate, only a single microcolony will develop (Hartman, 1960).

Abortive transductants remain stably diploid for the region donated and therefore are quite useful in testing dominance of alleles.

Specialized transduction / Although P22, the example of a generalized transducing phage that was chosen for discussion, happens to be a temperate phage (see Stanier, 1976), certain virulent phages also mediate generalized transduction. Specialized transduction, however, is mediated only by temperate phages, because the developmental aberration that generates specialized transducing particles occurs when the prophage is excised from its chromosomal location (Figure 16). Rather than a recombinational event between the homologous ends of the phage genome, illegitimate recombination between a point within

Figure 16. Mechanism of formation of specialized transducing particles of phage λ. A. On infection of a bacterial cell, phage λ DNA (represented as a pair of straight lines) enters the bacterial cell (bacterial DNA is represented as pairs of wavy lines) as a linear piece with complementary single-stranded regions (*m* and *m'*) 12 nucleotides in length and attached at the 5' ends. B. The complementary ends pair and the gaps are ligated, thus forming a covalently closed circle of phage DNA. The phage genome and the bacterial genome both contain identical 15-base pair regions of DNA designated *O*; these and surrounding (nonidentical regions of DNA) are the sites of pairing between phage and chromosome, followed by recombination and thereby (C₁) insertion of the phage genome into the bacterial chromosome. Because the regions surrounding *O* on the phage (designated P and P'; not shown) differ from those on the bacterial chromosome, they are distinguished by terming the attached site on the phage *att*λ*P* (POP') and that on the bacterial chromosome *att*λ*B* (BOB'). As a result of the integrative recombinational events occurring at *O*, the two attachment sites become hybrids of phage and bacterial DNA. The one on the left, designated *attL*, is composed of BOP' DNA. Recombinational insertion of λ is catalyzed by the protein product of the lambda gene, *int*. C₂. Occasionally, by the action of this λ gene product (and another, *xis*), recombination between the *O* regions of *attL* and *attR* occurs, a process that regenerates circular lambda and the original form of the bacterial chromosome. This process leads to vegetative replication of the phage, eventual cell lysis, and release of mature λ virions. D. Rarely, in approximately 1 out of every 10^5 such events, specialized transducing particles are generated as a consequence of an illegitimate recombination between a site within λ and one on the surrounding chromosome. There are restrictions as to where such recombinations can occur if a mature virion is to result: certain phage genes, including the packaging gene *cosA*, must be included in the particle and the size of the fragment must not be less than about 73% or greater than approximately 110% of the size of a normal λ genome. Represented here is a recombination that generates a prophage bearing *gal* genes. A lysate of this sort contains a small number

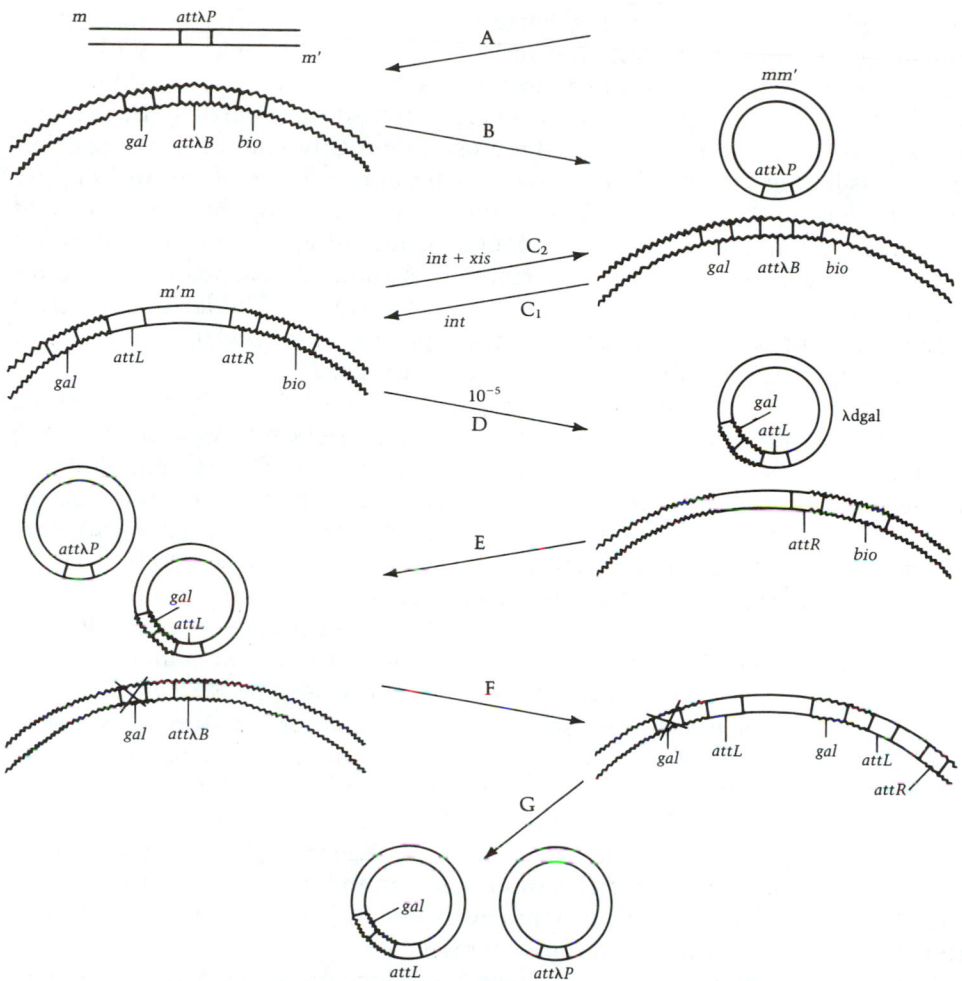

of specialized transducing particles (D) among a large number (C_2) of wild-type λ virions and hence is called a LOW FREQUENCY TRANSDUC-ING (LFT) lysate. E. The specialized transducing particles from such a lysate can be purified by using it to infect a strain that carries a defect (X) in one of the *gal* genes and plating the infected culture on a medium containing galactose as the only source of carbon and energy. F. Only those cells that have received functional *gal* genes from the specialized transducing particles will be capable of devel-oping into colonies. Usually, as depicted, integration of the λdgal occurs by two successive steps: a wild-type virion integrates (as pic-tured in C_1) and generates an *attL* site, where the specialized trans-ducing particle then integrates. Note that the chromosome contains two sets of *gal* genes: the defective one originally present and the functional one contributed by the specialized transducing particle. G. A lysate derived from such a culture would contain approximately 50% transducing particles and is termed a HIGH FREQUENCY TRANS-DUCING (HFT) lysate. (If a helper phage were not required, it could contain 100% transducing particles.)

the phage and the surrounding bacterial genome accounts for specialized transduction. Because the amount of DNA that can be packaged within the phage capsid is narrowly limited, only genes near the site of insertion of the prophage can become part of the specialized transducing particle. In the case of coliphage lambda (λ), which is the most thoroughly studied of the specialized transducing phages, genes encoding the biosynthesis of biotin (*bio*) are on one side of the attachment site and those encoding the catabolism of galactase (*gal*) are on the other. Although surrounding and intervening genes are also carried, by convention specialized transducing particles of λ are divided into those that carry *bio* and those that carry *gal*. The letter *d* is added if they lack sufficient phage genes to form plaques, that is, λdgal or λdbio. If they can form phages, the letter *p* is sometimes used.

Such aberrant excisions are rare, approximately 10^{-6}, so lysates obtained from induction of prophages are termed LOW FREQUENCY TRANSDUCING (LFT) lysates. However, if such a lysate is used to infect a Gal$^-$ culture and if clones are selected that are able to utilize galactase as a carbon source, one obtains a strain that carries λgal as a prophage (if the λgal were defective and therefore incapable of lysogenization or lytic growth, double lysogens would be selected that carry wild-type λ called HELPER PHAGE, as well as λdgal). On induction of such a lysogen, the lysate obtained—termed a HIGH FREQUENCY TRANSDUCING (HFT) lysate—contains specialized transducing particles at a frequency of approximately 0.5 because of the presence of the wild-type, helper phage.

As will be discussed later, specialized transduction has limited value in mapping the location of genes on the bacterial chromosome. It has been useful in the genetic analysis of λ because specialized transducing particles are deletion mutations in the phage. And it has been extremely useful in a variety of studies that are based on GENE AMPLIFICATION. An HFT lysate contains a high concentration of those bacterial genes carried on the transducing particle; and, therefore, a remarkable gene enrichment is effected. Moreover, during induction of the lysogen, these genes are expressed.

For these reasons, techniques have been developed to isolate specialized transducing phages that carry genes other than those adjacent to the normal attachment site of λ. Two techniques have been used. (1) Bacterial strains in which the attachment site has been deleted still can be lysogenized at low frequency, in which case the prophage is inserted at a variety of secondary sites. Thus, specialized transducing particles can be generated that carry genes adjacent to those sites. (2) The normal attachment site can be moved near the gene of interest. This can be accomplished by forcing illegitimate recombination between two plasmids, one of which carries the normal attachment site and the other, the gene of interest (Shimada, 1972). It can also be accomplished by transposing, via plasmids, the gene of interest to a site near the phage attachment site (Ippen, 1971) or vice versa (Press, 1971).

Conjugation and F plasmids

Conjugation, the direct transfer of genes from one bacterial cell to another, is mediated by plasmids termed CONJUGATIVE PLASMIDS. The plasmids are transferred at high frequency, and occasionally chromosomal

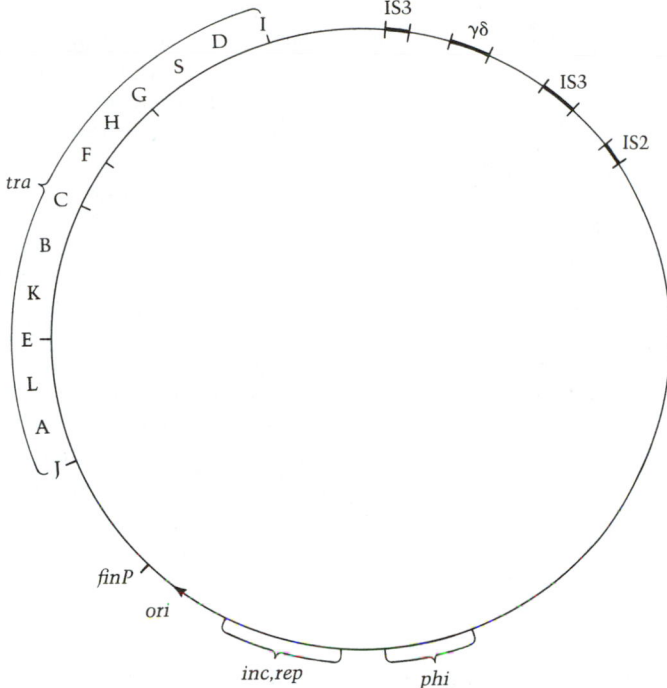

Figure 17. Genetic map of F plasmid. The relative position of genes encoding transfer functions (*tra*), fertility inhibition (*finP*), origin of transfer replication (*ori*), incompatibility (*inc*), replication (*rep*), and phage inhibition (*phi*), along with insertion sequences (IS2, IS3, and γδ) are shown. The length of the genome is 94.5 kilobases. (After Shapiro, 1977.)

genes are also transferred. The first-studied and most completely understood of the conjugative plasmids is the F plasmid, which can replicate in *E. coli* and certain closely related enteric bacteria, including *Salmonella typhimurium*. Like all plasmids, F is a self-replicating, circular molecule of dsDNA. Like all conjugative plasmids, it carries all the genes necessary to encode its transfer from one cell to another. These 13 transfer genes (*traA* through *traL*) form an operon (Figure 17) that makes up approximately ⅓ of the 94.5-kilobase genome. Outside the operon are a cluster of four IS sequences, certain other plasmid genes, and regions that amount to approximately ¼ of the genome and that encode no known functions. Apparently F encodes only its own replication and transfer. The *tra* genes of the F plasmid, and hence its ability to self-transfer are always expressed. However, *tra* gene expression in closely related plasmids (F-type plasmids) is normally repressed by the joint action of the products of two closely linked genes, *finO* and *finP*; F lacks *finO*. The *tra* genes of F control F-pilus formation and conjugal DNA metabolism.

If cells containing an F plasmid (designated F$^+$) are mixed with cells that lack it (designated F$^-$), conjugal pairs rapidly form by attachment of the end

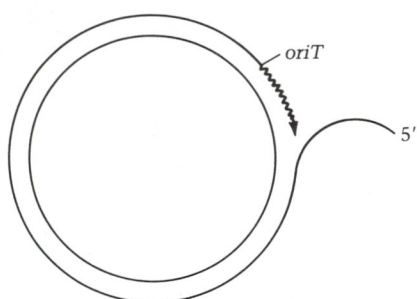

Figure 18.

Transfer of F plasmid from an F⁺ to F⁻ cell. Formation of a mating pair triggers transfer replication of F. By an F-encoded nuclease, one strand is cleaved (nicked) at *oriT*. Then replication (at arrowhead) occurs by a rolling circle mechanism. The newly synthesized DNA (wiggly line) displaces a preexisting single strand of F, which enters the F⁻ cell. Although the possibility exists that the single strand of F is passed through the pilus during conjugation, it appears more likely that it is passed from one cell to the other at sites of direct contact between them. Such cell-to-cell contact occurs after the cells have been linked through the F pilus.

of the F-pilus to the surface of the F⁻ cells. Attachment then triggers a series of events resulting in transfer of the F plasmid and very little else from the F⁺ to the F⁻ cell. At *ori* (Figure 18), a nick in a single strand is made in F; the nick serves as the starting point for replication of the unnicked strand (termed ROLLING CIRCLE REPLICATION) in the 5′ to 3′ direction. The 5′ end of the preexisting nicked strand enters the F⁻ cell as replication continues. It is not completely clear whether the single strand is conducted by the F pilus or is passed directly into the F⁻ cell at some other point where the two cells are in contact, but it is clear that replication and transfer occur simultaneously and that only a single strand is transferred to the recipient. There it is duplicated and recircularized. It will be noted from the form of replication that the donor cell retains a copy of the F plasmid.

Pair formation and plasmid transfer is extremely efficient between F⁺ and F⁻ cells. If two such cultures are mixed, all cells in the culture rapidly become F⁺—or, as they are termed because they serve as donors, male. At a much lower frequency (10^{-7}), some chromosomal genes are transferred to the F⁻ (female) cell along with the F plasmid.

As stated earlier, the F plasmid contains a cluster of IS sequences. The chromosome of *E. coli* also contains IS sequences. Homologous recombination between an IS on the F plasmid and one on the chromosome causes the F plasmid to be inserted into the chromosome. F genes in such a cell (characterized by its HIGH FREQUENCY OF RECOMBINATION, Hfr) continue to be expressed; the cell remains male. But certain of its properties change dramatically. When an Hfr contacts an F⁻ cell, conjugative transfer of F begins and, because it is covalently bound to the chromosome, transfer of one strand of the chromosome follows. However, the attachment between cells in a mating pair is a fragile one that breaks frequently, thus interrupting transfer. Transfer is a somewhat lengthy process, requiring in the case of *E. coli* approximately 100 minutes to complete at 37°C. So, complete transfer of the chromosome rarely occurs. The cluster of IS sequences on F (the site at which the chromosome is inserted in an Hfr) lies between the origin of transfer of the F plasmid and the *tra* genes (Figure 16). Thus, they are transferred only after chromosomal transfer is complete. Because this event is rare, the consequences of an Hfr $\times$ F$^-$ differ from those of an F$^+$ $\times$ F$^-$ mating. Chromosomal genes close to the site of F insertion are transferred at a high frequency (10^{-1}); the F$^-$ cell rarely becomes male.

Hfr strains are unstable; recombination between the homologous ends of the inserted plasmid regenerates an F$^+$ strain (Figure 19). Occasionally recombination occurs between a site within F and a site on the chromosome or between sites on either side of F. The plasmid product of the former (if enough F remains to preserve its properties as a replicon) is termed a TYPE I F′; the product of the latter is termed a TYPE II F′. F′ formation generates a chromosomal deletion; the genes carried on the F′ are not duplicated in the chromosome; *the cell remains haploid for all genes*. Such a cell is termed a PRIMARY F′. But the product of an F′ $\times$ F$^-$ mating, which usually results in the complete transfer of the F′ plasmid because of its relatively small size, is *a partially diploid cell*, a stable MEROZYGOTE, termed a SECONDARY F′ (Broda, 1965). Secondary F′ strains are quite useful in studies of dominance of alleles and, as discussed in Appendix D, in mapping the location of genes on the chromosome. They are also unstable unless they carry a *recA* mutation, owing to the considerable homology between the plasmid and the endogenote.

The utility of F′ strains is even greater than might be expected because of the existence of the process of HOMOGENOTIZATION. In *recA$^+$* secondary F′ strains, nonreciprocal crossovers occur frequently between the endogenote and the plasmid. Thus, transfer of a particular allele from one of these genetic elements to the other can be selected by establishing growth conditions that favor growth of the HOMOGENOTE, or enriching for cells with a phenotype expected from a desired homogenote. In practice, homogenotization becomes quite valuable in complementation or dominance studies. From a set of bacterial mutants that carry different mutant alleles of a particular gene, a set of F′ plasmids carrying these alleles can be easily constructed. Dominance or complementation tests can then be done by transferring the various mutation-bearing F′ plasmids to strains with various mutant alleles in the chromosomally located gene.

Other plasmids / A variety of other plasmids are conjugative and capable of mobilizing (facilitating the transfer of) the chromosome and any nonconjugative plasmids that might be present in the same cell. The molecular nature of mobilization is unclear other than that it seems to occur by a process analogous to Hfr formation. Only a portion of chromosome mobilization seen in F$^+$ strains can be attributed to the Hfr cells that they inevitably contain. The rest remains somewhat mysterious. Possibly transient single-strand connections between plasmid and mobilized DNA account for the phenomenon.

The vast variety of plasmids that occur in bacteria (Bukhari, 1977) are classified on the basis of the organism in which they were discovered and the INCOMPATIBILITY GROUP in which they fall, that is, on the basis of their ability

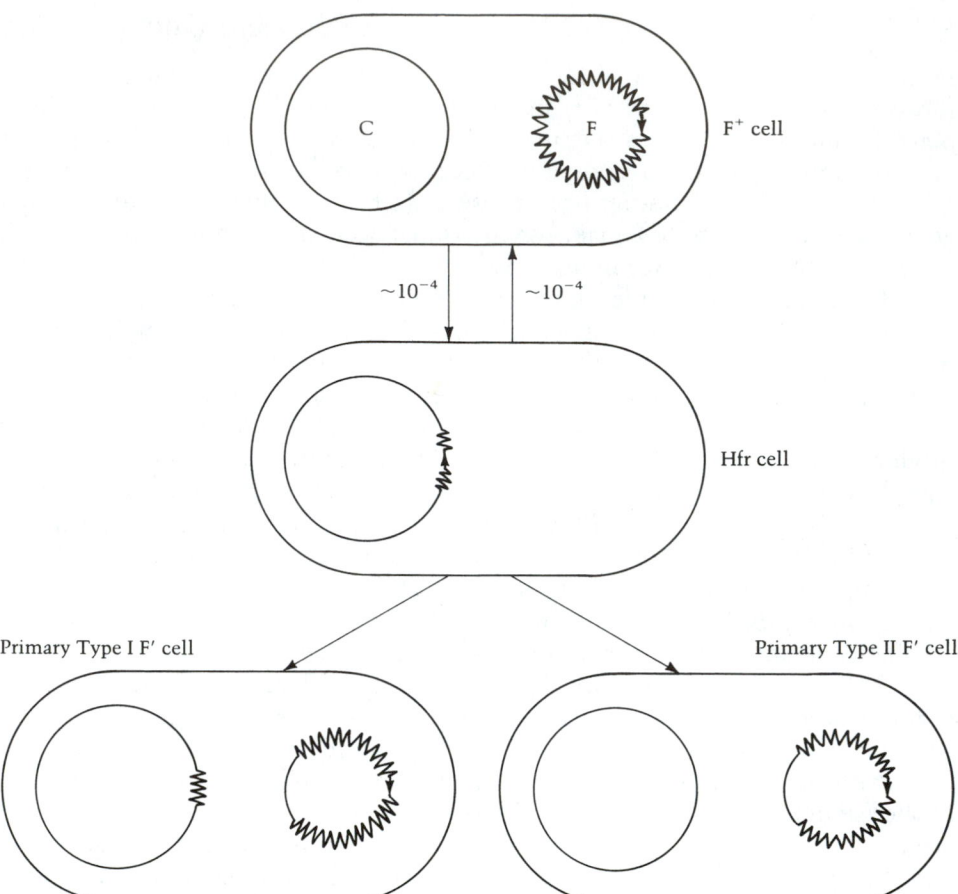

Figure 19. Formation of Hfr and F' strains. By homologous recombination (*recA*-dependent) between insertion sequences on the F plasmid and those in the chromosome (C), Hfr strains are formed from F⁺ strains at a frequency of approximately 10^{-4}. At approximately the same frequency, recombination between the homologous ends of the F plasmid DNA regenerates an F⁺ strain. Occasionally recombinations between the F plasmid and chromosomal DNA or between regions of chromosomal DNA on either side of the F plasmid occur, generating an F-prime (F') cell. In the former case, the product—termed a Type I F'—contains an altered F plasmid (an F' plasmid) that lacks a region of the F plasmid and contains certain chromosomal genes adjacent to the site of insertion of the F plasmid in the parental Hfr strain. The chromosome contains a fragment of the F plasmid and lacks the genes carried on the F' plasmid. In the second case, the F' plasmid (termed a Type II F') contains all the DNA of the F plasmid and chromosomal genes adjacent to both ends of the F plasmid DNA in the parental Hfr. The chromosome lacks these genes. Both of these cells (termed primary F' cells) are haploid for all chromosomal genes; they are carried on the F' plasmid or on the chromosome. If, by mating, the F' plasmid from a primary F' cell is transferred to an F⁻ cell, the product cell is termed a secondary F' cell. Secondary F' cells (both Type I and Type II) are merodiploid. Genes carried on the F' plasmid are also carried on the chromosome.

216

to coexist in the same cell. The phenomenon of incompatibility cannot yet be explained in mechanistic terms, but it can be easily described: certain closely related plasmids cannot stably coexist in the same cell. If two incompatible plasmids are introduced into a cell, only one will be passed to progeny cells. Bacterial cells can carry a number of different plasmids if they all fall in different incompatibility groups. Some plasmids occur in multiple copies in a cell (there are 20 to 40 copies of the plasmid ColE1 in each cell that carries it), but the copies are identical. If a change that affects its compatibility occurs in one of the copies, only one of them is inherited. Incompatibility has been chosen as primary criterion of relatedness, and therefore phenotypically quite different plasmids are grouped together if they are incompatible. If they are, they are placed in the same incompatibility group. Table 8 lists a number of bacterial plasmids and certain of their properties, including functions that they encode in addition to their common capacity for self-replication.

The plasmids of one incompatibility group, IncP in the enteric bacteria grouping or IncP1 in the pseudomonad grouping, deserve special attention. Whereas most plasmids can exist in only a limited number of closely related bacteria (NARROW HOST RANGE PLASMIDS), IncP plasmids can exist in almost any Gram-negative bacterium (BROAD HOST RANGE PLASMIDS). They also are capable of self-transfer among members of this broad group. (At first glance, this capacity seems startling, but those factors that confer a narrow host range on the majority of bacterial plasmids is equally inexplicable). IncP plasmids have the capacity to mobilize chromosomal genes of a number of species at low frequency (10^{-8}); plasmid variants have been isolated that mobilize them at higher frequencies (10^{-5}) (Holloway, 1979). By incorporating chromosomal genes into the plasmid to form R' particles (named to reflect their similarity to F' particles and the fact that the plasmid encodes drug resistance), they can be used to transfer genes of one species of Gram-negative bacterium to another.

R68.45 is a variant of the IncP plasmid R68 and has increased ability to mobilize chromosomal genes in several Gram-negative species, including *E. coli, Pseudomonas aeruginosa,* and *Rhizobium meliloti.* The development of this variant illustrates an approach of great utility in the construction of conjugation systems in species not naturally proficient. This variant was isolated from a TRANSCONJUGANT (product of a mating) of an R68-mediated cross that had received the plasmid and chromosomal genes. A few percent of such transconjugants contain plasmids with enhanced mobilization ability like R68.45. R68.45 is larger than R68 by 2400 base pairs. The added DNA appears to be the product of a tandem duplication of a fragment of R68. The duplication has the properties of an IS sequence; the original fragment does not. By an incompletely understood mechanism, R68.45 mediates the polarized transfer of small regions of the chromosome from a number of sites of origin. Transconjugants always contain the complete plasmid.

Conjugation in Gram-positive bacteria / Recent studies of the conjugation systems of the Gram-positive bacterium *Streptococcus faecalis* suggest that they will prove to be quite different from the better-studied Gram-negative systems (Clewell, 1981). Pili do not play a role in conjugation between strains of *Streptococcus faecalis;* mating signals between the cells are mediated by

Table 8. Properties of certain bacterial plasmids[a]

Incompatibility group	Host range	Plasmid	Original host	Phenotype[b]
IncFI	*Escherichia coli* *Proteus morganii* *Proteus vulgaris* *Salmonella typhimurium* *Rhizobium lupini*	F	*E. coli*	Tra^+ Dps (F1, F2, F17, MS2, μ2) Phi (T3, T7, Q11)
		ColV-K94	*E. coli*	Tra^+ Dps (F1, F2, μ2, MS2, M13)
	Enterobacter cloacae *Proteus mirabilis*	R453	*Proteus morganii*	Tra^+ Dps (F1, F2, (MS2) Phi (λ) Ap, Cm, Sm, Sp, Su, Tc
IncFII	*Shigella flexneri*	R100	*Shigella flexneri*	Tra^+ Fi^+ (F) Dps (F1, F2, MS2, M13) Cm, Fa, Sm, Sp, Su, Tc, Hg
IncA-c	*Providence*	R480	*Providence*	Tra^+ Fi^- (F) Dps (PRDI, PRRI)
IncP	Gram-negative bacteria	RP1	*Pseudomonas aeruginosa*	Ap, Km, Tc

[a]Selected from over 400 entries in a table compiled by A.E. Jacobs, J.A. Shapiro, L. Yamamoto, P.I. Smith, S.N. Cohen, and D. Berg describing plasmids studied in *Escherichia coli* and other enteric bacteria, p. 607, in Bukhari, 1977.

[b]Tra, Ability to promote self-transfer; Dps, plasmid encodes sensitivity of host to phages listed in parentheses; Phi, plasmid encodes resistance of host to phages listed in parentheses; Fi^+, fertility inhibition of F, produces product that inhibits expression of *tra* genes on F. Antibiotic resistance: Ap, ampicillin, Cm, chloramphenicol; Km, kanamycin; Sm, streptomycin; Sp, spectinomycin; Su, sulfanylamide; Tc, tetracycline; Hg, mercury resistance.

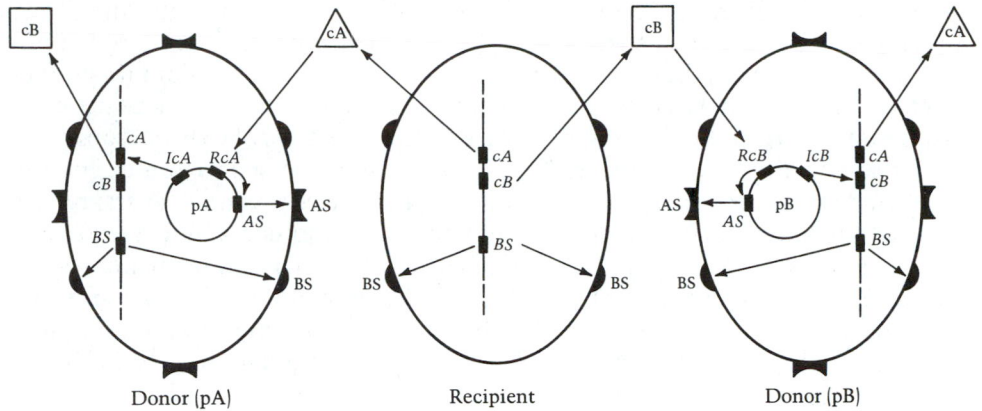

Figure 20. Conjugation of *Streptococcus faecalis.* Model shows the response of two donor cells to two chromosomally encoded pheromones (*cA* and *cB*) produced by a recipient. These react with plasmid-encoded genes (*RcA* and *RcB*), which signal another gene (*AS*) to produce aggregation substance (*AS*). AS causes producing cells to bind to other cells at the binding substance (*BS*) site. Cells that carry a particular plasmid (e.g., pA) do not produce the corresponding pheromone (*cA*) because the product of a plasmid gene (*IcA*) represses the structural gene (*cA*). (After Clewell, 1981.)

soluble molecules (PHEROMONES). These protease-sensitive, heat-resistant pheromones are released by cells that lack a particular plasmid; and the pheromones stimulate cells that have the plasmid to produce an aggregation substance on their outer surface. Cells with the substance form clumps with cells that lack the plasmid (Figure 20). Plasmids are transferred from one cell to another in these clumps.

MAPPING

A variety of techniques can be used to locate the relative position of genes on a bacterial chromosome and thereby to construct a genetic map. Some of these are more useful in determining the approximate position of a gene (ROUGH MAPPING), and others are useful in determining the relative position of closely linked genes or even various mutant alleles within a single gene (FINE STRUCTURE MAPPING). These procedures, which have become central to biochemical, structural, and physiological studies because of the power of mutant analysis in probing the living cell, are described in Appendix D.

IN VITRO RECOMBINATION

In recent years enzymatic techniques have been developed to join fagments of DNA *in vitro* using DNA LIGASE. Fragments can be generated by cutting DNA at specific sites by one or another RESTRICTION ENDONU-

CLEASE, as described in Chapter 2. Certain of these enzymes cut the DNA asymmetrically, thereby generating single-stranded ends that are complementary; they can anneal, thereby facilitating ligation (Table 9). Random fragments of DNA can be generated by mechanical shearing and complementary ends can be added enzymatically by a process termed TAILING. Such random fragments can be annealed and ligated by the same process used to join fragments generated by treatment with restriction endonucleases. Even DNA fragments lacking single-stranded complementary ends can be joined at low frequency. If these *in vitro* techniques are used to insert a DNA fragment from any source into a replicon (a phage or a plasmid) and if the new molecule is transformed into a bacterial cell, the DNA fragment becomes a part of the bacterial genome. Thus, in theory, any fragment of DNA can be made a part of a bacterium's

Table 9. Action of certain restriction endonucleases used in recombinant DNA technology

Class	Enzyme	Producing microorganism	Recognized DNA sequence[a]
Six base pairs recognized, complementary single-stranded ends produced	EcoRI	E. coli (R)[b]	G A A T C / C T T A A G
	HindIII	Haemophilus influenzae	A A G C T T / T T C G A A
Six base pairs recognized, blunt ends produced	HpaI	Haemophilus parainfluenzae	G T T A A C / C A A T T G
	HindII	Haemophilus influenzae Rd	C A / G T (T)(G) A C / C A (A)(T) T G / G C
Four base pairs recognized, complementary single strands produced	Hha	Haemophilus haemolyticus	G C G C / C G C G
	MboI	Moraxella bovis	G A T C / C T A G
Four base pairs recognized, blunt ends produced	HaeIII	Haemophilus aegypticus	G G C C / C C G G

[a]Arrow indicates site of single strand cleavage. Upper sequence of bases is written in the 5′ ⟶ 3′ direction.
[b]Encoded by genes that are plasmid-borne.

genome. This set of procedures and similar ones have been termed, collectively, RECOMBINANT DNA TECHNOLOGY or GENETIC ENGINEERING (Old, 1980).

IN VIVO GENETIC ENGINEERING

The relatively recent discovery of insertion sequences and transposons and the subsequent development of techniques to exploit their special properties have created a remarkable new capacity to manipulate the bacterial genome that in certain respects approaches the significance of in vitro recombinant DNA technology. In order to emphasize the utility of this set of techniques, its developers—with a certain competitiveness—have termed it IN VIVO GENETIC ENGINEERING. The procedures allow one to select positively for insertion of specific genetic elements into the bacterial genome, and the sites of insertion can be easily determined. The genetic elements that insert can carry a variety of genes and control elements, including those that encode resistance to antibiotics and other antimicrobial agents, those that encode ability to catabolize specific carbon sources, those that encode transcriptional start signals, and those that encode promoters. Through the ability to insert these genetic elements, a number of genetic tasks are simplified.

Transposable genetic elements (certain phages including Mu and λ as well as transposons and insertion sequences) encode their own ability to undergo transposition (or in laboratory jargon, hopping), and all, with the possible exception of insertion sequences, possess regulatory gene products that repress transposition. Thus, transposition occurs at greatest frequency when a transposable genetic element is first introduced into a cell—before there has been sufficient time for the repressor to rise to inhibitory concentrations.

Transposons—as discussed earlier—are extremely useful mutagenic agents. When they are inserted into a gene, they inactivate it and simultaneously introduce into the genome new genes, the phenotypes of which can be selected. The remarkable utility of the system can be illustrated by a procedure developed for *Salmonella typhimurium*. A hybrid phage P22 has been constructed that carries a transposon (Tn10, encoding resistance to the antibiotic tetracycline) and two amber mutations in vital phage functions; thus, it can be propagated only in amber-suppressing strains. If a nonsuppressing strain is infected by this phage, it does not replicate. The only tetracycline-resistant clones that develop from the infected culture are those in which transpositions of Tn10 from the phage genome to the bacterial genome have occurred; all are insertion-generated mutant clones. Thus, one can search for a desired mutant in a population (the tetracyline-resistant one) that is totally mutant, thereby greatly simplifying the search. Moreover, the mutation encodes resistance to tetracyline. As a consequence, one realizes the considerable advantage of having a set of mutants all of which can be selected directly. For example, if one isolates an auxotrophic mutation (which in itself expresses a negative phenotype) the mutation can be transferred to another strain by selecting for resistance to tetracycline (a positive phenotype). The opportunity of employing positive selection is a geneticist's dream.

This system of transposon mutagenesis can be effected by introducing the transposon into the recipient cell on any genetic element that is not itself a replicon (an IMMOBILIZED VEHICLE) in that cell. Immobilized vehicles include phages that can infect, but not replicate in the recipient cell (as we have discussed); plasmids that can be transferred to, but not replicate in, the recipient cell; and fragments of DNA transferred by genetic crosses (transductionally transformational or conjugation) into cells (*recA*) that cannot undergo homologous recombination.

In addition to their utility as insertion mutagens, transposons can be used to create areas of homology between different regions of the bacterial genome and thereby to cause certain recombinational events to occur at high frequency. For example, if a conjugative plasmid and a chromosome bear the same transposon, recombination will occur at high frequency between them, producing Hfr-type strains that can donate chromosomal genes commencing with those near the site of insertion of the transposon in the chromosome. By such techniques, systems of conjugation can be developed in a variety of bacteria. If two copies of a transposon are inserted into a genome and another is inserted into an accessory genetic element (a phage or a plasmid), two recombinational events will transfer the chromosomal genes between the transposons onto the accessory element.

As stated earlier, the bacteriophage Mu (named because it inserts randomly into the bacterial chromosome and therefore is *mu*tagenic) is a transposable element. This property of phage Mu has been exploited to construct a variant (Mud-1) that has proved to be an extremely valuable genetic tool (Casadaban, 1979). Mud-1 carries the genes from *E. coli* (*lacZ* and *lacY*) that encode the enzymes that catabolize lactose. These genes within Mu are associated with a translational start signal, but no signals for transcriptional start or stop are included. Thus, if Mud-1 is inserted within a bacterial gene in the correct direction, the lactose genes are transcribed from the gene's promoter and translated from the Mud-1-resident start signal. Such Mud-1 insertions are, in effect, gene fusions between the bacterial gene's promoter and the *lac* structural genes within Mud-1. The control of expression of the bacterial gene can thereby be followed by measuring the production of the easily assayed *lacZ* gene product, β-galactosidase (see Chapter 7).

IDENTIFICATION OF GENE PRODUCTS

As discussed, fragments of DNA can be inserted into phages or plasmids by recombinant DNA technology. They are also inserted into these replicons in the generation of F' particles, R' particles, and specialized transducing particles. A number of procedures can be used to identify which genes are carried on these recombinant molecules. Because the fragments carried are often small, identification often yields valuable information about gene location on the chromosome. Genes carried on phage λ, but not chromosomal genes, are expressed when they are injected upon infection into UV-irradiated bacteria. Thus, following infection, radiolabeled amino acids are incorporated into the products of the λ-carried genes. By identifying the radiolabeled proteins using gel chromotagraphy, the λ-carried genes are identified (Jaskunas, 1975).

This method cannot be applied directly to plasmids because transformation frequency of *E. coli* with plasmids is at most 10^{-2}. Therefore, alternative methods have been developed. The genes can be expressed and thus identified enzymatically *in vitro*. MINICELLS—tiny cells produced by certain mutants—can also be used. In some cases minicells contain the plasmids that normal cells of the culture contain. Thus, if minicells are separated from the culture of a minicell-producing strain and incubated with radiolabeled amino acids, all radiolabeled proteins produced will be plasmid encoded (Meagher, 1977). Alternatively, plasmids can be introduced into strains that carry two mutations—*recA* and *uvrA*—that confer hypersensitivity to UV. If such a cell is irradiated, extensive degradation of chromosomal DNA occurs and subsequently such a MAXICELL synthesizes only plasmid-encoded proteins (Sancar, 1979). As an alternative, use can be made of the fact that following chloramphenicol-treatment of cells that carry ColE1 plasmids or their derivatives (the most commonly employed plasmid vehicles for recombinant DNA), plasmid DNA is preferentially expressed (Neidhardt, 1980).

If radiolabeled DNA or messenger RNA that corresponds to some of the genes carried on a plasmid is available, it can be used as a PROBE to identify which bacterial colonies carry that plasmid (Grunstein, 1975). Colonies to be tested are replicated to Millipore filters and lysed by floating the filters on solutions containing lysozyme. Under these conditions, the DNA released from the lysing cells of the colonies is bound tightly to the nitrocellulose of the filter. When these filters are exposed to a solution of the radiolabeled probe, it will hybridize specifically to the area of the colony that carried genes corresponding to the probe. By exposing the filter to X-ray film, the radioactive areas are identified and the desired colonies can be located and picked from the replica plates.

SUMMARY

1. The bacterial genome is composed of a single circular chromosome of double-stranded DNA unassociated with basic proteins; in addition, bacteria may contain one or more copies of one or more types of self-replicating molecules called plasmids, which do not encode growth-essential functions.

2. The mutant technique of studying bacterial growth involves isolating mutant strains with specific functional losses and analyzing their phenotypes.

3. Isolation of a particular desired mutant strain is usually possible owing to the availability of effective mutagens, techniques for specifically enriching the proportion of desired mutants in a culture, and techniques for detecting them.

4. Genetic information is exchanged among bacteria by the processes of transformation, transduction, and conjugation.

5. By use of any of the mechanisms of genetic exchange, mutations can be located accurately on the bacterial chromosome or on a plasmid, and strains with a desired combination of mutations can be constructed.

PROBLEMS

List the steps and media you would employ to isolate the following mutant strains of *E. coli*. If necessary, you may start with a strain other than wild type.

1. A heat-sensitive mutant resulting from a lesion in *galK* (encodes galactokinase, which catalyzes one step of galactase catabolism): using penicillin to counterselect.
2. A cold-sensitive mutant resulting from a lesion in *argI* (encodes ornithine carbamyltransferase, which catalyzes one step in the biosynthesis of arginine): (a) using penicillin to counterselect; (b) using thymineless death to counterselect.
3. An *argI* mutant: using Tn10 as a mutagen.
4. A deletion in *galK*.
5. A duplication of *cod* (encodes cytosine deaminase; the products of the reaction catalyzed are uracil and ammonia).

REFERENCES

Adelberg, E. A., M. Mandel and G. C. C. Chen. 1965. Optimum conditions for mutagenesis by N-methyl-N'-nitro-N-nitrosoguanidine in *Escherichia coli* K-12. Biochem. Biophys. Res. Commun. 18:788.

Avery, O. T., C. M. Macleod and M. McCarty. 1944. Studies on the chemical nature of the substance inducing transformation of pneumococcal types. I. Induction of transformation by a DNA fraction isolated from pneumococcus type III. J. Exp. Med. 79:137.

Bachmann, B. J. and K. B. Low. 1980. Linkage map of *E. coli* K-12, edition 6. Microbiol. Revs. 44:1.

Beadle, G. W. and E. L. Tatum. 1941. Genetic control of biochemical reactions in *Neurospora*. Proc. Natl. Acad. Sci. USA 27:499.

Beck, C. F., J. Neuhard, E. Thomassen, J. L. Ingraham and E. Klecker. 1974. Mutants of *Salmonella typhimurium* defective in cytidine monophosphate kinase (*cmk*). J. Bacteriol. 120:1370.

Bonhoeffer, F. and H. Schaller. 1965. A method for selective enrichment of mutants based on the high UV sensitivity of DNA containing 5-bromouracil. Biochem. Biophys. Res. Commun. 20:93.

Broda, P., J. R. Beckwith and J. Scaife. 1965. The characterization of a new type of F-prime factors in *E. coli* K-12. Genet. Res. 5:489.

Bukhari, A. I., J. A. Shapiro and S. L. Adhya (eds.). 1977. *DNA: Insertion Elements, Plasmids and Episomes*. Cold Spring Harbor Laboratory, Cold Spring Harbor, New York.

Casadaban, M. J. and S. N. Cohen. 1979. Lactose genes fused to exogenous promoters in one step using a Mu-*lac* bacteriophage: *in vivo* probe for transcriptional control sequences. Proc. Natl. Acad. Sci. USA 76:4530.

Clewell, D. B. 1981. Plasmids, drug resistance, and gene transfer in the genus *Streptococcus*. Microbiol. Revs. 45:409.

Crick, F. H. C., L. Barnett, S. Brenner and R. J. Watts-Tobin. 1961. General nature of the genetic code for proteins. Nature 192:1227.

Davis, B. D. 1948. Isolation of biochemically deficient mutants of bacteria by penicillin. Amer. Chem. Soc. J. 70:4267.

DeLucia, P. and J. Cairns. 1969. Isolation of an *E. coli* strain with a mutation affecting DNA polymerase. Nature 224:1164.

Demerec, M., E. A. Adelberg, A. J. Clark and P. E. Hartman. 1966. A proposal for a uniform nomenclature in bacterial genetics. Genetics 54:61.

Folk, W. R. and P. Berg. 1971. Duplication of the structural gene for glycyl-transfer RNA synthetase in *Escherichia coli*. J. Mol. Biol. 58:595.

Griffith, F. 1928. Significance of pneumococcal types. J. Hyg. 27:113.

Grunstein, M. and D. S. Hogness. 1975. Colony hybridization: a method for the isolation of cloned DNA's that contain a specific gene. Proc. Natl. Acad. Sci. USA 72:3961.

Guerola, N., J. L. Ingraham and E. Cerdá-Olmedo. 1971. Induction of closely linked multiple mutations by nitrosoguanidine. Nature 230:122.

Hartman, P. E., Z. Hartman and D. Serman. 1960. Complementation mapping by abortive transduction of histidine-requiring *Salmonella* mutants. J. Gen. Microbiol. 22:354.

Hill, C. W., J. Foulds, L. Soll and P. Berg. 1969. Instability of a missense suppressor resulting from a duplication of genetic material. J. Mol. Biol. 39:562.

Holloway, B. W. 1979. Plasmids that mobilize bacterial chromosomes. Plasmid 2:1.

Horiuchi, T., S. Horiuchi and A. Novick. 1963. The genetic basis of hypersynthesis of β-galactosidase. Genetics 48:157.

Ippen, K., J. Shapiro and J. Beckwith. 1971. Transposition of the *lac* region to the *gal* region of the *Escherichia coli* chromosome: Isolation of λ*lac* transducing bacteriophages. J. Bacteriol. 108:5.

Jaskunas, S. R., L. Lindahl, M. Nomura and R. R. Burgess. 1975. Identification of two copies of the gene for the elongation factor EF-Tu in *E. coli*. Nature 257:458.

Kleckner, N., J. Roth and D. Botstein. 1975. Genetic engineering *in vivo* using translocatable drug-resistance elements. New methods in bacterial genetics. J. Mol. Biol. 116:125.

Lederberg, J. and E. M. Lederberg. 1952. Replica plating and indirect selection of bacterial mutants. J. Bacteriol. 63:399.

Mandel, M. and A. Higa. 1970. Calcium-dependent bacteriophage DNA infection. J. Mol. Biol. 53:159.

Meagher, R. B., R. C. Tait, M. Betlach and H. W. Boyer. 1977. Protein expression in *E. coli* minicells by recombinant plasmids. Cell 10:521.

Neidhardt, F. C., R. Wirth, M. W. Smith and R. VanBogelen. 1980. Selective synthesis of plasmid-coded proteins by *Escherichia coli* during recovery from chloramphenicol treatment. J. Bacteriol. 143:535.

Old, R. W. and S. B. Primrose. 1981. *Principles of Gene Manipulation*, 2nd Edition. University of California Press, Berkeley and Los Angeles.

Press, R. N., P. Glansdorf, P. Miner, J. DeVries, R. Kadner and W. K. Maas. 1971. Isolation of transducing particles of φ80 bacteriophage that carry different regions of the chromosome of *E. coli* K-12 genome. Proc. Natl. Acad. Sci. USA 68:795.

Riley, M. and A. Aniliones. 1978. Evolution of the bacterial genome. Ann. Rev. Microbiol. 32:519.

Sancar, A., A. M. Hack and W. D. Rupp. 1979. Simple method for identification of plasmid-coded proteins. J. Bacteriol. 137:692.

Schmieger, H. 1972. Phage P22 mutants with increased or decreased transduction abilities. Molec. Gen. Genet. 119:75.

Shimada, K., R. A. Weisberg and M. E. Gottesman. 1972. Prophage lambda at unusual chromosome locations. 1. Location of the secondary attachment sites and the properties of lysogens. J. Mol. Biol. 63:483.

Smith, H. O., D. B. Danner and R. A. Deich. 1981. Genetic transformation. Ann. Rev. Biochem. 50:41.

Speyer, J. F. 1965. Mutagenic DNA polymerase. Biochem. Biophys. Res. Comm. 61:410.

Zinder, N. D. and J. Lederberg. 1952. Genetic exchange in *Salmonella*. J. Bacteriol. 64:679.

Chapter Five

Growth of Cells and Cultures

INTRODUCTION

The coordinated summation of the biosyntheses described in Chapters 2 and 3 is termed GROWTH; its consequence is the production of new cells. The entire process of synthesizing a cell of *E. coli* from glucose and mineral salts, as occurs during growth in a minimal medium, takes only 40 minutes; and can be accomplished in half that time in a rich medium. Although some procaryotes grow quite slowly, requiring hours for each cell to synthesize a new one, rapid growth is a highly selected property in bacteria, particularly in those that exist in environments that are only transiently capable of supporting bacterial growth.

As observed under the microscope, the growth of a rod-shaped cell like *E. coli* appears quite simple. A freshly divided cell elongates with little or no increase in girth. Eventually a transverse wall is laid down near the center of the cell; when the cell reaches approximately twice its original size, it separates into two cells nearly equal in size. The products of the division are physiologically the same; each is equally young. Unlike yeast cells, which are able to produce only a certain number of buds (approximately 20) before they die, the concept of age of a bacterial cell applies only to its state of progress through the interdivision cycle. If the environment remains favorable, bacterial cells are capable of unlimited growth and division.

There is considerable variation among bacteria with respect to the manner in which the cross wall is synthesized, the distribution of cellular material to the daughter cells, and the pattern of association of the products of division (Stanier, 1976). Some replicate by budding rather than transverse fission; some (e.g., *Caulobacter*) produce morphologically distinct products of division; some produce cells aggregated in distinctive arrays.

In spite of the variations among bacteria in details of the division process, the growth of a population of bacterial cells (termed a CULTURE) dispersed in a liquid medium follows certain rules.

UNRESTRICTED GROWTH

If a bacterial culture is grown in a liquid medium in which the concentration of all nutrients (regardless of their qualitative value in supporting rapid growth) is sufficient to allow a maximum rate of growth in that particular medium—that is, if growth rate is not determined by the concentration of any component of the medium but rather by the overall composition of the medium—growth is said to be UNRESTRICTED. During unrestricted growth, the rate of increase of cells in the culture is proportional to the number of cells present at any particular time. In other words, unrestricted bacterial growth mimics an autocatalytic, first-order, chemical reaction—a response that might be anticipated from the fact that bacterial cells are self-duplicating units. Stating this in mathematical terms, one obtains the equation

$$dN/dt = kN \tag{1}$$

in which N is the concentration of cells (number of cells per unit volume), t is time, and the constant of proportionality k is called the SPECIFIC GROWTH RATE CONSTANT. Rearranging Equation (1) to the form

$$dN/N = kdt \tag{2}$$

one notes that the dimensions of N cancel; the dimensions of SPECIFIC GROWTH RATE k are reciprocal time, usually expressed as reciprocal hours, hr^{-1}.

Upon integration of Equation (1) between the limits of 0 and t and N_0 and N

$$\int_{N_0}^{N} dN/N = k \int_{0}^{t} dt \tag{3}$$

gives

$$\ln N - \ln N_0 = kt \tag{4}$$

and, converting to common logarithms,

$$\log_{10}N = kt/2.303 + \log_{10}N_0 \tag{5}$$

Equation (5) describes a straight line ($y = ax + b$). By plotting $\log_{10}N$ against t, one obtains a straight line that intercepts the ordinate at $\log_{10}N_0$ and has a slope of $k/2.303$.

If the antilogarithm of Equation (5) is taken, one obtains an exponential form:

$$N/N_0 = 10^{kt/2.303} \tag{6}$$

and, as a result, cultures in a state of growth that obey Equation (5) are said to be in the EXPONENTIAL PHASE OF GROWTH.

In practice, if one determines the number of bacterial cells present in a growing culture at various times and plots these data in the form indicated by Equation (4), a straight line is, indeed, obtained. From the slope of such a plot, one can calculate the specific growth rate k of the culture, which is a complete and reproducible description of how fast a particular bacterium grows in a particular environment; in practice and with the best modern cell counters, it can be determined with a precision of approximately 0.3%.

Other indices of growth rate are also commonly employed. Because the logarithm of the number of cells in a culture is a linear function of time, the number of cells always increases by the same factor during any given interval of time; hence, the time required for a culture to increase by a certain factor is an index of the rate of growth of a culture. The time required for the number of cells to increase by a factor of 2 is a commonly used index of growth rate, and is called the DOUBLING TIME, or g. Referring to Equation (4), t becomes equal to g when N_0 has increased to $2N_0$. Substituting these values for t and N into Equation (4), one obtains

$$\ln 2N_0 - \ln N_0 = kg \tag{7}$$

or

$$\ln 2N_0/N_0 = \ln 2 = kg \tag{8}$$

and doubling time can be defined as

$$g = \ln 2/k = 0.693/k \tag{9}$$

Doubling time is inversely related to the rapidity with which a culture grows; its reciprocal, termed μ (or doubling times per hour), is another commonly used index of the rate of growth of the culture. Thus,

$$\mu = 1/g = k/0.693 \tag{10}$$

Values of g can be determined by measuring the concentration of cells in a culture at various times, plotting the logarithms of the concentrations versus time, and reading the time required for the number of cells to double. Values of μ and k can be calculated from Equations (10) and (5).

Balanced growth

The question arises whether the same values of k would be obtained if measurements of components of the biomass of the culture other than cell concentration were made, for example, bacterial mass, bacterial protein, or bacterial DNA. This would be true if the average composition of the cells did not change—a condition termed BALANCED GROWTH, which Campbell (1957) defined as occurring over intervals of time if "every extensive property of the growing system increases by the same factor." Rigorous proof of the state of balanced growth according to the Campbell definition is, of course, impossible. However, there is a considerable body of data that is consistent with the Campbell definition, and most microbial physiologists accept that if sufficient care is taken to maintain constant environmental conditions the requirements of Campbell's all-inclusive definition can be met (see Chapter 6). Assuming balanced growth, Equation (1) can be extended to the form:

$$\left(\frac{dN}{dt}\right)\left(\frac{1}{N}\right) = \left(\frac{dx}{dt}\right)\left(\frac{1}{x}\right) = \left(\frac{dy}{dt}\right)\left(\frac{1}{y}\right) = \left(\frac{dz}{dt}\right)\left(\frac{1}{z}\right) = k \tag{11}$$

in which x, y, and z represent any constituent of the cell: protein, nucleic acid, carbohydrate, salts, water, etc. Similarly, Equation (5) can be extended so that a measurement of any extensive property of the cell can be substituted for cell numbers.

Balanced growth also requires that mean cell size (an extensive property of the growing system) remain constant, a fact that might at first glance appear paradoxical because during balanced growth individual cells grow by increasing in size and then dividing. However, the fact that the population shows balanced growth implies nothing about the growth of individual cells.

Measurements of growth

Because the rate of growth of bacterial cultures can be determined from measurements of any extensive property of the culture, the choice of measurement is determined by the ease and the accuracy with which it can be made, provided the measurement is compatible with the cultural condition employed.

The following procedures are some of the many that can be used to measure the growth of a bacterial culture: (1) counting viable cells; (2) counting the total number of cells; (3) determining the dry weight of cells; (4) estimating the weight of cells from their light scattering properties; (5) measuring some cellular constituent, for example, DNA, RNA, protein, or peptidoglycan.

Counting viable cells can be based on any property of live cells, for example, motility or resistance to staining, but usually colony-forming ability is the property used. One dilutes the culture appropriately, disperses a sample of the dilution on a solid medium, and counts the number of colonies that develop. In this determination there are errors that are difficult to avoid. An important error is associated with the determination of the time of sampling the culture. Although sampling time can be accurately determined, the culture will continue to grow at an unknown rate during the period of dilution in preparation for plating. There is no easy way to stop growth during the dilution period without severe risk of killing appreciable numbers of cells. As we will see later, diluting a culture into a medium that cannot support continued growth does not immediately stop the increase in cell numbers, and rapid chilling of the culture may kill significant numbers of the cells. Therefore, the effective sampling time is difficult to determine.

In addition to the error associated with events subsequent to sampling, there is an actual sampling error, that is, the number of bacteria in equal-sized samples differs by chance because the organisms are randomly distributed in suspension. The standard deviation (assuming no technical mistakes are made) is the square root of the total number of colonies counted.

Although the procedure is more time consuming, the number of viable cells in a culture also can be determined by the DILUTION END POINT method. The count thus obtained is usually termed the MOST PROBABLE NUMBER (MPN) because it is statistically derived. In this procedure, the culture is diluted appropriately so that a given volume of the dilution has a significant probability of not containing a viable cell. This volume is then added to a series of tubes containing a growth-supporting medium and incubated. If those tubes that remain clear are assumed to have received no viable cells and those that become turbid are assumed to have received one or more viable cells, the average number of viable cells (m) in the volume of the inoculum can be calculated from a simplified form of the Poisson distribution,

$$p(0) = e^{-m} \tag{12}$$

where $p(0)$ is the proportion of the tubes remaining clear. This procedure is usually employed when optimal growth conditions are more easily attained in liquid culture media or in closed culture tubes.

Most bacterial cultures contain a fraction—usually quite small—of dead cells. If growth is balanced, that fraction remains constant. Thus, the rate of increase of the total number of cells in a population is a valid index of growth rate. The total number of cells in a culture is usually determined microscopically, by counting the number of cells present in a COUNTING CHAMBER of known volume, or electronically, in a device called a COULTER COUNTER.

A counting chamber is a glass slide with a central depression of known depth, the bottom of which is ruled into squares of known area. The depression is filled with a bacterial suspension, covered with a ridged cover slip, and allowed to stand until the cells have settled to the bottom. The number of cells per unit area is counted using a microscope, and the total number of cells in the suspension is calculated.

Electronic counting is based on the principle that the electric conductivity of a bacterial cell is less than that of a saline solution. The device consists essentially of two chambers connected through a small pore (for counting bacteria, the pore is approximately 30 micrometers in diameter). Each chamber is provided with an electrode so that the electrical conductivity of the system can be monitored. In order to count bacteria, they are added to one of the chambers and pumped slowly into the other. Each time a bacterium passes through the pore, the conductivity of the circuit decreases momentarily and a pulse in voltage results; the number of pulses can be tallied electronically. Thus, by knowing the volume of suspension pumped through the pore, one has a rapid determination of the concentration of bacteria in that suspension. The size of the voltage pulse is proportional to the size of the bacterium traversing the pore; therefore, with proper electronic circuitry, the size distribution as well as the number of bacteria can be determined (Chapter 6, Figure 2). The device counts rapidly and accurately, but it is also expensive and subject to a number of artifactual complications. The pore can easily become plugged if the media and diluents are not carefully prepared.

The most direct way of measuring the dry weight of cells in a culture is to remove them by filtration or centrifugation, wash them with distilled water to remove adhering components of the medium, dry them at approximately

105°C, and weigh them. Although such measurements are accurate, they are time consuming and require the sacrifice of considerable culture volume. They are not suitable for routine monitoring of a growing culture.

The light scattering property of bacterial cells provides a rapid, simple method of estimating the dry weight of cells in a culture. It is the most commonly employed technique for following the growth of bacterial cultures. If a parallel beam of light is passed through a suspension of bacteria, it is diminished in intensity to an extent that is the sum of the light absorbed by the bacteria and that scattered by them. Absorption of visible light by bacteria is usually negligible because most bacteria are uncolored. Therefore, changes in light transmission through a bacterial suspension is determined largely by scattering.

The amount of light scattered is proportional to the ratio of particle size to wavelength of incident light. Therefore, in measurements of bacterial suspensions by scattering techniques, the measurement becomes more sensitive at shorter wavelengths; measurements are usually made between 420 and 660 nm. Two sorts of measurements can be made. One can measure the unscattered or the scattered light. If the scattered light is measured, the photodetection device is located at an angle other than 180° (usually 90°) to the incident light; the device is called a NEPHELOMETER. More commonly, however, the unscattered light is measured in an ordinary spectrophotometer (which has the photodetection device located at 180° to the incident beam) because of the general availability of these instruments. A spectrophotometer is an instrument designed to measure absorbance (A), that is, the logarithm of the ratio of the incident (I_0) to the transmitted (I) light through nonparticulate solutions. A is related to the concentration of the solution by the Lambert-Beer law:

$$\log (I_0/I) = \epsilon \ell c = A \tag{13}$$

where ϵ is the extinction coefficient, ℓ is the length of the light path in the liquid, and c is the concentration of the solute. Thus, knowing ϵ and ℓ and measuring A on any spectrophotometer, the concentration (c) of the solute can be determined. However, bacteria are in suspension, not in solution, and measuring A [or, as it is often called, OPTICAL DENSITY (OD)] of a bacterial suspension is not a direct measure of bacterial concentration. In fact, because scattering contributes so significantly to the determination, the measured value of A depends on the precise geometry of the instrument on which the measurement is made. The value of A of a bacterial suspension measured on one instrument will not be the same as that measured on an instrument with a different geometry. The particular instrument must be calibrated for the particular bacterium being studied by directly determining the dry weight of bacteria in a dense suspension and by measuring its absorbance A as well as the absorbance of known dilutions of the suspension. The results of such a calibration are shown in Figure 1. At low densities A is roughly proportional to the dry weight of the culture, but at higher densities there is significant deviation from linearity. Such deviation is a consequence of double scattering, that is, at high culture densities the probability of a scattered ray of light

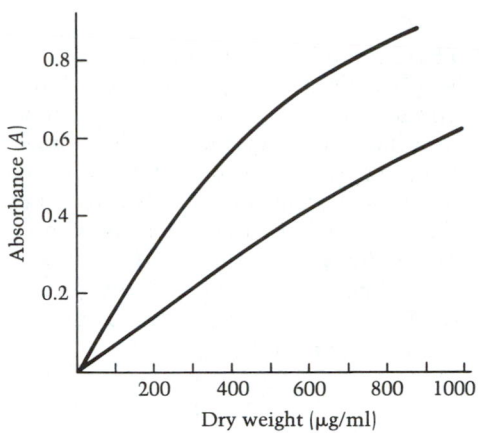

Figure 1.

Relationship between the absorbance (A) and biomass (dry weight) of a culture of *E. coli*. A was measured at 420 nm (upper curve) and 650 nm (lower curve). The relationship is linear only at low values of A.

being scattered back so that it strikes the photodetection system is increased. Once a calibration curve has been made, a rapid and reliable determination of dry weight of the culture can be made simply by measuring A and reading dry weight from the curve.

Commonly, uncorrected A measurements are used directly to follow the growth of a culture. Because such measurements are only reliable at low culture densities, it is necessary to dilute the culture to a value of A that lies in the linear range of the A versus dry weight plot. This value is then multiplied by the dilution factor.

Chemical measurement of any cellular constituent can also be used to follow the growth of a bacterial culture; measurements of protein, DNA, and wall constituents are most frequently employed, usually when cells tend to clump or settle rapidly, circumstances in which measurements of absorbance are unreliable.

Chemical constituents of the cell can also be estimated by determining the incorporation of radiolabeled components of the medium into cellular material, for example, incorporation of $^{35}SO_4$ or radiolabeled amino acids to estimate protein, $^{32}PO_4$ to estimate nucleic acid, or radiolabeled thymine to estimate DNA. If the radiolabeled component has been present in the medium for several previous generations, the amount incorporated into cellular material can be used directly to determine growth rate by plotting the logarithm of the amount of radioactivity in the cells versus time as would be done in any other method of calculating growth rate. However, for practical reasons, the radiolabeled component is usually added only after the culture has reached a significant density. Under these conditions, incorporated radioactivity is an index of the amount of the labeled constituent synthesized after the radiolabeled material was added, not of the total amount of that material. Under these conditions Equation (11) becomes

$$dz/dt = k \, (z^* + z_0) \tag{14}$$

where z^* is incorporated label in cellular component z, and z_0 is the amount

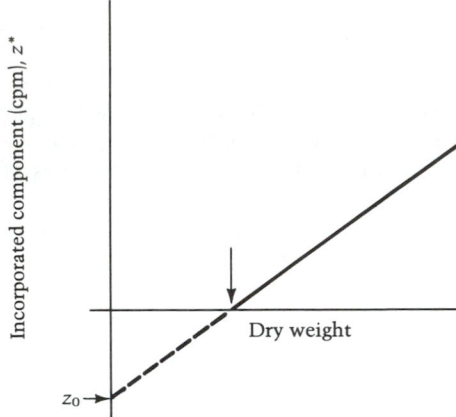

Figure 2.

Differential plot of incorporation of a radioactive component of the medium into cellular material. By plotting the counts per minute incorporated into the biomass (z) versus the mass of the culture (dry weight) a straight line is obtained, which when extrapolated to 0 mass gives an estimate of z_0, the amount of the cell component (measured as radioactivity) present when the radioactive component was added to the medium (↓).

of it in the biomass when the label was added to the medium. Hence, a plot of log z^* versus time does not yield a straight line but rather a curve of decreasing slope. The value of z_0 can be estimated from a DIFFERENTIAL PLOT in which z^* is plotted against mass of the culture (Figure 2). The extrapolation of this curve to 0 mass gives z_0 as a negative intercept. Plotting log ($z^* + z_0$) versus time yields a straight line with slope equal to $k/2.303$. By chemical analysis, the amount of component z can be related to counts of radioactivity.

Phases of growth

Unrestricted growth of a bacterial culture cannot continue indefinitely. The growing cells use nutrients from the medium and excrete certain waste products into it. Eventually some nutrient becomes exhausted or some waste product becomes toxic so that continued balanced growth is no longer possible; the culture stops growing and is said to enter the STATIONARY PHASE OF GROWTH. The transition from the exponential phase of growth to the stationary phase of growth is surprisingly abrupt when the culture depletes a limiting nutrient from a defined medium. However, in the case of a culture growing in a complex medium, declining O_2 tension and/or a shift to an unfavorable pH usually limits growth, causing a longer transition phase during which growth rate gradually declines. As was discussed earlier (Chapter 3), bacterial cells have elaborate mechanisms for maintaining a constant intracellular environment in spite of continually changing concentrations of nutrients in the external environment. Only when the limits of these mechanisms have been reached is the growth rate affected.

During the transition from the exponential phase of growth to the stationary phase, growth becomes unbalanced. The synthesis of certain components of the cell ceases before others.

Therefore, cells in the stationary phase of growth always differ both chemically and morphologically from cells in the exponential phase of growth, but the difference between them in terms of relative abundance of individual proteins depends on the particular nutrient that has been exhausted. However, stationary-phase cells are always smaller than those in the exponential phase, because cell division continues (see Chapter 6) after the synthesis of most macromolecules has slowed markedly.

If a culture in the exponential phase of growth is transferred to fresh medium of the same composition, exponential growth continues. On the other hand, transfer of a culture in the stationary phase of growth to fresh medium (of the same composition) is not followed immediately by exponential growth. Growth is delayed for a variable period of time depending on (1) the temperature, (2) the particular nutrient or toxic product that limited growth in the original culture, (3) the concentration of cells in the stationary-phase culture, and (4) the time during which the culture was held in the stationary phase of growth. Initially growth is unbalanced while the composition and size of the cells are adjusting to that characteristic of the exponential phase of growth. The period of time before exponential growth begins is called the LAG PHASE. The length of the lag phase is computed by extrapolating the curve of the exponential phase to the level of the measured component at the time of inoculation (Figure 3).

One might expect that the lag phase of growth is simply a period of unbalanced synthesis during which the chemical composition of cells necessary for exponential growth is reestablished. This correction of chemical composition does take place during the lag phase, but other factors play a major role in establishing the duration of the lag phase. For example, if a culture of *E. coli* is grown to stationary phase in a mineral-salts medium with a low concentration of glucose (around 0.02%), it will reinitiate growth (as measured by mass increase) with virtually no lag phase when transferred to a fresh medium after having been held in the stationary phase for periods up to several hours. If held for several days in the stationary phase, a lag time of several hours might result. On the other hand, if the glucose concentration were high (approximately 0.2%) in the initial culture, a lag period would result after only a brief holding period in the stationary phase, presumably because of the higher accumulation of metabolic by-products—principally acetate in the case of *E. coli*.

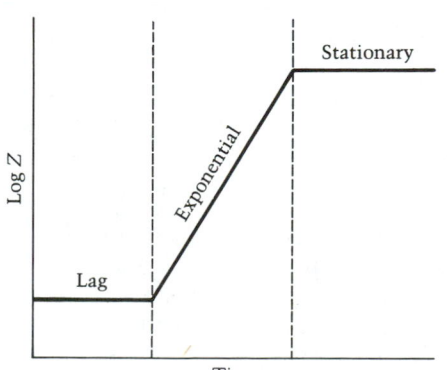

Figure 3.

Growth curve of a bacterial culture showing the various phases: lag, exponential, and stationary. The times of transition between the phases varies with the particular parameters of biomass (z) measured.

ABERRANT (LINEAR) GROWTH

As has been discussed, exponential growth is the normal consequence of the constantly increasing catalytic capacity of a culture of growing bacteria. A number of factors can intervene that arrest the increase in catalytic capacity without stopping growth. In some cases the aberrant growth continues for many hours. For example, if the amino acid analogue *p*-fluorophenylalanine, is added to a culture, it becomes incorporated in place of phenylalanine into newly synthesized protein. To a large extent such protein is not catalytically active. Thus, all cellular components continue to be synthesized, but the catalytic capacity to make new cells is fixed at the level present when the analogue was added to the culture. Growth under such conditions becomes LINEAR (or arithmetic), the rate of increase in mass being set by a fixed catalytic capacity, that is,

$$dx/dt = C \qquad\qquad (15)$$

where x is dry weight, t is time, and C is a constant. On integration of Equation (15), one obtains

$$x = Ct \qquad\qquad (16)$$

which predicts a linear increase of mass with time (Figure 4).

A number of other treatments cause arithmetic growth. Depriving an aerobic culture of iron fixes the active cytochrome content and hence ATP-generating capacity of the culture. Depriving a thiamin auxotroph of this vitamin fixes the pyridine nucleotide content of the culture. Shifting a conditionally expressed mutant of the class termed TEMPERATURE-SENSITIVE SYNTHESIS to nonpermissive temperature fixes the level of the affected gene product at the time of the shift.

YIELD COEFFICIENT

If we assume that the culture depicted in Figure 3 stopped growing because a component of the medium had been depleted, we could

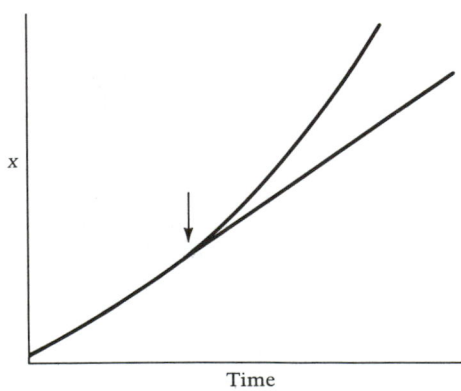

x

Time

Figure 4.

Linear growth of a bacterial culture. At the time indicated by the arrow (↓), the culture was treated (see text) in a manner that prevented further increase in the catalytic capacity of the culture, thus causing linear growth (increase in mass, x) to ensue. The upper line indicates the form of exponential growth that would have been obtained had the culture not been treated. Note that the slope of the line describing linear growth is set by the catalytic capacity of the culture at the time of treatment.

predict that the final yield of biomass would have been decreased by a factor of ½ if only ½ the concentration of limiting nutrient had been present in the medium. This relationship holds for almost all nutrients over wide ranges of concentration; that is, a medium that contains all nutrients in excess, except for one, supports a final crop of bacteria that is directly proportional to the concentration of the limiting nutrient. This relationship between the final crop of bacterial cells and the concentration of the limiting nutrient can be stated as

$$x_f - x_0 = Y(c_0 - c) \tag{17}$$

in which x_f is the final concentration of bacteria; x_0 is the initial concentration of bacteria; c_0 is the concentration of the limiting nutrient initially present in the medium; c is the concentration of the limiting nutrient when growth stops; and Y is the yield coefficient. Y is unitless, being the ratio of dry weight of cells produced per unit dry weight of limiting nutrient. The value of Y is usually approximately 0.5 for the source of carbon and energy—say, glucose, for an aerobe—and may be 100 or greater for a required amino acid or vitamin.

The constancy of the yield coefficient provides a relationship that can be used to determine the concentration of a particular nutrient in a particular medium. For example, an organism that requires methionine for growth could be used to assay the concentration of methionine in any unknown mixture. A medium that supplies all nutrients in excess except methionine is prepared. Then the yield coefficient for methionine is determined by adding various known amounts of methionine to the medium. By knowing the yield coefficient and the amount of growth that one obtains upon adding a certain amount of an unknown mixture to the methionine-deficient medium, one can calculate the amount of methionine in the unknown mixture. This procedure has been called a BIOASSAY and was used extensively during the 1940s and 1950s to determine the concentration of amino acids, vitamins, and other growth factors; to determine the potency of vitamin preparations; and to determine the nutritional value of foodstuffs. The procedure was also used extensively as an aid in isolating new growth factors. Today, with the availability of better physical and chemical methods for the separation and determination of nutrients, the bioassay technique is not as extensively employed, but it remains a valuable research tool in the study of biosynthetic intermediates.

The yield coefficient can also be used to deduce the probable pathways of bacterial fermentation because the ATP requirements for polymerization of macromolecules are remarkably constant among bacteria (Chapter 3). Thus, under proper conditions, cell yield is an index of ATP generation, which varies among different fermentations. Bauchop (1960) noted that if a culture medium contains all (or almost all) the monomers required for the synthesis of macromolecules, the molar yield (Y_m, grams cells produced per mole substrate utilized) is set by the number of ATP molecules produced per molecule of substrate utilized. They found that MOLAR ATP YIELD (Y_{ATP}, grams cells produced per mole ATP generated) measured under these conditions was roughly constant among microorganisms and was equal to 10.5. Thus, the measurement of Y_m can be used to verify the suspected pathway of fermentation of a substrate by a microorganism. For example, *Streptococcus faecalis* is thought to metabolize ribose to lactate and acetate via the phosphoketolase pathway, which would generate 2 moles of ATP per mole of substrate. The experimentally determined Y_m is 20, a result that indicates a production of 2.0 moles of ATP per mole of substrate ($20/10.5 = 1.9$) and is consistent with the hypothesis that the phosphoketolase pathway is the major route of metabolism of ribose by the organism.

In practice, measurements of Y_{ATP} are restricted to studies on cultures generating ATP by fermentation, because these are the only conditions for which the ATP yield per mole of substrate is known with precision. Much less is known about the energetics of aerobic growth, but recently Andersen (1980) made an interesting discovery that appears to be fundamental. He found that Q_{O_2} (millimoles oxygen consumed per hour per milligram dry weight) of batch cultures of *E. coli* does not vary significantly with growth rate (Figure 5). The values for growth with mannose as carbon source are low and those with pyruvate or lactate are high. But these are exceptions [for which plausible rationalizations are presented (Andersen, 1980)]; there is no pattern of change of Q_{O_2} with growth rate.

If we assume that ATP generated per mole of O consumed (P/O ratio) does not vary with growth rate (and there is no experimental or theoretical basis for questioning it), we are led to an interesting conclusion: *The total energy available for doubling the cell mass increases in proportion to doubling time (g).*

The energy required for growth can be subdivided into that required to

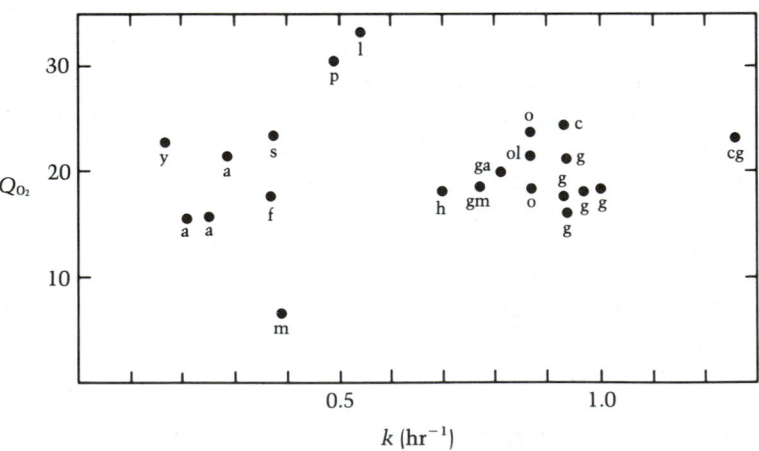

Figure 5. Effect of growth rate (k) on the specific oxygen consumption (Q_{O_2}) of batch cultures of *E. coli* B/r. Cultures were grown in a mineral salts medium with various carbon sources. Letters indicate the particular carbon source(s) employed: a, acetate; c, casein hydrolysate; f, fumarate; g, glucose; h, galactose; l, lactate; m, mannose; o, glycerol; p, pyruvate; s, succinate; y, glycolate. The units of Q_{O_2} are mmol/hr/mg dry weight of cells. (After Andersen, 1980.)

synthesize monomers, that required to polymerize monomers into macro-molecules, and maintenance energy (see later in this chapter). Energy required for maintenance is small and for present considerations can be neglected; the energy required for polymerization varies little with macromolecular com-position; so, with good approximation, the amount needed per doubling time can be taken to be constant. Thus, energy required to synthesize monomers is greater at low growth rates. This implies that the amount of energy needed to synthesize monomers from the available carbon source can only be provided if g exceeds by an appropriate amount the value necessary to sustain poly-merization. At least under the growth conditions of Andersen's experiments, growth rate appears to be set by the energy requirement for synthesis of monomers. It is not yet clear how general this relationship might be.

EFFECT OF CONCENTRATION OF NUTRIENTS ON RATE OF GROWTH

A reexamination of Figure 3 shows that a bacterial culture grows at a constant rate over the majority of the growth cycle in spite of the fact that substrate is constantly being consumed as the culture grows. The growth rate remains constant and then rather abruptly decreases to zero when the substrate is depleted.

Therefore, growth rate is a function of substrate concentration only at very low substrate concentrations. There are difficulties in measuring the relationship between growth rate and substrate concentration, because if the concentration of substrate is low enough to affect growth rate it will be rapidly utilized by an only moderately dense population of bacteria. Monod (1942) attempted measurements of this relationship by two techniques. (1) He used a nephelometer to measure growth in dilute cultures in which the concentra-tion of one nutrient was low enough to limit the growth rate. (2) He made estimates of the concentration of substrate as a bacterial population ceased to grow as a result of the limitation of one substrate; tangents to the growth curve (at times corresponding to those at which substrate determinations were made) were used to estimate growth rate. From his measurements, Monod proposed that the relationship between growth rate and substrate concentra-tion followed first-order kinetics, where some component of the system be-comes saturated (limiting) at low concentrations of substrates. It has the form

$$k = k_{max}c / (K_s + c) \tag{18}$$

in which k is the specific growth rate at a concentration c of limiting nutrient. The parameter k_{max} is specific growth rate at saturating concentrations of the nutrient; K_s is a constant numerically equal to the substrate concentration at which $k = k_{max}/2$. If growth rate is measured during exponential growth of a batch culture (unrestricted growth), as is discussed earlier in this chapter, the estimate of growth rate that one obtains is a measurement of k_{max}.

The form of the relationship that Monod found to exist between growth rate and substrate concentration is the same as the familiar relationship between the velocity of an enzyme-catalyzed reaction and the concentration of the substrate for the reaction, that is, the Michaelis-Menten equation:

$$v = v_{max} \, S/(K_m + S) \tag{19}$$

where v is the velocity of the reaction, S is the concentration of substrate, and K_m is the Michaelis constant.

Owing to the difficulties of making measurements of bacterial growth rate at low concentrations of substrate, the form of Equation (18) (which is so very important to studies of continuous culture of bacteria) has been questioned by some investigators. Shehata (1970) succeeded in measuring the specific growth rate of *E. coli* in sustained balanced growth with glucose as a limiting nutrient at concentrations as low as 10^{-7} *M*. These measurements were made on cultures kept below a density of 10^5 cells per ml by periodic dilution, and cell numbers were recorded by means of an electronic counter. Under these conditions, a significant increase in numbers of bacteria can be followed with only a minimum effect on substrate concentration. (Analyses have been made at even lower cell densities by the laborious procedure of counting colonies on agar plates; Novick, 1950.)

Figure 6 (insert) shows that saturation-type kinetics is obtained. As a test of the Monod equation, the reciprocal of growth rate is plotted against the reciprocal of substrate concentration as indicated by a rearranged form of Equation (18), that is,

$$1/k = (K_s/k_{max}) \, (1/c) + 1/k_{max} \tag{20}$$

which predicts that a plot of $1/k$ versus $1/c$ is a straight line, the slope of which is K_s/k_{max}, with ordinate intercept $1/k_{max}$. It can be seen (Figure 6) that the resulting plot is, indeed, linear over wide ranges of substrate concentration, thus proving the validity of the Monod equation and giving an estimate of K_s (6.4×10^{-7} *M*). Only at high substrate concentrations (low values of $1/c$) does the function depart significantly from linearity, an observation that indicates that at high substrate concentrations a second pathway for glucose utilization probably plays a significant role.

CONTINUOUS CULTURE OF BACTERIA

Continuous culture of bacteria, that is, maintenance of a culture in constant state of growth, can be readily accomplished in the sort of device sketched in Figure 7. One inoculates a vessel containing an appropriate medium with a culture of bacteria; after the culture has grown to a moderate density, fresh (sterile) medium is fed to it continuously and the volume of the medium in the culture vessel is kept constant by removing medium (containing cells) from the vessel at the same rate at which fresh medium is supplied to it. Vigorous mixing must be employed to ensure rapid mixing of the fresh medium with the vessel contents. Surprisingly, fresh medium can be added to the vessel over a wide range of flow rates, and the culture within the vessel maintains a constant density; that is, the rate of growth of the culture is just sufficient to balance the rate of loss of cells through the overflow. If the rate of addition of fresh medium from the reservoir is changed (the overflow rate follows this change precisely), the culture density may change slightly but

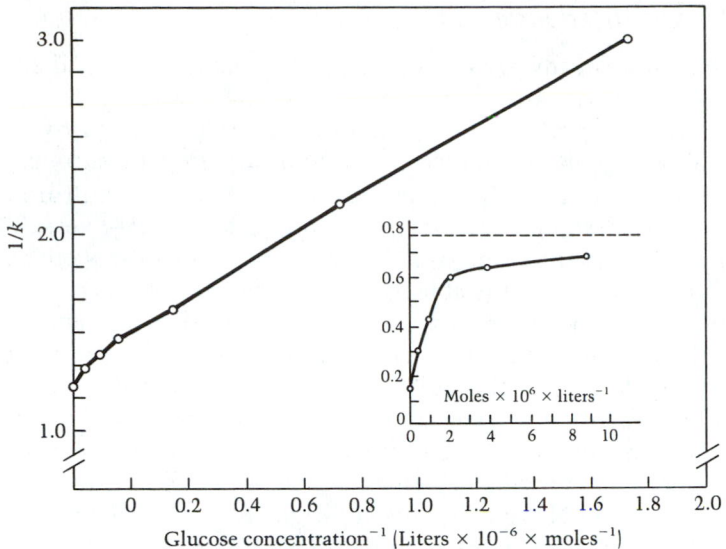

Figure 6. Effect of limiting concentrations of glucose on the growth rate of *E. coli*. In the insert, growth rate is plotted against substrate concentration; this method yields a hyperbolic curve. A plot of the inverse of growth rate $(1/k)$ against the inverse of substrate concentration (outer curve) yields a straight line, the slope of which is K_s/k_{max} [see Equation (20)]. (After Shehata, 1970.)

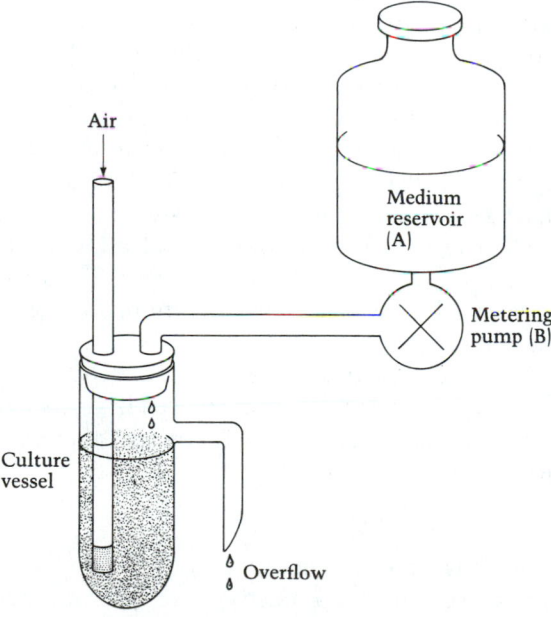

Figure 7. Schematic diagram of an apparatus for continuous culture of bacteria. Fresh medium from a reservoir is fed at a constant rate to the culture vessel through a metering pump. A constant volume of culture is maintained by means of an overflow through which culture passes out at the same rate at which fresh medium is added.

241

again a new steady state will be quickly established and the density of the culture again remains constant—that is, by changing the rate of addition of the fresh medium, the rate of growth of the culture in the vessel adjusts so that it again exactly balances the rate of loss of cells through the overflow.

The rate of growth of a culture in a continuous culture device seems to be set by the rate of addition of medium to the culture vessel. Of course there are limits. If the medium is added at a rate that requires growth of the culture at a higher rate than it is able to attain under conditions of unrestricted growth in the same medium, the culture density will decrease and eventually all the cells will be washed out of the vessel. Also, if the medium is added at an extremely low rate, the culture will sometimes wash out. The reasons for wash out at low flow rates are discussed later in this chapter.

The phenomenon of continuous culture was discovered simultaneously by two groups: Novick (1950) and Monod (1950). Monod called the device a "bactogen" and Novick and his colleague Szilard called their device the "chemostat." The latter name seems to have found more frequent use.

By studying the inset of Figure 6 one can see how the chemostat works. The chemostat operates in a region of the curve in which growth rate (k) varies with concentration of the limiting substrate (c). The system of the chemostat is stable because it is self-correcting: an increase or decrease in rate of addition of medium and of loss of culture through the overflow is matched by a corresponding increase or decrease in growth rate of cells in the culture. To illustrate how the system self-corrects, let us assume that a chemostat is operating in a steady state (culture density remains constant because growth rate is just sufficient to replace cells lost through the overflow) and that the rate of addition of medium is increased. The rate of loss of cells would increase transiently, exceeding their rate of formation in the vessel. As a consequence, the density of the cells in the culture would decrease. The less dense culture would utilize the limiting nutrient at a slower rate, causing the concentration of the nutrient in the vessel to increase. This would cause growth rate to increase (Figure 6) until it was sufficient to match the rate of loss of cells through the overflow. If the rate of addition of medium were decreased, an opposite series of events would ensue. Thus, growth rate of the culture in a chemostat adjusts to match the rate of loss of cells from it; the culture density remains constant.

This relationship can be arrived at mathematically. If V is the volume of culture in the vessel and FLOW RATE (f) is measured in units of culture volumes per hour, the MEAN RESIDENCE TIME (MRT), or the average time that a cell spends in the culture vessel, is

$$\text{MRT} = V/f \tag{21}$$

The reciprocal of MRT is DILUTION RATE (D).

If a chemostat is operating in a steady state, the number (N) or mass (x) of cells removed through the overflow is exactly matched by the number or mass of cells produced in the culture by growth, that is,

$$\frac{\text{rate of production of cells}}{\text{through growth}} = \frac{\text{rate of loss of cells}}{\text{through overflow}}$$

The rate of production of cells through growth of a culture can be described by the equation

$$dx/dt = kx \tag{22}$$

and the rate of loss of cells through overflow is

$$dx/dt = (f/V) \cdot (x) = Dx \tag{23}$$

so Equation (22) becomes

$$kx = Dx \tag{24}$$

or

$$k = D \tag{25}$$

Equation (25) is a restatement of the fact that the growth rate of a culture in a chemostat is determined by the flow rate and the volume of the culture vessel; it is exactly equal to the dilution rate. The equation is a statement of fact and does not help us to understand how the chemostat works. The key to understanding is the fact that at low concentrations of substrate the growth rate of a culture is dependent on the substrate concentration. As the culture grows in the chemostat, the various required nutrients are depleted until one of them limits the growth rate.

As we have established earlier, at low nutrient concentrations the specific growth rate (k) at any particular substrate concentration (c) is related to the maximum growth rate k_{max} (the growth rate if that nutrient is present in excess) by the relationship

$$k = k_{max}c/(K_s + c) \tag{18}$$

Using this expression, Equation (24) becomes

$$xk_{max}c/(K_s + c) = Dx \tag{26}$$

or

$$c = K_sD/(k_{max} - D) \tag{27}$$

thus giving us a fundamental relationship between substrate concentration (c) in the growth vessel and dilution rate (D).

Equation (27) was derived by considering the balance of bacterial dry weight during the operation of a chemostat; we can obtain another important relationship (between cell mass and substrate concentration) by considering the balance of substrate utilization. In the steady state, c remains constant, therefore,

$$\begin{array}{ccc} \text{substrate added from} \\ \text{the reservoir} \end{array} = \begin{array}{c} \text{substrate used} \\ \text{for growth} \end{array} + \begin{array}{c} \text{substrate lost} \\ \text{through overflow} \end{array}$$

or

$$Dc_r = (dx/dt)(dc/dx) + Dc \tag{28}$$

where c_r is the substrate concentration in the reservoir. Rate of cell production is given by Equation (22),

$$dx/dt = kx \tag{22}$$

The relationship between cells produced and limiting substrate utilized was given by Equation (17), which can be rewritten as

$$Y = \Delta x/\Delta c \tag{29}$$

or as the differential

$$Y = dx/dc \quad \text{or} \quad dx/dt = Y \, dc/dt \tag{30}$$

From Equation (30)

$$dc/dx = 1/Y \tag{31}$$

Hence Equation (28) becomes

$$Dc_r = (kx/Y) + Dc \tag{32}$$

and solving for x, we have

$$x = YD \, (c_r - c)/k \tag{33}$$

Because in the steady state $D = k$, we have

$$x = Y(c_r - c), \tag{34}$$

the second fundamental equation for the operation of the chemostat. From Equations (27) and (34), the relationships shown in Figure 8 can be plotted. The relationship between c and D can be seen from an inspection of Equation (27).

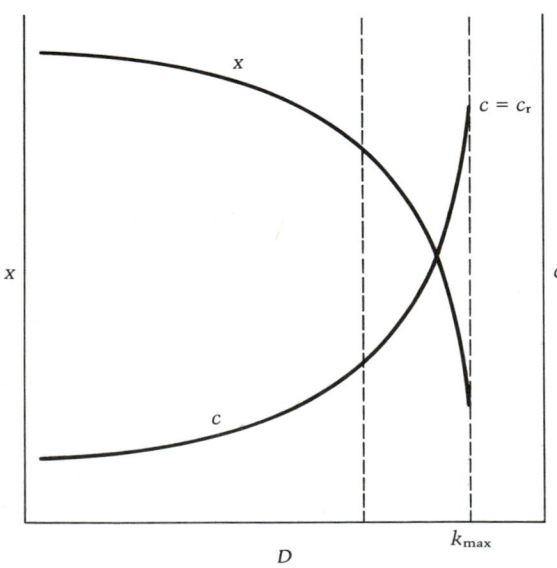

Figure 8. The relationship between biomass (x), concentration of limiting nutrient (c), and dilution rate (D) in continuous culture.

If D is small with respect to k_{max}, the denominator of the right side of Equation (27) is approximately equal to the constant (k_{max}) and hence c is approximately proportional to D, as indicated in the plot. But as D approaches k_{max}, the denominator decreases rapidly and c rises rapidly.

Knowing the relationship between c and D, from Equation (34) we can see that x would be approximately the inverse of this relationship to c.

Several important conclusions can be drawn from a consideration of Figure 8. First, over a wide range of dilution rates (D), the density of the culture (x) remains approximately constant, and this is the range over which a chemostat is most conveniently operated. At high dilution rates, where x changes rapidly with D, the chemostat becomes unstable because small (inadvertent) changes in D could cause wash out.

The turbidostat

In experiments with the chemostat, the operator regulates the flow rate, and the bacteria grow to a density sufficient to maintain the substrate concentration such that the growth rate equals the dilution rate. A similar instrument in which the cell density is regulated by the operator is called a TURBIDOSTAT. A turbidostat is provided with a device that constantly monitors the absorbance of the culture. If the absorbance exceeds the setting, the device automatically increases the flow rate, and if it is less, the flow rate is decreased. As is the case with a chemostat, a steady state is reached with a constant flow rate and a constant cell density. Although the two devices (chemostat and turbidostat) differ, the principle upon which they operate is identical, as can be seen by considering the following example. We are operating a turbidostat, that is, we preset a desired absorbance and the flow rate is adjusted automatically by the device to maintain this cell density. We then disconnect the absorbance sensor and steady-state operation continues. With the absorbance sensor disconnected, the device is a chemostat.

There are, however, practical differences between the operation of the chemostat and the turbidostat. For maximum stability (because it senses changes in absorbance), the turbidostat should operate at a dilution rate where cell density changes rapidly with dilution rate (at high dilution rates), whereas the chemostat should operate in a region where small changes in dilution rate should not have a major effect on cell density (lower dilution rates) (see Figure 8).

A turbidostat actually can be operated at k_{max}, that is, under conditions of unrestricted growth, by setting an absorbance level that maintains a culture density insufficient to allow any nutrient to fall to a growth rate-limiting value.

MAINTENANCE ENERGY

As mentioned earlier, a chemostat cannot be operated at very low dilution rates if the limiting nutrient is the source of energy for growth. This implies that if ATP is made available to the cell at a low rate, some cellular process(es) other than those leading to an increase in mass preferentially utilize ATP; growth can occur only when this nongrowth-associated demand for ATP, termed MAINTENANCE ENERGY, is met.

The existence of a maintenance requirement might be expected because there are a number of energy-consuming metabolic processes that are not directly coupled to growth. Motility requires energy; accumulation of substrates to a higher concentration within the cell than exists in the medium requires energy; the constant hydrolysis and resynthesis of macromolecules (termed TURNOVER) requires energy; maintenance of a potential across the cytoplasmic membrane requires energy. None of these expenditures results in increased mass.

In the eight decades since the concept of maintenance was introduced (Duclaux, 1898), a variety of indices of the magnitude of maintenance energy have been proposed. Principal among these are a, the SPECIFIC MAINTENANCE RATE (Marr, 1963) and m, the MAINTENANCE COEFFICIENT (Pirt, 1965). Specific maintenance rate is defined by a restatement of Equation (30) for the special condition in which the limiting substrate is the source of energy, namely,

$$dx/dt + ax = Y\, dc/dt \tag{35}$$

This equation states that over a period of time (dt) a certain amount of the substrate used (dc) contributes to growth (dx) and the rest (ax) is used for maintenance. The amount utilized for maintenance depends on the value of the specific maintenance rate (a) and on cell density (x). Thus, if a culture is fed a limiting substrate at a constant rate (a technique termed SLOW FEED) the rate of increase of biomass (dx/dt) will constantly decrease as cell density (x) increases when the limiting substrate is the source of energy. If the limiting substrate is not the source of energy (e.g., a required amino acid), growth rate remains constant (Figure 9). If Equation (35) is used in place of Equation (30) to derive the equation describing the operation of a chemostat [and if certain simplifications are made (Marr, 1963)], one obtains the equation

$$1/x = (a/x_{max})(1/D) + (1/x_{max}) \tag{36}$$

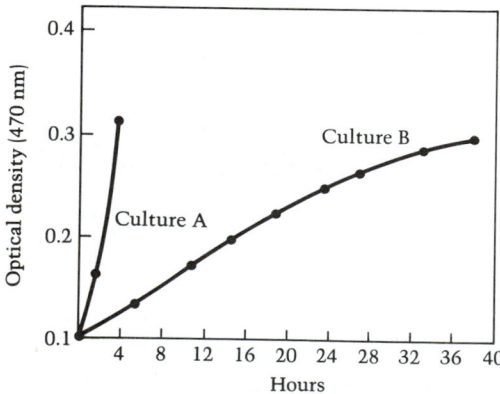

Figure 9.

Growth of *E. coli* strain PS limited by feeding glucose. Culture vessels containing 500 ml each of basal medium without glucose were inoculated with an exponential culture fed glucose at a rate of 0.038 mg/ml/hr until the optical densities were approximately 0.10. At this time, cultures A and B were fed a solution of glucose at a rate such that 0.377 mg of glucose was fed per milliliter of culture in 10 and 40 hours respectively. (After Marr, 1963.)

in which x_{max} is the cell density if $a = 0$. Thus, by plotting the reciprocal of cell density (x) versus the reciprocal of dilution rate (D) for various steady states of the chemostat, a straight line is obtained, the slope of which is a/x_{max}. By this method, the value of the specific maintenance rate for *E. coli* strain PS has been measured to be 0.028 hr^{-1} at 30°C.

Equation (35) introduces a certain complication into the definition of Y. We previously defined Y as the mass ratio of cells produced to substrate utilized. However, Equation (35) states that in the case of the source of energy, Y is the mass ratio of the sum of cells produced and substrate used for maintenance to substrate utilized. Rather than redefine Y for the special case of substrates that serve as an energy source, Pirt (1965) recommended the use of a new term Y_G that fits the original definition of Y, and he used this Y to describe substrate used for growth and maintenance. He defined the maintenance requirement at a rate of substrate utilization not contributing to growth, that is,

| overall rate of substrate utilization | = | rate of substrate utilization for maintenance | + | rate of substrate utilization for growth |

which becomes

$$ds/dt = ds/dt_m + ds/dt_G \tag{37}$$

and

$$kx/Y = mx + kx/Y_G \tag{38}$$

or

$$1/Y = m/k + 1/Y_G \tag{39}$$

predicting a linear relationship between $1/Y$ and $1/k$ with a slope m and ordinate intercept $1/Y_G$. Such a relationship is found (Figure 10).

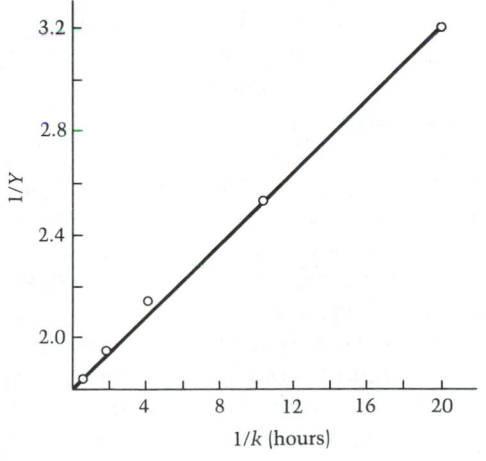

Figure 10.

Data derived from measurements on glycerol-limited growth of *Enterobacter aerogenes*. (After Pirt, 1965.)

The maintenance coefficient m and specific maintenance rate a are related by the equation

$$m = a/Y_G \tag{40}$$

Under many conditions of bacterial growth, particularly those encountered during unrestricted growth in laboratory media, maintenance energy might be considered negligible. A specific maintenance rate of 0.02 hr^{-1} for a culture growing at a rate 1.0 hr^{-1} does not constitute a major diversion of substrate utilization to nongrowth purposes. However, in many natural environments where substrate concentration is typically low, especially those anaerobic ones that further restrict the rate of generation of ATP, the maintenance requirement becomes an important determinant of cell growth.

SYNCHRONIZED CULTURES

Studies on bacterial cultures undergoing restricted or unrestricted growth provide composite information on all the cells in the population. Such studies are unsuitable for revealing the sequential changes that occur in individual cells during their interdivision cycle. Single cells could be studied, but their small size (10^{-12} g) severely restricts the types of physical or chemical measurements that can be made on them. A practical solution to this problem is the use of SYNCHRONOUS CULTURES, that is, cultures composed of cells that are all at the same stage of their normal interdivision cycle.

The very idea of working with synchronous populations of cells (not only bacteria) is fascinating. However, in their enthusiasm biologists have tended to forget that the properties of a synchronous population, and hence the problems that can be solved by studying it, depend utterly on the method by which synchrony has been obtained.

Several means of obtaining DIVISION SYNCHRONY have been devised, but in many cases it is evident that the division cycles produced are far from normal. Thus, if *Tetrahymena pyriformis* (a large protozoan cell) is exposed repeatedly to sublethal temperatures, several sharply defined cell divisions take place at the normal growth temperature. Interesting morphological changes have been studied with such cultures, but the division cycles produced by the temperature regime are definitely abnormal: during the pretreatment the cells grow to several times their normal size, and the synchronous divisions that follow take place at intervals shorter than the normal doubling time. With bacteria (*Streptococcus pneumoniae*, *Bacillus megaterium*, and *E. coli*), exposures to low temperatures for various lengths of time can result in synchronous division upon return to normal growth temperatures. In all these cases the mechanism responsible for synchrony is obscure and the division cycles are probably not normal. *E. coli* cultures growing in broth at 37°C have been caused to divide synchronously by short exposures to 25°C at regular intervals; however, it was later shown that this treatment upsets the normal DNA replication pattern. These ways of *inducing* division synchrony are related, and they are supposed to be effective because the cells accumulate at a certain stage of the interdivision cycle, enabling more or less simultaneous reinitiation of growth and division upon return to the higher temperature.

Attempts have been made to obtain synchrony also by "lining up" cells at the point in the cycle where they are ready to initiate chromosome replication. This technique rests on observations indicating that certain proteins must be synthesized shortly before a round of replication can be initiated; however, once started, the round runs to completion even if the cell is temporarily unable to synthesize protein. The experiment can be done by starving a culture for a required amino acid long enough for all cells to complete replication. The culture is then switched to a medium permitting protein, but not DNA, synthesis (lacking the required thymine). During the second starvation period protein synthesis is resumed, thus preparing the cell for replication. If thymine is now added back at the right time, REPLICATION SYNCHRONY ensues. However, little, if any, division synchrony obtains and that should not be a surprise because the protocol of the experiment indicates that the cells will be unevenly affected by the pretreatment. Thus, when protein synthesis is blocked, replication terminates very soon in some and much later in other cells; the DNA:mass ratio will, therefore, remain almost normal in the former, whereas it nearly doubles in cells that go through most of their replication cycle while starved of an amino acid. During the period of thymine starvation, the ratio will decrease, but it will not become normal throughout the population. The role of the DNA:mass ratio for the timing of initiation of replication is discussed in Chapter 6.

The purpose of this lengthy discussion of division and replication synchrony is to argue that methods that work because they interfere with one or another step in the normal division cycle probably affect cells of different ages at the time of induction to different degrees. It may, therefore, not be possible to obtain by induction a synchronous culture in which normal division cycles follow one after another. Nevertheless, induced division or replication synchrony has been used to advantage to study the processes of division and of replication and especially the initiation of the latter.

To a good approximation, an ideal synchronous culture can be obtained by *selection*: the smallest, and hence the youngest, cells can be removed from a culture in balanced growth by filtration or centrifugation. The quality of the synchrony in the samples thus selected varies from poor to quite good. An ingenious technique has been developed by Helmstetter (1963); it involves the continuous collection of young cells (baby cells) coming off of the lower surface of a filter onto which the parent population is adsorbed. This method has been used extensively to analyze the pattern of DNA replication, and it is described in more detail in Chapter 6. The quality of synchrony observed in the effluent from the filter apparatus is illustrated in Figure 11; the cell number is seen to remain constant for approximately 45 minutes, then doubles somewhat rapidly as the cells in the population divide. A second and third division can be discerned, but the period over which division occurs constantly increases: synchrony is rapidly lost. In other words, the interdivision times of various cells in the population are quite variable, and consequently experiments on synchronous cultures of this kind must be made soon after the culture is collected.

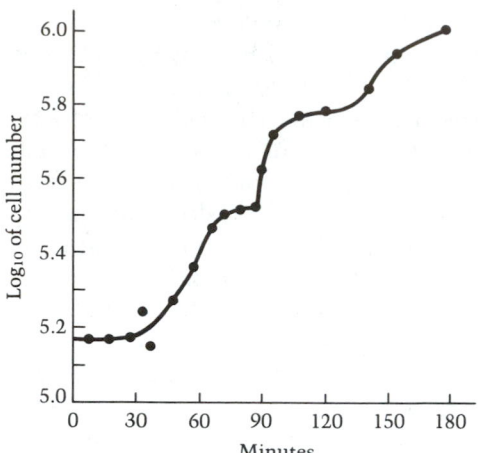

Figure 11.

Synchronous growth of *E. coli* ML30G following induction by the Helmstetter-Cummings technique. (After Shehata, 1970.)

From the rate of loss of synchrony of a culture, the frequency distribution of interdivision times can be calculated (Figure 12). Although the mean interdivision time of this culture is 53.9 minutes, significant numbers of cells divide after only 40 minutes, when they are relatively small, and others divide only after 60 minutes, when they are quite large. Evidence exists to support the contention that there is a negative correlation between the interdivision times of mother and daughter cells; cells produced by the division of a small cell have longer interdivision times than cells produced from a large cell.

As stated earlier, the kinetics of growth of an individual cell is not established by knowing the growth kinetics of a culture. One might expect it to be a logarithmic function of time, thus fitting the autocatalytic model of culture, or possibly a linear function because of some imagined constraint on individual rod-shaped cells that increase in length, not girth, as they grow. The best available evidence suggests that neither of these simple models is correct. Rather, the growth kinetics of an individual cell of *E. coli* is a relatively complex function that varies with time, fitting neither linear nor exponential kinetics within the cell cycle.

Finally, techniques are emerging that produce cultures of *E. coli* in which excellent division synchrony, once established, continues over 10 or more cycles without outside interference of any kind. Two procedures have been described: first, a minimal-medium culture is allowed almost to reach stationary phase; it is then diluted (approximately 100-fold) and, if the timing is right, the cell number now continues to double at intervals corresponding to the mass doubling time in an exponentially growing culture (Cutler, 1966). Second, a culture is conditioned by going through many cycles involving addition of enough sulfur at the beginning of each cycle to permit one mass doubling. After this pretreatment, surplus sulfur is added, and sustained division synchrony ensues (Képès, 1980). These results are surprising in that they contradict the established fact that the interdivision time is subject to considerable variation (see earlier, and Chapter 6). It is difficult to account for

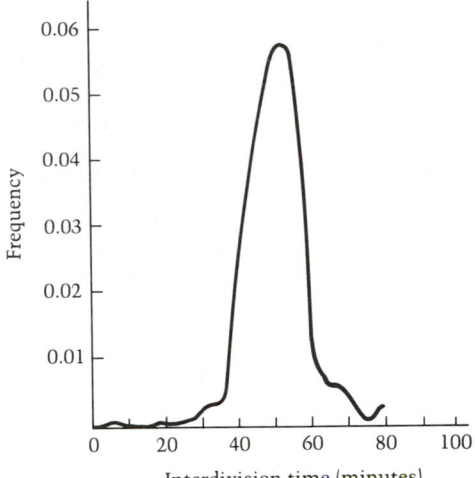

Figure 12.

Frequency distribution of interdivision time of cells of a culture of *E. coli* ML30G. The distribution was calculated from the rate of loss of synchrony of the culture depicted in Figure 11. (After Shehata, 1970.)

the results of Cutler and of Képès without postulating some sort of cell-to-cell interaction, but, whatever the explanation turns out to be, their methods hold considerable promise for future work with synchronous cultures.

EFFECT OF TEMPERATURE ON BACTERIAL GROWTH

As discussed, growth of bacteria is the consequence of a complex and highly regulated set of chemical reactions. Hence, the effect of temperature on the rate of growth might be expected to reflect its effect on the velocity of chemical reactions. The Swedish chemist Arrhenius discovered this relationship in the last century, finding it to be described by

$$v = Se^{-\Delta E^*/RT} \tag{41}$$

in which v is the velocity of the reaction, S is a constant, ΔE^* is the activation energy, R is the gas constant, and T is temperature in °K. Taking the logarithm, one obtains

$$\ln v = (-\Delta E^*/R)(1/T) + \text{constant} \tag{42}$$

Thus, the logarithm of the velocity of a chemical reaction is a linear function of the reciprocal of absolute temperature. If this type of plot (frequently termed an ARRHENIUS PLOT) is made for bacterial growth rate (Figure 13), a somewhat different response is seen. In the midrange of temperature, normal chemical kinetics seem to apply—a straight line is obtained (the slope of which is the TEMPERATURE CHARACTERISTIC, R). This range is termed the LINEAR or NORMAL RANGE. Above and below it, growth rate is less than the value predicted by extrapolation of the Arrhenius relationship. In the LOW RANGE, the slope of the plot increases, becoming vertical at the MINIMUM TEMPERATURE FOR GROWTH. At high temperatures, the plot becomes vertical at the MAXIMUM TEMPERATURE FOR GROWTH. The temperature at which growth rate is maximal

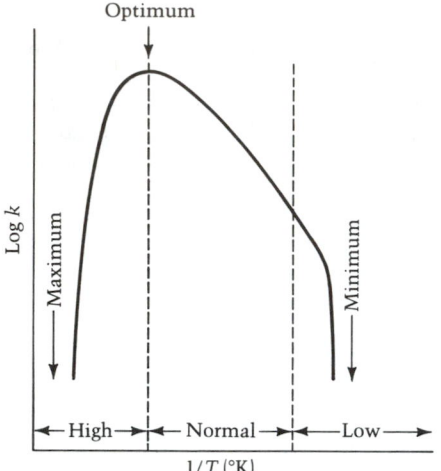

Optimum

Log *k*

Maximum

Minimum

←— High —→ ←— Normal —→ ←— Low —→

$1/T$ (°K)

Figure 13.

General form of an Arrhenius plot of bacterial growth. The cardinal temperatures (the maximum, optimum, and minimum for growth) and the growth temperature ranges (high, normal, and low) are shown.

is termed the OPTIMUM TEMPERATURE FOR GROWTH. Collectively, these three temperatures are called the CARDINAL TEMPERATURES.

The general form of the Arrhenius plot of growth rate is typical for all bacteria studied; but the absolute values of the defining parameters (temperature characteristic, and cardinal temperatures) vary widely among bacteria. Certain information in the generalized curve is worthy of emphasis. (1) Bacteria do not simply grow more slowly at high and low temperatures; there are precise temperature limits for the growth of any particular bacterium above and below which growth cannot occur. (2) Growth rate decreases rapidly at temperatures above the optimum: as a consequence, the optimum and maximum growth temperatures are not far apart.

An Arrhenius plot of the strain of *E. coli* analyzed in Chapter 1 is shown in Figure 14. Its maximum temperature of growth is approximately 48°C; its optimum temperature is approximately 37°C; its minimum is 8°C (data not shown in plot); its linear range extends from 37° to 21°C. One also notes from this plot that richness of the medium affects growth rate at all temperatures, but it does not affect the temperature characteristic, nor (in the case of *E. coli* and many other bacteria) does it affect the optimum and maximum temperatures.

Ratkowsky (1982) has made the very interesting observation that if the square root of growth rate is plotted as a function of growth temperature a straight line relationship is obtained over the normal and low temperature range (Figure 15). Although this relationship has no theoretical basis and fails by extrapolation to predict accurately the minimum temperature of growth, it should prove useful in predicting intermediate growth rates from limited data. Failure to predict the minimum temperature of growth suggests that the square root relationship between growth rate and growth temperature does not hold in the extreme low range.

The cardinal growth temperatures vary widely among bacteria. It is now established that some bacteria can grow at temperatures as high as 96°C and

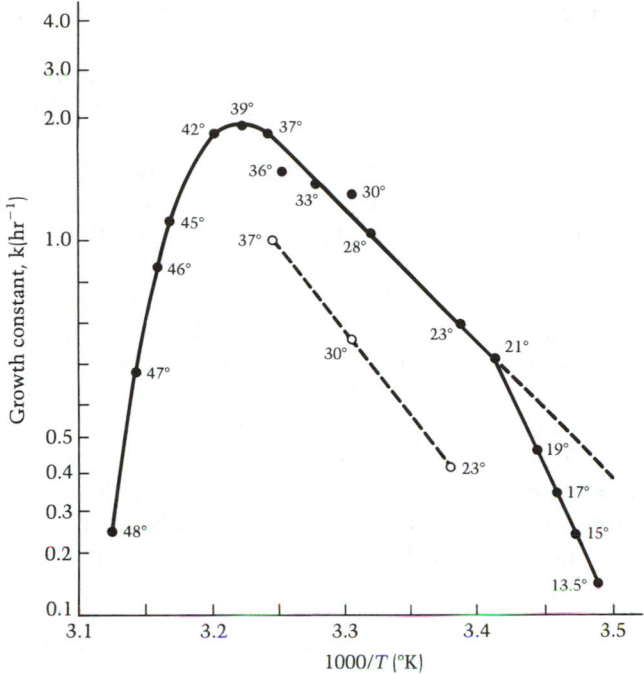

Figure 14. Arrhenius plot of growth rate of *E. coli* B/r. Individual data points are marked with corresponding degrees Celsius. *E. coli* B/r was grown in a glucose-rich medium (●) and a glucose-minimal medium (○). (After Herendeen, 1979.)

others at temperatures as low as −10°C. Indeed, bacterial growth seems to be limited only by the availability of liquid water, the upper limits being set by the boiling point and the lower being set by the bacterium's ability to withstand the osmotic pressure required to maintain water as a liquid below the freezing point of pure water. Although, there is not a direct relationship

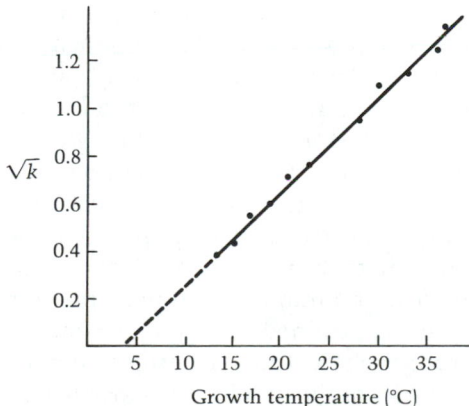

Figure 15.

Plot of data of Figure 14 according to the method of Ratkowski (1982). The square root of growth rate *k* is plotted as a function of growth temperature. The data points closely fit a straight line, but the extrapolation of the straight line to zero (at 3.5°C) does not accurately predict the actual minimum temperature for growth (8°C).

between maximum and minimum temperatures of growth of bacteria, most bacteria grow over a range of approximately 40°C. If the maximum and minimum temperatures of various bacteria for which this information is available are plotted versus one another, they define a straight line the intercept of which on the maximum temperature scale is approximately 40°C. Some points lie below the line indicating that certain bacteria grow only over a more limited range of growth temperatures, but none lies significantly above the line, indicating that the ability to grow over a temperature range significantly greater than approximately 40 Celsius degrees is a rare one. The reason for this natural restriction of growth-temperature range is not obvious and reflects natural selective pressures more than innate biological capacity; studies on mutant strains, termed TEMPERATURE-SENSITIVE MUTANTS (*ts*), that have more restricted growth-temperature ranges than their parents suggest that the maximum and minimum temperatures of growth are set by different genetic determinants. Those *ts* mutants, termed HEAT-SENSITIVE (*hs*), that have a decreased maximum temperature of growth do not have altered minimum temperatures of growth; and COLD-SENSITIVE (*cs*) MUTANTS in which the minimum temperature of growth is increased do not have altered maximum temperatures of growth.

On the basis of their temperature range of growth, bacteria are frequently divided into three broad classes: those that grow at 50°C or above, THERMOPHILES; those that grow best at approximately 37°C, MESOPHILES; and those that can grow at 5°C or below, PSYCHROPHILES or RHIGOPHILES. One notes that thermophiles are defined by their maximum temperature of growth, mesophiles by their optimum, and psychrophiles by their minimum.

To the bacterial physiologist, the observed effects of temperature on bacterial growth pose two important questions: (1) Does growth temperature affect the physiological state and/or chemical composition of bacteria? (2) What biochemical factors set the temperature limits of growth of a bacterium?

EFFECT OF TEMPERATURE OF GROWTH ON CELL COMPOSITION AND PHYSIOLOGY

That growth temperature can profoundly affect the physiological state of bacterial cells can be illustrated by a series of simple experiments (Ng, 1962). If the incubation temperature of an exponentially growing culture of *E. coli* is suddenly shifted within the normal temperature range, exponential growth continues at the rate characteristic of the new temperature. But if temperature shifts are made between the normal range and either the low or high ranges, transition growth rates often ensue before exponential growth at a rate characteristic of the new temperature begins. Transient growth periods are most marked in shifts to or from the low range. Shifting an exponentially growing culture of *E. coli* from 37° to 12°C causes it to stop growing for approximately 4 hours before exponential growth begins again. The reverse shift in temperature causes the culture to grow at a markedly reduced rate for approximately two doublings before full rate is attained. These experiments suggest that within the normal temperature range cellular reactions remain coordinated simply by modifications of enzyme activity, but full growth rate in the high or low range requires that composition of the cell be altered. Detailed analysis of the effect of growth temperature on the cellular content of 133 different proteins (constituting 70% of the cells' protein mass; Herendeen, 1979) lends support to this hypothesis. The concentrations of very few proteins change appreciably if growth temperature is varied within the normal range. But growth in the high or low range is accompanied by large changes (up to 25-fold) in the steady-state level of a number of proteins. These changes in level are brought about very rapidly; within a minute after a shift in temperature from, say, 30° to 42°C, rapid changes occur in the synthesis of many of the 1000 cellular proteins. There is a gradient of response extending from some proteins that are hyperinduced transiently 100-fold to some that for a time virtually cease being made (Lemaux, 1978). Proteins that are transiently induced by an increase in temperature are repressed by a decrease in temperature, and vice versa. Qualitatively the same transient pattern is observed for shifts of only a few degrees within the normal growth-temperature range.

The response of *E. coli* to a shift-up in temperature is analogous in many ways to the HEAT SHOCK RESPONSE that exists throughout the living world. Cells of all animals, plants, and microbes studied so far display a transient, nearly exclusive, synthesis at high rates of approximately a dozen proteins when exposed to temperatures above the normal for growth. Recent work (Neidhardt, 1981; Yamamori, 1982) indicates that a group of 13 *E. coli* proteins induced by a shift-up in temperature are under the control of a single gene, *htpR* (also called *hin*), located at a distance from the genes encoding the induced proteins. Mutational inactivation of this gene eliminates the heat induction, and the mutants die after a short period of growth at the elevated temperature. Evidence indicates that the *htpR* (*hin*) gene encodes a regulatory protein that controls transcription of the 13 heat-induced genes. These genes are said to constitute a REGULON, because they are not contiguous on the chromosome but share a common regulatory protein. The concentration of one or more of their products must be elevated to enable growth at high temperature, but the nature of this requirement is unknown.

The fatty acid composition of the phospholipids in bacterial membranes also varies with growth temperature. Like those of other bacteria and, indeed, of all other poikilothermic organisms that have been studied, the fatty acids of *E. coli* (Figure 16) are predominantly saturated (contain no double bonds) in cells grown at high temperature. Because the melting point of lipids decreases as their proportion of unsaturated fatty acids increases, the change in fatty acid composition acts as a homeostatic mechanism to preserve a more or less constant membrane fluidity.

FACTORS THAT DETERMINE THE TEMPERATURE LIMITS OF GROWTH

A number of studies establish that the maximum temperature for growth of most bacteria is set by the stability of proteins. Genes encode the primary structure of proteins, which determines their heat stability as well as their functional activity, be it catalytic or structural. The portion of

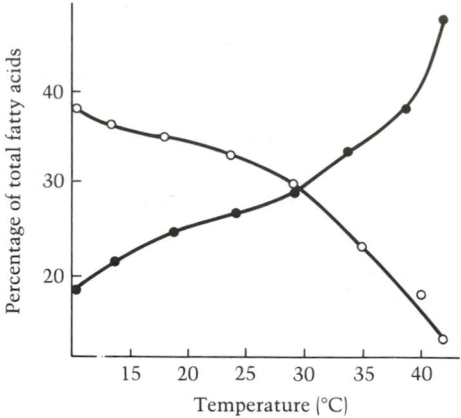

Figure 16.

Effect of growth temperature on the fatty acid composition of *E. coli* ML30. Cells were harvested during exponential growth in a glucose minimal medium. Results are presented as the percentage of the total fatty acid fraction (by weight) that is the saturated fatty acid palmitic acid (●) and the unsaturated fatty acid octadecanoic acid (○). (After Marr, 1962.)

the size and complexity of a protein that is required for heat stability as contrasted with functional activity remains unknown, but certainly heat stability is a particularly sensitive property of a protein, and it seems to be more readily changed by mutations affecting primary structure than is the property of functional activity itself. Several lines of experimental evidence support this contention.

The most detailed data is that of Langridge (1968), who examined 52 randomly altered mutant forms of the enzyme β-galactosidase from *E. coli*. The set of altered proteins was generated by isolating a number of nonallelic amber mutations in the gene encoding β-galactosidase and then introducing a suppressor mutation that introduced a new amino acid at the site of the amber codon. Thus, he was able to obtain a series of mutationally altered enzymes without selecting for loss of catalytic activity. He found that over 70% of these mutant proteins showed distinct loss of heat stability (Table 1) whereas only one showed detectable loss of catalytic function.

These results suggest that most naturally occurring mutations are counterselected on the basis of loss of heat stability rather than loss of function. It follows that a microorganism would, in the absence of the challenge of elevated temperature, rapidly accumulate mutations causing loss of heat stability. Bacteria capable of growing at elevated temperatures must contain heat-stable proteins, a fact that has been established by many studies. In contrast, bacteria that are incapable of growing at high temperature contain very few thermostable proteins (Koffler, 1957). Whereas most of the total cytoplasmic proteins from representative mesophiles are precipitated by an 8-minute heat treatment at 60°C, only a very small percentage of the proteins from representative thermophiles are precipitated by the same treatment.

The thesis that mutations conferring thermolability tend to accumulate in organisms not subjected to the challenge of exposure to high temperature is also supported by studies on marine psychrophiles. Whereas terrestrial psychrophilic bacteria that are unable to grow at temperatures as high as 20°C (obligate psychrophiles) are rare, these organisms are abundant in certain marine environments where the temperature *always* remains low, suggesting

Table 1. Frequency of half-lives in minutes at 57°C of 52 serine-substituted β-galactosidases

Half-life classes[a] (min at 57°C)	Frequency	Half-life classes[a] (min at 57°C)	Frequency
10	16	60–70	1
10–20	5	70–80	1
20–30	4	80–90	1
30–40	3	90–100	3
40–50	4	100–110	15
50–60	1		

[a]Wild-type enzyme has a half-life of 104 minutes.

again that in the absence of selective pressures, protein stability to temperatures even as low as 20°C is lost by random mutational events. Geneticists have come to accept, as a matter of course, that any missense mutation has a high probability of decreasing the heat stability of the product protein.

LOSS OF FUNCTION AT LOW TEMPERATURE

The chemical basis for loss of function of a protein at high temperature is self-evident; those chemical bonds that maintain the proper secondary and tertiary structure of proteins become weakened at elevated temperatures, a process that results in denaturation and loss of function of the protein. At low temperature, loss of function is more difficult to explain because most chemical bonds are strengthened as temperature decreases; but hydrophobic bonds, as a consequence of physical changes in the structure of the solvating water, do weaken at low temperature.

The selective pressures for growth at low temperature are quite different from those for growth at high temperature. We have discussed evidence indicating that exposure to the challenge of high temperature is essential if a microorganism is to retain its ability to grow at high temperature, but such does not seem to be the case for growth at low temperature. Enteric bacteria must have grown at temperatures quite close to 37°C for millions of years, yet both *Salmonella typhimurium* and *E. coli* have retained their ability to grow at temperatures as low as 8°C. The study of cold-sensitive mutants has provided examples of loss of function at low temperature. The typical growth response of a cold-sensitive mutant of *E. coli*, as affected by temperature, is shown in Figure 17. At 37°C the mutant strain grows almost as fast as its parent, but as temperature is lowered, the growth rate of the mutant decreases more rapidly than that of the parent, and it stops growing completely at approximately 20°C rather than at 8°C.

Although cold sensitivity is a frequent consequence of mutation—approximately as frequent as heat sensitivity—it appears that such mutations are found in a lesser number of genes. On analyzing 75 cold-sensitive mutants

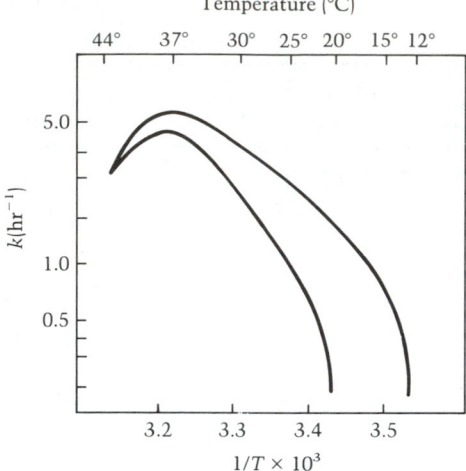

Figure 17.

Arrhenius plot of the specific growth rate (*k*) of an *E. coli* cold-sensitive mutant. Both the mutant (lower curve) and its parent (upper curve) are grown in a minimal medium. The ordinate is the specific growth rate (log scale) and the abscissa is the reciprocal of the absolute temperature times 1000.

of phage T4 (Scotti, 1968), it was found they were contained in only nine complementation groups. In contrast, mutations conferring heat sensitivity are randomly distributed among 37 genes.

Cold-sensitive mutants provide a means of analyzing the biochemical basis for the minimum temperature of growth. Because a single genetic change increases the minimum temperature of growth, the resulting protein alteration is the biochemical determinant of the minimum temperature of growth of the mutant. It can be argued that artificially induced defects might not be representative of the types of loss of function that account for the minimum temperature for growth of wild-type organisms in nature and that, therefore, knowledge concerning the genetic and biochemical basis for the lower temperature limits of growth can only be gained by the direct study of wild-type organisms. However, several practical considerations indicate that such a direct approach holds very little hope of success. In most microorganisms, many independent biochemical functions cease simultaneously at the minimum temperature of growth. Thus, there is not a single cause determining the minimum temperature of growth, and this fact makes biochemical analysis and even the distinction between cause and effect quite difficult. Moreover, there is not reason to believe that artificially induced mutations would preferentially affect different functions from naturally accumulated mutations that set the minimum temperature of growth of wild-type microorganisms. A study of cold-sensitive mutants gives information as to the types of lesions that prevent function at low temperature.

One such type of lesion is that which alters the sensitivity of regulated proteins to their small molecule effectors and hence results in a distortion of the regulated system such that function at low temperature is prevented. Cold-sensitive histidine mutants are an example of this type (O'Donovan, 1965). These mutants have a minimum temperature of growth approximately 12° higher than that of their parent; however, in the presence of histidine, the parent and mutant grow at identical rates at all temperatures. The functional block at low temperature lies in the first reaction of the histidine pathway—that catalyzed by phosphoribosyl-ATP pyrophosphorylase (PR-ATP ppase; ATP phosphoribosyltransferase). This enzyme is sensitive to feedback inhibition by the end product of the pathway, histidine. The enzyme produced by the mutant is almost 1000-fold more sensitive to feedback inhibition than is the enzyme produced by the parent (wild type). Moreover, mutant and wild-type enzyme are approximately 10 times more sensitive to feedback inhibition at 20°C than at 37°C. Thus, the increased sensitivity of the mutant enzyme to feedback inhibition, plus the increased sensitivity imposed by low temperature, creates conditions whereby the intracellular concentration of histidine that prevents its own further biosynthesis by feedback inhibition is insufficient to allow the biosynthesis of proteins. As a consequence, the mutant cannot grow at 20°C in the absence of exogenous histidine.

If the biochemical explanation of cold sensitivity of histidine biosynthesis is of any general significance, the degree of inhibition of many other allosteric proteins might be expected to be changed significantly by temperature. Considerable evidence indicates that changes in sensitivity of allosteric proteins to their effectors with temperature is the rule rather than the exception; but predictions are not possible as to the sense of change with temperature. Some proteins become more sensitive to inhibition as temperature is decreased and others become less sensitive.

As an example, we may consider the detailed data concerning the effect of temperature on the inhibition of fructose-1, 6-diphosphatase from rat liver by adenosine monophosphate (AMP). This enzyme has dramatically increased sensitivity to AMP at low temperature. In the absence of AMP, the effect of temperature follows normal chemical kinetics: the Arrhenius plot is completely linear over the temperature range of 46° to 2°C; however, in the presence of the inhibitor, the reaction is selectively inhibited at low temperature (Figure 18). Adenosine monophosphate at 10^{-4} has no inhibiting effect

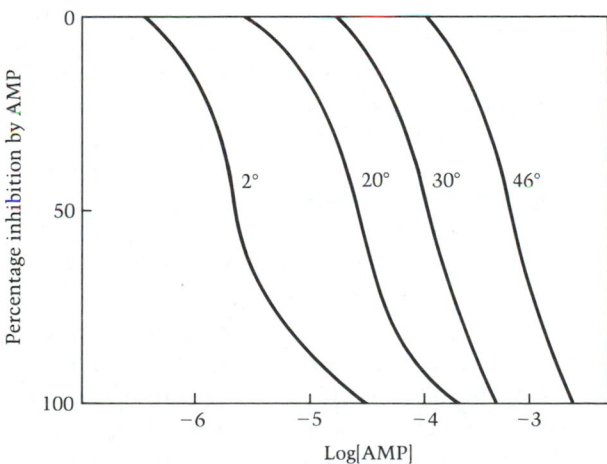

Figure 18. Percentage inhibition of fructose-1,6-diphosphatase by varying concentrations of AMP at different temperatures (°C). (After Taketa, 1965.)

on the enzyme at 46°C, but this concentration inhibits more than 90% at 20°C. From these data it is clear that because enzymes from homothermic organisms also exhibit strong temperature effects on the sensitivity to inhibition, such effects most probably reflect fundamental properties of allosteric proteins.

We can conclude that mutations that alter the sensitivity of allosteric proteins to their effectors would have a high probability of being expressed as a cold-sensitive phenotype.

One major class of cold-sensitive mutants is affected in an assembly process; they are unable to synthesize ribosomes at low temperature (Tai, 1969; Guthrie, 1969) because they are unable to assemble them.

Cold-sensitive mutants of both the regulatory and the ribosome assembly type probably share the same biochemical basis: a change in primary structure plus the weakening of hydrophobic bonds causes the protein to have a slightly altered conformation at low temperature. Those proteins the function of which is sensitive to conformational changes can, as a consequence, lose activity. The sensitivity of regulated functions to modulation of activity by effectors is known to be effected by conformation, and one would expect that the assembly of ribosomal proteins into ribosomes would be dependent on a precise conformational state. Other aspects of protein activity are also affected.

When shifted to a temperature slightly below the minimum for growth, polysomes do not form in *E. coli* (Friedman, 1969), probably as a consequence of a conformational change in a protein. As has been suggested, this defect might set the minimum temperature for growth of *E. coli*. But other bacteria are not so affected (Saruyama, 1980). It seems that one can generalize about the sorts of protein changes that set the minimum temperature of growth of bacteria but not the particular protein that is affected.

LETHAL EFFECTS OF TEMPERATURE

Bacteria are killed by exposure to high temperature, by freezing, and by sudden chilling. Although some bacteria die when held at temperatures above the freezing point and below the minimum for growth, there is no evidence that such killing is caused by low temperature rather than by the usual decline in bacterial populations that occurs in the absence of growth.

Lethal effects of high temperature have been studied extensively because sterilization by heat is simple and effective. In spite of complications such as the enormous variations among bacteria with respect to susceptibility to heat, the importance of the physiological state of bacteria, and the protective or synergistic effects of components of the medium, certain generalizations can be made. (1) Rate of killing of bacteria is directly related to their moisture content. (2) Rate of killing of bacteria is, in general terms, exponential. Although deviations (sometimes major ones) occur, a plot of the logarithm of the number of surviving cells in a population exposed to a lethal high temperature is usually a linear function of the time of exposure. Thus, sterility is an absolute state, but the design of lethal heat treatments to attain it are based on probability calculations, even if the initial population and the slope of the killing curve are known precisely. The more common indices of susceptibility to killing by heat of a particular suspension of bacteria are the following:

1. THERMAL DEATH POINT (TDP) is the lowest temperature that results in the organism's being unable to reproduce under optimum conditions after 10 minutes or some other fixed exposure time.
2. THERMAL DEATH TIME (TDT or F value) is the time of exposure to a particular temperature necessary to sterilize a definite concentration of organisms in a particular medium.
3. D VALUE is the time of exposure necessary to reduce the number of survivors by a factor of 10.

Often the logarithms of both D values and F values are linear functions of temperature over relatively narrow ranges. The slope of the curve relating the logarithms of F values to temperature, that is, the number of degrees required for the line to transverse one log cycle, is termed the Z value.

Freeze-killing of bacteria is dependent on the same factors that affect susceptibility to heat-killing. In addition, the rate at which the culture is chilled markedly affects the number of bacteria that survive the treatment. The degree of killing by freezing can be separated into two components: the killing (IMMEDIATE EFFECT) that occurs at the time of freezing and the slow loss of viability (STORAGE EFFECT) that occurs in the frozen state. The immediate effect of freezing is to kill a certain portion of the population. Thus, the logarithm of the number of survivors of a particular suspension of bacteria is usually a linear function of the number of times it has been frozen. The rate of loss of viability of a culture held in frozen state is usually directly related to the temperature at which it is held.

A number of compounds, when added to the medium in which a bacterial suspension is to be frozen, are remarkably protective. These include glycerol, dimethyl sulfoxide, dimethylacetamide, dimethylformamide, *N*-methylpyrrolidinone, acetamide, and polyvinylpyrrolidone. Other compounds confer somewhat less protection, including milk proteins, meat extract, sucrose, glucose, and lactose.

Many bacteria exhibit a curious susceptibility to death when a culture is rapidly chilled. This phenomenon, known as COLD SHOCK, can, if unappreciated, cause huge errors in certain procedures, such as the making of dilutions prior to determining the number of viable cells in a growing culture. For example, the rapid chilling of growing cultures of *E. coli* from 37° to 5°C can kill over 90% of the population. Growing cultures are most susceptible to cold shock. Slow chilling of a culture protects it completely, as does the presence of optimal concentrations of glycerol. In general, Gram-positive bacteria are more susceptible to cold shock than are Gram-negative bacteria.

SUMMARY

1. When growing in a medium in which all essential nutrients are present in nonlimiting concentrations—a condition termed unrestricted growth—the rate of increase of biomass mimics an autocatalytic, first-order chem-

ical reaction: the rate of increase of mass is proportional to the amount of mass present at any particular time.

2. Under conditions of unrestricted growth, the rate of increase of cell numbers can be predicted from the equation

$$\log_{10}N = kt/2.303 + \log_{10}N_0$$

where N is a number of cells, t is time, and k is the specific growth rate constant.

3. The value of k or the related parameter g (doubling time) is sufficient to describe completely the growth rate of a bacterial culture. These parameters are related by the equation $k = \ln 2/g$.

4. After a period of unrestricted growth, balanced growth ensues, that is, all components of the biomass increase at the same rate. As a consequence, any measurement of biomass can be used to compute growth rate.

5. Upon exhaustion of an essential nutrient or accumulation of toxic products of metabolism, growth of a bacterial culture ceases and the culture is said to leave the exponential phase and enter the stationary phase of growth. On transfer of such a culture to a fresh medium, growth is delayed for a period—termed the lag phase of growth.

6. The final crop of cells that a medium can support is linearly related to the concentration of the growth-limiting nutrient. The constant of proportionality is termed the yield coefficient, Y.

7. Bacterial cultures can be grown continuously by feeding medium to a culture and withdrawing a portion of that culture at the same constant rate. If the rate of dilution of the culture by this process does not exceed the unrestricted growth rate of the bacterium, culture density remains constant. This phenomenon is a consequence of growth rate being set by the concentration of limiting nutrient in the culture.

8. A certain portion—termed the maintenance requirement—of the utilization of the nutrient from which ATP is derived does not contribute directly to growth. As a consequence, continous cultures of bacteria are unable to grow when this limiting nutrient is added slowly.

9. Most bacteria are able to grow over a temperature range of about 40°C. Some can grow at temperatures over 90°C, and some can grow at temperatures as low as −10°C.

10. Thermal instability of proteins sets the maximum temperature of growth of most strains of bacteria.

11. Those factors that set the minimum temperature of growth of bacteria seem to derive from the weakening of hydrophobic bonds at low temperatures.

PROBLEMS

1. The following data were collected using a culture of *Salmonella typhimurium*. A sample (100 ml) of a growing culture was collected, centrifuged, resuspended in distilled water, centrifuged again, dried at 105°C, and weighed. This sample weighs 103.5 mg. A second sample of the same

culture was taken at the same time. The A_{420nm} of this culture and dilutions of it were measured with the following results.

Dilution*	A_{420}	Dilution*	A_{420}
10 + 0	1.017	5 + 5	0.670
9 + 1	0.972	4 + 6	0.568
8 + 2	0.919	3 + 7	0.445
7 + 3	0.844	2 + 8	0.300
6 + 4	0.763	1 + 9	0.147

*dilution indicates milliliters of culture + milliliters of uninoculated medium

By taking a number of samples of the same culture of *Salmonella typhimurium* growing in a minimal medium with glucose as the total source of carbon and energy, you obtained the following data. At the time of inoculation (0 minutes), the glucose concentration was 0.2% (w/v). Assume that the culture stopped growing because glucose was exhausted in the medium.

Time of sampling (min)	A_{420}
0	0.091
20	0.092
40	0.111
60	0.152
80	0.208
100	0.303
120	0.430
140	0.577
160	0.732
180	0.902
200	1.001
220	1.080
240	1.080

a. Plot dry weight and log of dry weight in the form of a growth curve on millimeter paper.
b. Plot the data on semilog paper.
c. Calculate k.
d. Calculate g.
e. Calculate Y.
f. Calculate Y_m.
g. Although monomers were not present during growth, assume that the value of $Y_{ATP} = 10.5$ holds and calculate the number of moles of ATP produced per mole of glucose metabolized.
h. If you wanted a stationary culture containing 200 µg cells/ml, what concentration of glucose would you have the medium contain?
i. If the average dry weight of a *Salmonella typhimurium* cell in exponential phase is 1×10^{-12} g, how many cells per milliliter do you think are in the culture at 240 minutes?

2. A culture of *E. coli* is growing in a minimal medium at a rate $k = 1.0$ hr^{-1}. At the time that the cell density is 300 μg/ml (dry weight), you add *p*-fluorophenylalanine to the culture. (a) What would the cell density in μg (dry weight)/ml be 2 hours later? (b) What would it be if *p*-fluorophenylalanine had not been added?

3. A glucose-limited chemostat is operated at a dilution rate of 0.4 hr^{-1}. If K_s for glucose is 1×10^{-7} M, the glucose yield constant (Y) is 0.48 and if the same medium supports growth in batch culture with $k = 0.8$ hr^{-1}, (a) what concentration of glucose could you add to the reservoir to maintain a steady-state cell density in the growth vessel of 600 μg/ml? (b) After resetting the flow rate of the chemostat, you find that the steady-state concentration of glucose is 2.0×10^{-7} M glucose. At what dilution rate is it now operating?

4. Assume that you have a glucose-limited chemostat operating in a steady state. (a) List three changes you could make to increase the steady-state cell density in the growth vessel. (b) What change could you make to increase the growth rate (k)?

5. You have isolated a mutant of *E. coli* that has lost a certain function. When mixed with its parent and grown at a low dilution rate in a glucose-limited chemostat, the mutant persists and the parent disappears. How could you explain this phenomenon?

REFERENCES

Andersen, K. B. and K. von Meyenburg. 1980. Are growth rates of *Escherichia coli* in batch culture limited by respiration? J. Bacteriol. 144:114.

Bauchop, T. and S. R. Elsden. 1960. The growth of microorganisms in relation to their energy supply. J. Gen. Microbiol. 23:457.

Bergey's Manual of Determinative Bacteriology. 1974. R. E. Buchanan and N. E. Gibbons, eds. Williams and Wilkins Co., Baltimore.

Campbell, A. M. 1957. synchronization of cell division. Bact. Rev. 21:263.

Cutler, R. G. and J. E. Evans. 1966. Synchronization of bacteria by a stationary-phase method. J. Bacteriol. 91:469.

Duclaux, E. 1898. Traite de Microbiologie 1:208. Mason, Paris.

Friedman, H., P. Lu and A. Rich. 1969. Ribosome subunits produced by cold sensitive initiation of protein synthesis. Nature 223:909.

Guthrie, C., H. Nashimoto and M. Nomura. 1969. Structure and function of *E. coli* ribosomes. VIII. Cold-sensitive mutants defective in ribosome assembly. Proc. Natl. Acad. Sci. USA 63:384.

Helmstetter, C. E. and D. J. Cummings. 1963. Bacterial synchronization by selection of cells at division. Proc. Natl. Acad. Sci. USA 56:707.

Herendeen, S.L., R.A. VanBogelen and F.C. Neidhardt. 1979. Levels of major proteins of *Escherichia coli* during growth at different temperatures. J. Bacteriol. 139:185.

Képès, F. and A. Képès. 1980. Synchronisation automatique de la croissance de *Escherichia coli*. Ann. Microbiol. 131A:3.

Koffler, H. and G. O. Gale. 1957. The relative thermostability of cytoplasmic proteins from thermophilic bacteria. Arch. Biochem. Biophys. 67:249.

Langridge, J. 1968. Thermal responses of mutant enzymes and temperature limits to growth. Mol. Gen. Genetics 108:116.

Lemaux, P. G., S. L. Herendeen, P. L. Bloch and F. C. Neidhardt. 1978. Transient rates of synthesis of individual polypeptides in *E. coli* following temperature shifts. Cell 13:427.

Marr, A. G., E. H. Nilson and D. J. Clark. 1963. The maintenance requirement of *Escherichia coli*. Ann. N.Y. Acad. Sci. 102:536.

Monod, J. 1942. *Recherches sur La Croissance de Cultures Bacteriennes*. Hermann et Cie, Paris.

Monod, J. 1950. La technique de culture continue. Théorie et applications. Ann. L'Inst. Pasteur, Paris 79:390.

Neidhardt, F. C. and R. A. VanBogelen. 1981. Positive regulatory gene for temperature-controlled proteins in *Escherichia coli*. Biochem. Biophys. Res. Comm. 100:894.

Ng, H., J. L. Ingraham and A. G. Marr. 1962. Damage and derepression in *Escherichia coli* resulting from growth at low temperature. J. Bacteriol. 84:331.

Novick, A. and L. Szilard. 1950. Experiments with the chemostat on spontaneous mutations of bacteria. Proc. Natl. Acad. Sci. USA 36:708.

O'Donovan, G. A. and J. L. Ingraham. 1965. Cold-sensitive mutants of *Escherichia coli* resulting from increased feedback inhibition. Proc. Natl. Acad. Sci. USA 54:451.

Pirt, L. J. 1965. The maintenance energy of bacteria in growing cultures. Proc. Roy. Soc. B 163:224.

Ratkowsky, D. A., J. Olley, T. A. Meekin and A. Ball. 1982. Relationship between temperature and growth rate of bacterial cultures. J. Bacteriol. 149:1.

Saruyama, H., N. Fukunga and S. Sasaki. 1980. Effect of low temperature on protein-synthesizing activity and conservability of polysomes in bacteria. J. Gen. Appl. Microbiol. 26:45.

Scotti, P. D. 1968. A new class of temperature conditional lethal mutants of bacteriophage T4D. Mutation Res. 6:1.

Shehata, T. E. and A. G. Marr. 1970. Synchronous growth of enteric bacteria. J. Bacteriol. 103:789.

Tai, P. -C., D. P. Kessler and J. L. Ingraham. 1969. Cold sensitive mutations in *Salmonella typhimurium* which affect ribosome synthesis. J. Bacteriol. 97:1298.

Yamamori, T. and T. Yura. 1982. Genetic control of heat-shock protein synthesis and its bearing on growth and thermal resistance in *Escherichia coli* K12. Proc. Natl. Acad. Sci. USA 79:860.

Chapter Six

Growth Rate as a Variable

INTRODUCTION

An experiment in bacterial growth physiology usually begins with observations made on cells sampled from a batch culture. It is therefore necessary to become familiar with populations of growing bacteria and with the criteria that define the STATE OF GROWTH one wants to study. Unless growth is carefully monitored throughout an experiment, the results obtained may not be reproducible, whether in your own or in somebody else's laboratory. In fact, an experiment done with a poorly characterized culture is all but useless.

It is commonly accepted that measurements made on samples from exponentially growing cultures are reproducible and, furthermore, that they are relatively easy to interpret. This is correct, but it implies that sound criteria are applied to define the state of exponential growth and, beyond that, the state of balanced growth. Figure 1 illustrates this point. It shows exponential growth curves in terms of cell mass and cell numbers (see Chapter 5) in a broth medium inoculated with the small, resting-phase cells typical of a fully grown culture. Both mass and cell number can be seen to increase at the same rate, but not over the same period of time. In fact, the linear portions of the curves are parallel only during a short period approximately 1.5 hours after inoculation. By this time the vertical distance between the curves shows that the average cell mass has increased approximately fourfold. If the culture is now diluted with fresh, prewarmed broth *and* if the two curves remain parallel, we conclude that the definitive value for mass per cell, characteristic of growth in broth, at 37°C, has been established, that is, that there is balance between the rates of mass increase and of cell division. The upper part of the curves on Figure 1 shows how the initial state is gradually reestablished when

267

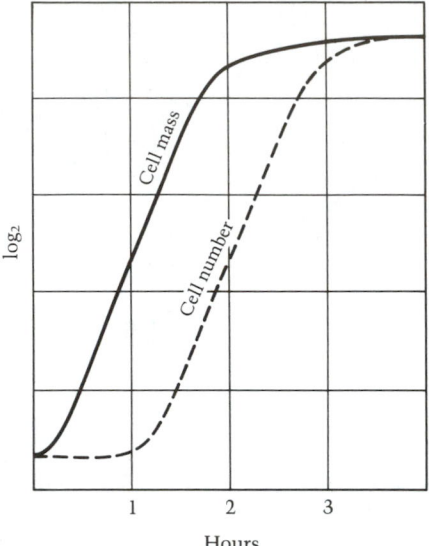

Figure 1.

Growth in terms of mass and cell number. Ordinate: mass in arbitrary units (solid line) and cell number (dashed line). The log unit represents one doubling. The curves correspond to growth in broth at 37°C with a doubling time of 25 minutes, or 2.4 doublings per hour. Note that the mass:cell ratio first increases and then decreases about fourfold. (From Maaløe, 1966.)

the cell density becomes so high that the metabolic activities of the cells begin to affect the medium, say, by changing the pH. During the transition to the resting phase, cell division outstrips the increase in mass. Figure 2 shows the impressive difference in volume between very slow-growing, virtually resting cells and those of a fast-growing culture. Note also that, in both cell populations, the biggest cells are more than twice as big as the smallest. Considering that the interdivision time varies considerably (see Chapter 5), this should be no surprise. The large variations in cell size and in division time both derive from the fact that the cycle of elongation and division is far from perfect.

BALANCED GROWTH IN THEORY AND PRACTICE

The concept of balance can be extended to all cell components, and in the ideal state of balanced growth every one of them must, by definition, increase exponentially and at the same rate. This description applies to a genetically homogeneous population of cells in equilibrium with an unchanging environment. Despite the great complexity of the individual cell, extremely simple experiments can, in principle, be done with cultures in balanced growth. In the first place, samples for optical and chemical measurements can be taken without disturbing the balance. If the culture is diluted at intervals, growth can be followed indefinitely or at least for as long as the culture remains genetically stable. Second, such measurements can be used to identify the state of balanced growth that corresponds to a given set of external conditions. And, third, the transition between two states can be studied by suddenly changing the composition of the medium or the temperature. Thus, a culture in balanced growth can in many ways be handled as if

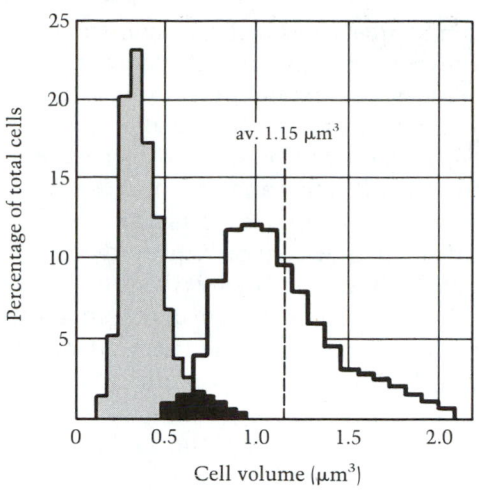

Figure 2.

Distribution of cell sizes in balanced growth of *Salmonella typhimurium*. Shaded area represents cells from a glucose-limited chemostat, growth rate 0.13 doublings per hour. White area shows cells grown in glucose minimal medium enriched with casamino acids, growth rate 2.0 doublings per hour. (From Ecker, 1963.)

it were a simple dialysis system with a large number of tiny dialysis bags—except, of course, that the bags grow and divide. (Before molecular biology changed our views so completely, some biochemists actually referred to bacteria as "bags of enzymes.")

When we pass from theory to practice, absolute terms like "unchanging environment" and "equilibrium" must yield to terms such as "essentially unchanging environment" and "sufficient approximation to balance, or equilibrium." It must be realized that these terms can be defined only empirically: the first by finding out to what density a culture can be allowed to grow before the medium no longer remains "essentially" the same; and the second by applying common sense arguments based on the process through which the ideal state of balanced growth is approached.

Some cell properties change early in the growth of a culture, that is, long before a culture density is reached at which the rate of mass increase can be seen to decrease. This was very striking in a study of lysogenization by a temperate phage: in broth cultures with cell densities ranging from 2×10^7 to 10^8 cells/ml, a constant fraction of the cells became lysogenic under standard conditions. However, at densities higher than 10^8 cells/ml, the lysogenization frequency dropped sharply, despite the fact that the rate of mass increase remained unchanged for at least one more doubling time. This rather special case is reviewed here to emphasize that it is unsafe to assume that the physiological state of the cells remains unchanged if the culture comes close to the point where the growth rate can actually be observed to decrease. Reference is often made in the literature to cultures in "early" or "late" log phase; such statements are disturbing because they indicate that the growth rate might still be increasing or, conversely, that it was about to decrease; furthermore, we are left with the impression that no great care was taken to establish exponential growth.

The simplest indication that the process of growth has become balanced is that readily measurable quantities, such as mass and cell number, increase exponentially and at the same rate. True balanced growth obviously is approached gradually and asymptotically, and strict exponential growth must therefore be observed for a considerable time before we can accept that the state of growth is "essentially" balanced. A good rule is not to start experiments in growth physiology unless both optical density and cell number have been observed to increase exponentially and in parallel by a factor of 10 or more. There are two reasons for advocating such stringent requirements: in the first place, the slope of the growth curve often increases very slowly, and a long observation time with many readings is necessary to make sure the definitive growth rate has been reached. Second, the final adjustment of parameters such as cell dimensions and number of ribosomes requires several generations (Chapter 5).

It was not obvious, a priori, that cultures satisfying the criteria just described could be maintained at cell densities suitable for biochemical and radiochemical studies. Thus, at densities of, say, 10^8 cells/ml substrate might be consumed or waste products might accumulate at rates such that the medium could not be assumed still to remain "essentially" unchanged. As it turned out, in most media and for most purposes, various strains of *E. coli* and related enteric bacteria can safely be used at densities up to approximately 2×10^8 cells/ml. This is true also of some unrelated bacteria, such as *Pseudomonas* and *B. subtilis*. With a good spectrophotometer and an electronic cell counter, it is possible to monitor growth in terms of mass and of cell number. The introduction of one- and two-dimensional gel separation techniques has greatly increased the scope of experimentation at these cell densities.

A convenient property of a system in balanced growth should be noted: when the rate of synthesis of any given protein i is expressed relative to the overall rate of protein synthesis, the ratio obtained equals the amount of i as a fraction of total protein, which we refer to as α_i.

The advantages of balanced growth—reproducibility and ease of interpretation—have their price. The system appears simple because the process of growth is treated as continuous in all respects, ignoring the fact that it consists of discrete cycles of growth and cell division. Thus, when the rate of synthesis per milliliter of culture of some cell constituent is measured, an average rate is obtained. In practice, the average is usually taken over at least 10^7 cells. This does *not* mean that such measurements are in any sense invalid, but it *does* mean that if the synthesis of some cell constituent is associated with a particular event during the cell cycle, this relationship will not be detected.

We have seen in Chapter 5 that several techniques have been developed for obtaining division synchrony in mass cultures. In principle these methods open the way to study events as functions of cell age. However, the criteria we have chosen for accepting that growth is "essentially" balanced are no longer sufficient; each synchronization technique must be critically examined before accepting it as producing "essentially" normal division cycles (see "Experiments with Synchronized Cultures" in the second part of this chapter).

GROWTH RATE-RELATED MEASUREMENTS

The concept of balanced growth is simple and offers the best possible conditions for making reproducible experiments with bacterial cultures. This is essential, but two additional features, neither of which could have been anticipated, have added enormously to the usefulness of the results of such experiments. In the first place, balanced growth can be established in batch cultures in various, chemically defined media at rates that span an almost ten-fold range; and, second, *several key parameters of growth turn out to be either independent of, or proportional to, the growth rate.* The measurements that have been made to establish this are chiefly chemical and radiochemical determinations of protein and nucleic acids. Some common features of these experiments will be discussed as background for the next section, in which the key parameters are presented.

A model experiment with generally labeled [^{14}C]glucose was described in Chapter 1 (Figure 1) to illustrate the time course of labeling of different classes of cell constituents. It was argued that the precursor pools (e.g., of amino acids) quickly reach the specific activity of the [^{14}C]glucose because the flow through these pools of newly synthesized ^{14}C-labeled amino acids is rapid relative to the pool size. In contrast, highly stable components like the proteins become labeled pari passu with growth. Thus, as total protein per milliliter increases exponentially, the fraction that was present as unlabeled protein when [^{14}C]glucose was added must decrease exponentially. A few unstable products—notably, mRNA—are labeled less rapidly than the amino acid pools and more rapidly than protein.

Choice of label

A class of macromolecules may be labeled by adding a precursor specific for that class. Thus, [^{3}H]- or [^{14}C]thymidine will be found in DNA and [^{3}H]- or [^{35}S]methionine in protein almost exclusively. Labeled adenine, on the other hand, will be incorporated as adenine *and* guanine in RNA *and* in DNA; similarly, uracil winds up as thymine *and* cytosine in DNA and as uracil *and* cytosine in RNA. The time course of labeling of RNA, whether with ^{3}H, ^{14}C, or ^{32}P, is complex because for quite some time after adding, say, labeled adenine to the medium the pool of ATP receives not only radioactive adenine from outside but also nonradioactive adenine derived from the breakdown of mRNA. This complication has been analyzed by Salser (1968) and by Winslow (1969).

Duration of labeling

Addition of a radioactive precursor—say, [^{35}S]methionine—to the medium marks a discontinuity followed by rapid labeling of the pool; and when its specific activity has come close to that in the medium, each methionine passing from the pool into protein will contribute the same amount of radioactivity. It is commonly found that the content of an amino acid pool is equivalent to the consumption of that amino acid during 10–20 seconds of protein synthesis. Thus, 1 minute after addition of [^{35}S]methionine, the pool has been renewed 5–10 times and its specific activity will be very close to that in the medium. After 10 more minutes, the protein synthesized during the labeling period will have a methionine specific activity that is 90% of that in the medium. For even longer periods, the pool effect can be neglected and the radioactivity in the acid insoluble material will be a good measure of the *amount* of protein synthesized after addition of [^{35}S]methionine.

On the other hand, the *rate* of protein synthesis—*dP/dt*—can be estimated by measuring incorporation during a short period—PULSE LABELING. At the end of the pulse, a large surplus of nonradioactive precursor may be added, thus reducing the specific activity in the medium by a factor of 100 or 1000. The latter procedure allows the radioactivity present in the pools at the end of the pulse to become incorporated into protein, whereas little activity will be taken up from the medium where the specific activity is now very low. This is called a PULSE-CHASE LABELING.

However, with short labeling periods the endogenous pool of nonradioactive precursor delays the incorporation of label. In principle this can be taken into account by determining the mean specific activity in the pool of precursor during the pulse, but this difficult measurement is usually unnecessary. In a culture in balanced growth, the absolute rate of synthesis of a stable macromolecule is best obtained by determining its amount (e.g., the number *N* of molecules per unit volume of culture) and applying the equation:

$$\frac{dN}{dt} = kN = \mu \ln 2N$$

where *t* = time, *k* = specific growth rate constant, *μ* = doublings per hour [see Equations (1)–(10) in Chapter 5]. Pulse-labeling is therefore necessary only in situations where the balance has been deliberately disturbed, and in such cases we are more concerned about relative than about absolute rates. A case in point would be measurements of the gradual reduction of the rate of protein synthesis following addition of rifampicin (see following).

Double labeling

The main virtue of double labeling is that one can correct for unequal recovery among samples. The culture under study may first be labeled for a period with [^{3}H]methionine, which may be allowed to be totally assimilated some time before pulse-labeling with [^{35}S]methionine is begun. After total assimilation, each species of ^{3}H-labeled, stable macromolecule will be represented by a fixed radioactivity per milliliter. When the relative rate of synthesis of a given protein is to be estimated, the ratio of ^{35}S to ^{3}H counts is determined on material extracted from a gel after one- or two-dimensional electrophoresis. Variations in sampling volume and in recovery of the protein are thus eliminated. (See Problem 1.)

Auxotrophs versus prototrophs

When a prototroph is labeled with a radioactive amino acid, there is no simple way of telling to what extent endogenous production of that amino acid will continue to contribute to protein synthesis. As long as balanced growth is maintained, we may assume the contribution from endogenous synthesis to remain constant; however, if the experiment involves an operation disturbing the overall balance, the balance between endogenous and exogenous amino acid supply may also change. This uncertainty is eliminated if the culture is pregrown with a supply of the amino acid in question large enough for its concentration to remain "virtually" constant during growth. Amino acids are available at very high specific acitivities, and an adequate amount of radioactivity may be added without increasing the concentration of the amino acid significantly. It is even simpler to use a mutant strain unable to synthesize the particular amino acid. However, both procedures impose limitations. On the one hand, the exogenous supply of unlabeled amino acid reduces the specific activity and hence the intensity of protein labeling; and second, it may not be possible to raise the amino acid concentration in the medium sufficiently to terminate a pulse by an effective chase.

Useful antibiotics

Many antibiotics inhibit macromolecular synthesis in bacteria, but only those for which the mechanism of action is well understood are useful for studying cell function. Five antibiotics with very different actions in the bacterial cell will be briefly discussed in terms of target, effect, and typical research application.

Rifampicin blocks the initiation of transcription by DNA-dependent RNA polymerase (RNA-P). Ongoing transcription runs to completion. An experiment with rifampicin typically consists of adding the antibiotic together with a labeled precursor for either RNA, protein, or DNA and then watching the RUN-OUT of synthesis of the labeled material. Because the target for rifampicin is the RNA-P, the direct effect of this antibiotic is observed if the precursor chosen is one that labels RNA. Specific experiments of this type will be presented later (see Figures 9–11). If the precursor used is a labeled amino acid, we observe the run-out of protein synthesis. This is an indirect effect of rifampicin due to the instability of mRNA, which must be continuously renewed by de novo transcription if protein synthesis is to continue. Figure 11A illustrates how the protein run-out curve can be used to estimate the half-life of mRNA. Finally, we may choose to label with a DNA precursor (usually thymidine), in which case we observe a very protracted run-out. The reasons for this are, in the first place, that de novo RNA synthesis is required at the site of initiation for a new round of DNA replication to be started and, second, that a single round of replication lasts 40–50 minutes (at 37°C). It therefore takes that long for replication to run out in those cells in which a round of replication had just been initiated when rifampicin was added.

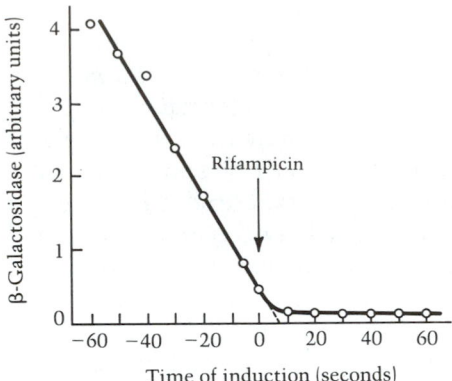

Figure 3.

Estimation of the lag before induction of β-galactosidase is inhibited by rifampicin. The drug was added at $t = 0$, and IPTG was added at the times shown on the abscissa. Strain AS19 was grown in glycerol minimal medium ($\mu = 1.1$); after induction, time was allowed for expression of enzyme synthesis. (From Pato, 1970.)

A necessary and obvious condition for doing "clean" experiments with an antibiotic is that the drug enters and becomes effective in the cell very rapidly; ideally, all the cells in the culture experience the full effect of the drug instantaneously. In practice, this means within a few seconds after addition of the antibiotic. Figure 3 illustrates a control experiment with rifampicin which shows that the full effect was expressed within 5–10 seconds. However, this favorable condition does not normally obtain with enteric bacteria, and most of the antibiotic experiments discussed here were therefore done with a mutant (AS19) of the B strain of *E. coli*. AS19 cells are highly permeable to several antibiotics, but they grow as rapidly as do cells of strain B in many media. In general, Gram-positive bacteria, such as *Bacillus subtilis*, are at least as permeable as cells of strain AS19, whereas most Gram-negative bacteria, including the enterics, are inhibited only by very high concentrations of many antibiotics.

Streptolydigin also interacts with the RNA-P, but in contrast to rifampicin it stops ongoing RNA synthesis. Streptolydigin can be used to estimate the transcription time for the mRNA corresponding to an inducible enzyme.

Chloramphenicol binds to ribosomes, and in adequate concentrations it stops protein synthesis very rapidly by inhibiting peptidyltransferase. It was mentioned that this effect produces a run-out of DNA synthesis. The three antibiotics so far described are included in the experiments of Problem 2.

Fusidic acid is interesting because it slows down the movement of ribosomes along mRNA by blocking the function of elongation factor G (EF-G). This antibiotic is not ideal for in vivo experiments because it penetrates slowly into *E. coli* cells, even into the otherwise highly permeable cells of the mutant strain AS19. In experiments to be described later, fusidic acid was used to show how *E. coli* responds to a specific reduction of the efficiency of its ribosomes.

Nalidixic acid can be used to stop DNA synthesis; it acts rapidly by inhibiting DNA gyrase. In most bacteria, nucleic acid precursors are incorporated into DNA as well as RNA, but in some situations it is desirable to exclude label from DNA. Figures 10 and 11A show run-out experiments with rifampicin, in which the results would have been distorted if the labeled precursor had entered DNA as well as RNA. Nalidixic acid was used to prevent this.

PARAMETERS CHARACTERIZING BALANCED GROWTH

In Chapter 1 (Table 1) we presented the chemical composition of an average cell from a culture in balanced growth at 37°C in glucose minimal medium. The macromolecules total approximately 96% of the dry weight, and the specific components of the envelope—the lipids, lipopolysaccharides, and peptidoglycan—amount to 17–18%. Thus, protein, RNA, and DNA together account for nearly 80% of the dry weight, and the first thing now to be considered is how the *relative* amounts of these major classes of macromolecules vary with the growth rates. For the time being we will disregard the fact that slow-growing cells are small and fast-growing cells big and will concentrate on the amounts of protein, RNA, and DNA per milliliter, normalizing to unit culture density ($A = 1$).

A particularly complete set of chemical measurements is presented in Figure 4. First we note that for $\mu > 0.6$–0.7, the RNA and protein measurements can be combined to show that the RNA:protein ratio, R/P, is proportional to μ, that is, $(R/P) = \mu(R/P)_{\mu=1}$. In the same range, the tRNA:total RNA ratio remains constant. We shall now show that the $R:P$ ratio permits us to estimate the efficiency of the ribosomes at different growth rates. This efficiency is expressed in terms of the number of amino acids added per second to a growing polypeptide chain—for short, the CHAIN GROWTH RATE (cgr_p). Consider balanced growth of a bacterial system with an arbitrary number N of amino acids in all of its proteins. In this system the number of amino acids incorporated per second into protein is $N\mu \ln 2/3600$. This process of polymerization is sustained by a number of ribosomes that can be calculated from the $R:P$ ratio, the known average molecular weight of an amino acid residue in protein (110), and the molecular weight of the RNA of a ribosome (1.45×10^6)—provided we also know the fraction of total RNA that resides in active ribosomes (i.e., in polysomes). This fraction is between 0.7 and 0.75 and is obtained by subtracting from the total RNA 10–15% for the pool of ribosomes not engaged in protein synthesis, 12% for tRNA, and 3–4% for mRNA. The pool of free ribosomes and ribosomal subunits has been determined by Forchhammer (1971) and shown to vary little, if at all, with μ. Though of minor importance at this point, the amount of mRNA plays a major role in a different set of calculations to be presented later.

The number of active ribosomes in our system is found by dividing the total weight of the RNA in active ribosomes $[N\mu(R/P)_{\mu=1}] \times (110 \times 0.7)$ by the total molecular weight of the 4565 nucleotides in one ribosome (1.46×10^6). (In our notation the subscript ($\mu = 1$) identifies the $R:P$ ratio as that obtained at a growth rate of μ equal to 1; thus, the whole term $\mu(R/P)_{\mu=1}$ stresses the proportionality between R/P and μ.) If $(R/P)_{\mu=1} = 0.23$, $N\mu \times 1.23 \times 10^{-5}$ is the number of active ribosomes in a system with N amino

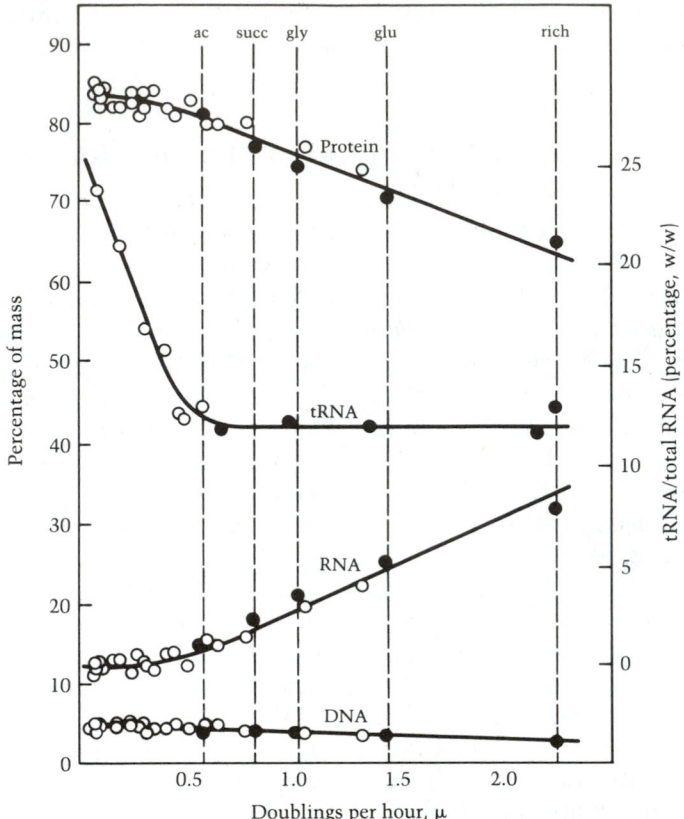

Figure 4. Cell composition at high, intermediate, and low growth rates. Filled circles refer to balanced growth in acetate, succinate, glycerol, glucose, and enriched media, as indicated on top of the frame. Open circles represent growth in a glucose-limited chemostat. (From Jacobsen, 1974.)

acids. Finally, the number of amino acids incorporated per second, $N\mu \ln 2/3600$, is divided by the number of active ribosomes. At this point, N and μ cancel; and for μ-values higher than approximately 0.7, we obtain a constant cgr_p of approximately 16 at 37°C. In other words, *the efficiency of an active ribosome appears to be constant.*

A second notable result of the experiment in Figure 4 is that macromolecular composition is the same at a given μ-value whether the cells utilize glucose in limited supply (as in a chemostat) or a poor carbon source. We shall return to this point when discussing the actual source and energy consumption during growth in different media.

The constant efficiency of the ribosomes over a fourfold range of growth rates (μ between 0.7 and 2.8) is remarkable, if only because it makes such obvious sense. As we have seen (Chapter 3), most of the carbon and energy consumed during the growth of bacteria go to make protein, and the overall

PROTEIN-SYNTHESIZING SYSTEM (PSS) is large and complex. It comprises not only the ribosomes, tRNA, and mRNA, but, in addition, a whole set of indispensable proteins, such as RNA-P, aminoacyl–tRNA synthetases, and a variety of factors; by itself the PSS makes up as much as 50% of the mass of rapidly growing cells (Maaløe, 1979). Therefore, the evolutionary advantages of high growth rates point to the PSS as perhaps the most important subsystem to be effectively regulated. Ideally, it should never be bigger than necessary to process the precursors that can be supplied from the raw materials in the growth medium; here and in Chapter 8 we shall examine how this is achieved. We can anticipate two levels of control: one at which the size and capacity of the PSS is balanced against the capacity of the combined biosynthetic and fueling systems that produce precursors and energy, and a second level at which the production of the many individual components within the PSS itself are balanced.

Size of the protein-synthesizing system

Estimates of the size of the PSS and its variation with growth rate are based essentially on *measurements of the rate of synthesis of total ribosomal proteins relative to the total rate of protein synthesis* (α_r) *during balanced growth*. To this is added the relative rate of synthesis of those nonribosomal proteins that belong to the PSS. A technique for determining α_r was introduced by Schleif (1967), and it was modified and made more generally useful by Gausing (1974). Figure 5 shows an almost complete set of her α_r measurements. The nonribosomal proteins were scored by Pedersen (1978a) on the O'Farrell type of two-dimensional gels (Figure 3 in Chapter 1).

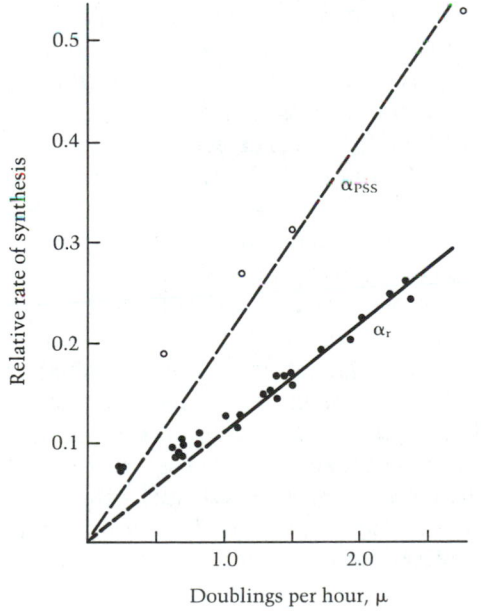

Figure 5.

Relative rates of synthesis of ribosomal proteins, α_r, and of all the proteins of PSS, α_{PSS}, as determined by Gausing (1974) and by Pedersen et al. (1978a), respectively. Measurements were on cultures in balanced growth; relative rates therefore equal relative abundance of the proteins. (From Maaløe, 1979.)

The α_r values of Figure 5 were determined by the pulse-chase method and are therefore rate measurements; however, they were obtained from samples of cultures in balanced growth, and these values can therefore be read also as relative amounts of the ribosomal proteins. They vary from approximately 7 to 28 percent of the total protein of the cells (this set of α_r values does not extend to $\mu = 2.8$, the maximum growth rate, but additional measurements show that the trend continues). The α_r data show a definite resemblance to the R/P values we derived from Figure 4; in both cases the measured parameter is proportional to μ, with notable deviations at low growth rates. We shall therefore use the same notation as before and write $\mu \cdot \alpha_{r,\mu=1}$ to stress the proportionality between α_r and μ. We can calculate the cgr_p the same way we did on the basis of the $R:P$ ratio. Using α_r, we can express the number of active ribosomes, r_{act}, in a system with a total of N amino acids in the proteins as follows:

$$r_{act} = (\mu \alpha_{r,\mu=1} N)\,(0.85/8000)$$

where the number in parentheses gives the amino acids in active ribosomes, and 8000 is the number of amino acids in one set of ribosomal proteins. When the number of amino acids incorporated into protein per second ($N\mu \ln 2/3600$) is divided by r_{act}, μ and N again cancel; and for an $\alpha_{r,\mu=1}$ of 0.10 to 0.11, we obtain a cgr_p of 17–18 amino acids per second. Thus, two rather different approaches—one in which the number of ribosomes is based on their RNA and the other in which it is based on their protein content—give virtually identical values of cgr_p and show it to be constant for μ values higher than 0.7–0.9.

Inefficiency of PSS at low growth rate

In the range of growth rates within which R/P and α_r both are proportional to μ, the size of the PSS is effectively regulated and correctly adjusted to the growth rate the medium permits. However, both parameters deviate from proportionality at the lowest growth rates. For example, cgr_p for a μ of 0.3 is approximately 8 amino acids sec^{-1}. However, the simple experiment of Figure 6 shows that extremely slow-growing cells ($\mu = 0.1$) respond to induction of β-galactosidase synthesis as fast as do rapidly growing cells. The lag in synthesis agrees well with the time needed to make a β-galactosidase polypeptide chain at the normal cgr_p (Figure 6), and we therefore conclude that the rate of chain elongation cannot be reduced significantly even at extremely low growth rates. The rate of protein synthesis, dP/dt, can be written as the product of two terms $r_{act} \times cgr_p$; and if cgr_p is not reduced at low growth rates, the term r_{act} must go down. This could be because part of the ribosomes synthesized at low growth rates are defective or temporarily inactivated. The latter is obviously the case because the "excess" ribosomes carried by slow-growing cells can become active in protein synthesis in a very short time. This is demonstrated by the experiment in Figure 7, in which cells growing very slowly because of glucose limitation are supplied with an excess

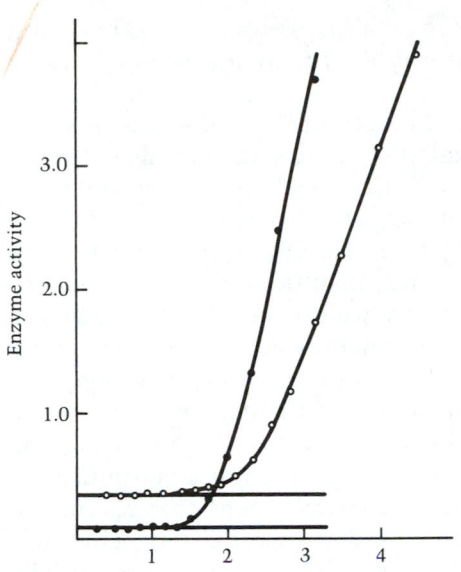

Figure 6.

Induction lags at extreme growth rates. At $t = 0$, IPTG was added to induce maximal synthesis of β-galactosidase. Filled circles represent glucose minimal medium ($\mu = 1.5$), and open circles refer to a glucose-limited chemostat ($\mu = 0.1$). (From Jacobsen, 1974.)

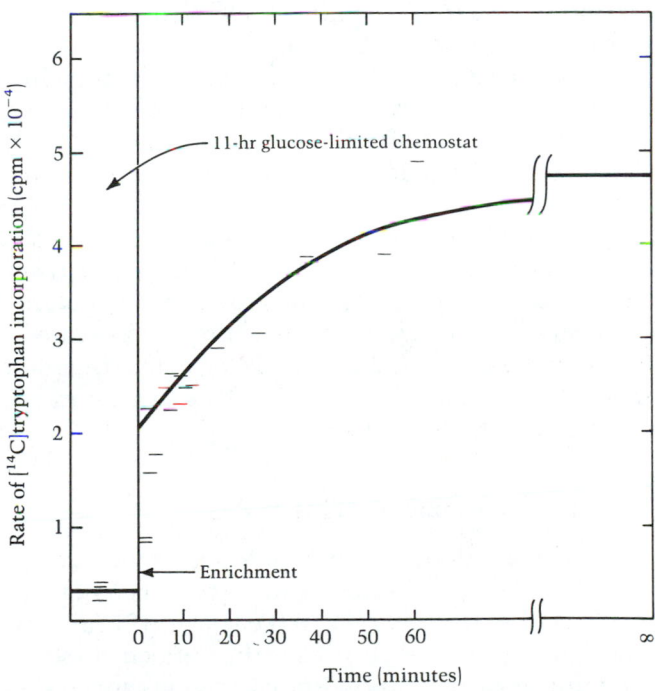

Figure 7. Relative rates of protein synthesis before and after a dramatic upshift, measured by 2-minute pulses of radiolabeled tryptophan. (From Maaløe, 1979.)

of glucose and rapidly increase their dP/dt by a factor of 5 or 6. This jump occurs within such a short time that it cannot be attributed to de novo ribosome synthesis.

This brief analysis of the apparent inefficiency of the ribosomes in slow-growing cells does not resolve the dilemma. We are left with an unanswered question: Why do these cells "overproduce" ribosomes? As will be explained in Chapter 8, the adjustment of the number of ribosomes to the growth conditions, or to μ, can be accounted for by two different modes of regulation, one "passive," the other "active." At this point, it suffices to say that according to the former the mechanism involved may well be incapable of reducing α_r below the observed minimal value of approximately 0.07. The active mode of regulation, on the other hand, might have evolved to do better but may not have done so because it would be a disadvantage to the cells ever to reduce the number of ribosomes too severely. The argument here is that it may be important for the cells, not only to survive under starvation conditions, but also to be able to start growing rapidly once the environment changes for the better. Cells with a very efficient mechanism of regulation and hence left with few ribosomes at the end of a period of semistarvation would resume growth very slowly when external conditions improved, and they would compete unfavorably with cells that started growing more rapidly because they had more ribosomes. This ecological problem was discussed by Koch (1971) as the dilemma of "feast and famine." We shall return to the problem of passive versus active control in Chapter 8.

Great importance has been attached to measurements of the average elongation rate of polypeptides (cgr_p); one reason, of course, is that the average probably also applies to individual proteins, because once the synthesis of a polypeptide chain has been initiated, all steps are the same except for the identity of the amino acid to be added next. A few experiments have been made to verify this. Figure 8 shows measurements by Engbaek (1973) of the time it takes for the NH_2-terminal amino acid of β-galactosidase to appear in finished molecules. The cgr_p at various growth rates, obtained in this study, agrees with the average, or global, values already presented. The kinetics of ribosomal protein synthesis is likewise in agreement, as is the synthesis of the RNA-P subunits β and β'. The cgr_p of the latter can be estimated from pulse-chase experiments (Problem 3).

Nonribosomal proteins of PSS

The nonribosomal proteins can, as we have seen, be displayed on two-dimensional gels, and some of them have been quantitated at growth rates ranging from $\mu = 0.55$ to $\mu = 2.8$ (Pedersen, 1978a). Approximately 200, including the r-proteins, have been identified (Bloch, 1980; Phillips, 1980). A considerable number of the nonribosomal proteins increase in relative abundance with the growth rate; that is, their individual α_i values behave more or less like α_r. Among the proteins identified on the gels, all those that belong to the PSS (the RNA-P, the aminoacyl-tRNA synthetases, and the elongation

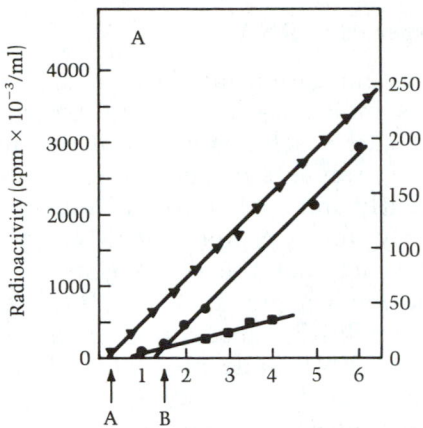

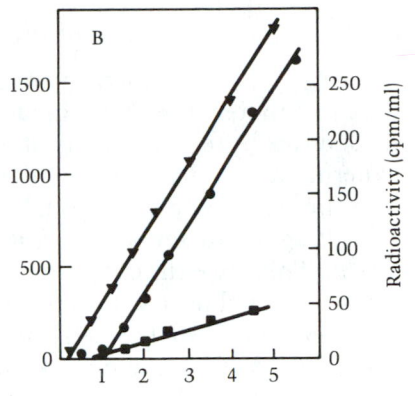

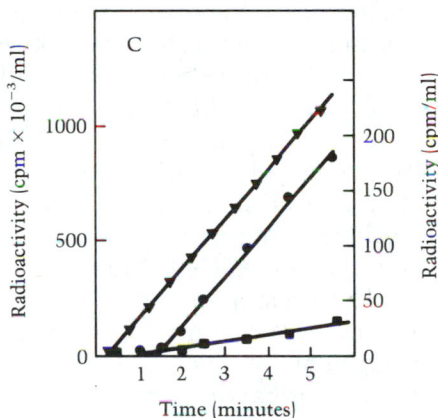

Figure 8.

Incorporation of [^{14}C]threonine into total protein and into NH$_2$-terminal threonine in purified β-galactosidase. Fully induced, exponentially growing cultures were labeled from $t = 0$; counts per minute and per mass unit in total protein and in NH$_2$-terminal threonine are shown by (▼) and by (●) respectively (left-hand scales). Counts in NH$_2$-terminal threonine from uninduced cultures (■) are shown on the right-hand scales. A. Supplemented glucose medium; $\mu = 1.9$. B. Glycerol medium; $\mu = 1.0$. C. Acetate medium; $\mu = 0.5$. (From Engbæk, 1973.)

factors G, Tu, and Ts) behave in this way. By extension, all the proteins that exhibit an α_r-like relationship to μ are, provisionally, assigned to the PSS; their total mass is not very different from that of the 50-odd r-proteins. All proteins of the PSS are, by definition, essential to protein synthesis, and their collective mass, relative to total protein, is designated α_{PSS}. In Figure 5 our present crude estimates of α_{PSS} are compared with the α_r curve. As can be seen, during the most rapid growth a little more than half of the cells' protein is devoted to protein synthesis; add to this the RNA of the cells, all varieties of which serve the same purpose, and it is clear that at maximum growth rate the PSS accounts for more than half of the total mass of an *E. coli* cell. Evidence from two-dimensional gels will be used extensively in the section on amounts and instability of mRNA and in Chapter 8, when discussing promoter strength.

Synthesis of the stable species of RNA

For polypeptide synthesis, elongation rate (cgr_p) is an important parameter; the same holds for RNA molecules. An elegant experiment to determine the cgr_{rRNA} is shown in Figure 9. This is a typical rifampicin experiment, as described earlier; the drug was added together with labeled adenine, and samples were taken frequently and applied to one-dimensional gels designed to separate 5 S RNA clearly from 4 S RNA, or tRNA (Molin, 1976). This experiment is based on the fact that the rRNA genes form a transcriptional unit with 5 S RNA at the end (Chapter 2). Thus, all the RNA-P molecules in all the cells of the culture finish transcription of the rRNA genes by synthesizing one of the small 5 S RNA molecules. These function as "tags" at the 3' end of the transcripts, and the measurement consists in determining the time at which label no longer appears in the 5 S RNA. In these experiments it is immaterial that pools label only gradually, because all that matters is the time at which RNA synthesis stops (Chapter 2). The cgr_{rRNA} measured by this technique is 50–60 nucleotides per second at all the growth rates examined ($\mu > 0.8$). We shall see presently that very similar estimates are obtained for the cgr_{mRNA} for some large proteins (β-galactosidase and the β and β' subunits of the RNA-P); and, within the accuracy of the experiments, RNA transcription appears to proceed at an elongation rate just about three times that of the cgr_p, that is, at a rate that would permit the first ribosome attaching to the 5' end of a nascent mRNA molecule to proceed not far behind the transcribing RNA-P molecule.

The rifampicin experiment just described is a special case of the RNA RUN-OUT METHOD introduced by Pato (1970). Figure 10 shows the outcome of experiments of this type at three different growth rates. A great deal of information can be gleaned from these curves; each of them is composed of an ascending part during which RNA synthesis still goes on, a descending part during which the labeled mRNA decays, and, finally, a horizontal part representing label in stable RNA. Two complicating features should be noted. In

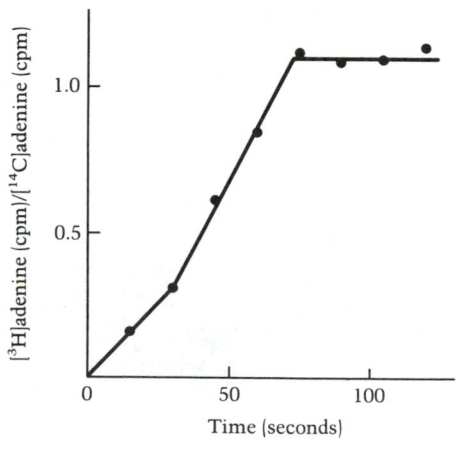

Figure 9.

Kinetics of 5 S rRNA synthesis after addition of [³H]adenine and rifampicin at $t = 0$. The culture was grown in glucose minimal medium and was prelabeled with [¹⁴C]adenine. Samples were collected into SDS at 90°C; and 5 S rRNA was separated on 10% acrylamide gels. (From Molin, 1976.)

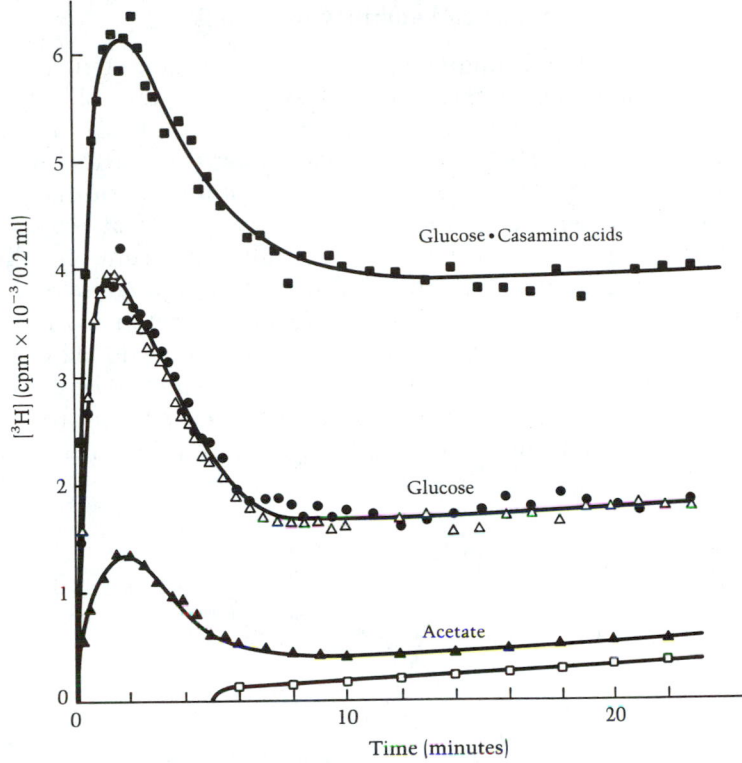

Figure 10. Residual RNA synthesis determined by incorporation of [^{3}H]uridine added together with rifampicin and nalidixic acid at $t = 0$ (see also Figure 3). The cultures of strain AS19 in enriched medium, in glucose, and in acetate media were in balanced growth at μ values of 2.4, 1.4, and 0.5, respectively. The bottom curve ($\square$) shows incorporation when [^{3}H]uridine was added 5 minutes after the drugs; this control did not differ between experiments. (From Pato, 1970.)

the first place, the initial slope of the curves does *not* reveal the full rate of synthesis of RNA at $t = 0$; to measure that, it is necessary to correct for the increase in pool specific activity (see Problem 4). Second, because run-out of synthesis and onset of mRNA decay overlap, the peak value and the time it is reached do not express accurately either total RNA synthesis or its duration, respectively. However, although the initial slope underestimates the initial rate of RNA synthesis, the relatively small uncertainties attaching to height and position of the peak do not obscure the important features of the experiment: first, that the peak positions roughly coincide, in agreement with the notion of a *cgr*$_{RNA}$ that is independent of growth rate; and second, that mRNA accounts for a large fraction of total RNA synthesis, as indicated by the difference between the peak value and the final plateau; this fraction increases as μ goes down. We note also that at all three growth rates the plateau value is reached at more or less the same time.

Amount and stability of mRNA

In a semiquantitative way the rifampicin run-out experiment identifies the two parameters needed to estimate the mRNA content in cells during balanced growth: the overall rates of synthesis and of decay of the many species of mRNA. We shall discuss decay of mRNA first. The run-out of protein synthesis has been examined in parallel with that of RNA, and the results are shown in Figure 11A. We first note that very little protein is synthesized after the initial 12–14 minutes, the time at which all mRNA seems to have decayed. Figure 11B illustrates the progress of decay and shows that, after a lag of 1 to 1.5 minutes, the rate of protein synthesis (the slope of the run-out curve) falls exponentially with time. The lag can be interpreted as the average transcription time, that is, the time during which still unfinished messenger molecules are completed. The rate of protein synthesis remains unchanged during the lag because newly finished mRNA molecules

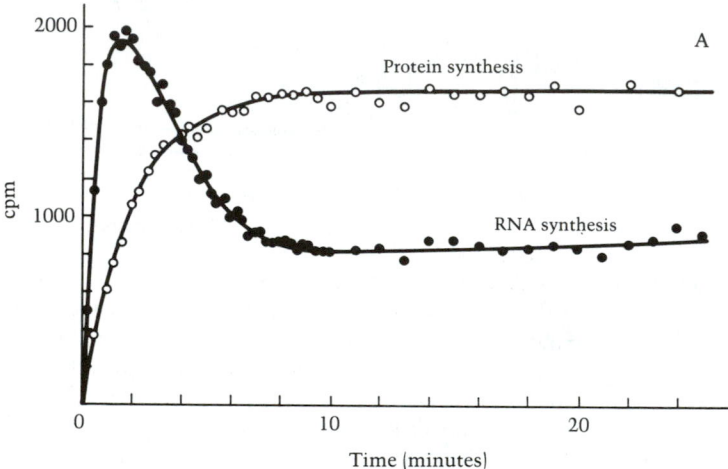

Figure 11.

Residual RNA and protein synthesis in the presence of rifampicin and naladixic acid. A. Filled circles repeat the glucose experiment of Figure 10; open circles show simultaneous measurements of protein synthesis by incorporation of [^{14}C]proline (cumulative). B. The decline of the rate of protein synthesis with mRNA decay; the points represent derivatives of the protein curve in A, obtained from pulse-labeling (20-second pulses) with proline. (From Pato, 1973.)

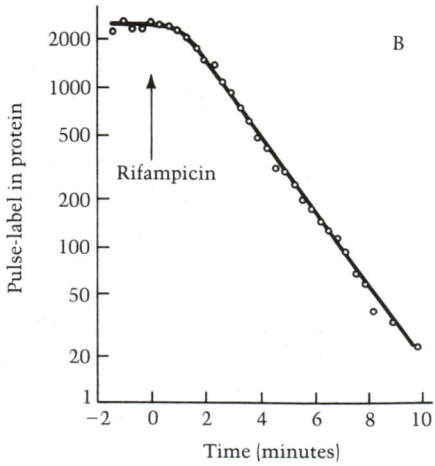

replace those that decay. The slope of the curve indicates that the bulk of the mRNA has a functional half-life of approximately 1.3 minutes. Other experiments indicate that the chemical half-life is very similar.

Surprisingly little is known about the decay process, except that it starts at the 5′ end of the mRNA. It is generally assumed that decay proceeds behind the last ribosome to attach to the messenger. The signal to initiate decay presumably resides in a base sequence near the 5′ end; and, for all we know, this critical, but as yet unidentified sequence may differ from one species of mRNA to another and cause their half-lives also to differ. This has been tested by conducting run-out experiments in which the proteins were pulse-labeled at various times after adding rifampicin and were subsequently separated on two-dimensional gels (Pedersen, 1978b). The half-life of a particular species of mRNA is then estimated by plotting the radioactivity in the corresponding protein spot on the gels, as a function of the pulse time. The result is that most half-lives fall between 0.5 and 2.0 minutes, but that rare species of mRNA exist that have much longer half-lives. As pointed out earlier, the protein run-out curve of Figure 11A shows that only a very small fraction of the protein of *E. coli* can originate from very long-lived mRNA.

The relative rate of synthesis of mRNA, $d(\text{mRNA})/d(\text{total RNA})$, can be estimated in different ways; one approach requires measurement of the specific activity of the precursors in the pool (Problem 4). However, this complication can be circumvented by hybridizing pulse-labeled RNA to an excess of DNA encoding one set of rRNA genes. In this way, and by taking the known contribution of tRNA into account, the fraction of stable RNA contained in the pulse-labeled material can be estimated, and the $d(\text{mRNA})/d(\text{total RNA})$ calculated. Extensive measurements of this type have been made by Gausing (1977). In the first place, she found that the relative rate of synthesis of rRNA increased with growth rate (between μ values of 0.15 and 2.0), but that it increased *less* than the rate of accumulation of ribosomes. The conclusion is that although rRNA in ribosomes is indeed stable, some of the newly synthesized rRNA is broken down, and more so the slower the cells grow; at very low μ values approximately 60–70% and at high μ values approximately 10% of the total rRNA synthesized seems to be broken down. The possible significance of this will be discussed in Chapter 8. Second, hybridization to DNA encoding a group of r-proteins shows that the rates of r-protein mRNA synthesis correlated well with the rate of r-protein synthesis, α_r, indicating that the latter is regulated at the level of transcription. Finally, the mRNA:total RNA ratio during balanced growth was calculated, and the result, which agrees with measurements by others, was that mRNA amounts to 3–4% of total RNA at all the growth rates examined. In this calculation the global half-life of mRNA is assumed not to vary with μ; the RNA run-out curves of Figure 10 provide good support for this.

The life cycle of a polysome

The polysome life cycle begins with the initiation of transcription of a gene, soon after which a ribosome attaches to a specific binding site near the 5' end of the nascent mRNA molecule. The cycle ends when the last ribosome to attach at this site has completed its round of translation and the last protein molecule derived from that particular transcript is finished. The point of this analysis is to estimate the average number of protein molecules produced per transcript and to see how this number relates to the growth rate. First we note that at μ values above 0.7–0.9 total RNA is accounted for as follows: 3–4% mRNA, 12% tRNA, and approximately 85% rRNA; of the latter, approximately 85% are in active ribosomes, that is, in polysomes. It follows that each active ribosome with its approximately 4600 nucleotides can be allotted a piece of mRNA with approximately 225 nucleotides. The latter is calculated by multiplying 4600 by the ratio between the amounts of RNA in mRNA and in active ribosomes [0.035/(0.85)(0.85)]. This is the average spacing of ribosomes in a polysome; it takes approximately 4 seconds for a ribosome to cover this distance and for a nascent polypeptide to grow by some 75 amino acids. Second, we recall that the global messenger half-life is approximately 1.3 minutes and varies little, if at all, with μ; this half-life corresponds to a mean lifetime of 1.9 minutes (1.3/ln 2). The average number of ribosomes that will translate the messenger is obtained simply by dividing the mean lifetime (114 seconds) by the average time (4 second) between successive ribosome attachments at the 5' end. Thus, on the average, *initiation of transcription of a gene gives rise to the synthesis of 25–30 molecules of the corresponding protein* (Figure 12).

It must be kept in mind that this estimate is based entirely on average values. However, not only does the lifetime vary considerably between different species of mRNA, but, as shown by Steitz (1979), the sequences defining the RIBOSOME ATTACHMENT SITES are similar, but not identical. A given species of mRNA may, therefore, combine a relatively long lifetime with an unusually effective ribosome binding site, and the average number of protein molecules synthesized from such a messenger would exceed considerably the figure of 25–30. This distinction between averages and individual values is important in the final step of this analysis of polysomes—the calculation of frequencies of initiation of transcription at different growth rates.

The general question can be stated this way: How many mRNA transcripts must be initiated per second to sustain protein synthesis at a given growth rate? In a model system with N amino acids in its proteins, the number of protein molecules of average size (400 amino acids) that must be produced per second equals $(N\mu\ln 2)/(400 \times 3600)$. If each gene were transcribed from its own promoter, the number of polysomes needed would be approximately four times the number of protein molecules to be synthesized, because each polysome, on the average, would deliver a newly finished protein every 4 seconds. However, several genes are often grouped behind one promoter, and a polygenic transcript would form a polysome from which, every 4 seconds, a copy of each of the proteins encoded on the mRNA would be finished. The average number of genes transcribed on one mRNA is not known, but the best estimates put the mean size of an mRNA at 2000–3000 nucleotides, that is, equivalent to 2–3 genes of average size (Molin, 1974). The number of polysomes, therefore, equals the number of protein molecules synthesized per second, multiplied by 4, and then divided by 2 or 3 for the number of gene

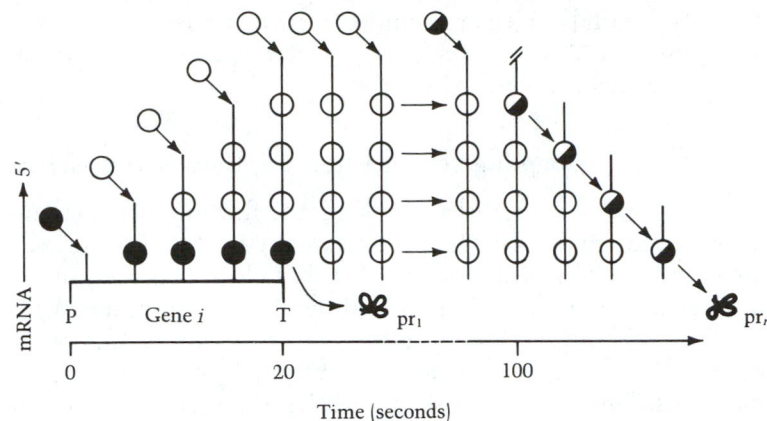

Figure 12. The "life cycle" of a polysome of average lifetime, whose mRNA is transcribed from a gene *i* of average size. P and T denote promoter and terminator; pr₁ and prₙ denote the first and the last protein molecule to be synthesized; the filled and the half-filled circles represent the first and the last ribosome, respectively, to attach to the mRNA.

transcripts per polysome. To maintain this standing population of polysomes, all the 5′ ends of mRNA that are eliminated by the onset of decay must be replaced by a new 5′ end created by an act of transcription. With an average mRNA half-life of approximately 80 seconds, nearly 1% of the polysomes must be replaced per second. To arrive at a meaningful initiation frequency, we introduce the number (N) of amino acids (10^9) in the proteins of the average *E. coli* cell growing in glucose minimal medium at 37°C (Table 1 in Chapter 1). The requisite number of initiations of mRNA transcription is approximately 10 per second. This frequency is proportional to μ at μ values higher than 0.7–0.9 and maybe also at low growth rates, as neither the mRNA:total RNA ratio nor the mRNA half-life seem to change. We do not yet understand how slow-growing cells manage to use relatively few of their ribosomes, albeit with full efficiency (normal *cgr*ₚ).

In order to estimate the frequency of initiation (F_i) at the promoter of a given gene (i), consider again a cell with N amino acids in its proteins and growth rate μ. In general, the rate of synthesis of protein molecules of type *i* (dn_i/dt) equals the rate of total protein synthesis (dP/dt) multiplied by the fractional yield of protein *i* (α_i) and divided by the number of amino acids in protein *i* (aa_i). The calculation then proceeds as above. As an example, we may calculate the frequency of initiation at the *lac* promoter necessary to maintain β-galactosidase ($aa_i = 1000$) at 1% of total protein in a fully induced culture growing at μ = 1.0. In a cell with 10^9 amino acids in its proteins, we find that two monomers must be synthesized per second; and knowing that the half-life of β-galactosidase message is close to the average mRNA half-life, we calculate that the requisite number of initiations at the *lac* promoters of the cell is 0.08 per second. Because the cell we consider contains approximately three copies of the *lac* gene, an initiation must occur approximately every 30 seconds at each promoter. If the ribosome attachment site on the *lacZ* gene were "better" or "worse" than average, our estimate would be too low or too high, respectively.

Coupling between transcription and translation

As already mentioned, the first ribosome to translate a nascent mRNA binds to the attachment site at the 5' end very soon after initiation of transcription. The best evidence for this close coupling *in time* between the two starts is provided by the revealing electron micrographs of Miller (1970).

Throughout this section on the parameters characterizing balanced growth, the emphasis has been on protein synthesis. As a consequence, we have considered only mRNA molecules that took up ribosomes and participated in protein synthesis by forming polysomes. In other words, our calculations of frequencies of mRNA transcription relate exclusively to events that lead to the synthesis of *productive* mRNA. Considering that the cells carry a surplus of RNA-P (see Chapter 8), *abortive* mRNA synthesis might be a frequent event. This is difficult or impossible to detect by labeling experiments. Thus, if an act of mRNA transcription were followed by decay of the nascent RNA chain at a distance of no more than 200–300 nucleotides behind the RNA-P molecule, the time resolution of a pulse-chase experiment would be insufficient to detect it. However, abortive mRNA transcription would, in principle, be revealed on electron micrographs of the type shown in Figure 20 in Chapter 1. If excess transcription not leading to protein synthesis were at all frequent, RNA-P molecules would be seen on DNA segments *between* the points where polysomes branch off. In high resolution micrographs (negative stain), structures are revealed at the root of polysomes that can be interpreted with confidence as RNA-P molecules. Isolated objects of this kind are not seen on pictures such as Figure 20 in Chapter 1; however, they are not easy to identify and might be overlooked on pictures showing sufficiently long stretches of DNA.

In the absence of positive indications of abortive transcription, a functional coupling between transcription and translation becomes a possibility. Stent (1966) conjectured that, in vivo, transcription might require the formation of a complex between the RNA-P and the leading ribosome on the mRNA in order to raise the cgr_{mRNA} above the very low values that were observed in vitro. However, transcription in vitro can now be made to proceed about as fast as in vivo, and it has been shown that the cgr_p can be reduced (e.g., by fusidic acid) without slowing transcription. In this and other cases we must conclude that the distance between the RNA-P and the leading ribosome grows as transcription progresses. The resulting exposure to nuclease action of a stretch of a "naked" mRNA is thought to cause the phenomenon called INTRAGENIC POLARITY (Pato, 1973). For all these reasons, Stent's original model cannot be correct, but it may contain an important truth: the leading ribosome, or some other factor of initiation of translation, could be required to release RNA-P from a position near the promoter but not to sustain transcription once activated. Furthermore, translation and transcription are interrelated in the process by which transcription is terminated at specific attenuation signals (Chapters 3 and 7). Here, however, we are not concerned with these specific control devices.

Coupling between translation and transcription would account for mRNA being synthesized in proportion to the number of ribosomes. The remarkable constancy of the mRNA:total RNA ratio at all growth rates would thus be assured. This would also be the case if abortive transcription occurred but the transcripts were kept very short by trimming; it would not be the case if they followed the decay pattern of the unstable fraction of the rRNA.

When we try to imagine how the complicated process of initiation of transcription and translation proceeds, it is pertinent to ask what the role of diffusion of large entities like ribosomal subunits or RNA-P molecules may be. It is probably not possible to calculate rates of diffusion for such particles in the crowded cytoplasm of a bacterium, but the constancy of the mRNA:total RNA ratio permits us to set certain limits. We know that approximately 15% of the ribosomal material forms a pool of 30 S, 50 S, and, perhaps 70 S particles from which the subunits are recruited to initiate translation. (In this context it does not matter whether a ribosome leaves a polysome as a 70 S particle or as a pair of 30 S and 50 S subunits.) It is also know that over a fairly wide range of growth rates (μ between 0.7 and 2.8) the average time between the arrival of successive 30 S subunits at a ribosome binding site at the 5' end of a messenger is always approximately 4 seconds. The following reasoning shows that diffusion accounts for only a small fraction of this time: the number of ribosomes and the fraction of them residing in the pool, as well as the number of binding sites on polysomes, are proportional to μ; and, if diffusion were rate limiting, the number of ribosomes binding to one of the polysomes in the cell per second would be proportional to μ^2. This would imply that the number of polysomes, and hence dP/dt, is proportional to μ^2 and not to μ as, in fact, it is.

With an average of approximately 400 amino acids per protein molecule, an act of translation lasts 20–25 seconds; and with approximately 1/6 of the ribosomes in the pool, the average time they reside there is 3–4 seconds. This is an independent estimate of the previously calculated time interval of 4 seconds between ribosomes moving along an mRNA strand, but it does not add to our understanding of what happens during the relatively long time ribosomes stay in the pool. Because diffusion to a new binding site probably takes a fraction of a second, the 3–4 seconds of residence in the pool may be needed to "process" a ribosome for a new run of translation and/or to complete the many steps in the overall process of ribosome reassociation and initiation of translation.

GROWTH RATE AND THE CELL CYCLE

We have until now treated a growing culture as a homogeneous chemical system, deliberately ignoring that its mass is concentrated in large numbers of cells that grow and divide as individual entities. Now we discuss the cell, emphasizing cell size and the pattern of DNA replication as they relate to the growth rate.

In the light microscope, the small, rodlike cells of *E. coli* can be seen to increase their volume and mass by elongation and their number by division. As pointed out in the introduction to Chapter 1, the time course of this process can be followed if the cells are spread on the surface of an agar block mounted on a microscope stage. For observations extending over long periods, ideal growth conditions can be maintained by replacing the agar block by a semipermeable membrane below which fresh, prewarmed medium is circulated. With such an arrangement the statistics of the division cycle have been studied extensively with several species of bacteria. It is important to know that the interdivisional time is subject to considerable variation; the standard deviation usually amounts to 10–20% of the mean (Powell, 1956).

It has also been emphasized that a pure culture is quite homogeneous with respect to viability and growth rate. Thus, when cells from such a culture are spread on agar, very few (usually less than 1%) fail to divide, and microcolonies of remarkably uniform size develop. Genetically and phenotypically the cells are equally homogeneous. Except for rare mutants, they all exhibit the same growth requirements and the same patterns of sensitivity to agents such as antibiotics or bacteriophages; *thus, growth and division are organized so as to ensure genetic continuity and phenotypic homogeneity.*

In a sense, this is an elaborate way of stating a fact on which bacteriology itself relies, that is, if a suitable medium is inoculated from a single colony, a pure and stable culture is usually obtained.

The genetic continuity we observe has definite implications concerning the organization of DNA replication in relation to the cell cycle. The reasoning is as follows: at division, genetically identical sister cells are born, and for this to be possible a round of replication must have terminated, creating two identical DNA complements some time before division. Each sister cell receives one of these identical and separate DNA complements. An act of termination is obviously conditioned by an act of initiation that must have occurred one replication time earlier. This tells us that maintenance of a 1:1:1 relationship between cell division, termination, and initiation of replication will ensure genetic continuity. The time interval between termination and division is commonly referred to as the D-time, and the interval between initiation and termination, that is, the duration of a round of replication, is called the C-time. Note also that during the C period genes close to the origin of replication are duplicated early and that the order in which individual genes are duplicated can be read off the circular genetic map of *E. coli* (Figure 1 in Chapter 4) keeping in mind that replication begins at a defined position (*oriC*) at 83.5 units and proceeds bidirectionally, to terminate in a region 180° from the origin.

CELL SIZE AND INITIATION OF DNA REPLICATION

In 1928, Henrici wrote a book titled *Morphologic Variation and the Rate of Growth of Bacteria*. Referring to the changes that *E. coli* cells could be seen to undergo between the inoculation of a culture and the time when net growth ceased, Henrici wrote, "In this work I shall show that, contrary to the orthodox teaching, the cells of bacteria are constantly changing in size and form and structure; but that instead of these changes occurring in a haphazard or meaningless fashion, or instead of being phases in a rather vague and complex life cycle, they occur with great regularity and are governed by simple laws which, after more data have been accumulated and analyzed, may probably be very precisely formulated." We are still working toward this goal, and considerable progress has been made.

The size variation cells undergo as they pass from the lag phase through the most active growth phase and into the resting state was illustrated in Figure 1. The average cell mass was shown to increase and again decrease approximately fourfold; Figure 2 shows that a similar volume (or mass) difference exists between slow- and fast-growing cells when they are examined during balanced growth. It is clear that the differences in composition noted earlier in this chapter cannot account for anything like a fourfold increase in mass (or volume); thus, the PSS, which is a major cell constituent, increases from approximately 15 to 55% of the mass, respectively, between μ values of 0.3 and 2.8. By itself, this increase could account for no more than a 40% change of mass. The pattern of cell division and DNA replication is now known to be very rigid, and the cells are, so to speak, forced to increase their mass in step with the growth rate, in order to accommodate the process of replication.

To understand this, let us look at the size of the genome and compare it with the replication time C. With approximately 3×10^6 base pairs per genome and with bidirectional replication the number of base pairs replicated per second (cgr_{DNA}) equals 1.5×10^6 divided by the C-time in seconds. Assuming for the moment that replication progresses at the same rate as transcription, that is, with a cgr_{DNA} of approximately 50, we obtain $C = (1.5 \times 10^6)/50$ or 3×10^4 seconds (approximately 8 hours). In itself, this looks unreasonable considering that bacteria can double their mass, including their DNA, in just over 20 minutes, and it is now accepted that replication is organized quite differently from transcription. Great efforts have, therefore, been made to estimate the true C-time and to find out how it is related to μ. This has turned out to be difficult.

Estimation of C-time and D-time

It is natural, as in earlier chapters, to begin with intact cells. In a given state of balanced growth, a definite sequence of events can be assumed to be repeated in successive division cycles, and, in particular, *division* will occur D minutes after *termination* of replication, which in turn is conditioned by an act of *initiation* that took place C minutes earlier. This trivial statement tells us that whenever $C + D$ exceeds the shortest known doubling time (approximately 22 minutes at 37°C), the round of replication to which the cell responds by dividing $C + D$ minutes later must have been initiated in an earlier division cycle. It has actually been known for a long time that the C-time in cells growing in a glucose–salts medium at least equals the doubling time of approximately 45 minutes. This was shown by pulse-labeling the DNA and then preparing autoradiograms (Schaechter, 1959).

All cells had incorporated [^{3}H]thymine into their DNA, and the C-time must therefore have been $\geq$ 45 minutes.

Less direct experiments show that the C-time probably varies little, if at all, with the growth rate for $\mu > 1.5$. These are the protracted run-out experiments referred to in the paragraph about rifampicin and illustrated in Problem 6. There are reasons for being cautious when interpreting this kind of experiment. Setting the run-out time equal to the C-time is based on two assumptions: first, that a new round of replication cannot be initiated without de novo synthesis of RNA (rifampicin experiments) or of protein (amino acid starvation or chloramphenicol experiments); and second, that the drug, or treatment, takes effect at once without changing the cgr_{DNA}. The assumption about de novo synthesis is plausible and it has been very useful; it received strong support from the genetic experiments to be described below. The point about the cgr_{DNA} is instructive; it calls attention to the troublesome fact that different strains of a species, such as *E. coli*, can behave quite differently. Thus, it has been shown by Pato (1975) that in some (but not all) of the commonly used strains the cgr_{DNA} is strongly reduced in chloramphenicol run-outs.

Genetic analysis of replication during balanced growth

We will now discuss genetic experiments that support the assumption about de novo synthesis and even go a long way to prove that replication is bidirectional and that the C-time is very nearly constant for doubling times of less than 1 hour. The idea is to use genetic markers in known positions and to measure their relative copy numbers under various conditions (Bird, 1972). As a fixed marker, phage λ integrated in its normal attachment site at 17 units on the chromosome was used; and phage Mu, which behaves much like a transposon, was employed as a variable marker. The position of Mu in the different strains constructed for these experiments are indicated on the abscissa of Figure 13A and B. The measurements were made by extracting the DNA from cells in balanced growth and using prelabeled λ and Mu DNAs as probes. The radioactivity specifically hybridizing to the cellular DNA was then converted into numbers of λ and Mu residues per unit of DNA; the ratio between these numbers (Mu/λ) appears on the ordinate of Figure 13A and B. The authors describe experiment A as follows: "The different strains were grown in L-broth or in minimal medium supplemented with casamino acids, with average doubling times of 25 and 38 minutes. The ratio of Mu to λ shows a maximum value near *ilv* and a minimum near *trp*. The most striking features are that maximum and minimum are 180° apart on the circular genetic map and that the gradient of phage Mu-1 frequency decreases symmetrically from the maximum to the minimum points in both directions. The ratio of maximum to minimum is 1.6:1. Our interpretation is that replication of the chromosome is initiated near *ilv* and proceeds simultaneously in both directions to a terminus near *trp*." We now know that *oriC* is, in fact, close to *ilv*. Other experiments showed that if the C-time and hence the number of replicating forks is increased by shifting from high to low

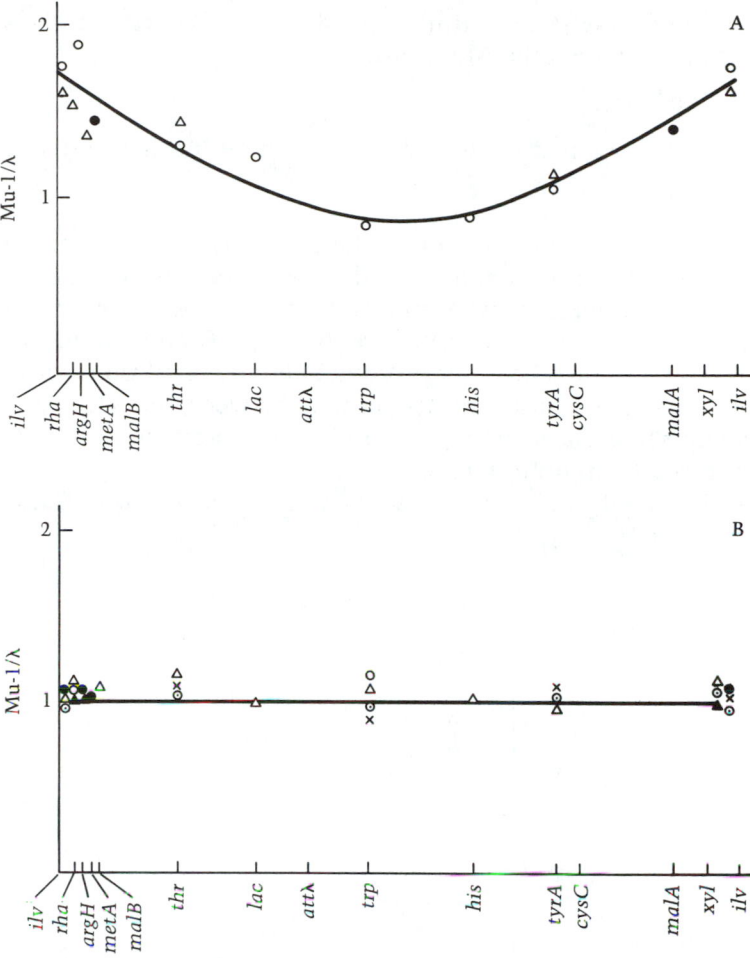

Figure 13. Frequencies of phage Mu-1 genomes, inserted into the *E. coli* genes indicated on the abscissa, relative to genomes of phage λ at its normal attachment site. (See text for more detail.) A. Frequencies in a broth culture; μ = 2.4. Each symbol represents an independent culture. B. Frequencies as observed after 80–100 minutes of amino acid starvation of a culture in minimal medium with casamino acids; μ = 1.6. The different symbols represent removal of casamino acids or of required amino acids. (From Bird, 1972.)

thymine concentration, a symmetrical curve like that of Figure 13A is again obtained, except that it is much steeper (maximum to minimum ratio 3.6:1). Conversely, if the cells are prestarved for a required amino acid, the result is precisely as predicted from the theory based on the DNA run-out experiment; namely, replication runs to completion as shown by all markers having attained the same frequency (Figure 13B).

Finally, this MARKER FREQUENCY technique has been used to test the notion of a constant C-time (Chandler, 1975). It can be shown that the relationship between the Mu:λ ratio and μ is

$$Mu/\lambda = 2^{lC\mu}$$

Where *l* is the relative map distance. Where Mu is inserted near *ilv*,

$$Mu/\lambda = 2^{0.64C\mu}$$

For a constant value of *C*, the results shown in Figure 14 are predicted. Note the discrepancy at low μ-values betewen the Mu/λ curve and the dashed line, which refers to experiments done with a different strain (and a different method). The strains involved are the major *E. coli* strains (K and B) in current use; the K strains host phage λ and under all conditions grow slower than the B strains. The significant discrepancy between the K12 data shown in Figure 14 and the dashed line representing experiments with B cultures can perhaps be traced to strain differences.

In the beginning of this section on cell size and initiation of DNA repli-

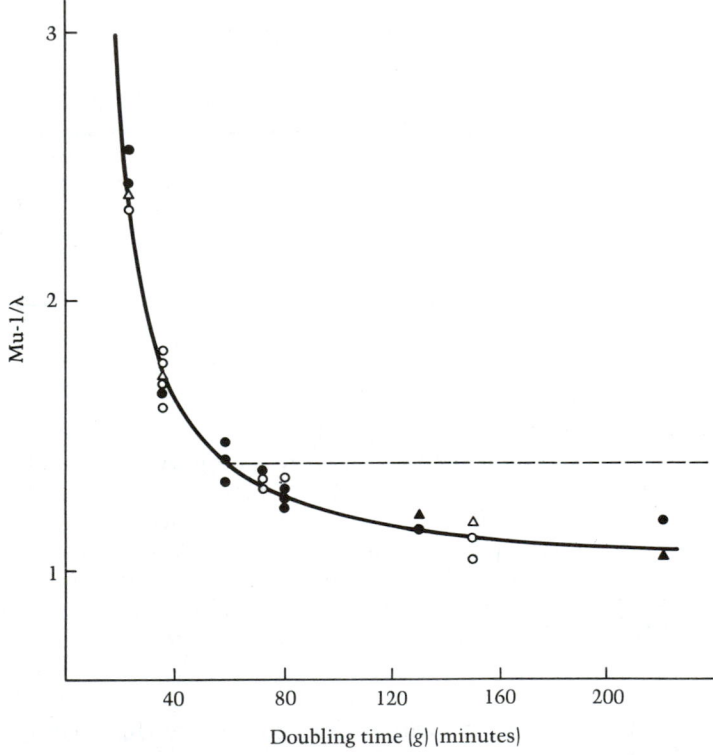

Figure 14. Frequency of phage Mu-1 integrated near *ilv*, relative to λ (see Figure 13), for different growth rates. The solid curve is calculated as described in text, assuming the replication time *C* is independent of μ. Filled and open symbols refer to strains with and without requirement for thymine (low concentration). The dashed line shows the ratios expected if, as is sometimes assumed, *C* increases with *g* at low growth rates. (From Chandler, 1975.)

cation, we introduced the thesis that cells are bigger the faster they grow because the pattern of DNA replication is very rigid. Now that we have seen that C-time is more or less constant, at least for $\mu > 1$, this argument can be finished: in an *E. coli* chromosome on which replication is being initiated, with the associated mass M_i, C minutes later replication terminates and the mass will now be $M_t = M_i 2^{C\mu}$, or ln $(M_t/M_i) = C\mu \ln 2$. Cell division can follow D minutes later, and it is clear that if C, and possibly D, are rigidly fixed, the mass increase during $C + D$ minutes will depend entirely on μ. So far, M_i and D are both unknown, but we shall see presently how they can be estimated.

Physical demonstration of bidirectional replication

Bidirectional replication has also been demonstrated by entirely independent, visual means (Prescott, 1972). Cultures of a thymine-requiring strain, growing exponentially, were labeled in two stages: first with a relatively low specific activity of [^{3}H]thymine for 15–20 minutes, followed immediately by labeling for a few minutes at a greatly increased specific activity. The DNA was then gently stretched out on a suitably prepared glass surface and autoradiograms such as those in Figure 15A and B were obtained. Numerous cases were found in which short, dense tracks, corresponding to the high specific activity, are joined by less dense tracks representing lightly labeled DNA. These pictures come close to a direct proof of bidirectional replication.

Experiments with synchronized cultures

The parameters we now want to study are characteristics of the normal division cycle. One parameter is the age (which runs from 0 to 1) at which DNA replication is initiated; the other is when it is terminated. The D-time is the interval between termination and division.

With this definition in mind, we consider division synchrony in batch liquid cultures as the method of choice. But not just any method will do because the division cycles must be "essentially" normal (see earlier discussion of "essentially" balanced growth). Two criteria may be used to guide the choice among the techniques available: first, the synchronously dividing cells must emerge from a culture in balanced growth without the balance having been disturbed; and second, more than one division cycle should be produced, and any event observed at a particular age in the first cycle should repeat itself at the same age in the next cycle. Cell division itself can be used to select "baby cells" from a population of cells adhering to the lower surface of a filter disk through which fresh, warm medium is passing. The best indication that the attached cells grow quite normally is probably that for hours after they were put onto the filter they divide and give off a constant stream of newborn cells that grow and divide with essentially the same rates as were observed in the batch culture with which the experiment was begun. This elegant method has been used extensively to study the division and replication cycles. A very useful technical point should be mentioned: the parent culture may

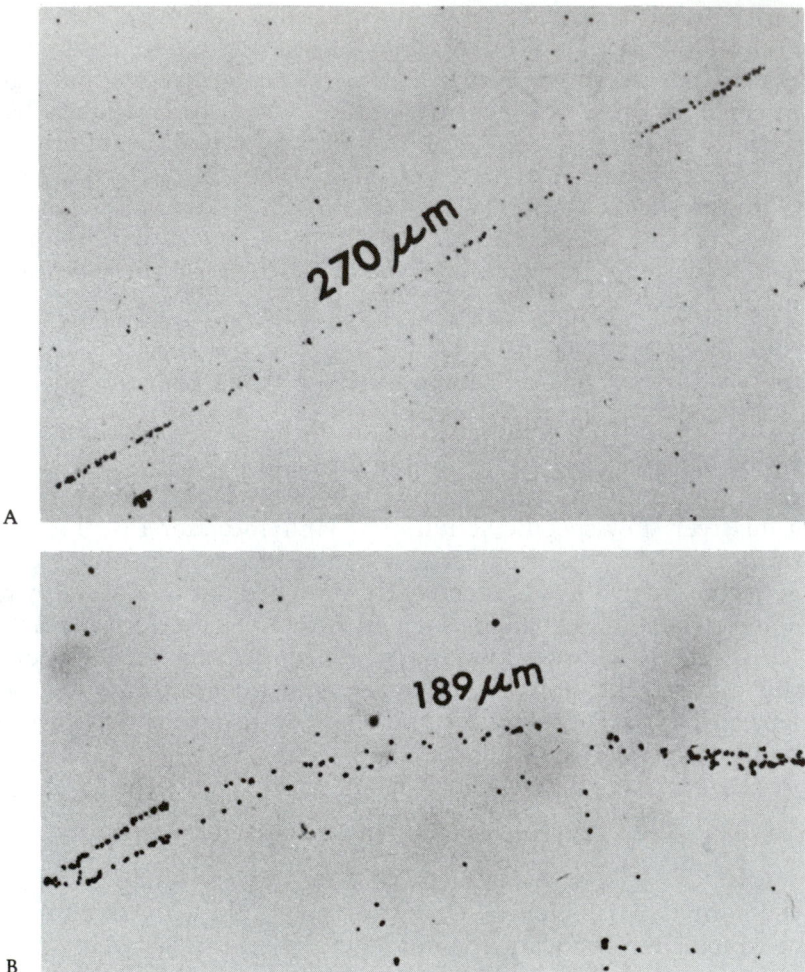

Figure 15. Autoradiograms of replicating *E. coli* DNA labeled successively at low and high specific activity of [³H]thymine. A. Portion of *E. coli* chromosome after labeling for 19 minutes with [³H]thymine followed by labeling for 2.5 minutes with [³H]thymine and [³H]thymidine. B. Cells labeled as in A; the daughter duplexes connecting the replicating zones happen to be well separated. (From Prescott, 1972.)

be pulse-labeled just before the cells are put onto the filter. The cells in a sample of baby cells that are collected in the effluent between, say, 10 and 12 minutes originate from the age class of cells on the filter that were labeled 10–12 minutes before they would have divided if left in the batch culture (Figure 16). Thus, the filter setup can be employed as an "age discriminator."

Figure 16. Cells were pulse-labeled with [¹⁴C]thymidine during balanced growth, adsorbed to membranes, and eluted as described in text. The stepwise increases of μ (frames A through F) were obtained by supplementing glucose minimal medium with, respectively, 2, 6, 12, and 15 amino

Cell age (fractions of a generation)

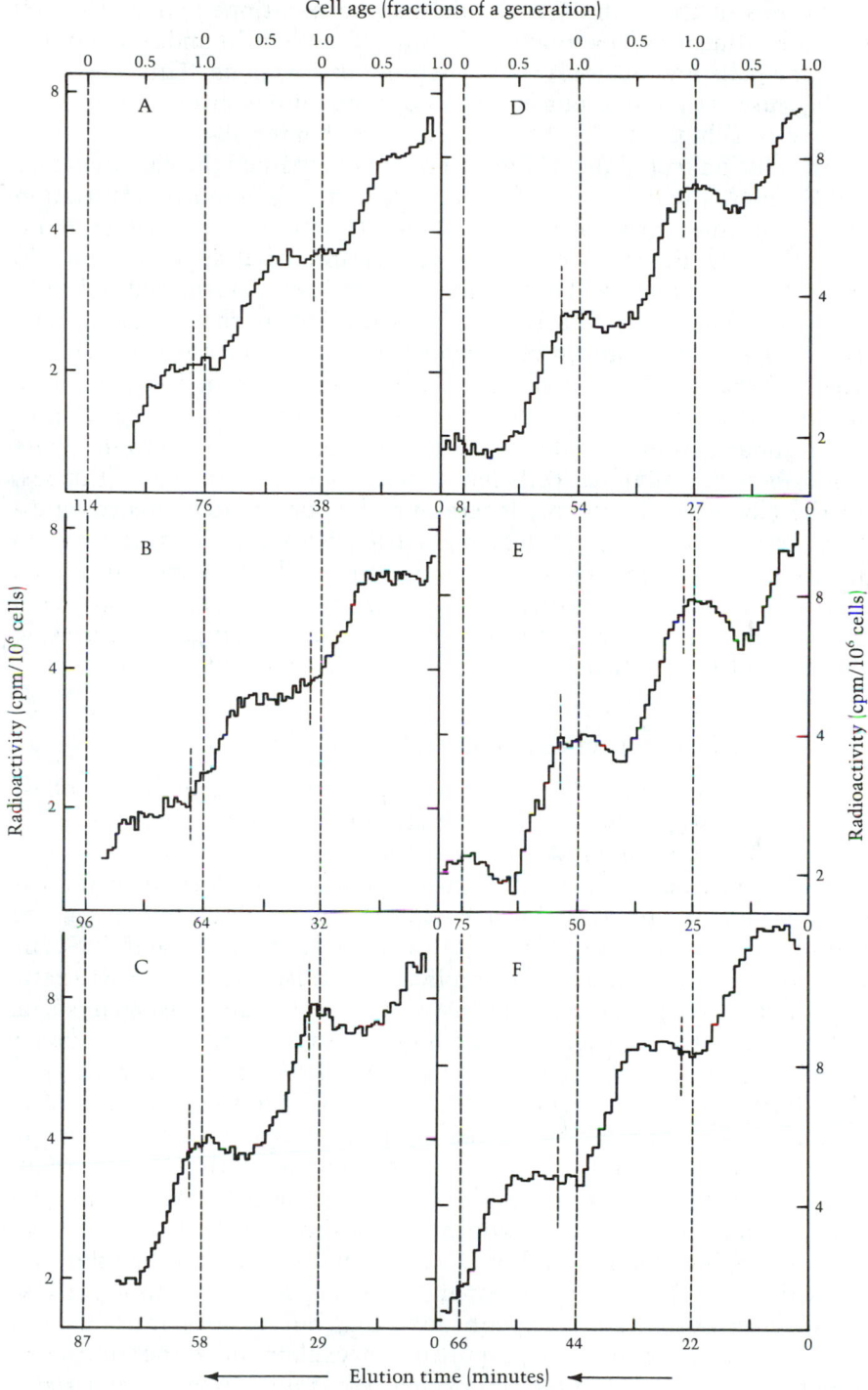

acids (A–D) and with casamino acids (E) and the same plus nucleosides (F). Sampling times were 1 or 1.5 minutes. (From Helmstetter,
1968.)

Two limitations of the technique should also be mentioned: first, the cell density in the effluent is low (maximally 2×10^6 cells/ml); and, second, the collected baby cells produce only 2–3 synchronous divisions. This was to be expected because, as pointed out earlier, the standard deviation of the inter-division time is as high as 10–20% of the mean (Chapter 5).

With the elution technique, Helmstetter (1968) obtained the data in Figure 16; Figure 17 shows the interpretation of the experiments in terms of initiation and termination times (or ages) at different growth rates (Cooper, 1968). From this extensive study, the C- and D-times were estimated at approximately 40 and 20 minutes, respectively. The D-time has not been determined independently, and it is subject to the same large variation as the division time. Estimates of D are, therefore, to be accepted with caution. Moreover, it has been suggested that the D-time varies even between different lines of the *E. coli* B strain. Virtually nothing is known about the processes that lead up to cell division some 20 minutes after replication has terminated. Measurements on large numbers of individual cells have been made to estimate D (Koppes, 1978). In the electron microscope, length and the constriction preceding division were recorded, and pulse-labeling with [³H]thymidine was used to distinguish between replicating and nonreplicating cells. Unfortunately, this distinction applies only when growth is very slow, so this interesting approach cannot, in principle, give much information in the upper range of the μ-scale where both C and D are assumed to be more or less constant.

Cell mass at initiation of replication

Having estimated the termination time and thereby D as best we can, we shall now consider M_i, the initiation mass, which appeared in the expression $\ln (M_t/M_i) = C\mu\ln2$. As we have just seen, the age at which *E. coli* cells initiate replication varies with the growth rate, and so does the cell mass (Figure 18). It may therefore be asked whether these observations can be combined in such a way that common features emerge that characterize cells at the time, or age, they initiate DNA replication, irrespective of growth rate. The question has been examined by Donachie (1968) and his analysis is presented in Figure 19. He suggested the following very simple rule: *when a defined mass (M_i) per origin of replication is reached during growth, replication is initiated at every origin (oriC) in the cell.* Like the estimates of D above, this neat formula must be viewed with caution, and a few comments are in order. First, it should be noted that constancy of the M_i per origin depends on the slope and not on the absolute values of mass (Figure 18) and on the match between this slope and a $C + D$ value of approximately 60 minutes. This implies that if C and/or D vary with μ (as may well be the case at low growth rates) M_i would not remain constant. Second, a more or less constant M_i does not in itself explain anything, but it does introduce an important restriction on models purporting to explain how the timing of initiation of replication within the cell cycle is controlled. And, third, parameters such as M_i, C, and D may be altered by mutation, by drugs, or by special growth conditions.

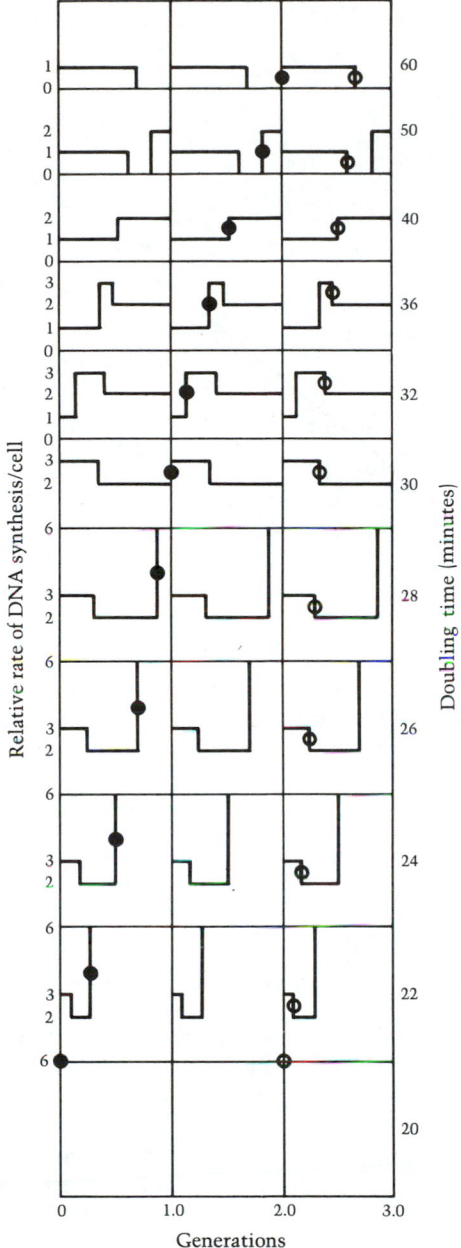

Figure 17. Proposed relationship between a round of chromosome replication and the division cycle. Rationalization of experiments of the type in Figure 16. Rate of DNA synthesis per cell during three successive division cycles with filled circles indicating starts and open circles ends of rounds of replication. (From Cooper, 1968.)

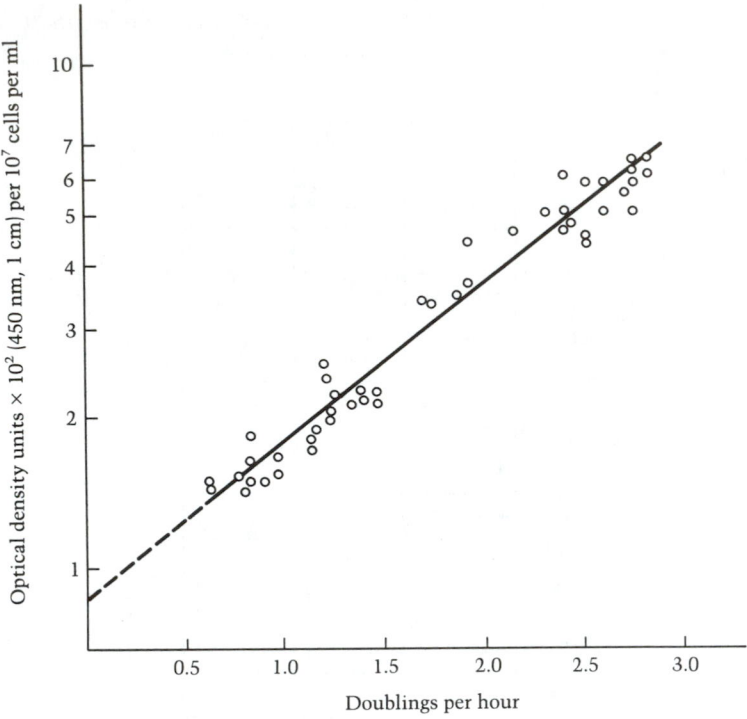

Figure 18. Dependency of cell mass on growth rate of *Salmonella typhimurium* at 37°C. From the optical density (mass) measurements and the viable counts in the different media, values for optical density per 10^7 cells per ml were calculated. The logarithm of these values is plotted against the growth rate, expressed as doublings per hour. (From Maaløe, 1966.)

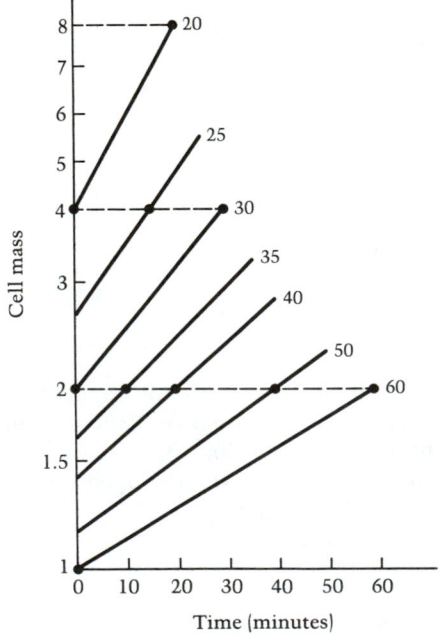

Figure 19.

Proposed mass, and mass increase, in individual cells at different growth rates. A cell is assumed to grow exponentially and to double its mass in each division cycle. Each line therefore shows mass increase of individual cells from division at 0 minutes to the next division, assuming that an act of initiation of chromosome replication always precedes division by $C + D = 60$ minutes. The graph shows that initiation must occur at a fixed cell mass (M_1) or a multiple thereof. (From Donachie, 1968.)

300

THE BACTERIAL NUCLEUS AS AN ACTIVE CENTER

The major polymerization processes—DNA replication, transcription, and translation—were surveyed in Chapter 3. Two aspects of these processes of special significance to growth physiology will be treated here: the first is the *frequencies of initiation*, whether of DNA, RNA, or polypeptide chains; and the second is the topology within growing cells that creates specific problems not encountered in in vitro systems from which much of our information about the enzymology of initiation and elongation of chains has been derived.

It has been stressed more than once that cgr_{RNA} and cgr_p are independent of μ; and we have seen that cgr_{DNA} is constant, at least for $\mu > 1$ (Table 1). The primary significance of these observations is the strong case they present for *efficiency*. Thus, chain growth rates much below the maximum rates defined by the physical properties of the system would result in ribosomes spending unnecessarily long times synthesizing protein molecules; hence, there must have been strong selection for such features of the overall system as would permit near-maximum cgr_p and cgr_{RNA}. Similarly, the fact that at high μ values new rounds of replication begin before the previous round terminates strongly suggests that each growing point operates at near-maximum efficiency.

Note that to allow polymerization of one or another kind to proceed at near-maximum rate the cell must be organized so as to maintain the substrates (all the charged tRNAs and the ribose and deoxyribose triphosphates) at concentrations high enough almost to saturate the various enzyme reactions.

In a system with the properties just summarized, *the rates of DNA, RNA, and protein synthesis are set by the frequencies with which chain initiations take place.* The implications of this are analyzed below and in more detail in Chapter 8.

In vitro studies of initiation processes

A common trend in bacterial biochemistry is to extract selected subsystems for study by refined biophysical, biochemical, and genetic techniques. Programs of this kind range all the way from structural work on individual proteins to the use of complex in vitro systems in which DNA-dependent RNA and protein synthesis are achieved. The degree to which the latter have been perfected is illustrated by the fact that DNA carrying an operon that encodes the two big subunits of RNA-P at the end of a large operon can be transcribed and translated so effectively in vitro that the β and β' subunits (molecular weight approximately 150,000 each) are synthesized in equal amounts, as in fact they are in vivo (this particular operon with its complex regulatory mechanisms will be discussed in Chapter 7).

In vitro studies of the initiation processes say that we are a long way from understanding them in detail. Thus, at least 10 different proteins, most of which have only been identified through mutations in various genes, are involved in the initiation of replication. A considerable number of proteins and low-molecular-weight effectors interact with RNA-P when transcription

Table 1. Some growth parameters as functions of μ

Parameter	Function	Figure references	Comments
cgr_p	Invariant	6, 8; see also problems 2 and 3	Measured by several methods, in several labs
cgr_{RNA}	Invariant	9, 10; see also problem 2	No evidence that different RNA species differ in cgr
mRNA/total RNA	Invariant		Estimated in several ways; see text
Half-life of mRNA	Invariant	11	The *average* half-life seems not to vary with μ; species of mRNA variations known; see text
cgr_{DNA} (C-time)	Constant at $\mu \geq 1(37°C)$	17; problem 6	Difficult to estimate at low μ; strain difference may be important
tRNA/total RNA	Constant at $\mu > 0.7(37°C)$	4	
Fraction of ribosomes in polysomes	Constant at $\mu > 0.7(37°C)$		Not measured at low μ; recovery uncertain
Average spacing of ribosomes	Constant at $\mu > 0.7(37°C)$		Varies among mRNA species
Average number of proteins per polysome	Constant at $\mu > 0.7(37°C)$	12	Varies with mRNA half-life
R/P	Proportional at $\mu > 0.7(37°C)$	4; problem 5	
α_r	Proportional at $\mu > 0.7(37°C)$	5	Cf. Figure 7 and text
α_{PSS}	Proportional at $\mu > 0.7(37°C)$	5	Much less well defined and less accurately known than α_r

is initiated at different promoters. This may sound discouraging considering the prime importance of the initiation events, and it would be discouraging indeed if analysis of the overall system in terms of initiation frequencies required that the mechanisms were understood in detail. However, the complicated events that trigger polymerization can be thought of as happening in a *black box*. A car is a good example; it is not necessary to know how the engine works in order to analyze traffic patterns. All we need to know is what cars are designed to do and when and where they start and stop. In Chapter 7 we shall discuss the mechanisms that permit starts and stops at specified times and places (promoters and terminators). The initiation events themselves are largely treated at the black-box level of understanding. In Chapter 8, we shall examine the regulatory networks that make the process of bacterial growth so remarkably efficient.

Spacing of DNA duplexes in the bacterial nucleus

The anatomy and composition of the *E. coli* cell have been described in Chapter 1, and it is clear from electron micrographs that ribosomes never diffuse into the nuclear region, which contains most, if not all, the DNA. We thus deal with a two-phase system. We know nothing about the forces that cause the DNA to condense and form the nucleoid without benefit of a separating membrane. Nevertheless, it is instructive to calculate the spacing between neighboring segments of the genome. We may assume that in small portions of the nuclear region the long, coiled genome takes the shape of a multistranded cable composed of parallel DNA duplexes. The nuclear region accounts for roughly 10% of the total cell volume, and the mean distance between the axes of neighboring DNA duplexes is therefore close to 5 nm; if they are viewed as cylinders, 2.2 nm in diameter, this leaves only approximately 3 nm between them. (If the DNA occupied 20% instead of 10% of the cell volume, these distances would increase to approximately 7 and 5 nm, respectively). Figure 20 shows an imagined cross-section through a region in which DNA duplexes are cut at right angles to the axis; the regular hexagonal packing used to illustrate the spacing must not be construed to

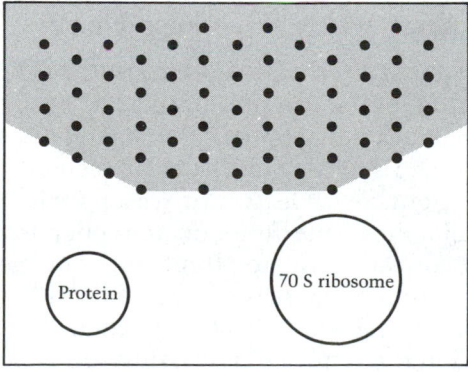

Figure 20.

Spacing of the DNA in the bacterial nucleus. The shaded area represents the nucleus; the black dots illustrate cross sections through DNA rods, 22 Å in diameter and arranged in a hexagonal pattern with 50 Å between centers. The two circles outside the nucleus illustrate the dimensions of a 70 S ribosome (reduced to a sphere with a diameter of 160 Å), and a spherical protein 100 Å in diameter. (From Maaløe, 1966.)

Protein

70 S ribosome

mean that the nucleus is a rigid, crystallike body. Outside the nucleus two circles indicate the dimensions of a 70 S ribosome and of a spherical protein molecule with a molecular weight of approximately 10^6.

It is in a physical system of this description that replication and transcription take place. Two questions should therefore be examined, neither of which applies to homogeneous in vitro systems: Are all parts of the genome accessible to transcription at a given instant? How does a newly synthesized RNA molecule disengage itself from the DNA template?

Accessibility of genes in the nucleus

The arguments developed here do *not* imply that a large part of the genome is more or less permanently inaccessible to transcription. At a given instant most of the strands in a bundle of, say, 150 are obviously "internal"; however, we visualize the DNA in the nuclear region as being in constant thermal motion such that any part of the genome would expose itself at the interface at relatively short intervals (probably seconds). With the synchronization technique described earlier, Cummings (1965) has actually demonstrated that the lactose operon is accessible to transcription throughout the normal division cycle.

Note that accessibility is not a problem as far as replication is concerned because this process takes place at special sites—the growing points through which the whole genome is thought to be threaded. It is easy to see that the two sister strands can be separated by spinning the DNA molecule around its axis; and the energy liberated by hydrolysis of the deoxytriphosphates in the course of replication would, in fact, suffice to spin the entire genome so as to unwind the parental duplex and rewind the two sister duplexes generated at the replication fork. However, this apparently simple solution is fraught with difficulties: in the first place, it takes no account of the fact that, at all times— even during an interim between two rounds of replication or in the presence of nalidixic acid—transcription is in progress at many points along the genome; and in the second place, it does not explain how nascent transcripts disengage themselves from the DNA template to form polysomes without being spun around. These topological problems are by no means trivial.

Replication and transcription advance at incompatible rates

It has been thought that replication and transcription might be coupled mechanically and that separation of the parental strands in the act of replication and the disengagement of nascent transcripts from the templates were aided by the same process of rotation, or spinning, of the entire genome. However, this unitarian concept is ruled out for at least two reasons. First, the cgr_{DNA} and the cgr_{RNA} are incompatible; the former amounts to replication of more than 500 base pairs per second, corresponding to 50 full turns of the double helix per second, whereas transcription progresses at less than 1/10 of this rate. Second, and perhaps more convincing, there is strong genetic evidence that all operons are *not* transcribed in the same direction; thus, uniform rotation of the entire genome would have positive and negative effects, re-spectively, on transcription from operons with opposite orientation.

This last statement requires explanation. During transcription, the RNA-P must interact with successive sites along the DNA duplex. The fact that a few hydrogen bonds between bases are opened in the region with which the RNA-P interacts does not affect the general topology; and if we imagine a DNA duplex at rest, transcription would require that the polymerase move from one site to the next, winding its way around the DNA template; the RNA transcript would therefore be laid down as a third strand parallel to those of the DNA double helix. In a relatively dilute, homogeneous, in vitro system, the nascent mRNA, with ribosomes attached, separates from the template by diffusion, and presumably does so without steric complications. In the cell it is different. Here all the templates being transcribed are part of a covalently continuous genome that, as Figure 12 in Chapter 1 shows, is largely confined to the central portion of the cytoplasm. Also, as Figure 20 in Chapter 1 illustrates, a nascent mRNA takes on ribosomes and forms a poly-some while still attached to the template by the RNA-P at the point where transcription is proceeding. We must therefore try to visualize how the ge-nome can remain confined to the nuclear compartment while the RNA-P, with the nascent polysome(s) attached, is reading the DNA template, which, as pointed out, requires that it make contact with successive sites disposed along a helical path. It has been suggested that the steric problems would be solved if the DNA being transcribed looped out into the cytoplasm; in this configuration, the entire complex of RNA-P and the growing polysome(s) might spiral around the DNA; this would allow for the necessary contacts between enzyme and template.

Figure 21 shows examples of autoradiograms of thin sections of pulse-labeled *E. coli* cells viewed in the electron microscope (Ryter, 1975). These elegant experiments have been taken to support the notion of transcription taking place in the cytoplasm on DNA loops. They are certainly compatible with this notion, but in fact they neither prove nor disprove it. What the pictures clearly show is that newly synthesized RNA is *not* seen in the nuclear regions; however, the *E. coli* cells in Figure 21 were labeled with [^{3}H]uracil for 2 minutes, which is long enough to produce transcripts several cell di-ameters in length. The grains seen in the autoradiograms therefore do not point to the sites where the newly synthesized RNA was made. To identify these sites, only very short RNA chains should be labeled (pulses of 1 or a few seconds), but this would hardly allow incorporation of enough ^{3}H for autoradiography. Experiments with *B. subtilis* (30-second pulses) show that the pulse-labeled RNA is found outside, but not far from the nuclear region.

Transcription may be confined to interface between nucleus and cytoplasm

Because the newly synthesized RNA appears to be excluded from the nuclear region, where could it be made except in the cytoplasm? The possibility we shall consider is that it is made at the interface between

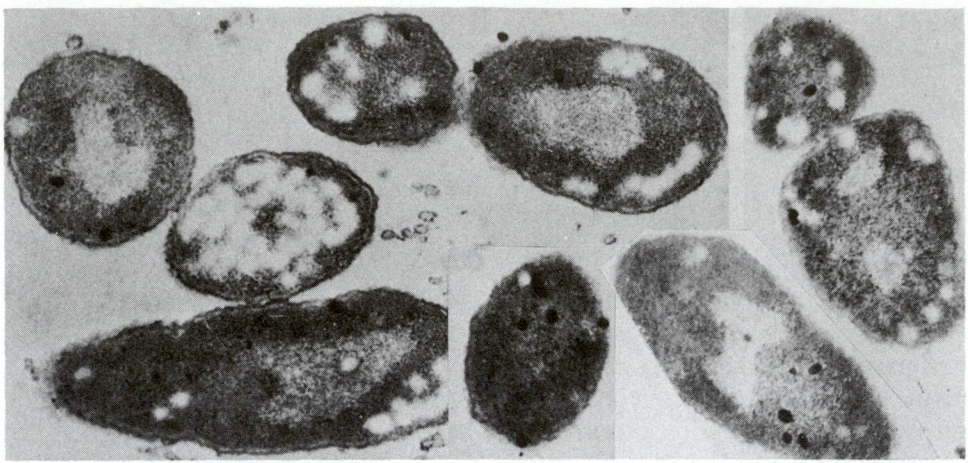

Figure 21. Autoradiograms of thin sections of *E. coli* cells grown in glucose minimal media and labeled for 2 minutes with [^{3}H]uracil. (From Ryter, 1975.)

the nuclear region and the cytoplasm. As illustrated in Figure 21, this boundary, which is clearly visible in micrographs of thin sections, seems to exclude structures as large as a ribosome or an RNA-P molecule. The absence of newly synthesized RNA in these regions indicates that the polymerase does not penetrate the interface; and neither single ribosomes nor polysomes are seen in the nuclear region.

In our model the RNA-P attaches to a promoter site exposed at the interface; as transcription progresses the DNA segment involved is rotated to permit the polymerase to interact with successive bases. Such a mechanism would unwind the nascent mRNA chain and allow polysome formation to occur in the cytoplasm in step with transcription. Thus far the model is adequate, but rotation through a great many turns, and in the sense required by the orientation of the gene(s) being transcribed, is only possible if the segment of the genome concerned can rotate independently of neighboring segments. The model therefore demands that single-strand breaks (SWIVEL POINTS) be created in front of as well as behind the advancing RNA-P. This assumption is less arbitrary than it may seem; as we have seen (Chapters 2 and 3), proteins are known that recognize positively or negatively supercoiled DNA and upon binding to such segments nick one of the two strands. The same protein may close the nick when the stress is relieved.

In general, complex processes like DNA replication and protein synthesis turn out to require a large number of accessory proteins with highly specialized functions, such as primases and gyrases or initiation and elongation factors (Chapter 3). The gene for a protein with the function envisaged in our model for transcription at the interface might be looked for among mutants with temperature-sensitive RNA synthesis.

Segregation of newly synthesized DNA duplexes

Finally, an even more elusive mechanism of great significance should be mentioned: the process by which the twin duplexes created in the act of replication of the genome segregate to form the nuclear bodies of two sister cells. We know that the replicating genome is attached to the cell envelope, and it is easy to visualize envelope growth as the process separating two attachment sites between which the septum is to be formed. It is much harder to imagine how two genomes—each approximately 1000 cell diameters in circumference—are kept apart and induced to coil up separately as replication progresses. A reasonable guess is that each sister duplex winds up on itself, each coil being linked somehow to the preceding one. We know nothing about this process, except that it occurs, and the one feature that eventually may lead to an understanding is the fact that the nascent sister duplexes are chemically distinct, despite their having identical base sequences: initially, the two newly synthesized DNA strands of opposite polarity are unmodified. Until modification (methylations, etc.) is completed, each duplex can be distinguished and could be linked, at intervals, to itself. This hypothetical linking together of successive coils of a chromosome could be viewed also as the means by which DNA is maintained separate from the cytoplasm without being surrounded by a membrane.

SUMMARY

1. Balanced growth of *E. coli* and similar bacteria at densities below approximately 2×10^8 cells/ml can be maintained as long as genetic stability permits. Constant growth rate is the most useful criterion of balance, and this condition is sine qua non for reproducible experiments in growth physiology.

2. Balanced growth at 37°C at rates between 0.2 and 2.8 doublings per hour can be achieved with *E. coli* in chemically defined media. Chemical and radiochemical methods and the effects of antibiotics are used to characterize and analyze growth.

3. The ratios between macromolecular species (total RNA:protein; rRNA:tRNA; DNA:protein, etc.) are known for *E. coli* to be functions of μ. These and other measurements show that RNA and polypeptide chain growth rates (cgr_{RNA} and cgr_p), half-life of mRNA, and the mRNA:total RNA ratio are invariant with μ. Active ribosomes are equally effective at all growth rates (cgr_p unchanged); at the lowest rates, the active fraction decreases.

4. Differential rates of synthesis of individual proteins (α_i) and of groups of proteins (r-proteins or PSS proteins) characterize states of balanced growth and shifts between growth rates. The α_i of constitutive operons decreases, whereas α_{PSS} increases with μ.

5. Rifampicin run-out and pulse-chase experiments show cgr_{RNA} to match cgr_p (one codon transcribed per amino acid added). The "life-cycle" of

polysomes and the average yield of protein per polysome are independent of μ. The constant mRNA:total RNA ratio may reflect obligatory coupling between transcription and polysome formation. Diffusion of 30 S particles is too rapid to account for the kinetics of polysome build-up. When measuring protein synthesis, only transcription coupled to polysome formation counts.

6. The genotypic stability of cell populations implies 1:1:1 relationships between cell division, termination, and initiation of DNA replication. The minimum doubling time (22 minutes) is much shorter than the replication time (40–45 minutes). A complex replication pattern maintains the 1:1:1 ratios and accounts largely for the μ-dependent increase in cell volume and mass.

7. Replication is initiated at a fixed origin and proceeds bidirectionally, as shown by physical (autoradiography) and genetic analyses on cultures in balanced growth. Synchronized cultures enable estimation of the time required for replication (C-time) and the time between termination of replication and cell division (D-time) and of the cell age at initiation of replication at different growth rates.

8. The bacterial chromosome occupies 10–20% of the cell volume; the average free space between neighbor duplexes is 3–5 nm. Large structures (RNA-P and ribosomes), therefore, do not enter the nucleus despite the absence of a membrane. Transcription probably takes place at the interface between nucleus and cytoplasm; new transcripts are passed directly into the cytoplasm. Sister duplexes coil up to form separate nuclear bodies; the mechanism involved is not known.

PROBLEMS

1. A culture was grown for many generations at $\mu = 1.4$, with uniformly ^{14}C-labeled glucose as the carbon source. Total protein was extracted and separated on a two-dimensional gel. The spots representing ribosomal protein S6 and the α subunit of the RNA-P were cut out in toto and their radioactivity measured; these two proteins accounted for 0.30 and 0.25%, respectively, of the radioactivity applied to the gel. All S6 molecules (molecular weight 16,000) are assumed to reside in 30 S subunits, with one in each; all α molecules (molecular weight 40,000) are assumed to be part of the $\alpha_2\beta\beta'$ core of the RNA-P. Total protein per genome is equivalent to 4×10^8 amino acids.

Calculate (1) the numbers of ribosomes and of RNA-P per genome. Calculate (2) the fraction of RNA-P engaged in synthesizing rRNA. The cgr_{rRNA} is assumed to be 60, and there are 4565 nucleotides in one ribosome.

It can be difficult to cut the spot corresponding to a given protein free and measure its total radioactivity. Labeling with more than one isotope may then be used. Consider four cultures in balanced growth labeled in the following manner:

    A. Succinate minimal medium, $\mu = 0.70$, label with [^{35}S]methionine
    B. Glucose minimal medium, $\mu = 1.40$, label with [^{35}S]methionine
    C. Glucose-enriched medium, $\mu = 2.80$, label with [^{35}S]methionine
    D. Glucose minimal medium, $\mu = 1.40$, label with [^{3}H]methionine

Suitable volumes of cultures A, B, and C were mixed with ^{3}H-labeled cells and analyzed by two-dimensional gel electrophoresis. The mole percent of methionine is the same in all cases. The results were as follows:

Culture	Counts per minute in total protein applied to gel		Counts per minute in sample from center of S6 spot	
	^{35}S	^{3}H	^{35}S	^{3}H
A + D	1.0×10^6	2.5×10^5	1.0×10^3	5.0×10^2
B + D	5.0×10^5	1.0×10^6	1.0×10^3	2.0×10^3
C + D	2.5×10^6	1.0×10^7	5.0×10^3	1.0×10^4

    Calculate (3) the relative amounts of S6 in A, B, and C. (4) Use the answer to (1) to find the number of ribosomes per genome in A, B, and C. (5) Calculate cgr_p for three μ values.

2. A culture of *E. coli* AS19 in balanced growth with glycerol as the carbon source produces a low basal level of β-galactosidase; however, immediately after adding the inducer IPTG, the frequency with which the *lac* operon is transcribed is increased 100- to 1000-fold. Three antibiotics, rifampicin (rif), streptolydigin (stl), and chloramphenicol (cam) are used as follows: At time $t = 0$, IPTG is added; and every 2 minutes samples are transferred to tubes containing rif, stl, and cam, respectively. All three antibiotics exert full inhibition within 5–10 seconds, and the experimental tubes are kept at 37°C for 20–30 minutes before the cells are killed and the β-galactosidase activity measured. The 20–30 minutes the samples are held at 37°C is enough time to enable expression of the capacity for further β-galactosidase synthesis that exists when the antibiotic is added. Typical curves for the three antibiotics are shown on Figure P2A.

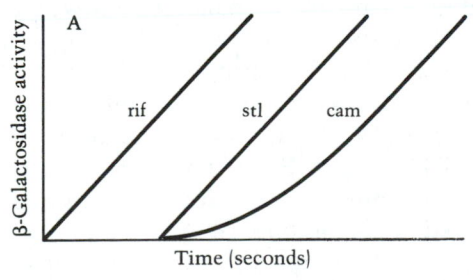

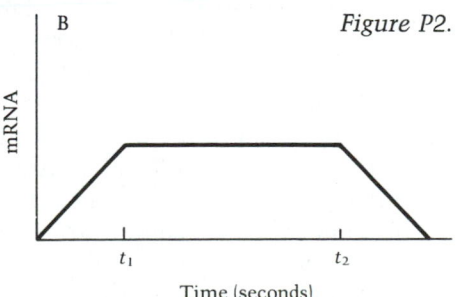

Figure P2.

Figure P2B is a simplified version of Figure 12; t_1 is the time it takes to transcribe the gene, and t_2 is the average life time of the mRNA.

For Figure P2A, explain the three curves in terms of β-galactosidase synthesis; for example, what do the intercepts on the abscissa signify? Why is the initial part of the cam curve *not* linear? What do the vertical distances between the parallel, linear curves mean?

For Figure P2B, illustrate by geometric construction and express in terms of t_1 and t_2 the residual β-galactosidase synthesis expected when rif or stl, respectively, are added in totally inhibiting concentration to a culture induced with IPTG more than 3 minutes earlier. The background level of synthesis can be neglected.

3. The cgr_p of a protein can be measured by a pulse-chase experiment; and if the polypeptide chain is long the experiment is technically easy. A culture in balanced growth may be labeled for 3 seconds, say, with [^{3}H]leucine and the pulse terminated by adding a large excess of cold leucine. At short intervals samples are lysed and the extracted protein is applied to a one-dimensional gel and electrophoresed. Large proteins like β-galactosidase and the β and β′ subunits of the RNA-P form separate bands near the top of the gel. They can be stained and cut out for measurement of the radioactivity. (For the experiment to work for smaller proteins, a two-dimensional gel would have to be run for each time point.)

Consider that label is incorporated into growing peptide chains and show on a graph how the labeling of the finished polypeptide chain progresses with time. What time scale would you use on the graph if the β and β′ subunits, each of which contains approximately 1400 amino acids, have been measured?

How would your curve look if (a) the labeling amino acid were represented only near the NH$_2$-terminal end of the polypeptide? (b) the pulse were extended to 30 seconds?

4. The total rate of RNA syntheses (unstable and stable species) can be measured by radioactive labeling if the observation time is so short that the breakdown of labeled mRNA can be neglected. In such experiments, the specific activity (cpm/mol) of the precursor used to label RNA increases greatly during the first few minutes. Correction for this is necessary.

The precursor used is [^{3}H]adenine and the specific activity of the ATP pool is measured by double labeling. A culture in balanced growth is labeled with $^{32}PO_4^{3-}$ of known specific activity, and for long enough to equilibrate all P-containing pools. At $t = 0$, [^{3}H]adenine is added and at short intervals (e.g., 30 seconds), two samples are taken; one to determine counts per minute of [^{3}H]adenine in RNA (Figure P4B), the other to estimate the specific ^{3}H activity of ATP (Figure P4A). The latter is estimated from the ^{3}H/^{32}P ratio in ATP extracted from a thin-layer chromatogram. The ^{32}P counts are converted to picomoles of ATP (1 pmol = 10^{-12} mol). The curves in Figure P4A and B are used to fill in columns 2 and 3, respectively, of the table. The corrected values (column 4) are used to draw a curve showing the actual incorporation of adenine into RNA

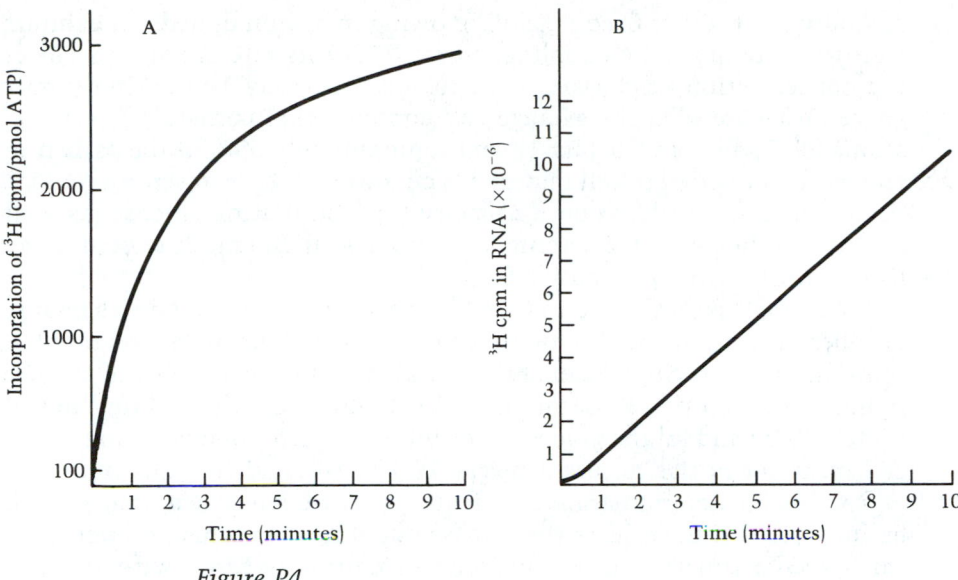

Figure P4.

Time intervals (minutes)	Average specific activity of ATP pool	Incorporation of [³H]adenine	Corrected values for adenine incorporation
0 – 0.5			
0.5 – 1			
1 – 1.5			
1.5 – 2			
2 – 3			
3 – 4			
4 – 5			
5 – 6			
6 – 7			
8 – 9			
9 – 10			

as a function of time. Estimate the slopes of this curve at $t = 0$ and at $t = 10$ minutes and explain why they differ. How can the ratio of the slopes be interpreted in terms of the synthesis of stable and unstable RNA?

5. A wild-type strain of *E. coli* (Kmt$^+$) growing in minimal medium exhibits a growth rate (μ) of 1.4 doublings/hr at 37°C; this rate is independent of the concentration of glucose down to approximately 10 mg/liter (upper curve on Figure P5). The average cell contains approximately 5×10^9 C atoms (cf. Table 1 in Chapter 1), and approximately 50% of the carbon in glucose is converted to cell material. The lower curve represents a mutant (Kmt$^-$) with a greatly reduced capacity for assimilating glucose (as well as several other sugars and amino acids). Thus, at 100 mg/liter of glucose, Kmt$^-$ grows with a μ of only 0.5.

We shall assume that the Kmt$^+$ and Kmt$^-$ cells contain the same number of C atoms at all growth rates and that their metabolism is the same except for the uptake of glucose. Calculate (1) the number of glucose molecules taken up per minute in a Kmt$^-$ culture with 100 mg/liter of the C source and (2) the glucose concentrations after one doubling of the cell mass, for initial culture densities of 10^6, 10^7, and 10^8 cells/ml.

A reliable determination of μ requires, as we have seen, that growth be followed during at least three mass doublings. At what cell densities can a good estimate of μ be obtained in a culture of Kmt$^-$ with 100 mg/liter of glucose? Present your arguments, including choice of technique, and suggest a simple method for maintaining balanced growth over, say, 24 hours.

The mutant Kmt-7 is known to revert. How would this affect the growth curve? How would you proceed in order to obtain a culture of Kmt-7, at 100 mg/liter of glucose, with zero, or at least few, revertants to begin with?

Figure P5 shows that a Kmt$^+$ strain grows very slowly at 1 mg/liter of glucose. Explain how you would proceed to obtain a good estimate of μ at this low concentration.

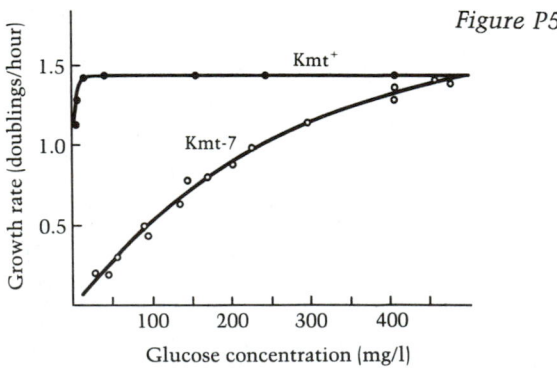

Figure P5.

6. Figure P6 shows the run-out of DNA syntheses in cultures in balanced growth at $\mu = 1.5$ and at $\mu = 2.5$, respectively. At $t = t_1$, protein synthesis was stopped at once with chloramphenicol. Both cultures had been labeled with [^{3}H]thymine for several generations before $t = t_1$.

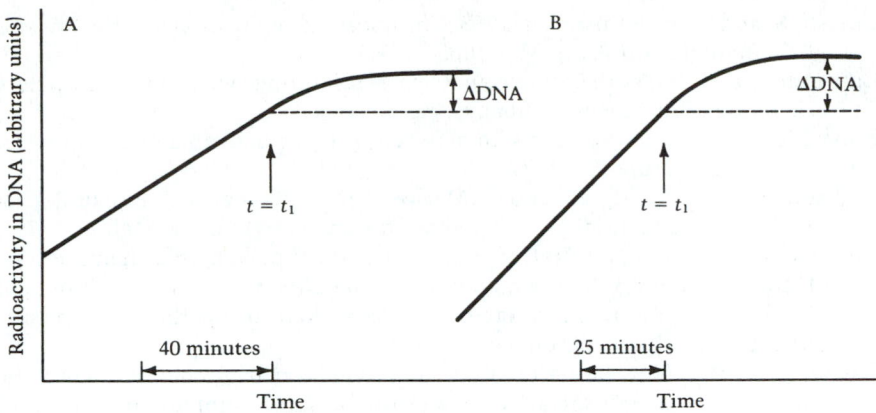

Figure P6. The "run-out" of DNA replication after addition of chloramphenicol at $t = t_1$. The doubling times are indicated for the two cultures: A, 40 minutes; B, 25 minutes.

A replicating genome can be illustrated as in (a), and if the circle is viewed edge on it is reduced to (b):

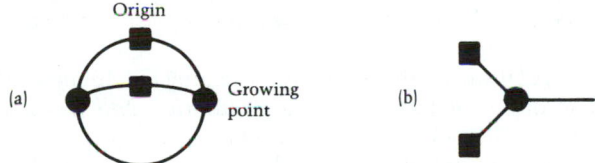

This simplified representation can be used to draw a "time-lapse" series illustrating the run-out. Show in this qualitative manner why the ΔDNA increases with growth rate. Indicate how the expected DNA could be obtained by calculation.

The two run-out curves suggest that the C-time is more or less the same at $\mu = 1.5$ and at $\mu = 2.5$, respectively. How would you modify the experimental design if the purpose had been to estimate the C-time more precisely?

REFERENCES

Bird, R. E., J. Louarn, J. Martuscelli and L. Caro. 1972. Origin and sequence of chromosome replication in *Escherichia coli*. J. Mol. Biol. 70:549.

Bloch, P. L., T. A. Phillips and F. C. Neidhardt. 1980. Protein identifications on O'Farrell two-dimensional gels: locations of 81 *Escherichia coli* proteins. J. Bacteriol. 141:1409.

Chandler, M., R. E. Bird and L. Caro. 1975. The replication time of the *Escherichia coli* K12 chromosome as a function of cell doubling time. J. Mol. Biol. 94:127.

Cooper, S. and C. E. Helmstetter. 1968. Chromosome replication and the division cycle of *Escherichia coli* B/r. J. Mol. Biol. 31:519.

Cummings, D. J. 1965. Macromolecular synthesis during synchronous growth of *Escherichia coli* B/r. Biochim. Biophys. Acta. 85:341.

Donachie, W. D. 1968. Relationship between cell size and time of initiation of DNA replication. Nature 219:1077.

Engbaek, F., N. O. Kjeldgaard and O. Maaløe. 1973. Chain growth rate of β-galactosidase during exponential growth and amino acid starvation. J. Mol. Biol. 75:109.

Forchhammer, J. and L. Lindahl. 1971. Growth rate of polypeptide chains as a function of the cell growth rate in a mutant of *Escherichia coli* 15. J. Mol. Biol. 55:563.

Gausing, K. 1974. Ribosomal protein in *E. coli*: Rate of synthesis and pool size at different growth rates. Mol. Gen. Genet. 129:61.

Gausing, K. 1977. Regulation of ribosome production in *Escherichia coli*: synthesis and stability of ribosomal RNA and of ribosomal protein messenger RNA at different growth rates. J. Mol. Biol. 115:335.

Helmstetter, C. E. and S. Cooper. 1968. DNA synthesis during the division cycle of rapidly growing *Escherichia coli* B/r. J. Mol. Biol. 31:507.

Henrici, A. T. 1928. *Morphologic Variation and the Rate of Growth of Bacteria.* Microbiology Monographs. Bailliere, Tindall and Cox, London.

Koch, A. L. 1971. The adaptive response of *Escherichia coli* to a feast and famine existence, p. 147. In *Advances in Microbial Physiology*, Vol. 6, A. H. Rose and J. F. Wilkenson, eds. Academic Press, London.

Koppes, L. J. H., C. L. Woldringh and N. Nanninga. 1978. Size variations and correlation of different cell cycle events in slow-growing *Escherichia coli*. J. Bacteriol. 134:423.

Maaløe, O. 1979. Regulation of the protein-synthesizing machinery—ribosomes, tRNA, factors, and so on, p. 487. In *Biological Regulation and Development*, Vol. I, R. F. Goldberger, ed. Plenum Press, New York.

Miller, O. L. Jr., B. A. Hamkalo and C. A. Thomas Jr. 1970. Visualization of bacterial genes in action. Science 169:392.

Molin, S. 1976. Ribosomal RNA chain elongation rates in *Escherichia coli*, p. 333. In *Control of Ribosome Synthesis*, Alfred Benzon Symposium IX, N. O. Kjeldgaard and O. Maaløe, eds. Munksgaard, Copenhagen.

Molin, S., K. von Meyenburg, K. Gulløv and O. Maaløe. 1974. The size of transcriptional units for ribosomal proteins in *Escherichia coli*. Mol. Gen. Genet. 129:11.

Pato, M. L. 1975. Alterations of the rate of movement of deoxyribonucleic acid replication forks. J. Bacteriol. 123:272.

Pato, M. L., P. M. Bennett and K. von Meyenburg. 1973. Messenger ribonucleic acid synthesis and degradation in *Escherichia coli* during inhibition of translation. J. Bacteriol. 116:710.

Pato, M. L. and K. von Meyenburg. 1970. Residual RNA synthesis in *Escherichia coli* after inhibition of transcription by rifampicin. Cold Spring Harbor Symp. Quant. Biol. 35:497.

Pedersen, S., P. L. Bloch, S. Reeh and F. C. Neidhardt. 1978a. Patterns of protein synthesis in *Escherichia coli*: A catalog of the amount of 140 individual proteins at different growth rates. Cell 14:179.

Pedersen, S., S. Reeh, and J. Friesen. 1978a. Functional mRNA half lives in *E. coli*. Mol. Gen. Genet. 166:329.

Phillips, T. A., P. L. Bloch and F. C. Neidhardt. 1980. Protein identification on O'Farrell two-dimensional gels: locations of 55 additional *Escherichia coli* proteins. J. Bacteriol. 144:1024.

Powell, E. O. 1956. Growth rate and generation time of bacteria, with special reference to continuous culture. J. Gen. Microbiol. 15:492.

Prescott, D. M. and P. L. Kuempel. 1972. Bidirectional replication of the chromosome in *Escherichia coli*. Proc. Natl. Acad. Sci. USA 69:2842.

Ryter, A. and A. Chang. 1975. Localization of transcribing genes in the bacterial cell by means of high resolution autoradiography. J. Mol. Biol. 98:797.

Salser, W., J. Janin and C. Levinthal. 1968. Measurement of the unstable RNA in exponentially growing cultures of *Bacillus subtilis* and *Escherichia coli*. J. Mol. Biol. 31:237.

Schaechter, M., M. W. Bentzon and O. Maaløe. 1959. Synthesis of deoxyribonucleic acid during the division cycle of bacteria. Nature 183:1207.

Schleif, R. 1967. Control of production of ribosomal protein. J. Mol. Biol. 27:41.

Steitz, J. A. 1979. Genetic signals and nucleotide sequences in messenger RNA, p. 349. In *Biological Regulation and Development*, Vol. I, R. F. Goldberger, ed. Plenum Press, New York.

Stent, G. S. 1966. Genetic transcription. Proc. Roy. Soc. Ser. B. 164:181.

Winslow, R. M. and R. A. Lazzarini. 1969. The rates of synthesis and chain elongation of ribonucleic acid in *Escherichia coli*. J. Biol. Chem. 244:1128.

Chapter Seven

Molecular Genetics of Selected Operons

INTRODUCTION

We have seen how the fueling enzymes convert any of the many usable carbon sources to a set of twelve intermediates and how these in turn are converted by the biosynthetic enzymes to amino acids and nucleotides that are polymerized by the protein synthesizing system (PSS). Figure 1 shows that the different groups of genes all feed transcripts, either mRNA or stable RNA, to the PSS.

The number and variety of fueling enzymes define the capacity of the organism for utilizing different nutrients. It is often found that the presence of a particular substrate in the medium will induce the synthesis of enzymes necessary for assimilating and metabolizing that substrate. The classic example is the *lac* operon, which remains practically silent (repressed) unless a substrate, such as lactose, is available to the cell, in which case the synthesis of β-galactosidase, acetylase, and lactose permease are induced. This system is understood in great detail. On the whole, *E. coli* is not especially rich in catabolic pathways. It is among *Pseudomonas* species that enzymes are found that will break down almost any organic compound (Stanier, 1976). As in the case of β-galactosidase, most catabolic enzymes are produced in very small amounts unless they are needed for growth. This repression/induction mechanism is one way by which bacteria effectively adjust their metabolism to growth in a particular medium.

Many reactions catalyzed by catabolic enzymes are of great practical importance. Thus, a large number of fermentation products, and of more or less purified enzymes, are manufactured on an industrial scale for a variety of uses (Aunstrup, 1979). Conversely, the presence or absence in growing cells of characteristic catabolic functions, such as the ability to ferment different

317

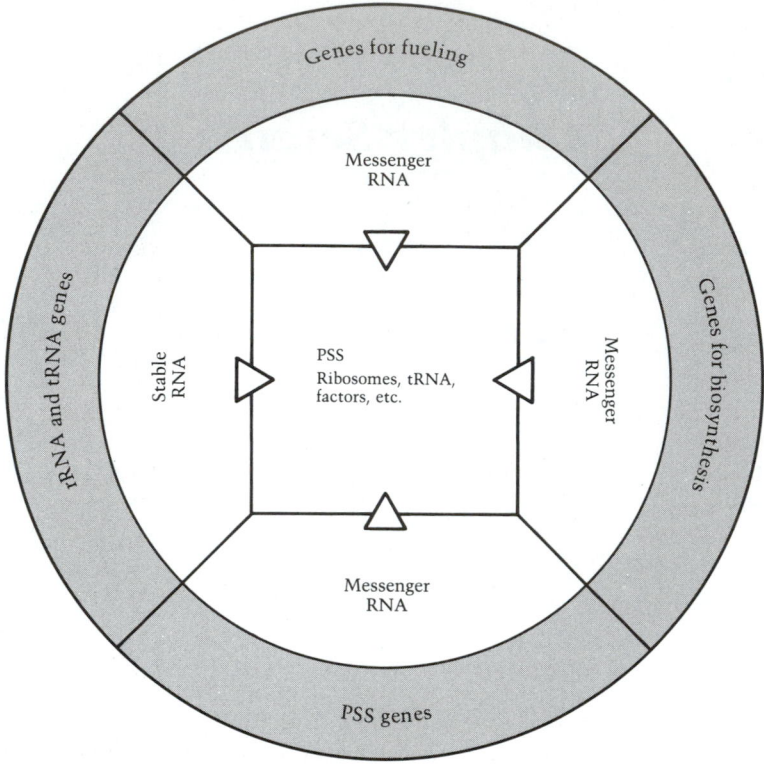

Figure 1. Diagram illustrating how transcripts of genes with different functions all flow to the protein-synthesizing system (PSS). (From Maaløe, 1979.)

sugars, is widely used to identify bacterial strains for medical and veterinarian diagnostics, as well as for general taxonomic purposes (Buchanan, 1974).

The biosynthetic domain is made up of genes for a fairly well known set of enzymes that form the pathways for the synthesis of building blocks (see Chapter 3). In contrast to the catabolic enzymes, many of which are produced and utilized only under particular growth conditions, *all* the biosynthetic enzymes are necessary for growth in minimal medium. Even so, control mechanisms of several types exist that enable the cells to adjust the rates of synthesis and the activity of the biosynthetic enzymes, thereby avoiding excessive synthesis of either the enzymes or their products. If one of these products (say, an amino acid) is added to the medium, synthesis of the relevant enzymes will be repressed and the capacity of the cell for synthesizing other proteins will increase correspondingly.

Finally, there is the protein synthesizing system, in which we place all the proteins that are involved in making protein by polymerizing amino acids (see also Chapters 3 and 6) as well as r- and tRNA. Together these components make up a functional entity that plays a dominant role in the overall process of growth. We shall focus our attention on the PSS and leave aside other systems that are equally essential for growth, such as those responsible for synthesizing the outer layers of the cell and for replicating its DNA. In Figure 1 the PSS genes are subdivided into those that produce mRNA—as nearly all genes do—and the small set of genes that produce the stable species of RNA. The unifying characteristic, of course, is that *all* the PSS functions are indispensable under *all* growth conditions. With a few exceptions, the same is true of the individual PSS genes that encode these functions. The known exceptions are the protein elongation factor (EF-Tu), which is encoded by two separate genes (*tufA* and *tufB*), lysyl-tRNA synthetase, also encoded by two separate genes (*lysS* and *lysU*), and the rRNA encoded by seven almost identical but noncontiguous genes (*rrnA* through *rrnG*).

Probably the most striking feature of the growing bacterial cell is that the processes of fueling, biosynthesis, and polymerization are linked by a network of controls and regulations which, as we have seen, renders the overall system highly efficient. Thus, in media supporting greatly varying growth rates, all the small molecules necessary for protein and nucleic acid synthesis appear to be produced at rates just sufficient to ensure that an act of polymerization, once initiated, will proceed about as fast as possible at the prevailing temperature. In other words, the substrates for polymerization (amino acids and nucleotides) are always maintained at concentrations sufficient to support high and constant chain growth rates. Moreover, the flexibility of the control network is such that this high efficiency of polymerization is maintained over a considerable range of temperature (approximately 20°–40°C, in the case of *E. coli*).

The exchange of genetic material between related bacteria was discovered late, but it gave rise to incredibly rapid progress in understanding the cell. The main reasons, of course, are that mutants are easy to obtain in a haploid organism, that very large populations of bacteria can be screened for the presence of new phenotypes, and that every step in these procedures takes a short time because the cells grow so rapidly. This development enabled detailed mapping of the genomes of *E. coli* and a few other bacteria (Chapter 4). There now exists a remarkably rich and precise map of many hundreds of the genes of *E. coli* and many thousands of cell lines carrying mutations in one or more of these genes. The enormous potentialities of this material are gradually being revealed as, one after another, entirely new technical developments widen the scope of molecular genetics.

SELECTED EXAMPLES OF REGULATION

In the following sections, we shall examine a few thoroughly analyzed systems: the *lac* and *ara* operons, which exemplify NEGATIVE and POSITIVE CONTROL, respectively, of the synthesis of some catabolic enzymes; the *trp* operon, which illustrates two independent control mechanisms—REPRESSION and ATTENUATION—that act on the synthesis of a set of biosynthetic enzymes; and, finally, from the PSS, an operon containing r-protein genes only, and a "mixed" operon composed of four r-protein genes and the genes for the large subunits, β and β', of RNA polymerase (RNA-P). In addition, the control exerted by guanosine tetraphosphate (ppGpp) will be described. The strategy developed by Beckwith and his group for a detailed analysis of complex promoters appears as Appendix D.

Each of the control systems we shall describe is of great interest in itself, and the groups working on them tend to become highly specialized. Therefore, in Chapter 8 we shall consider some general features of the control mechanisms and shall see to what extent they can help us understand the behavior of the whole cell in which approximately 1000 promoters and a large number of control circuits coexist.

LACTOSE OPERON

Background

A vast amount of information has been and continués to be gathered by studying the synthesis of the enzyme β-galactosidase. The reasons for this are instructive. Enzyme induction was a central subject in the early days of bacterial physiology, and among the many systems studied was that of lactose fermentation. The strong focus on this system in *E. coli* came with the discovery that genetic analysis was possible in this organism and with the introduction of a rapid and sensitive assay for β-galactosidase activity. Both the major discovery and the minor contribution are due to Lederberg (1948).

Work on β-galactosidase induction reached a climax when Jacob and Monod presented the operon model in 1961. This event marked the beginning of a new era, and any modern textbook of biochemistry, genetics, or microbiology offers its own description of the model. An interesting account by Stent (1978) emphasizes the logic and the history of the model; in our version, we shall stress the questions posed by the model, many of which have been solved by the invention of ingenious new methods. In this process, molecular genetics has been enriched by techniques such as DNA-dependent protein synthesis in vitro, nucleic acid hybridization and sequencing, and the construction of hybrid genes and plasmids.

When the *lac* operon model was first put together, strong evidence had just been obtained that growing polypeptide chains were associated with ribosomes and that the increase in β-galactosidase activity measured when an INDUCER has been added reflects de novo synthesis of the enzyme (and not, as some had speculated, activation of a hypothetical proenzyme). These experiments, by themselves, suggested that induction somehow released the genetic message contained in the *lac* gene and made it available at the sites where the new enzyme molecules were to be synthesized. Furthermore, the kinetics of induction had been carefully studied, and it was clear that the then hypothetical message was released quickly by the inducer and that it disappeared quickly when the inducer was removed.

Finally, several β-galactosides had been synthesized in which the oxygen atom linking the glucose and galactose moieties of lactose had been substituted by sulfur. Some of the artificial galactosides [notably, TMG and IPTG (Figure 2)] are efficient inducers, but they are *not* hydrolyzed by β-galactosi-

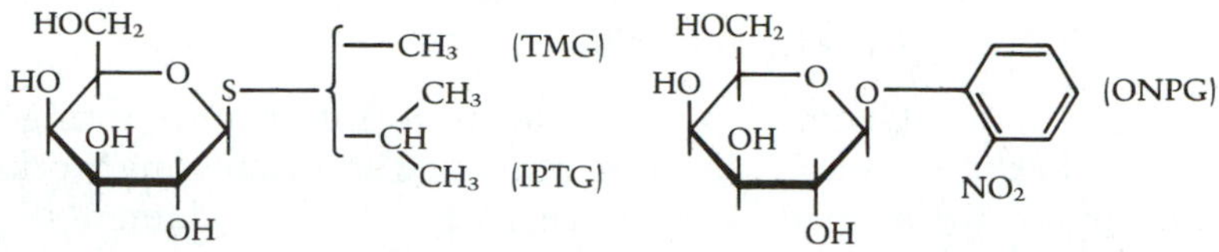

HOCH₂ ... —CH₃ (TMG) ... HOCH₂ ... (ONPG)

Figure 2. Structure of two nonmetabolizable, or gratuitous, inducers of *E. coli* β-galactosidase—thiomethylgalactoside (TMG) and isopropylthiogalactoside (IPTG)—and of the chromogenic enzyme substrate *O*-nitrophenylgalactoside (ONPG) used in the enzyme assay. (From Stent, 1978.)

dase. This is important for two reasons: first, it means that the induction process can be studied in a virtually unchanging state of balanced growth—say, with glycerol as a carbon source and IPTG as inducer. With lactose instead of IPTG, the induced synthesis of β-galactosidase would enable the cells to utilize the inducer itself as a carbon source with the result that the growth rate and cell composition would shift gradually to those characteristic of a glucose culture. Monod introduced the term GRATUITOUS for the condition in which induction does not affect metabolism. Second, these artificial inducers led to the discovery that no correlation exists between their effectiveness as inducers and their affinity for β-galactosidase; this finding suggests that the enzyme itself might not be the target of the inducer. Considering the high specificity of induction, it was natural to think that the target might be a different protein, as indeed it turned out to be.

Such were the important physiological and biochemical observations that had to be accounted for by any model claiming to explain the phenomenon of induction. Also, the fact that the lactose permease and the acetylase are invariably induced together with β-galactosidase must be accounted for. All this is accomplished by the operon model shown in Figure 3.

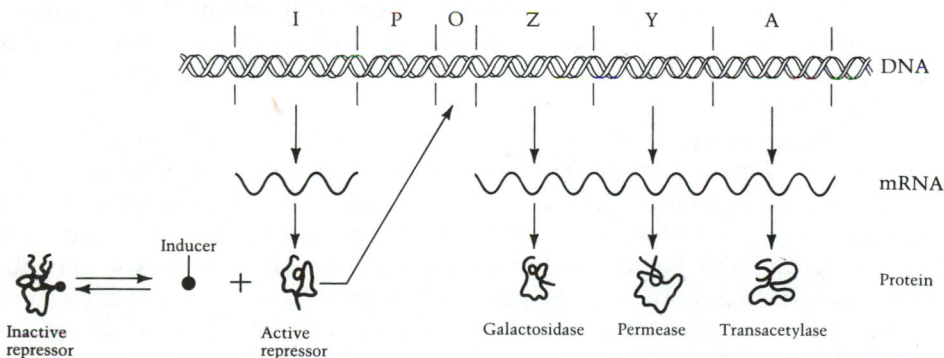

Figure 3. The original operon model of Jacob and Monod, as proposed for the regulation of the *lac* genes of *E. coli* in 1961. (From Jacob, 1961.)

Genetic analysis

Genetics played a decisive role in promoting the operon be-yond what might otherwise have remained an interesting working hypothesis. The point to be made is at once simple and fundamental. With a model involving specific interactions, on the one hand, between proteins and small molecules and, on the other hand, between proteins and nucleic acids, clear predictions can be made about the effects to be expected if one or another reactant in the system changes its specificity vis-a-vis its target. We have already seen that the small molecule involved—the inducer—can be changed (in this case by organic synthesis) and can lose its substrate character while retaining inducer activity. The macromolecular reactants—proteins as well as DNA target areas—can be changed by mutation. The functioning of a system like the *lac* operon can therefore be analyzed genetically by selecting mutants in which both phenotype and genotype are changed in the ways predicted by the model. The art of this game lies in devising suitable selection procedures. A typical and very simple technique by which *lac* CONSTITUTIVE mutants (that is, mutants producing β-galactosidase in the absence of inducer) have been isolated consists of spreading large numbers of wild-type *E. coli* cells on agar plates with phenyl-β-D-galactoside as the only carbon source. In contrast to IPTG, this galactoside is a substrate for β-galactosidase but not an inducer, which means that it is readily hydrolyzed by *lac* constitutive mutants, which then grow, now using galactoside as their carbon source.

Table 1 is reproduced from the original presentation of the operon model, and it can be read together with Figure 3. The symbols I^s and O^c refer to types of mutants, predicted by the model and isolated by special selection proce-dures: I^s is a cell line producing a "super repressor," that is, an *I*-product modified in such a way that it is not released, or not readily released, from the operator site (*O*). O^c stands for *operator-constitutive*, and this phenotype results from a mutation within the *O*-site that prevents the normal *I*-product (the active repressor) from binding to its target. Finally, the genes following the letter *F'* in the table are introduced into the cells on various derivatives of the *F fertility factor*, resulting in a cell that is diploid for the set of chro-mosomal genes carried on the F' factor. A manual by Miller (1980) contains detailed descriptions of a number of specific selection procedures and also explains how partial diploids, such as those presented in the table, are con-structed (Chapter 4).

One example will suffice to indicate how the entries in Table 1 relate to the model in Figure 3: consider line 3, with the genetic constitution *lacI⁻*, *Z⁺*, *A⁺* (chromosomal)/F' *lacI⁺*, *Z⁺*, *A⁺* (plasmid). This strain is inducible for β-galactosidase as well as acetylase (an enzyme with no known function in cells), just like the haploid wild-type strain (line 1); this shows that the chromosomal *lacZ* gene is repressed as effectively as the plasmid *lacZ* gene by the active *I*-product originating from the episomal *I⁺* gene. Thus, as the model indicates, the *repressor* must be released into the cytoplasm and by diffusion reach the chromosomal *O*-site as well as the *O*-site on the plasmid carrying the *I⁺* gene: it is said to act in the TRANS as well as in the CIS position.

Table 1. Relative concentrations of galactosidase and galactoside trans-acetylase in *E. coli* variants with various haploid and diploid genomes[a]

Genome[b]	Galactosidase (*lacZ*)		Galactoside transacetylase (*lacA*)	
	Noninduced	Induced	Noninduced	Induced
$lacI^+, Z^+, A^+$	0.1	100	1	100
$lacI, Z^+, A^+$	100	100	90	90
$lacI^-, Z^+, A^+/F'\ lacI^+, Z^+, A^+$	1	240	1	270
$lacI^s, Z^+, A^+$	0.1	1	1	1
$lacI^s, Z^+, A^+/F'\ lacI^+, Z^+, A^+$	0.1	2	1	3
$lacO^c, Z^+, A^+$	25	95	15	100
$lacO^+, Z^-, A^+/F'\ lacO^c, Z^+, A^-$	180	440	1	220
$lacI^s, O^+, Z^+, A^+/F'\ lacI^+, O^c, Z^+, A^+$	190	219	150	200

[a]From Jacob (1961).

[b]*lacZ*+, Inducible for β-galactosidase; *lacZ*−, noninducible for β-galactosidase; *lacA*+, inducible for galactoside transacetylase; *lacA*−, noninducible for galactoside transacetylase; *lacI*+, I-product produced; *lacI*−, inactive I-product produced or no I-product produced; *lacI*s, modified I-product produced; *lacO*c, operator-constitutive phenotype; *lacO*+, normal operator; F′, episome carrying *lac* genes.

We shall see later that chromosomal material also can be introduced onto plasmids that are present in the cell, not in 1–2 copies per chromosome as is the F plasmid, but in as many as 50 copies. In such cases, and if the repressor is produced from a single gene on the chromosome, there is apparently not enough repressor in the cell to block all 50 O-sites; therefore, the phenotype of the cells will be constitutive despite the presence of a normal I^+ gene. A common way of describing this situation is to say that the many O-sites "titrate out" the repressor.

All eight entries in Table 1 should be carefully examined, and, in particular, the genetic demonstration that the O^c character acts in cis only should be appreciated. However, it is also to be noted that, initially, there was no evidence for the involvement of unstable RNA (mRNA) as a go-between on the way from a gene to its protein product. The proof came somewhat later, when the DNA–RNA hybridization technique was applied to the Lac system. It was then shown that mRNA appeared and disappeared quickly as the kinetics of induction demanded and that the quantity of the specific messenger present in the cells paralleled the rate of synthesis of β-galactosidase (Adesnik, 1970).

Crude bacterial extracts contain a large number of different mRNA species, which are very difficult to separate. To measure the amount of a particular mRNA in an extract by hybridization, a "selector" DNA is prepared that is greatly enriched in the gene(s) one wants to study. The methods that enable recovery of a specific DNA segment from the bacterial chromosome and then preparation in bulk are at the root of all experiments in molecular genetics.

Isolation of the *lac* repressor

The *lac* operon model generated intense efforts to isolate the *lacI* gene product and to study its binding properties. Eventually, the base pair sequences that are responsible for the critical and highly specific DNA--protein interactions that make the system work were determined.

As implied in our earlier reference to "titration," the *lac* repressor molecules are normally present in small numbers. To isolate the repressor, various protein fractions were tested in an equilibrium dialysis experiment for their ability to bind and thus to retain radioactive IPTG. However, this assay is insufficiently sensitive at very low concentrations of repressor, and a genetic "trick" was applied to overcome this difficulty. The procedure is very instructive. It was assumed that certain mutations in the *lacI* gene might result in a repressor that would bind IPTG more effectively than does the wild-type repressor. It was then argued that such tight-binding mutants (with genotype *lacI^t*) would behave as constitutive mutants if presented with an inducer concentration too low to be effective in a *lacZ$^+$,Y* cell. The anticipated *lacI^t* mutants were found and the way was paved for isolation of the *lac* repressor. The active form proved to be a tetramer with identical monomers, each with a molecular weight of approximately 38,000 and each of which can bind one molecule of inducer. It appears that cells normally contain 5–10 tetramers, a quantity that corresponds to the yield to be expected from an average of one or two transcripts from a *lacI* gene per doubling time.

Analysis by DNA sequencing

Once the repressor had been isolated, it was soon demonstrated that the *lac* operator is indeed a small DNA segment (Gilbert, 1966, 1967). This implies that at points along a sequence of approximately 20 DNA base pairs (6–7 nm) the repressor makes contacts that result in firm and specific binding. How this and many other protein–nucleic acid interactions are achieved is a major problem in molecular biology. To solve it (as some enzyme–substrate complexes have been solved), a highly ordered, preferably crystalline, preparation of the particular protein–nucleic acid complex is needed.

Figure 4 illustrates the next big advance in our ability to characterize structures such as the *lac* operon. With the methods described in Chapter 4, almost any genetically defined DNA segment can be recovered from the chromosome, prepared in the requisite amount, and characterized by means of restriction enzymes. Finally, methods have been worked out for determining the sequence of base pairs along a DNA segment with precisely identified ends. Segments with several thousand base pairs have been sequenced (the genes for the 16 S and 23 S rRNA and the whole genome of the phage φX174). The much bigger genome of phage λ (with 48 kilobases) is now completely

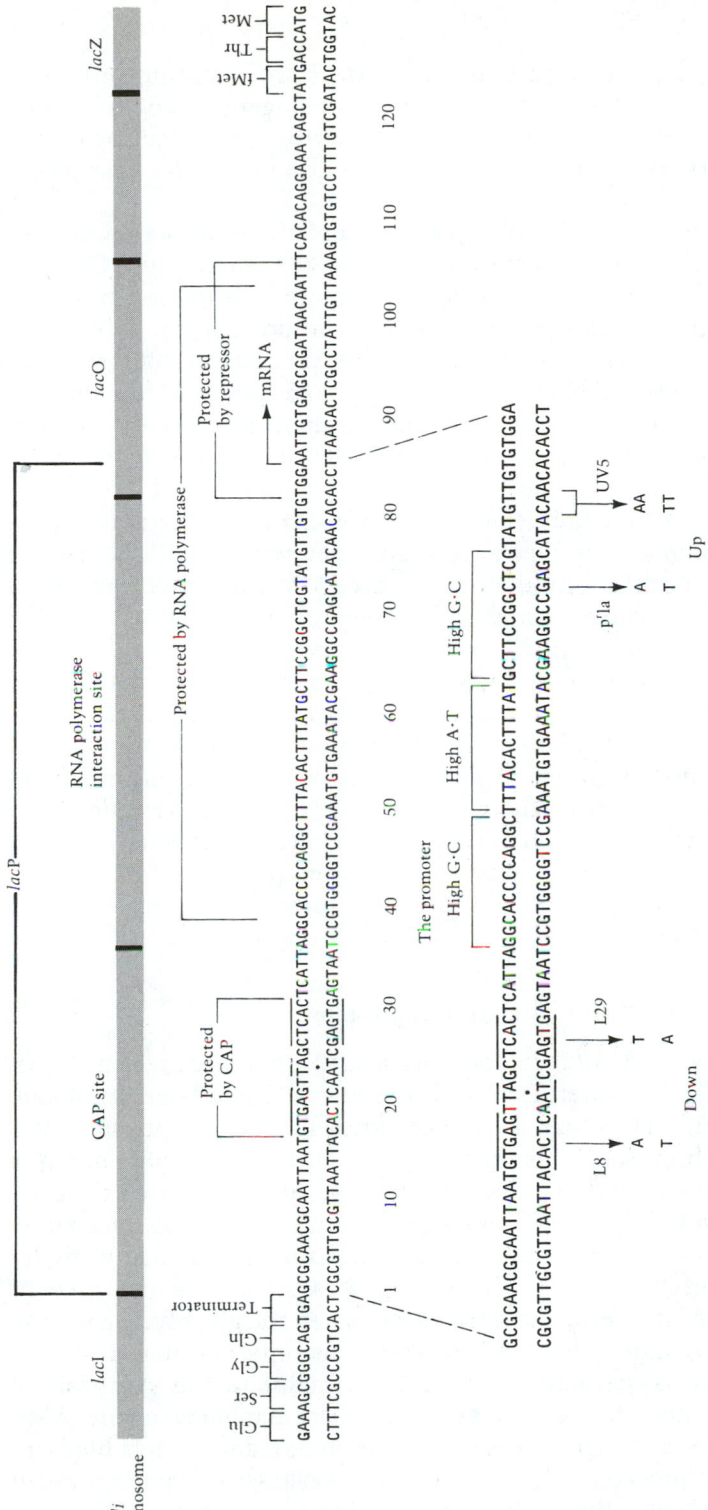

Figure 4.

DNA base sequence of the regulatory region of the *E. coli lac* operon. Regions of twofold rotational symmetry are indicated by horizontal lines above and below the symmetrical base pairs and by a dot at their axis of symmetry. The base-sequence changes associated with promoter mutations are indicated below the vertical arrows. (From Stent, 1978.)

known, and serious people believe that the entire *E. coli* chromosome may be sequenced in the foreseeable future. As longer and longer sequences become available, the aid of computers is essential in identifying special features, such as putative promoters, ribosome-binding sites, or the targets for restriction enzymes.

Many bacterial genes have been sequenced, and their ribosomal attachment sites and all the codons from start to stop can be read. Because DNA or RNA splicing is not known to occur in bacteria, more information has been obtained about the amino acid sequences of bacterial proteins via DNA base sequences than by direct analysis of the proteins themselves. But the two methods complement each other, and even very large proteins [notably, β-galactosidase (1023 amino acids) and the β and β' subunits of RNA-P (approximately 1500 and 1600 amino acids, respectively)] have been sequenced by direct analysis.

The presence of translatable genes can be looked for by established in vivo and in vitro methods. The DNA to be analyzed will usually be carried on a plasmid, which is then introduced into a cell without a chromosome (minicell assay) or into a cell whose chromosome is inactivated by UV light (maxicell assay) as described in Chapter 4. Alternatively, the plasmid may be used to prime an in vitro system in which transcription is coupled with translation. As noted in Chapter 6, this type of system has reached a high degree of perfection, but it must be set up with meticulous attention to detail.

Once the start and stop codons of actual genes carried on a plasmid have been identified, the sequences of the INTERGENIC REGIONS are revealed, and that is where the regulatory functions usually operate. Mutations affecting these functions can be definitively located and related to the regions that are protected against DNase digestion when RNA-P or one of the control proteins are attached (Figures 4 and 7).

Cyclic AMP and catabolite repression

The cyclic AMP receptor protein (CRP), which has been shown to interact with the *lac* promoter, is an addition to the original operon model, but the phenomenon it relates to had long been known to growth physiologists (Figure 5). What had been seen was that in media with glucose plus one of a number of other sugars (including galactose, maltose, and arabinose), the cells used glucose preferentially; and synthesis of the enzymes necessary for metabolizing the second sugar were induced only after the glucose had virtually disappeared. It had been observed also that glucose itself did not prevent induction and that the effect emanated from a step(s) in the glycolytic pathway beyond glucose-6-phosphate. The "glucose effect" is only one manifestation of a general regulatory system that cells use to modulate the synthesis of many catabolic and other fueling enzymes. It is mediated by cyclic AMP (cAMP), which serves as a signal of carbon sufficiency and which binds to CRP. This regulatory process, called CATABOLITE REPRESSION, helps the cell achieve a balance between the fueling and biosynthetic domains. Glycerol does not elicit the "glucose effect" and it is often used as the carbon source in induction experiments.

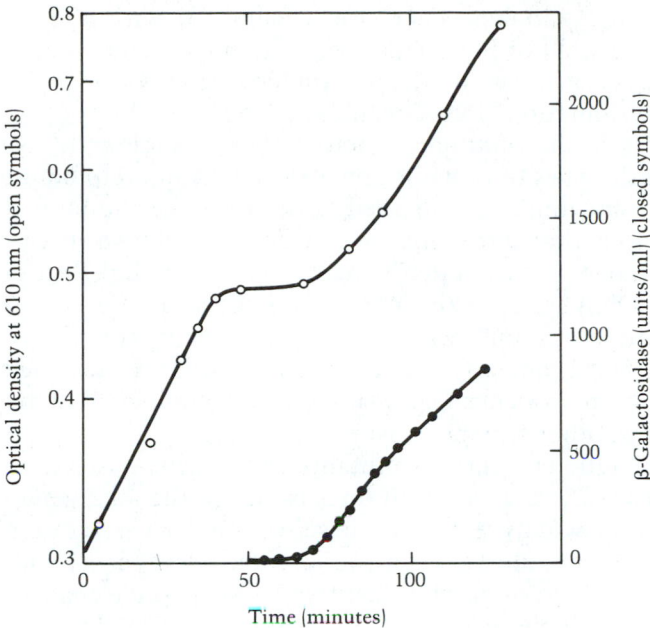

Figure 5. Delayed appearance of β-galactosidase in *E. coli* growing in a medium containing initially 0.4 mg/ml of glucose and 2 mg/ml of lactose. The left-hand ordinate indicates the cell density of the growing culture. (From Stent, 1978.)

As was the case with the *lac* repressor itself, the CRP protein was isolated by way of affinity for its effector, cAMP. That the latter was implicated had been suggested by an ingenious genetic approach. It was argued that if one and the same mechanism were involved in stimulating induction of several operons, mutants should exist that in one step had lost the ability to ferment several sugars. Such pleiotropic mutants were looked for on indicator plates on which cells that ferment either of the sugars added to the agar—say, lactose and arabinose—form red colonies. A few white colonies appeared; and on further analysis some were found that lacked the activity of the enzyme ADENYLATE CYCLASE and consequently contained no cAMP. Other white colonies, as anticipated, were mutants lacking functional CRP.

Is the *lac* system complete?

With the new components, CRP and cAMP, the *lac* system appears to be complete, in the sense that transcription and translation can be reproduced in vitro. The in vitro system is usually based on extracts containing both repressor and CRP; inducer and cAMP must be added. On the other hand, at least one more *E. coli* protein plays a role in the *lac* system. It has been identified as the product of the *nusA* gene, and it is present in most cell extracts. In the absence of this protein, transcription of the *lac* operon may terminate at a *rho*-dependent site (Chapter 3) within the *lacZ* gene, thus reducing the yield of β-galactosidase. The *nusA* gene product seems to function by associating with RNA-P, thereby modifying its specificity toward different termination sites (Greenblatt, 1981).

To end this account of the lactose operon, some reflections will be made about certain types of mutants and about DNA–protein binding studies. First, note the point mutations marked with arrows at the foot of Figure 4. They are mutations that either increase or decrease the strength of the promoter many fold; one mutation may cancel or augment the effect of another. Thus, the entire DNA segment that constitutes the promoter is a unit within which many point mutations can change dramatically the characteristics of the promoter. This is not easy to understand, considering that the many promoters that have been sequenced show imperfect homology in some regions and no homology in other regions (Chapter 3).

A very different class of mutants arises when a deletion brings about a fusion between two genes. Cell lines in which the *lacZ* gene is fused to other genes are now widely used, and these will be discussed later. The fusions we shall discuss now are between the *lacI* and *lacZ* genes, and the point to be made concerns the surprising plasticity of polypeptide chains. A *lacI::Z* fusion in which very little of the COOH-terminal part of the repressor and of the NH$_2$-terminal part of β-galactosidase are missing gives rise to hybrid polypeptides that exhibit the activity of both proteins. This is the more surprising because the two activities are associated with tetrameric molecules. In the same vein, it is interesting that some *lacZ* deletions make a polypeptide lacking the first 60–70 amino acids of the β-galactosidase; this product is inactive, but it can be complemented in vitro by a fragment (auto-α) consisting of the NH$_2$-terminal portion of the enzyme. Thus, the long loose peptide tails attached, say, to an active repressor tetramer formed by a *lacI::Z* fusion do not exclude repressor function; nor does β-galactosidase activity necessarily require that the monomer be in one piece (Miller, 1978).

Finally, a comment on DNA–protein binding studies. They serve not only to identify sites on the DNA but have been used successfully to characterize the physicochemical properties of the protein–DNA interaction (von Hippel, 1979). Note also that RNA-P and CRP protect segments of DNA that are separated by a short length of unprotected DNA (Figure 4). However, it is entirely possible that, in situ, the two proteins touch and that the CRP may promote transcription by improving the fit of RNA-P to the promoter.

ARABINOSE OPERON

Background

Arabinose has long been known to induce the synthesis of a set of three catabolic enzymes encoded by the contiguous genes, *araB*, *araA*, and *araD* (Figure 6). A fourth gene, *araC*, is involved in regulation. However, its product (the C-protein) actively promotes transcription of the *ara* operon, in contrast to the *lac* repressor, which must be removed from the *lac* operator to allow transcription. Therefore, the *lac* repressor is said to act negatively

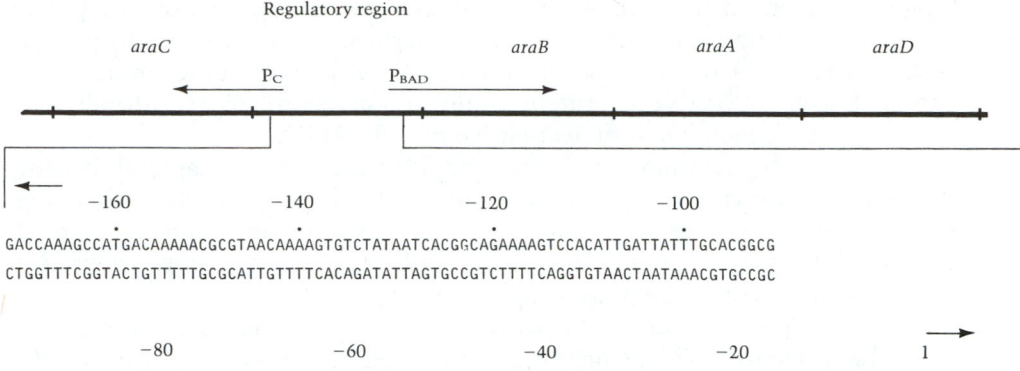

Figure 6. Structure of the L-arabinose operon, with origins and direction of the genes, nucleotide sequence, and position notation. (From Ogden, 1980.)

and the *araC* protein positively on transcription from the respective operons.

After the *lac* operon had been successfully analyzed, great interest was focused on the *ara* system, which could be expected to reveal a control circuit that had a different design but served the same purpose, that is, to regulate enzyme synthesis by controlling *access to transcription* of the operon. There-fore, it is not surprising to find that the elements of the *lac* system have their opposite numbers in the *ara* system. However, the latter is constructed such that it enables interactions not possible in the *lac* operon.

Initially, the special features of the *ara* operon were deduced from genetic studies (Englesberg, 1974). Eventually, all the recent technical innovations in the field of molecular genetics were applied, and the functioning of the operon is now understood in considerable detail (Ogden, 1980; Lee, 1981). The ana-logue of the *lac* repressor is the *araC* gene product; however, the genetic elements are arranged differently in the two operons and the C-protein has two functions: *araB*, *araA*, and *araD* genes are transcribed to the right from promoter p_{BAD}, which requires the C-protein–arabinose complex and CRP. This is where the C-protein acts positively. However, without the inducer (arabinose), the same protein is in a different configuration, with a different specificity and function. It binds to another site in the regulatory domain of the operon and blocks transcription of the *araC* gene from the leftward pro-moter (p_C) and thus *represses its own synthesis*. This is the phenomenon of AUTOREPRESSION, and in the case we now consider it is not very tight; even in the absence of the inducer, some C-protein is being synthesized.

Isolation and function of the C-protein

This brief account should suffice to indicate that the C-protein itself had to be isolated to pave the way for a better understanding of the biochemistry of the system. Cell lines carrying the *araC* gene on a transducing λ phage were used to increase the yield, and the C-protein was assayed by measuring its stimulating effect on the synthesis of L-ribulokinase (the product of the *araB* gene) in an in vitro system. A sufficiently pure product was obtained, and the active C-protein could be characterized as a dimer with monomers that each had a molecular weight of 29,000.

Progress became more rapid when *araC::lacZ* fusions (Chapter 4) became available to substitute for the laborious in vitro assay. With such fusion strains, the activity of p_C can be measured by the rapid and accurate β-galactosidase assay. The genetic analysis could therefore be complemented by physiological experiments requiring large numbers of assays.

Figure 7A is a schematic drawing of the *ara* operon in *three* different physiological states. This is one more state than can be visualized for the *lac* operon because of the dual activity of C-protein and the different arrangement of the elements in the two cases. All the *lac* genes are transcribed to the right and the *lacI* gene intervenes between its own promoter and that of the *lac* operon; the *araC* gene and the *araB, araA, araD* group of genes are transcribed

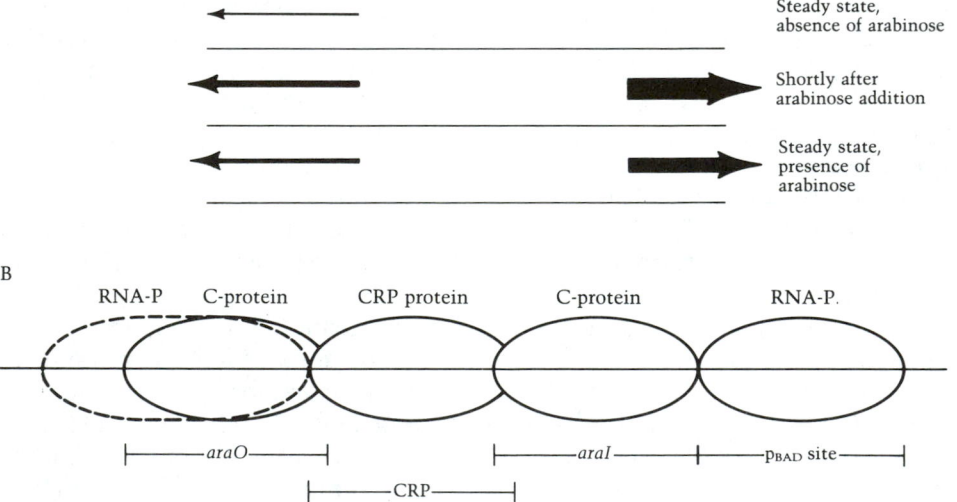

Figure 7. A. Relative activities at p_C and p_{BAD} before addition of arabinose, shortly thereafter, and during steady-state growth in the presence of arabinose. The high, transient activity at p_C was demonstrated experimentally. The thick arrows pointing to the right denote high, but not necessarily equal, activity at p_{BAD}; no rate measurements were made, but the definitive activity can be assumed to be lower than during the transient. B. Proposed CRP, C-protein, and RNA-P binding sites. See text for details. (Data from R. Schleif, pers. comm., 1982.)

in opposite directions, and the p_C therefore lies within the regulatory region (Figure 6). This enables interactions not possible in the *lac* system. The two domains C^{ind} or C^{rep} are assigned according to binding and protection experiments and represent the genetically defined *araI* and *araO* sites, respectively.

The CRP and RNA-P binding sites were located by in vitro protection experiments (Figure 7B). The combined physiological and genetic data suggest the following scenario (R. Schleif, 1982, pers. comm.):

> The *araC* product, the C-protein, in the repressing state binds to the *araO* sequence about 85% of the time. In this position C-protein blocks access of RNA-P to p_C, allowing expression of this promoter at about 15% of its maximal value; p_{BAD} is hardly expressed at all in the absence of C-protein in its inducing form (C-ind). Addition of arabinose reduces the afinity of C-protein for *araO* and hence increases the activity of P_C; at the same time, CRP may bind at its site from which it stimulates transcription from p_C, on the one hand, and assists binding of C-ind to the *araI* sequence on the other. The greater occupancy of this site by C-ind allows RNA-P to bind and transcribe from p_{BAD}.

Note that this mechanism accounts for positive as well as negative control by the *araC* protein and for the fact, known from genetic studies, that the p_{BAD} promoter is repressed from a site upstream relative to the site from which it is activated. It also accounts for the role of CRP as activator of both promoters.

Transient activation of the *araC* gene promoter

The model presented in the preceding section further suggests that arabinose should act as an inducer of both the p_{BAD} and the p_C promoter; an effect on the latter was looked for but was not observed. The reason for this failure is that the effect is transient. With the *araC::lacZ* fusion strain, the activity at p_C could be followed closely before and after addition of arabinose. During the first 10–15 minutes after induction, the activity was increased approximately fivefold; it then declined and eventually fell back to approximately the preinduction level (Figure 7A). Control experiments indicate that p_C is sensitive to catabolite repression. The transient character of the stimulation of the synthesis of the C-protein by arabinose may therefore be explained as a result of the induced synthesis of the B, A, and D proteins and the onset of arabinose catabolism, which is known to elicit catabolite repression. Autorepression may also be involved in turning down synthesis of the C-protein.

Finally, the model can account for the paradoxical behavior of diploids of the C^c/C^+ constitution. These cell lines do not display the constitutive phenotype characteristic of the C^c genotype; the normal C-product dominates, and no enzymes are produced without induction. The explanation that can now be offered is that, if the O-site is occupied by C-protein in the repressor conformation, the activity at *both* promoters is blocked because the CRP–cAMP complex cannot function. This means that the C^c product, although present in the cell, cannot work.

Physiological importance of catabolite repression

In terms of growth physiology, the mechanisms that regulate transcription serve to balance the overall system; and to understand how this is achieved, it is not enough to know how individual mechanisms are constructed. We must also know how they are operated. For example, when cells grown on glycerol are transferred to a medium with arabinose as the sole carbon source, the *ara* operon will be fully induced. Without a check on the synthesis of the *araB, araA,* and *araD* enzymes, they would continue to be produced at the highest possible rate. A high initial rate is clearly an advantage because the enzymes needed to metabolize the new carbon source can quickly reach optimal concentrations in the cell. Beyond that time, two adverse effects can be anticipated: first, the cell might produce more of these enzymes than it needs for balanced growth. This would be a minor disadvantage, but with enzymes in excess more arabinose would be metabolized than can be used for growth. Heat would be generated as well as lower grade products, which would pass into the medium. This need not affect growth rate, but it could seriously reduce the yield of cell mass per milligram of arabinose and thus be a major disadvantage from the point of view of evolution.

In the *ara* system, induction leads to a high initial rate of enzyme production followed by a decline. It is probably the metabolism of arabinose itself that acts as a buffer and prevents runaway synthesis of the *ara* enzymes by modulating the cAMP level in the cells. The same reasoning applies to growth with carbon sources such as lactose or galactose, both of which depend for their utilization on specific processes of induction subject to catabolite repression. It is characteristic of all these operons that the inducer, or effector, is present outside the cell in a more or less fixed concentration, and it is therefore not possible to arrive at the optimal *degree* of induction by manipulating the effector concentration. The secondary effector, cAMP, which is produced inside the cell as a by-product of catabolism, adds flexibility to the individual processes of induction.

TRYPTOPHAN OPERON

In this section we shall consider operons in the domain of biosynthesis. Most of these operons display the pattern of the *lac* system. A promoter–operator complex is placed in front of the gene(s) for the biosynthetic enzymes, and a regulatory gene (*R*) exists somewhere on the chromosome. Its product is a repressor protein analogous to the *lacI* gene product, but the repressor protein is constructed such that its affinity for the operator is greatly *increased* when it combines with the effector (in contrast to the *lac* repressor whose affinity for *lacO* is reduced when it combines with an inducer).

The *trp* operon has been analyzed in more detail than any other biosynthetic operon, and it is an ideal example of a repressible system. In addition, the *trp* operon offers by far the best example of a novel and entirely different mode of regulation known as attenuation. We shall consider these two entirely independent control mechanisms separately.

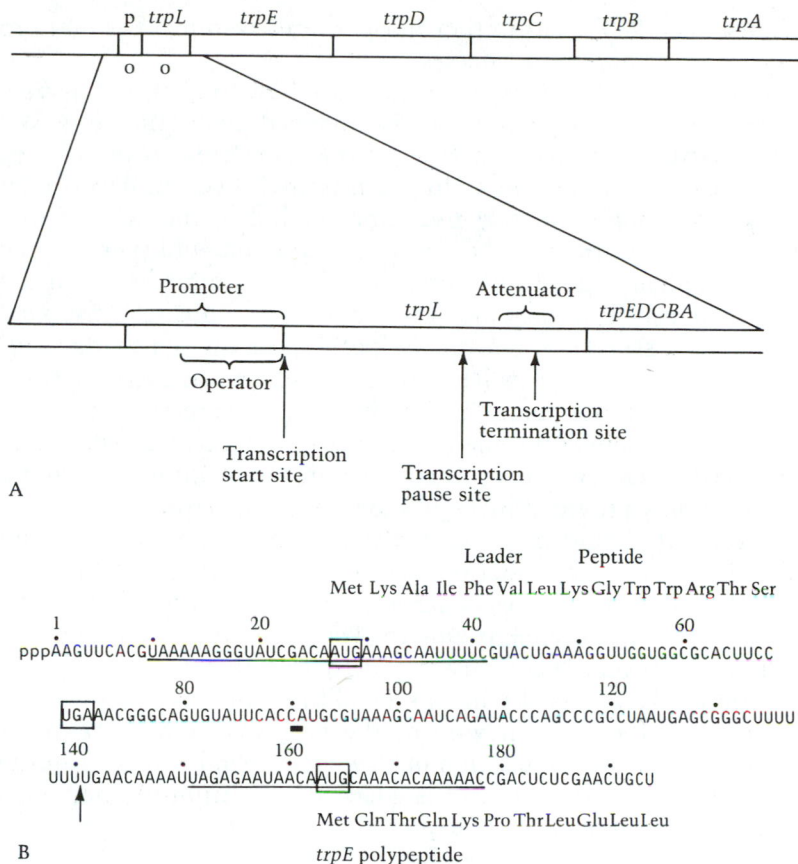

Figure 8. A. The regulatory and structural gene regions of the *trp* operon of *E. coli*. Transcription initiation is controlled at a promoter-operator. Transcription termination is regulated at an attenuator, in the transcribed 162-base pair leader region, *trpL*. All RNA polymerase molecules transcribing the operon pause at the transcription pause site before proceeding further. B. The nucleotide sequence of the 5′ end of *trp* messenger RNA. The nonterminated transcript is presented. When transcription is terminated at the attenuator, a 140-nucleotide transcript is produced. Its 3′ terminus is marked by an arrow. The 3′ terminus of the pause transcript, at nucleotide 90, is underlined by a bar. The two AUG-centered ribosome binding sites in this transcript segment are underlined. The boxed AUGs are where translation starts and the boxed UGA is where it stops. The predicted amino acid sequence of the *trp* leader peptide is shown. (From Yanofsky, 1981.)

Repression by tryptophan

Figure 8A shows the familiar arrangement of promoter and operator followed by the genes encoding the five enzymes of the operon. The *trpR* gene specifying the repressor is not shown; it maps far from the *trp* operon and, like the *araC* product, the *trp* repressor regulates its own synthesis

(Gunsalus, 1980). As in other biosynthetic systems, the end product, in this case tryptophan, is the effector.

In a system like this, it is easy to see how the optimal *degree* of repression is reached, because the effector is produced inside the cells. Wild-type and a *trp* constitutive mutant of *E. coli* (*trpR*$^+$ and *trpR*, respectively) grow practically at the same rate in a glucose minimal medium, but the level of the *trp* enzymes is approximately five times higher in the cells of the totally derepressed mutant than in the normally regulated wild type. Evidently, the flow of tryptophan needed for protein synthesis is maintained in both cases, and the relatively low enzyme level in the *R*$^+$ culture must be the result of INTERNAL REPRESSION due to the buildup of the appropriate, relatively high concentration of tryptophan. In this case, the activity of the *trp* operon is reduced to approximately 20% of its fully derepressed level by adjustment of the effector concentration in the cells; the further reduction to 1 or 2%, observed when tryptophan is supplied in the medium, does not add greatly to the economy provided by regulation on that operon.

Here it should be reiterated that nearly all *E. coli* proteins are stable.[1] This implies that a reduction in the *rate* of synthesis of an enzyme by repression does not change the *concentration* of that enzyme except as it is gradually diluted out by subsequent growth. Therefore, it is important to keep in mind that, for the overall economy of growth, the effector of a biosynthetic operon—its end product—usually has a second function: it causes *feedback inhibition* directed at the first enzyme of the pathway. This is a *fast* reaction that prevents wasteful production of, say, tryptophan and the intermediates in its synthesis if that amino acid is added to the minimal medium in which the cells have been growing (Chapter 3).

Attenuation in the *trp* operon

During unrestricted growth, tryptophan synthesis is regulated almost exclusively by repression and feedback inhibition; the mechanism of attenuation only plays a role when the cells are starved of tryptophan. The *trp* operon is thus "opened up" in two stages: when the concentration of tryptophan in the cell goes down, the operon is first derepressed; and when starvation conditions are reached, attenuation is relieved, with the result that complete *trp* mRNA is made approximately 10 times more frequently than in a fully derepressed *trpR* mutant.

Like induction and repression, attenuation essentially controls mRNA production, but the mechanism is entirely new: no specific protein is involved, and the DNA–protein interaction that controls access to the promoter according to the classic operon concept is replaced by interactions between the

---

[1] This is true, but exceptions exist. The best known probably is the antiterminator factor produced from the *N* gene of phage λ. In a survey of approximately 300 *E. coli* proteins displayed on two-dimensional gels, all except one small acidic protein were stable. We shall therefore not discuss the proteases known to exist in bacteria. These otherwise interesting enzymes seem to serve mainly as scavengers eliminating faulty proteins and, possibly, fragments of normal proteins, such as signal peptides (Swamy, 1981).

nascent transcript, ribosomes, tRNA, and termination factors. In other words, the whole translation machinery is involved. Very briefly, the *trp* promoter is separated from the *trpE* gene, which encodes the first of the five enzymes, by a LEADER SEQUENCE (*trpL*) of 162 base pairs (Figure 8A). Near position 140 of the transcript of *trpL* is a termination site that is characterized by a sequence of eight uracils (Figure 8B). Except during *trp* starvation, approximately 9 out of 10 transcripts stop here; when they continue to full length, *trp* mRNA is produced. ATTENUATION refers to this dramatic reduction in the number of transcripts that extend beyond the early terminator (Yanofsky, 1981).

It is a remarkable feature of the mechanism of attenuation that the movement of ribosomes translating the short *trpL* transcript determines whether or not the RNA-P further downstream will pass the termination point. The first indications that translation was involved came with the isolation of mutants with abnormal synthesis of *trp* or *his* aminoacyl-tRNA. It was observed that mutations in the synthetases themselves, in tRNA, or in modifying enzymes could affect the production of the cognate species of mRNA (Singer, 1972). However, the detailed analysis of the phenomenon of attenuation had to await the development of the sequencing techniques by which we identify the secondary structures that a particular nucleic acid sequence can assume.

Leader sequence and leader peptide

The early part of the transcript of the *trp* operon contains two AUG start codons with corresponding ribosomal binding sites (Figure 8B). The leader peptide of 14 amino acids has not been isolated; it is assumed to be unstable, and it would have remained hypothetical had it not been for an elegant proof that translation actually is initiated effectively from the AUG codon of the leader transcript. This was shown with deletion mutants that were missing a segment between a late codon in the leader sequence and an early codon in the *trpE* gene. Such mutants produce stable fusion polypeptides that begin with the early part of the leader peptide and continue from the end of the deletion in the *trpE* gene. (The *trpE* gene product is an ordinary, stable enzyme.) This proof that a leader peptide in fact is synthesized with the amino acid sequence indicated by the base sequence of the transcript turned out to be essential.

The next step in the analysis concerns the stem-and-loop structures that the leader transcript can be imagined to form (Figure 9). Computer constructs of this kind must be viewed with caution; they are easy to make, but their correspondence to real structures is by no means assured. In the case of the *trp* leader sequence, however, digestion with RNase T1 produces fragments that migrate together in nondenaturing gels and that support the base-pairing pattern illustrated in Figure 9.

The model that has emerged suggests that as transcription proceeds along the *trpL* segment a ribosome quickly occupies the binding site centered on the first AUG codon and that the subsequent positions of this ribosome along the leader transcript determines which alternate stem-and-loop structures are

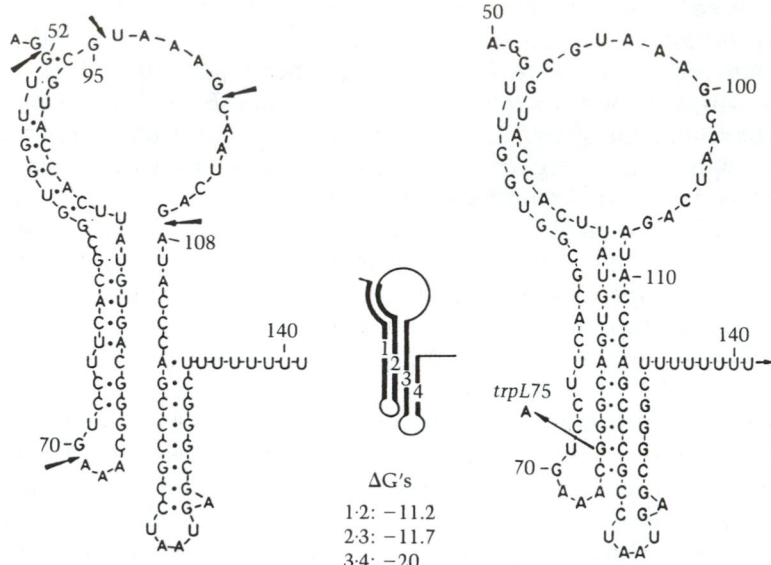

Figure 9. Secondary structure alternatives in the *trp* leader transcript. In the large structure on the left, the two base-paired structures that are detected in vitro are presented. The arrows indicate the sites of RNase T1 attack. The G-bonds in the hydrogen-bonded regions are not cleaved, presumably because the Gs are base-paired. In the large structure on the right, an alternative secondary structure is shown. The center insert is explained in the text. (From Yanofsky, 1981.)

allowed to form. The decisive feature of the leader peptide is the occurrence of two neighboring *trp* codons (a rare event considering that in *E. coli* only approximately one amino acid residue in a hundred is tryptophan). As translation advances and the ribosome faces a *trp* codon, further progress will depend on the availability of a charged tRNA^{Trp}; in this situation, starvation for tryptophan can be assumed to cause translation to pause. In the position now occupied by the ribosome, the stem formed by segments 1 and 2 in Figure 9 will have dissolved, permitting the alternative stem, between segments 2 and 3, to form. This, in turn, prevents the formation of the structure presumed to be the terminator, namely, the 3:4 stem with the U-tail.

In summary, delaying the ribosome at the *trp* codons favors the 2:3 stem formation and prevents the simultaneous existence of the terminator; consequently, the RNA-P transcribing ahead of the ribosome may pass through the "danger" region because this segment cannot assume the terminator configuration. Conversely, if the tRNA^{Trp} is more or less fully charged, translation can proceed apace; this prevents formation of the 2:3 loop and allows the 3:4 loop to persist, with the result that transcription is terminated in most cases.

A number of mutants with known base substitutions in the leader sequence have been isolated, and their properties bear out the correctness of the model just described. One mutant harbors a G → A change in the stem formed by pairing strands 2 and 3 (Figure 8B); the mutation (*trpL75*) greatly weakens this stem, which in turn stabilizes the 3:4 termination structure. In agreement with the model, starvation for tryptophan does *not* relieve attenuation in the *trpL75* mutant. In contrast, approximately 30 mutants in which transcription termination is reduced harbor base changes clustering around the middle of the stem formed when strand 3 pairs with strand 4. In all these mutants the structure held responsible for termination is weakened and termination accordingly is reduced. Further support for the attenuation model and a comprehensive list of references are found in a review by Yanofsky (1981).

Attenuation and growth physiology

The simultaneous presence in the *trp* operon of two independent control systems invites speculation as to their respective places in evolution and in the growth physiology of present-day *E. coli*. We know that during balanced growth the expression of the operon is regulated by repression and that attenuation is relieved only when tryptophan is in very short supply. Now consider a culture of wild-type *E. coli* that has grown for a long time with tryptophan supplied in the medium: when this amino acid is exhausted, derepression will, by itself, increase the frequency of transcription from the *trp* promoter approximately 100-fold, and a further tenfold increase may result from relieving attenuation. Therefore, the question is how much a cell gains by being able to increase the potential of the *trp* operon 1000, rather than 100, times and thus shorten the period necessary to build up an adequate supply of the *trp* enzymes. We do not know the answer; but it is quite conceivable that the *trp* attenuator in fact is an evolutionary relict with little or no physiological significance today. If so, this relict has been the source of very precious information.

If there can be doubt about the importance of attenuation in regulating synthesis from the *trp* operon, this uncertainty does not apply to the amino acid biosynthetic operons that appear to be controlled by attenuation alone (Figure 10). Each of the predicted leader peptides is excessively rich in the amino acids thought to be the specific effectors, and an extensive search has failed to turn up repressor-negative mutants. It is therefore natural to assume that repression has been replaced by attenuation in the five operons illustrated in Figure 10, and we shall consider briefly what the consequences of this may be.

When an effector amino acid is added to a minimal medium culture, operons regulated by attenuation will respond qualitatively as do repressible operons; synthesis of the relevant biosynthetic enzymes is reduced. It is not obvious, however, that enzyme concentration can be adjusted effectively by attenuation during growth in minimal media. This new uncertainty stems from the fact that the *level* of charging of tRNA, which determines the *degree* of attenuation, is not a free and independent variable. The reason is that the high and constant cgr_p observed at all growth rates (Chapter 6) requires that certain minimum concentrations of charged tRNA are maintained at all times. We shall return to this problem in Chapter 8 when discussing the "dual role" of amino acids as effectors and as substrate for protein synthesis.

Finally, we note that control of the expression of an operon by modulating

Operon	Leader sequence
pheA	Met Lys His Ile Pro <u>PHE PHE PHE</u> Ala <u>PHE PHE PHE</u> Thr <u>PHE</u> Pro
his	Met Thr Arg Val Gln Phe Lys <u>HIS HIS HIS HIS HIS HIS HIS</u> Pro Asp
leu	Met Ser His Ile Val Arg Phe Thr Gly <u>LEU LEU LEU LEU</u> Asn Ala Phe Ile Val Arg Gly Arg Pro Val
	Gly Gly Ile Gln His
thr	Met Lys Arg <u>ILE</u> Ser <u>THR THR ILE THR THR THR ILE THR ILE THR ILE THR THR</u> Gly Asn Gly Ala Gly
	Gly Ala Ala Leu Gly Arg Gly Lys Ala
ilv	Met Thr Ala <u>LEU LEU</u> Arg <u>VAL ILE</u> Ser <u>LEU VAL VAL</u> Ile Ser <u>VAL VAL VAL ILE ILE ILE</u> Pro Pro Cys

Figure 10. Leader sequences in amino acid operons. (From Yanofsky, 1981.)

the translation of a leader transcript probably applies to amino acid biosynthetic operons exclusively. On the other hand, attenuation can be an effective element in autoregulation; thus, an "effector protein" temporarily present in excess in a cell may interact with a specific sequence in its own mRNA and terminate transcription (or interfere with translation). These modes of regulation will be discussed in the next section.

OPERONS IN THE PROTEIN SYNTHESIZING SYSTEM

Introduction

When a bacterial cell grows, the dominating activity is protein synthesis, and we have seen how efficiently this process is organized. We have also presented examples of the refined control mechanisms that regulate the flow of energy and matter that feeds the PSS. It is time now to discuss the set of controls that serve to maintain balance within PSS.

There are good reasons why this set of controls should be treated as something special. In the first place, the problem of regulation is bigger and more complex than that of any single operon, because the PSS consists of at least 100 different proteins that map all over the chromosome. In addition, the r- and tRNA species belong to the system and they are transcribed from their own special promoters and do not serve as templates for protein synthesis. The magnitude of the problem is perhaps best envisaged when we recall that the high efficiency of the PSS demands that all its components be present in more or less fixed ratios and that the overall size of the PSS varies with the growth rate (Chapter 6). Here we shall discuss our limited knowledge of the mechanisms that maintain these ratios; the regulation of the size of the PSS is the main subject of Chapter 8.

Both in aggregate mass and in complexity, the ribosomes predominate, and it is obvious that we face two main problems: first, balancing the production of 50-odd r-proteins originating from approximately 20 widely scattered transcriptional units, and second, adjusting the production of rRNA (and tRNA) to that of the r-proteins.

Before we examine the balancing mechanisms that have been shown to exist, consider what the situation would be if the output of different r-proteins was not regulated at all. If the promoters involved were of equal strength, all the r-protein genes would issue mRNA at the same rate, but the different genes would not be present in the same number. The extreme case would be the L34 gene (*rpmH*), now mapped less than one unit from the origin, and the L32 gene (*rpmF*) mapping close to the terminus of replication. The ratio between origins and termini equals $2^{C\mu/60}$ (Chapter 6), and to a good approximation this would also be the ratio between the rates of synthesis of L34 and L32. At maximum growth rate ($\mu = 2.8$), the ratio between the rates would be slightly less than $2^{1.9} = 3.7$; and in a relatively slow-growing culture ($\mu = 0.7$), the ratio would drop to approximately 1.4. This dependence on μ could not be eliminated by having promoters of different but fixed strengths, and because one of each of the r-proteins (except the product of the *rplJ* gene) is needed to assemble a complete ribosome, the rate of synthesis of L32 would limit the rate at which finished ribosomes could be made. In rapidly growing cells, L34 would therefore be overproduced by approximately 250%.

The pools of free r-proteins are small fractions of the total protein bound in the ribosomes (Gausing, 1974); however, if r-proteins are synthesized in the presence of rifampicin, some remain free, and they are fairly stable (Dennis, 1974). This suggests that mechanisms must have evolved to balance the rates of synthesis of the individual r-proteins; but the nature of these controls cannot be deduced from what we already know about the regulation of operons. First of all, the r-proteins are not enzymes catalyzing individual reactions, and therefore no conventional effector, such as the substrate or product of a reaction, exists. Second, a common effector, if it were found, could not possibly serve to balance the synthesis of the different r-proteins because the gene ratios vary with μ.

Autoregulation and its effectors

The r-proteins are themselves the substrates in the process of ribosome assembly (Chapter 2), and it has turned out that some of them are effectors as well. The type of control that results is known as AUTOREGULATION, of which we have seen two examples: the *araC* protein, which in one of its configurations acts as an autorepressor, and the *trpR* protein. It is a general feature of this mode of regulation that it is capable of maintaining in the cell a nearly constant concentration of the products of the genes it controls (Sompayrac, 1973).

The problems we are about to discusss have only begun to be analyzed, and there are many loose ends. Here we have selected a few case studies to illustrate what has been achieved and by what means. [A more comprehensive survey may be consulted for further details (Lindahl, 1982).]

Most of the experiments done to analyze operons involve adding or withdrawing the specific effector. If it is a protein molecule that acts as the effector, other methods must be sought. Two lines are regularly pursued: the first is to introduce into the cell a vector (usually a plasmid) carrying the gene(s) for the effector to be tested. This gives an in vivo system in which the product of this gene(s) is overproduced; one then looks for the effect of an excess of the putative effector. Second, the problem of getting the effector into the cell can be bypassed by working in vitro. Clean experiments of this kind require that the protein to be tested be available in adequately purified form. In vitro experiments in which the *lac* or *ara* repressor is added are good examples.

The proof that autoregulation works as predicted must be that it can be shown to operate in vivo as well as in vitro. In vivo, the major difficulty is that cells carrying a plasmid with extra copies of a particular gene will adjust to this new condition, and the result of that may be quite different from the response to an immediate increase in the intracellular concentration of the product of that gene. This problem has been circumvented most elegantly by making the extra r-protein gene(s) carried on the plasmid depend on IPTG for expression (Lindahl, 1982). This was achieved by constructing a number of fusions between the proximal end of the *lac* operon and various parts of the r-protein operons (Figure 11). Note that the purpose of these fusions was to put the gene(s) to be tested under *lac* operon control; no effort was made to fuse the *lacZ* gene directly to a gene in the r-protein operon. Thus, we deal with purely TRANSCRIPTIONAL FUSIONS, and the r-protein(s) originating from the plasmids are produced by ribosomes that enter at the natural binding site for an r-protein but on a messenger transcribed from the *lac* promoter. This situation must be clearly distinguished from the TRANSLATIONAL FUSIONS discussed under the *ara* operon.

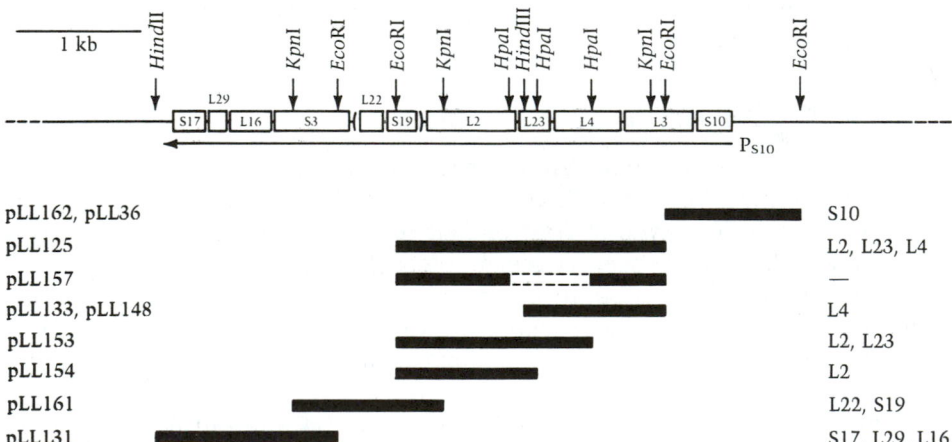

Figure 11. S10 operon and fragments of the operon cloned on plasmids. In the top part of the figure, the genes in the S10 operon are identified by the names of their products; the boxes are approximately proportional to the molecular weights of the corresponding proteins. P_{S10} designates the expected position of the promoter, and the arrow gives the direction of transcription. The solid bars below the map indicate the DNA fragments carried by the plasmids named on the left. The proteins synthesized from the plasmids upon induction with IPTG are named on the right. (From Zengel, 1980.)

S10 operon

The operon in Figure 11 was analyzed by adding IPTG and measuring, 10 minutes after the addition, the rate of synthesis from the chromosomal genes *not* carried on the plasmid. All 11 r-proteins in the operon were severely and coordinately repressed when L4 was overproduced; moreover, hybridization of pulse-labeled RNA to probes representing various segments of the operon showed that transcription along the entire operon was reduced in roughly the same degree as translation. These in vivo results seem very clear; autoregulation at the level of transcription is exerted by L4 and not by any other protein in the operon. The term *repression* used above suggests a mechanism such as in the *lac* operon, but there is no evidence that L4 blocks initiation of transcription at the promoter; L4 is an RNA-binding protein and it might bind to an early site on the mRNA and terminate transcription. It is in fact observed that an excess of L4 causes attenuation some 150 bases downstream from the S10 promoter. In this region a stem and loop structure can form which shares a nine-base sequence with 23 S RNA near the presumed binding site for L4. This stem and loop structure may be the actual terminator (Lindahl, personal communication, 1982). At least one loose end remains: the clear cut in vivo results have been only partially confirmed in vitro. In a coupled system programmed by DNA containing the operon and little else, addition of L4 represses synthesis from the proximal but not the distal genes. It is not known why the in vitro system fails to reproduce the results observed in vivo; perhaps the S10 operon is regulated at the translational as well as the transcriptional level.

A complex case: the β operon

Autoregulation in a pure r-protein operon seems to be relatively simple, in the sense that a single effector, presumably interacting with a single target, balances the output of the set of r-proteins in one operon against the output of other sets from other operons. However, in several cases, r-protein genes occur together with the genes of other proteins of the PSS. For example, EF-G is synthesized as a UNIT PROTEIN (one copy per ribosome) and would be classed as an r-protein were it not that it is only a part-time constituent of the ribosome. Another set of genes, often referred to as the β operon, comprises four r-protein genes in two groups and the *rpoB* and *rpoC* genes coding for the large subunits (β and β') of the RNA-P. We shall discuss this operon in detail because it exhibits several modes of regulation.

Figure 12A shows the entire region from which mRNA for four r-proteins (L11, L1, L10, and L7/L12) plus the β and β' subunits is transcribed. L7 and L12 are identical except that the former is acetylated at the NH_2-terminal end. The relative importance of promoters p_K and p_1 (also called p_β) is not clear, and no indication of a termination site can be seen near the end of *rplA*; p_2, p_3, and p_4 are minor promoters and may not be active in vivo. Figure 12B indicates length in units of 1000 base pairs and identifies the restriction sites relevant to the experiments in Figure 14.

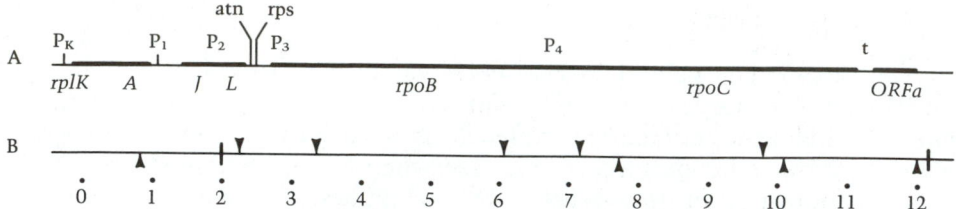

Figure 12. A. Positions and relative sizes of the genes in the β operon. The promoters and control elements are discussed in the text. B. *Eco*RI (▼) and *Bgl*II (▲) restriction sites; the segment between the bars is used for the deletion analysis in Figure 14. The numbers below the line show distance in kilobase pairs. (From Squires, 1981.)

The two leftmost genes, *rplK* and *rplA*, appear to form a separate operon autogenously regulated by r-protein L1. Evidence for this has been obtained in vivo as well as in vitro, but the mechanism has not been analyzed. However, if it turns out that the whole β operon is transcribed from the promoter p_K in vivo, L1 would have to act at the translational level in order not to interfere with expression of genes further downstream.

Role of L10 and L12 proteins

The second pair of r-protein genes, *rplJ* and *rplL*, are preceded by a leader sequence of approximately 400 base pairs (Figure 13A). The control again seems to act at the level of translation, but in this case the mechanism is beginning to be understood. All the r-proteins are unit proteins except L7/L12, which is present in four copies. These four, and one L10 molecule, form a complex that binds to the 23 S rRNA at an early time during assembly of the 50 S subunit. A plasmid carrying the complete *rplJ* but only part of the *rplL* gene inhibits growth almost completely; but if both genes are present, the plasmid is tolerated.

With both genes on the plasmid, the r-proteins L10 and L12 are synthesized from the same number of genes; the *rplJ* and *rplL* on the chromosome, and one pair on each plasmid. In this situation, L10 and L12 will be produced in the natural ratio (1:4) whatever the degree of autocontrol due to the high copy number of the genes. If the plasmid carries a complete *rplJ* gene while all or part of *rplL* is missing, L12 can be supplied only from the chromosomal *rplL* gene. Therefore, the argument is that L10 will be overproduced and that it will reduce its own production from the multiple *rplJ* genes on the plasmids, but also, and to the same degree, the production of L12 from the one and only *rplL* gene on the chromosome. Thus, the system designed to control the output from genes *rplJ* and *rplL* will reduce the supply of L12 much more severely than that of L10, and the cells will grow very slowly (Fiil, 1980).

Mutant studies and in vitro experiments support this notion. On a lawn of cells that grow very little because they harbor a plasmid with the *rplJ* and

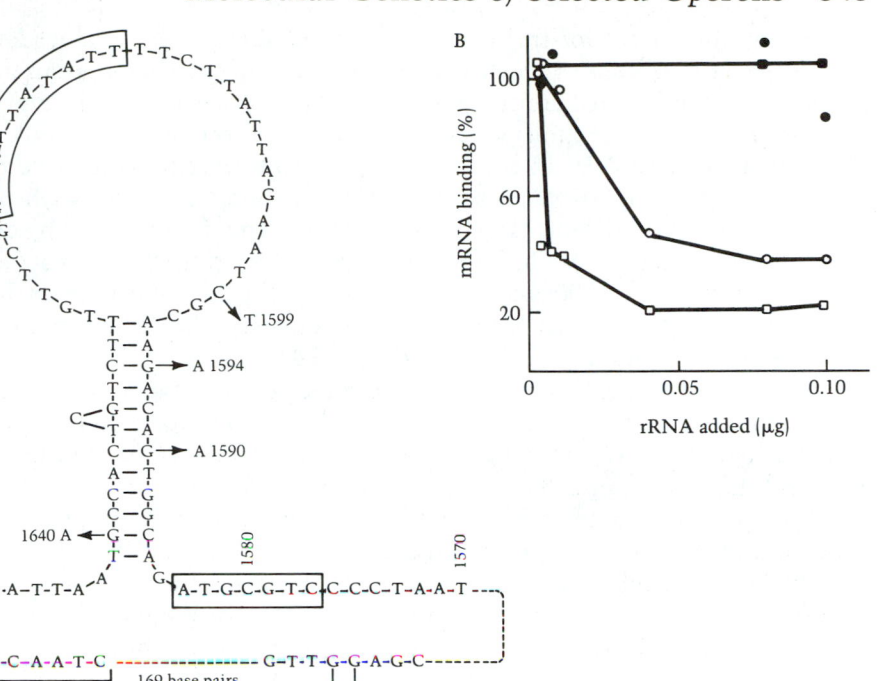

Figure 13. A. Base sequence and possible secondary structure in the segment of the β operon beginning with the Pribnow box of promoter P₁ (see Figure 12A). Arrows with letters and base numbers indicate mutations affecting expression of *rplJ* and *L*; boxes contain sequences common to this control region of the β operon and 23 S rRNA. (From Fiil, 1980.) B. Binding of r-protein L10 (●, ○) or the L10/L12 complex (■, □) to labeled mRNA transcribed from the early promoters of the β operon. Filled and open symbols refer to addition of increasing amounts of yeast RNA and of *E. coli* ribosomal RNA, respectively. (From Johnsen, 1982.)

only part of the *rplL* gene (see earlier), a few big colonies come up, and some of them turn out to carry mutations in the *rplJ* leader. In Figure 13A, six point mutations are indicated; these and partial deletions of the stem-and-loop structure (not shown) all greatly reduce translation of the *rplJ* and *rplL* mRNA. In vitro studies complement this picture and show that addition of L10 or of L10 and L7/L12 together strongly depresses the synthesis of both proteins from plasmid DNA carrying the *rplJ* and *rplL* genes. The model that emerges is far from clear in detail, but the evidence implies that the secondary structure of the *rplJ* leader transcript is involved as a positive element in the initiation of translation and that its efficiency can be reduced either by mutations that affect the stability of the structure or its interaction with the effector, which apparently can be either L10 or the L10–L7/L12 complex.

A possible target for this control was looked for in the *rplJ* leader, and a very suggestive feature turned up. Figure 13A shows that a stem-and-loop structure can be formed around position 1600 and that runs of 10 and 7 bases (boxes) are repeated around positions 590 and 560 on the 23 S RNA. If, at a time when the L10–L7/L12 complex happens to be in excess relative to other r-proteins, and/or to rRNA, surplus complexes may combine with the *rplL* leader, and this could depress translation. This model is supported by evidence that the complex actually binds to the leader transcript and protects a region between base pairs 1500 and 1600, including the box with seven bases in Figure 13A. Moreover, 23 S rRNA competes with the leader transcript in binding experiments (Figure 13B; Johnsen, 1982).

There are several loose ends here. The relative importance of promoters p_K and p_1, and the relative role of L10 and the L7/L12 complex must be defined. But that is not all. We do not know what makes the joint transcript of genes *rplL* and *rplJ* yield four copies of L7/L12 for each copy of L10. The entire DNA sequence has been determined, and the possibility of tandem repeats of *rplL* is ruled out. Also, the NH_2-terminal acetylation that turns L12 into L7 cannot be invoked to explain the unequal yields because mutants in which this modification does not take place grow perfectly well. We emphasize the very high yield of L7/L12, relative to the unit protein L10, because some new thinking may be required to explain it.

β and β' subunits of RNA polymerase

We shall now look at the distal part of the β operon with the *rpoBC* genes. The structures to be discussed are indicated in Figure 12A as attenuator (atn), target site for the RNase III (rps), and terminator (t), each of which permits the respective transcripts to assume characteristic stem-and-loop structures. An elegant functional analysis of these sites and their location is illustrated in Figure 14. A plasmid was constructed to contain the segment between bars in Figure 12B fused to the *ara* control region on the left and the *lac* genes on the right. Cells carrying this plasmid produce β-galactosidase upon induction with arabinose, and the rate of enzyme production will be modified if the inserted piece of the β operon contains functional promoters, attenuators, or terminators.

Figure 14 shows that a very effective terminator is close to the end of the *rpoC* gene. It also shows that the induced yield of β-galactosidase is reduced approximately fivefold if the region between the *rpoJL* and *rpoBC* genes, which contain the attenuator-like structure, is on the plasmid. In addition, the small segment on the left appears to harbor a secondary promoter that allows some β-galactosidase synthesis, provided the strong terminator has been removed (Barry, 1979).

The data in Figures 12–14 illustrate the intensive work being done on the β operon. Hybridization experiments confirm that transcription is partially terminated (attenuated) in front of the *rpoBC* genes. The role of the rps structure is uncertain, but the indications are that the *rpoBC* mRNA is more effectively translated once it is severed by RNase III from the early part of the

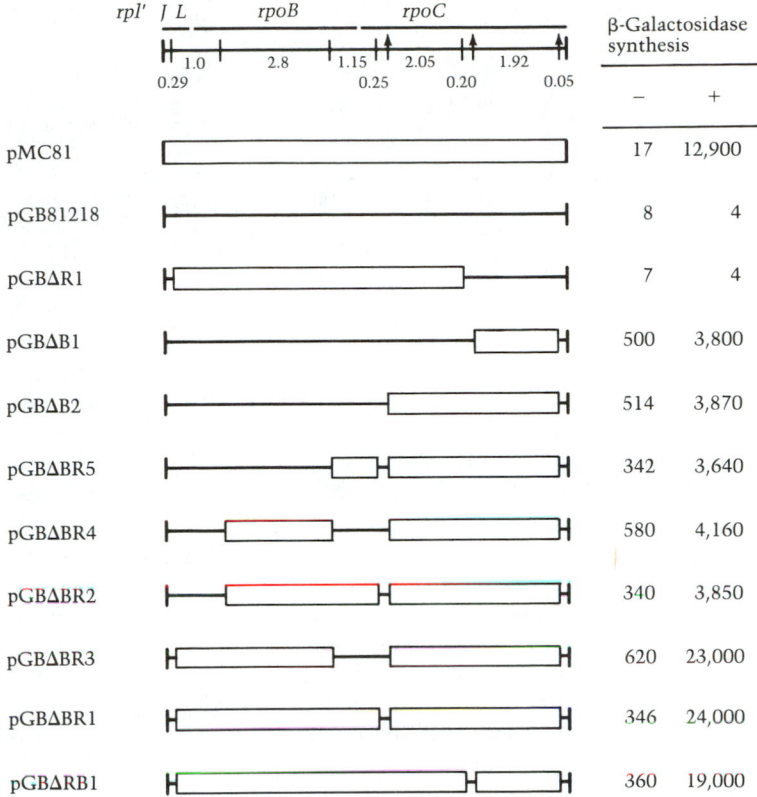

Figure 14. Deletion analysis in the β operon. Plasmid pGB81218 carries the intact segment between bars in Figure 12B; pMC81 is a control lacking this segment. Segments removed by in vitro deletion are represented by open bars. Differential rates of β-galactosidase synthesis are shown for each plasmid with (+) and without (−) induction by arabinose (see text). (From Barry, 1979.)

β operon transcript. This operon has now been shown to carry two terminators in tandem, only one of which seems to require the rho factor.

Translational versus transcriptional regulation

Autoregulation at the level of translation seems to be the rule rather than an exception in the r-protein operons (Nomura, 1980; and Figure 15). This mode of regulation poses a new problem. In the case of repression or induction, all the genes in an operon are expressed equally, or in fixed ratios, when a transcript has been produced. When regulation is exerted at the level of translation, the situation is different. Here it is the yield from an existing mRNA that is modulated, and it is not clear how a damping effect on the translation of the first gene on an r-protein mRNA is propagated to the downstream genes. This requires some sort of translational coupling, an in-

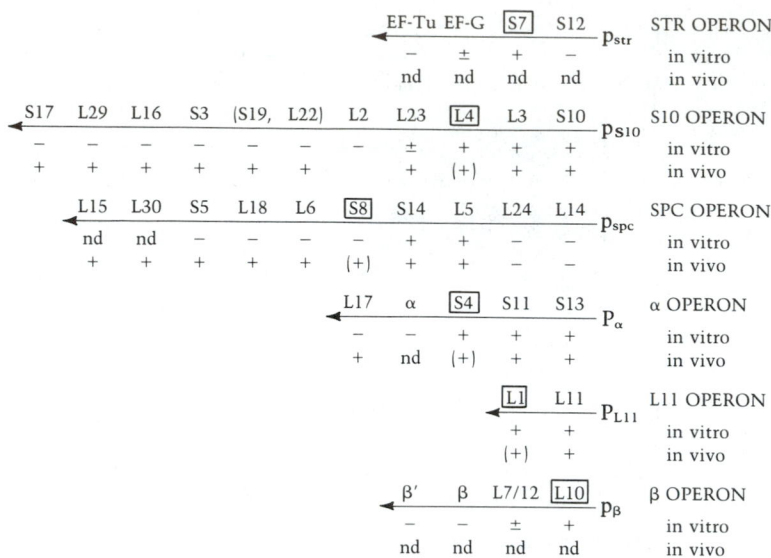

Figure 15. Summary of in vivo and in vitro experiments showing autogenous regulation of r-protein operons. Genes are identified by their products, with the regulatory proteins (effectors) boxed in. Arrows show direction of transcription; degrees of inhibition by the effector are indicated by +, ±, and −. (+) indicates a protein not tested but expected to be inhibited; nd means not determined. (From Nomura, 1980.)

stance of which has been observed in the *trp* operon where translation of the first two genes, *trpE* and *trpD*, is linked in the sense that the entire *trpE* region must be translated before the *trpD* can be expressed (Oppenheim, 1980). However, this may be a special case; the *trpE* and *D* genes are separated by only three base pairs. The S10 operon (Figures 11 and 14) has been partially sequenced and the known distances between genes are more "normal" (30–40 base pairs).

SUMMARY

1. Bacterial genes fall into three groups that serve fueling, biosynthesis, and polymerization. The first encodes enzymes of great practical and diagnostic value; the last includes the genes whose products constitute the PSS (the protein synthesizing system).

2. *Fueling reactions.* The *lac* operon model posed questions that stimulated the development of powerful techniques of molecular genetics: DNA-dependent protein synthesis in vitro, nucleic acid hybridization and sequencing, and construction of hybrid genes and plasmids. The complex *ara* operon illustrates autorepression of the *araC* gene function, and it shows that the effect of a primary inducer present in a high, fixed con-

centration outside the cell can be modulated by a secondary, internal effector (cAMP).

3. *Biosynthesis*. In balanced growth, the activity of the *trp* operon is reduced by internal repression; the repressor protein is autoregulated. Starvation for tryptophan releases repression *and* reduces transcription–termination in the leader sequence (attenuation). The latter involves alternating stem-and-loop structures in the transcript and synthesis of a leader peptide with a pair of tryptophan residues. Attenuation alone regulates the synthesis of several amino acids.

4. *Protein synthesis*. Coordinate production of the PSS components (r-proteins, etc.) requires autoregulation of many individual operons that are located throughout the chromosome. Translational as well as transcriptional regulation are encountered, and attenuation occurs inside at least one complex operon. The PSS appears to constitute a unit within which all syntheses are coupled by controls that ensure balanced production of its components.

REFERENCES

Adesnik, M. and C. Levinthal. 1970. The synthesis and degradation of lactose operon messenger RNA in *E. coli*. Cold Spring Harbor Symp. Quant. Biol. 35:451.

Aunstrup, K. 1979. Production, isolation and economics of extracellular enzymes, p. 27. In *Applied Biochemistry and Bioengineering*, Vol. 2, Enzyme Technology, L. B. Wingard, E. Katchalski-Katzir and L. Goldstein, eds. Academic Press, New York.

Buchanan, R. E. and N. E. Gibbons (eds). 1974. *Bergey's Manual of Determinative Bacteriology*, 8th Edition. Williams and Wilkins, Baltimore.

Dennis, P. P. 1974. Synthesis and stability of individual ribosomal proteins in the presence of rifampicin. Mol. Gen. Genet. 134:39.

Englesberg, E. and G. Wilcox. 1974. Regulation: positive control, p. 219. In *Annual Review of Genetics*, Vol. 8, H. L. Roman, A. Campbell and L. M. Sandler, eds. Annual Reviews Inc., Palo Alto, CA.

Fiil, N. P., J. D. Friesen, W. L. Downing and P. P. Dennis. 1980. Post-transcriptional regulatory mutants in a ribosomal protein-RNA polymerase operon of *E. coli*. Cell 19:837.

Gausing, K. 1974. Ribosomal protein in *E. coli*: rate of synthesis and pool size at different growth rates. Mol. Gen. Genet. 129:61.

Gerhard, B., C. L. Squires and C. Squires. 1979. Control features within the *rplJL-rpoBC*, transcription unit of *Escherichia coli*. Proc. Natl. Acad. Sci. USA 76:4922.

Gilbert, W. and B. Müller-Hill. 1966. Isolation of the *lac* repressor. Proc. Natl. Acad. Sci. USA 56:1891.

Gilbert, W. and B. Müller-Hill. 1967. The *lac* operator is DNA. Proc. Natl. Acad. Sci. USA 58:2415.

Greenblatt, J., M. McLimont and S. Hanly. 1981. Termination of transcription by *nusA* gene protein of *Escherichia coli*. Nature 292:215.

Gunsalus, R. P. and C. Yanofsky. 1980. Nucleotide sequence and expression of *Escherichia coli trpR*, the structural gene for the *trp* aporepressor. Proc. Natl. Acad. Sci. USA 77:7117.

Hippel, P. H. von. 1979. On the molecular bases of the specificity of interaction of transcriptional proteins with genome DNA, p. 279. In *Biological Regulation and Development*, Vol. 1, R. F. Goldberger, ed. Plenum Press, New York.

Jacob, F. and J. Monod. 1961. Genetic regulatory mechanisms in the synthesis of proteins. J. Mol. Biol. 3:318.

Lederberg, J. 1948. Gene control of β-galactosidase in *E. coli*. Genetics 33:716.

Lee, N. L., W. O. Gielow and R. G. Wallace. 1981. Mechanism of *araC* autoregulation and the domains of overlapping promoters, P_C and P_{BAD}, in the L-arabinose regulatory region of *Escherichia coli*. Proc. Natl. Acad. Sci. USA 78:752.

Lindahl, L. and J. M. Zengel. 1982. Expression of ribosomal genes in bacteria, p. 53. In *Advances in Genetics*, Vol. 21, E. Caspari, ed. Academic Press, New York.

Miller, J. H. 1980. *Experiments in Molecular Genetics*. Cold Spring Harbor Laboratory, Cold Spring Harbor, New York.

Miller, J. H. and W. S. Reznikoff (eds.). 1978. *The Operon*. Cold Spring Harbor Laboratory, Cold Spring Harbor, New York.

Nomura, M., J. L. Yates, D. Dean and L. E. Post. 1980. Feedback regulation of ribosomal protein gene expression in *Escherichia coli*: structural homology of ribosomal RNA and ribosomal protein mRNA. Proc. Natl. Acad. Sci. USA 77:7084.

Ogden, S., D. Haggerty, C. M. Stoner, D. Kolodrubetz and R. Schleif. 1980. The *Escherichia coli* L-arabinose operon: binding sites of the regulatory proteins and a mechanism of positive and negative regulation. Proc. Natl. Acad. Sci. USA 77:3346.

Oppenheim, D. S. and C. Yanofsky. 1980. Translational coupling during expression of the tryptophan operon of *Escherichia coli*. Genetics 95:785.

Singer, C. E., G. R. Smith, R. Cortese and B. N. Ames. 1972. Mutant tRNA^{His} ineffective in repression and lacking two pseudouridine modifications. Nature New Biol. 238:72.

Sompayrac, L. and O. Maaløe. 1973. Autorepressor model for control of DNA replication. Nature New Biol. 241:133.

Stanier, R. Y., E. A. Adelberg and J. Ingraham. 1976. *The Microbial World*, 4th Edition. Prentice-Hall, Englewood Cliffs, NJ.

Stent, G. S. and R. Calendar. 1978. *Molecular Genetics*, 2nd Edition. W. H. Freeman, San Francisco.

Swamy, K. H. S. and A. L. Goldberg. 1981. *E. coli* contains eight soluble proteolytic activities, one being ATP dependent. Nature 292:652.

Yanofsky, C. 1981. Attenuation in the control of expression of bacterial operons. Nature 289:751.

Chapter Eight

Regulation at the Whole Cell Level

INTRODUCTION

In this chapter we shall examine two major problems in bacterial growth physiology:

1. Does the overall design of the cell explain why bacteria utilize their energy and carbon sources with very high efficiency under most growth conditions?
2. What determines the growth rate in a given medium?

The provisional answers we can offer are based on these lines of evidence: the data on growth rates and molecular genetics discussed in Chapters 6 and 7 and a new set of experiments, SHIFTS BETWEEN MEDIA (or, more generally, between growth conditions).

In Chapter 6 we developed the life cycle of the polysome based on three μ-independent parameters: the cgr_p, the mRNA half-life, and the average amount of mRNA per active ribosome. The conclusion was that a typical polysome produces 20–25 protein molecules per gene and approximately 50 per operon during its lifetime of approximately 100 seconds, irrespective of growth rate. Thus, the average yield per initiation of transcription of a productive mRNA is nearly constant. In an exponentially growing culture with N amino acids per milliliter in its proteins, an estimate of the frequency of initiation of productive transcripts is obtained by dividing dN/dt ($= \mu \ln 2 N$) by 50. Like dN/dt, *the global initiation frequency, F_μ, is therefore proportional to μ.* The partitioning of initiations between promoters inside and outside the PSS determines the relative size of the PSS and thus the efficiency of protein synthesis.

349

In Chapter 7 we observed that inductions and repressions control the frequencies with which many individual promoters in the domains of fueling and biosynthesis are utilized. The biosynthetic operons controlled by attenuation essentially fall into the same category; at their promoters, transcription is probably initiated at maximum frequency, but only a controlled fraction of the transcripts proceed past an early terminator. From the point of view of control of the synthesis of the proteins of an operon, transcription that is terminated very early might as well not have been initiated. Commonly, induction or repression can change the transcriptional frequency of an operon by factors of the order of 100.

In the *domain of the PSS* the picture is different, but so is the purpose of the controls. Here we are not dealing with operons to be turned off or on individually, as the efficiency of growth demands. The entire PSS is necessary for protein synthesis (Chapter 7), and the great efficiency of growth, for which the high and constant cgr_p is the most direct evidence, requires balanced synthesis of the components of the PSS. For this purpose, autoregulation of the individual operons is an effective design. A priori it is not obvious that translational rather than transcriptional regulation should prevail, and as the PSS becomes better known, we shall probably find that the network that ties its components together involves regulation at the transcriptional as well as the translational level. The scant evidence so far available indicates that control at the level of translation operates with a relatively small amplitude, say, a factor of 10 or less.

As the name indicates, the PSS is defined by its function. This complex system is not separated from the surrounding cytoplasm by a membrane in the manner nuclei or mitochondria are. Even so, the regulatory network we have discussed confers on the PSS the properties of a structure, not in the rigid anatomical sense of the word, but in a more fluid, dynamic sense. The most important characteristics of the PSS "structure" are: *constancy of composition and of efficiency.*

SHIFTS BETWEEN GROWTH CONDITIONS

Shifts from one growth condition to another is a common experimental design (Maaløe, 1966), and it was actually employed in two of the experiments discussed earlier (Chapter 6, Figure 7, and Chapter 7, Figure 5). When the definitive postshift growth rate is lower than the preshift rate, we talk about a DOWNSHIFT; an UPSHIFT is the reverse of this.

The number of shifts that can be performed between media of different compositions or between temperatures or other relevant parameters is evidently very large; of all the shifts possible, those involving rich and minimal media have been most extensively studied. We shall begin with the downshift.

Shifts from rich to minimal medium

When cells that have grown for long periods in a rich medium are transferred to minimal medium (e.g., from broth to a glucose–salts medium), mass increase stops for at least 30 minutes (at 37°C); and it takes 1–2 hours to reach the definitive rate of growth in the postshift medium. This is understandable; in the rich medium all biosynthetic operons were repressed and the corresponding enzymes were reduced to low levels by dilution. Thus, growth in the postshift medium requires de novo synthesis of many biosynthetic enzymes under very unfavorable conditions (low concentrations of the enzymes necessary to supply the substrate for protein synthesis). It is quite possible that protein turnover initially contributes amino acids to de novo protein synthesis, but it is clear that the sequence of events that precedes the establishment of balanced growth in the postshift medium is complicated and very difficult to unravel. To stress this point, a shift of this kind is designated as a SLUGGISH SHIFT.

Stimulons and regulons

In its natural environment, sluggish shifts are *E. coli*'s way of life. More often than not, shifts in growth conditions require extensive adaptive response if growth (and even survival) is to be possible. For many years appreciation of these adaptive responses was built up by study of the induction and repression of various particular enzymes and enzyme pathways—enzymes that enable use of alternate carbon and energy sources or of alternate electron acceptors; enzymes that equip the cell to obtain N, P, or S from organic compounds; enzymes that catalyze reactions essential for repairing damaged DNA; and enzymes that enable growth at temperatures above or below normal. The number of metabolic capabilities that are known to be produced only under conditions that require their use is very large indeed, and the adaptability of bacteria to an environment they can do little to control has become, along with rapid growth, a hallmark of these cells.

The true complexity of a bacterium's response to common environmental stimuli was not fully revealed until the late 1970s. For example, when a culture of *E. coli* depletes a limited supply of inorganic PO_4^{3-}, the response of each cell is dramatic—the synthesis of approximately 85 cellular proteins becomes affected. Many are induced (or derepressed) greater than tenfold and rise from previously undetected low levels before the shift to quite high levels in a brief time after phosphate limitation. This information comes from the use of two-dimensional gels (Chapter 1) to resolve individual cellular proteins; comparison of autoradiograms of gels made from cells labeled during phosphate starvation with cells in normal growth reveals the magnitude of the cell's response to this stimulus. And the same technique can be employed to observe the response to any shift in the chemical or physical nature of the growth medium. In this way STIMULONS are being defined, that is, as sets of genes that become active and produce their protein products in response to particular environmental stimuli (Table 1).

Not all of the 85 genes–proteins that constitute the phosphate starvation-induced stimulon are likely to be controlled by one and the same regulatory protein. In fact, it is known that this stimulon consists of several REGULONS—each of which is a set of genes sharing a regulatory element. One of the tasks for bacterial physiologists is to analyze these stimulons (and others yet to be defined) to learn the molecular nature of the regulatory mechanisms of which

Table 1. Examples of stimulons in *E. coli*

Stimulon	Approximate number of genes–proteins	Number of regulons
Phosphate starvation-induced (PSI)	85	Several
Ammonium starvation-induced (NSI)	43	Several
High temperature-induced (HTI)	16	>1
Low temperature-induced (LTI)	6	?
SOS	10	1
O_2-induced (AER)	20	Several
Anaerobiosis-induced (ANA)	18	Several

they are constructed. Genetic techniques such as gene fusions are already proving useful in this analysis (Wanner, 1982). No less of a task, of course, is discovering the biochemical identity and cellular function of each of the protein members of a stimulon.

Two aspects of stimulon analysis deserve additional comment. First, there are preliminary indications that stimulons overlap; one protein may belong, for example, to both the phosphate starvation-induced and the nitrogen starvation-induced stimulon. Sorting out these responses can therefore be expected to reveal novel patterns, if not novel mechanisms, of gene regulation. Second, these studies continue to reveal proteins that are not detected on gels made from cells in balanced growth in glucose minimal medium. (Recall that only one-third of the cell's DNA is needed to encode the proteins found in the cell under any one growth condition.) It will be interesting to see what fraction of the cell's potential protein-coding capacity is devoted to proteins made only during shifts that tax the capacity of the cells for reorienting their metabolism.

The role of ppGpp in regulation

Shifts of several kinds have been used to study the production and the effects of the unusual guanosine phosphates (pppGpp and ppGpp) first described by Cashel (1969). We introduce this theme by describing a particular *temperature downshift*: with an *E. coli* strain carrying a temperature-sensitive valyl-tRNA synthetase, the supply of charged tRNA^{val} can be reduced instantaneously, and in a controlled manner, by raising the temperature. This immediately lowers dP/dt and elicits rapid synthesis of the tetra- and pentaphosphates (often called Magic Spots I and II). Shifts with these characteristics are designated IMMEDIATE-RESPONSE SHIFTS; a term implying that an immediate effect is produced that can be related to a clearly identified, primary cause. This type of shift lends itself more readily to detailed analyses than do sluggish shifts.

Figure 1A illustrates the inhibitory effects of temperature and ppGpp on RNA synthesis. As the level of the effector rises (Figure 1B), the corresponding levels of RNA synthesis falls (Figure 1A). From such data the relationship between the level of ppGpp (at 15–20 minutes after raising temperature) and the rate of RNA accumulation was established (Figure 2). With increasing temperature, this rate falls from unity (the value measured in a glucose-minimal culture, at 30°C, of the parent strain) to less than 0.20, which corresponds to growth with acetate as the carbon source. The ppGpp concentration required to effect this reduction is approximately 400 units, which is nearly 10 times the level measured during balanced growth in an acetate culture (Figure 9). It was concluded that ppGpp cannot be the main effector controlling rRNA synthesis (as had long been thought), but that it probably acts "as a fine control by modulating the effect of . . . [another] factor" (Fiil, 1972). This other factor remains hypothetical.

The response to temperature in cultures of the temperature-sensitive strain just described is elicited whenever the supply of one or more amino acids is curtailed. Accumulation of ppGpp above the basal level can therefore be interpreted as a sign of suboptimal charging of one or more species of tRNA, that is, as a sign of substrate limitation in protein synthesis. Cells that respond in this way are said to exert STRINGENT control of RNA synthesis. Mutants that fail to produce excess ppGpp and do not reduce their rate of RNA accumulation when starved of amino acids are said to exhibit a RELAXED phenotype (Figure 3).

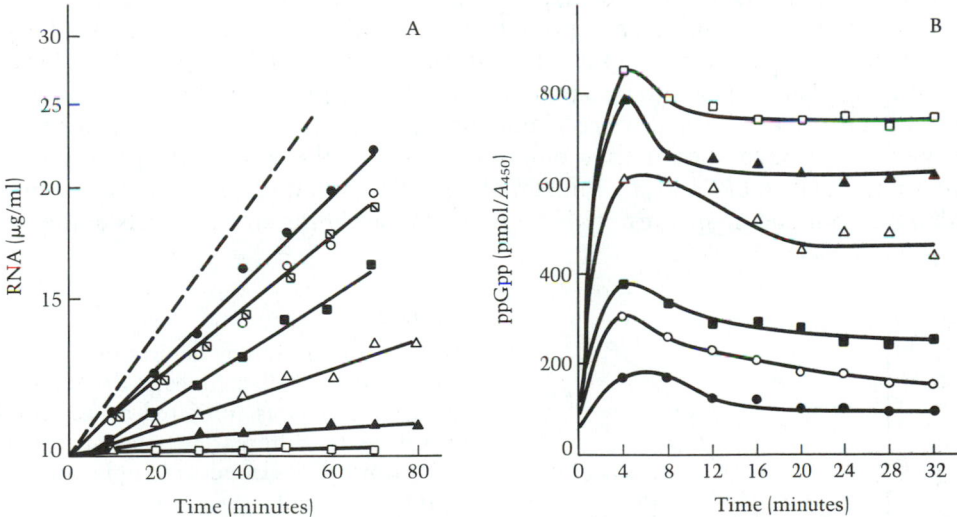

Figure 1. A. RNA accumulation as determined by chemical measurements (orcinol) in a temperature-sensitive valyl-tRNA synthetase mutant at 30°C (●), 35.5°C (○), 36.5°C (◨), 37.2°C (■), 38°C (△), 38.8°C (▲), and 39.5°C (□). The dashed line represents wild-type growth at 37°C. B. Time course of ppGpp accumulation by radiochemical determinations; symbols are as in A. (After Fiil, 1972.)

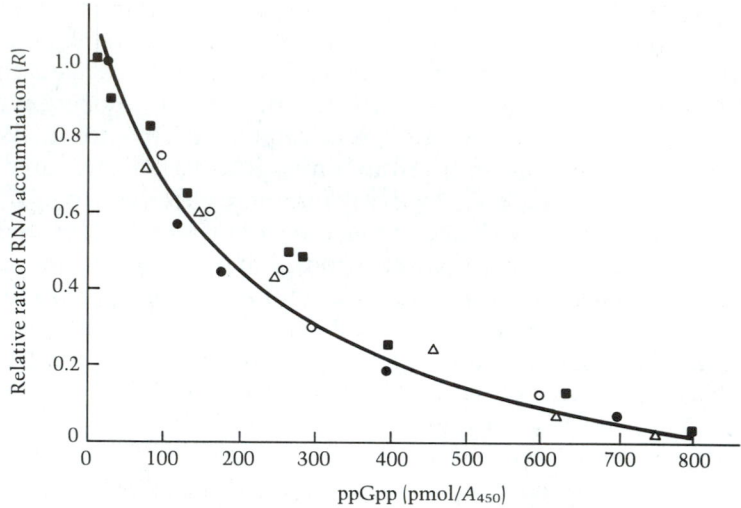

Figure 2. Relative rate of RNA accumulation (R) plotted against concentration of ppGpp measured 15–20 minutes after raising the temperature. Data are from experiments similar to that illustrated in Figure 1; different symbols represent different experiments. (After Fiil, 1972.)

The mechanism underlying the stringent response is now largely understood. Typical relaxed cell lines carry a mutation in the *relA* locus, and they fail more or less completely to produce the so-called STRINGENT FACTOR. This protein is involved in the rapid synthesis of pppGpp and ppGpp (in *E. coli* the pentaphosphate appears to be a precursor of the ppGpp, and the tetraphosphate is normally 10–20 times more abundant than the pentaphosphate). In vitro experiments have shown that rapid synthesis of the guanosine phosphates requires GTP, ATP, and an intact PSS and that it takes place when an uncharged, but codon-specific, tRNA enters the acceptor site on a ribosome to

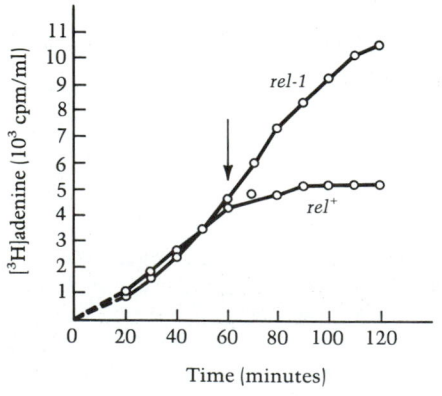

Figure 3.

RNA accumulation as determined by incorporation of [³H]adenine. At $t = 60$, protein synthesis was inhibited by addition of valine to differentiate the stringent (rel^+) from the relaxed (rel-1) phenotype. (After Fiil, 1969.)

which a molecule of the stringent factor is attached. Actually, a $relA^+$ cell contains only one molecule of the factor per 1–300 ribosomes. Nevertheless, upon amino acid starvation, $relA^+$ cells quickly produce large amounts of ppGpp and maintain concentrations greatly in excess of the levels during normal growth (Figure 1). Evidence, which is presented later (Figure 9), shows that ppGpp turns over rapidly and that continued synthesis is therefore required to maintain the high levels seen in Figure 1B.

Protein synthesis and ppGpp

High levels of ppGpp have been shown to affect protein as well as RNA synthesis. This was tested with a $relA^+/relA$ pair of cell lines, *both* harboring the temperature-sensitive valyl-tRNA synthetase. At 35.5°C (that is, under the condition of partial valine starvation), the relative rates of r-protein synthesis (α_r) for many individual r-proteins were measured (Dennis, 1974). The results were clear: at 35.5°C all the α_r values were reduced to approximately two-thirds of their values at the permissive temperature (29°C) in the $relA^+$ strain, and they increased to approximately 1.5 times these values in the $relA$ strain. In terms of ppGpp, the shift to 35.5°C elicits a slow reduction in the ppGpp level in the $relA$ strain, but a big increase in the $relA^+$ strain. At the time samples were taken for the α_r measurements, the ppGpp level in the $relA^+$ culture was at least 10 times above the basal level characteristic of moderate to high growth rates. Thus, ppGpp at such high concentrations not only reduces the rate of synthesis of RNA, *but also reduces the rate of synthesis of the r-proteins and of a number of other PSS proteins* (Reeh, 1976).

The shift of the $relA$ strain to 35.5°C reveals a paradox: it is a downshift as far as growth rate is concerned; but in terms of α_r, it is an upshift of the kind produced when amino acids are added to a minimal medium culture (Figure 8 and accompanying text). The explanation is straightforward: when the flow of amino acids into protein is suddenly cut short by reducing the activity of the valyl-tRNA synthetase, the amino acid pools will swell exactly as they do if the amino acids are added from outside. In the $relA$ strain, the high level of uncharged valyl-tRNA prevailing at 35.5°C will not elicit rapid synthesis of ppGpp and the cell responds as if amino acids had been supplied in the medium.

An analogous situation is created when fusidic acid is added to a culture (Bennett, 1974). This drug slows down the translational movement of the ribosomes by interacting with EF-G (protein synthesis elongation factor G). The efficiency of PSS is thus lowered, and therefore the flow of amino acids into protein is reduced. In this case, the ppGpp concentration in the cells actually falls, as in a bona fide upshift (Figure 9). The temperature shift with $relA$ cultures described earlier and the fusidic acid experiments are instructive; they demonstrate that in at least two different situations a downshift in terms of μ is accompanied by an INTERNAL UPSHIFT in terms of α_r. Our simplistic classification of shifts is thus not unambiguous.

Stringent factor is not essential

Stringent and relaxed strains behave alike, except when challenged by amino acid shortage to overproduce ppGpp. In fact, they maintain the same low basal level of ppGpp during balanced growth in all media tested. It has been commonly thought that residual stringent factor activity was responsible for the low level of ppGpp synthesis in relaxed cells. We now know this is not the case; nonsense mutations and insertions in the *relA* gene have been identified, and, as a final proof, a cell line has been isolated in which a chromosomal deletion has removed *relA* and its neighboring genes (Atherly, 1979). The implication of this for ppGpp synthesis in relaxed cells will be discussed in the section on upshifts.

The fact that stringent factor is not essential for normal balanced growth in no way implies that it is unimportant. Most bacteria carry, and have preserved, a functional *relA* gene; this suggests that the stringent phenotype has selective advantages over the relaxed type.

The characteristic stringent response to starvation (Figure 1) also is important for inducibility. It has long been known that when a culture of *relA* cells is deprived of a required amino acid, RNA–protein particles accumulate that do *not* contain r-protein. Apparently the excess rRNA synthesized during starvation associates with basic proteins already present in the cells. When the missing amino acid is returned to the culture, the presence of these abnormal particles does not prevent the resumption, after a short lag, of protein synthesis at the prestarvation rate. As long as the particles persist, however, induction of at least two enzymes, β-galactosidase and tryptophanase, is blocked. Thus, the stringent response to starvation—inhibition of RNA accumulation—also prevents a temporary loss of inducibility. These phenomena have been described collectively as the "relaxed syndrome" (Maaløe, 1979).

The relaxed phenotype can also result from mutations in the *relB* or the *relC* genes. The latter is identical with the *rplK* gene, whose product is r-protein L11. The phenotype of *relC* mutants probably reflects poor binding of stringent factor to ribosomes; *relB* mutants exhibit a delayed relaxed phenotype, and the mechanism involved is not completely understood.

ENERGY AND CARBON SOURCE DOWNSHIFTS

Different carbon sources have been used not only to establish balanced growth at various rates (Chapters 5 and 6), they have also been employed in shift experiments. The characteristics of such shifts naturally depend on the need for changing the enzymatic equipment of the cells during the transition period following the shift. For example, a culture of *Salmonella typhimurium* with glutamate as the sole carbon source grows at $\mu = 0.9$; when proline is substituted for glutamate, a long lag precedes the gradual establishment of the definitive postshift growth rate at $\mu = 0.7$.

This very sluggish downshift may be compared to the shift from rich to minimal medium discussed earlier. In that case, de novo synthesis of many biosynthetic enzymes was necessary; in the glutamate-to-proline shift, the lag probably reflects the need to synthesize the enzymes required to utilize proline and to synthesize glutamic acid. This would be a difficult and slow process in the virtual absence of a usable energy and carbon source.

Some strains of *S. typhimurium* exhibit a curious and interesting behavior in carbon source shifts. The compounds included in this study fall into two classes: those that are part of or feed into the tricarboxylic acid (TCA) cycle (e.g., glutamate, succinate), and those that feed into the glycolytic and related pathways (e.g., various sugars, glycerol). In general, shifts between members of the same class produce little or no lag, whereas shifts between members of opposite classes are followed by a pronounced lag. The few exceptions are cases in which de novo protein synthesis is required to metabolize the second carbon source (as in the glutamate-to-proline shift). This suggests that in *S. typhimurium* the energy yielding pathway *not* used (the TCA cycle or glycolysis) is suppressed (Richmond, 1962). In the *E. coli* strains tested, this phenomenon of mutual repression is much less pronounced.

Immediate-response downshifts

A downshift with virtually ideal experimental characteristics can be effected by suddenly reducing the availability of the carbon and energy source without in any other way changing the preshift minimal medium. Mutants with reduced transport capacity can be used. Some transport mutants require high concentrations of the carbon source (e.g., lactose or one of several amino acids) to grow at the same rate as does the wild type at much lower concentrations (von Meyenburg, 1971). Reducing the concentration of the carbon source by dilution produces an immediate-response downshift and eventually leads to balanced growth at a lower μ value. Cell densities of 10^7 to 10^8 per milliliter can be used without significantly lowering the concentration of the carbon source during an experiment, but the genetic stability of the cultures must be watched. The potential usefulness of transport mutants for studies in growth physiology is far from exhausted.

Meanwhile, a different approach has turned out to be extremely useful. The uptake of glucose by *E. coli* cells is competitively inhibited by the analogue α-methylglucoside (αMG). This compound accumulates in the cells as the metabolically inert αMG-6-phosphate, and this sugar phosphate can reach very high concentrations in the cells without being toxic.

The effects of the αMG:glucose ratio on growth rate and on ppGpp production have been established in both stringent and relaxed cultures (Figure 4). Shifts were produced by adding αMG in a tenfold excess over a glucose concentration of 0.1% in the medium. Figure 5 shows that both stringent and relaxed cultures respond immediately and identically by reducing the rate of protein synthesis (dP/dt) to one-half the preshift value (Johnsen, 1977). The reduced rate remains constant for at least 60 minutes, giving rise to an apparent steady state. It will become evident, however, that this response comes about by different routes, depending on the state of the *relA* gene. Table 2 presents measurements of the rates of synthesis of the main RNA fractions,

Figure 4.

A. Growth rate as a function of the
αMG:glucose ratio at glucose concentrations of 0.02% (○) or 0.1% (●). The insert
shows the growth curves in media containing 0.1% glucose (●) and, after addition at 10 minutes, of αMG at 0.12% (○),
0.50% (□), and 2.20% (■). B. Accumulation of ppGpp in cultures with 0.1% glucose of a stringent and a relaxed strain. At
$t = 0$, αMG was added at 2.6%. (After
Hansen, 1975.)

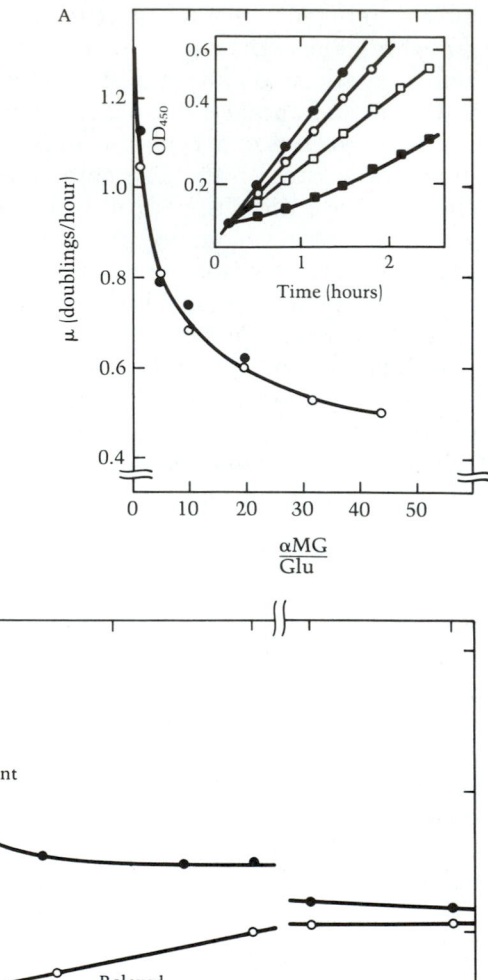

of RNA accumulation, and of ppGpp levels; note that the typical rapid increase
in the ppGpp level in the stringent cells is paralleled by quick and sustained
reductions in the rates of synthesis of *stable as well as messenger RNA*. In
the relaxed cells, there is no immediate response, but the rates fall gradually;
after 20–30 minutes, there is little difference between the two cultures (Molin,
1977).

Our problem now is to reconcile the uniform effect of the shift on dP/dt
with the different responses in terms of RNA synthesis. First, we recall that
dP/dt (expressed as number of amino acids incorporated into protein per

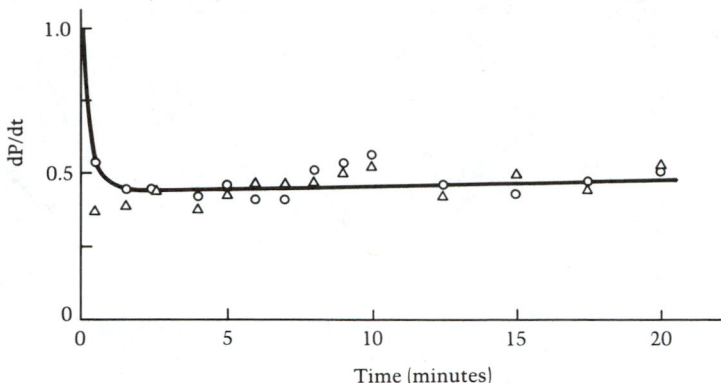

Figure 5. Rate of protein synthesis in a stringent (○) and a relaxed (△) strain following addition of αMG at 1.0% to induced cultures with 0.1% glucose. Rates were measured by 30-second pulses of [³H]proline. (After Johnsen, 1977.)

second) by definition equals the number of active ribosomes (r_{act}) times the cgr_p; note that the latter can be estimated with good precision from induction kinetics (Figure 6). Table 3 shows that the cgr_p is reduced immediately after the shift, but in NF541 ($relA^+$) it returns to normal within 2 minutes; in NF542 ($relA$), the cgr_p remains at half-normal values for 10–20 minutes before

Table 2. RNA synthesis and accumulation after addition of α-methylgluco-side to a glucose culture[a]

Cell line	Time after shift (min)	$dRNA_t/dt$	$drRNA/dt$	$dmRNA/dt$	$dRNA_{acc}/dt$	ppGpp
$relA^+$ (NF541)	−2	1.0	1.0	1.0	1.0	30
	2	0.45	0.26	0.56	0.1	210
	5	0.45	0.22	0.58	0.1	160
	10	0.45	0.24	0.56	0.1	135
	20	0.50	0.30	0.63	0.1	115
	30	0.65	0.46	0.61	0.1	112
$relA$ (NF542)	−2	1.0	1.0	1.0	1.0	35
	2	0.95	0.30	1.02	0.25	43
	5	0.85	0.65	0.95	0.25	56
	10	0.65	0.45	0.74	0.25	75
	20	0.65	0.40	0.77	0.25	117
	30	0.65	0.48	0.79	0.25	115

[a]Total rates of RNA synthesis ($dRNA_t/dt$) as determined by measurements of the incorporation of [³H]adenine into RNA (corrected for specific activity of the intracellular ATP); $drRNA/dt$ is the product of $dRNA/dt$ and the relative rate of rRNA synthesis. The rate of mRNA synthesis, $dmRNA/dt$, was calculated from the difference between total RNA and rRNA synthesis (the rate of tRNA synthesis is contained in this value and no separate estimates were made). The ppGpp pool is expressed as pM, at $OD_{450} = 1$.

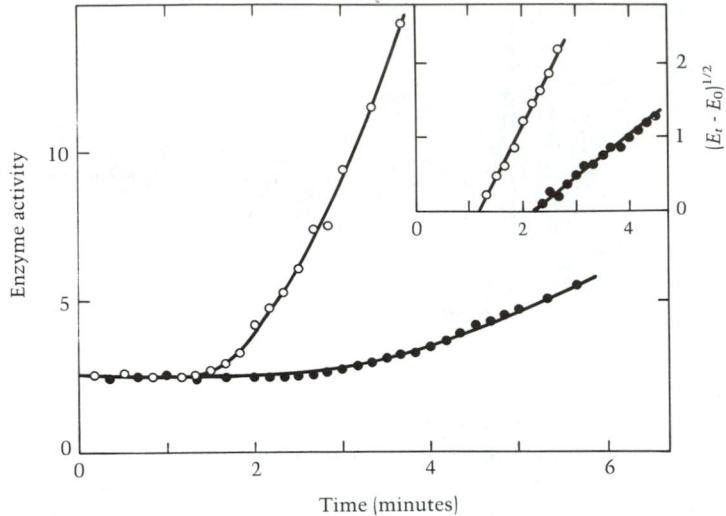

Figure 6. Induction of β-galactosidase synthesis in the relaxed strain before and after the downshift shown in Figure 5. ●, IPTG added 5 minutes after αMG; ○, control without αMG. The insert shows estimation of the induction lag by plotting the square root of the increment in enzyme activity [$(E_t - E_0)$, i.e., E at time t minus background, E_0] against time.

Table 3. Polypeptide chain growth rates (cgr_p) in amino acids per second[a]

Time of IPTG addition (minutes after shift)	cgr_p	
	NF541 or NF1133 ($relA^+$)	NF542 or NF1134 ($relA$)
−5.00	17,16,16,16	17,17,16,16
0.00	10,10,10	10,10,9
0.33	11	
0.67	10	
1.00	11	
1.33	13	
2.00	15	9
5.00	16,15	9,8,8
10.00		8,7
20.00		10
40.00		14
60.00		15

[a]Calculated from induction lags of β-galactosidase (measured as illustrated in Figure 6) and the number of amino acids in the β-galactosidase monomer.

slowly returning to the normal value. In the same experiments, the cgr_{rRNA} was monitored, using the technique described in Chapter 6. It was found *not* to be affected by the downshift. The changes in the rates of RNA synthesis shown in Table 2 must, therefore, solely reflect changes in the frequencies with which transcription is initiated.

We can sum up the results in a simple way be assigning unit values to dP/dt, cgr_p, and r_{act} in the preshift condition and by comparing them to the experimental values taken between 2 and 10 minutes after the shift. This gives:

$relA^+$ and $relA$, preshift $dP/dt = 1$, $cgr_p = 1$, $r_{act} = 1$
$relA^+$, postshift $dP/dt = 0.5$, $cgr_p = 1$, $r_{act} = 0.5$
$relA$, postshift $dP/dt = 0.5$; $cgr_p = 0.5$, $r_{act} = 1$

Evidently, the reduction of dP/dt from 1 to 0.5, which results from limiting the flow of carbon and energy to the PSS of the cells, brings about corresponding reductions in r_{act} in the $relA^+$ culture and in cgr_p in the $relA$ culture. It seems reasonable to relate these results to the inhibitory effect of ppGpp on RNA synthesis observed in vitro. We are therefore led to propose the following model: the immediate reduction of the flow of glucose into the cells caused by the addition of αMG sets a new and lower limit on dP/dt; and, because the PSS in both $relA^+$ and $relA$ cells is geared to sustain the high, preshift dP/dt, the only parameter that can "give" is the cgr_p. We interpret the quick return to the normal cgr_p in the $relA^+$ cells to mean that ppGpp inhibits not only the synthesis of r- and tRNA but also of mRNA and thereby reduces the number of ribosomes that can be accommodated in polysomes. This view obviously agrees with estimates of $dmRNA/dt$ (Table 2).

Further support for the notion that high levels of ppGpp reduce mRNA synthesis in vivo comes from an elegant DOUBLE SHIFT experiment. Here the effect of αMG, added at time zero, was offset by raising the glucose concentration sufficiently at 5 minutes (Johnsen, 1977). Figure 7A shows that in the $relA^+$ culture reversal of the downshift by glucose is followed by a lag of approximately 2 minutes before the preshift rate of protein synthesis is reestablished; in the $relA$ culture, no significant lag is observed (Figure 7B). This is exactly what our model predicts; if in the $relA^+$ culture the available mRNA is reduced so that r_{act} is down to one-half its preshift value, a return to the normal value, and hence to the preshift dP/dt, should not be possible without building up a normal complement of polysomes. The time required to do this would approximately equal the transcription time for an average operon, which cannot be very different from the observed lag.

Energy charge during a downshift

An intriguing feature of the immediate-response downshifts deserves attention. When faced with a sudden and severe reduction in the supply of carbon and energy, the cells do *not* deplete themselves of nucleoside triphosphates. Despite the fact that these key compounds turn over very rapidly, the pool levels of the nucleoside triphosphates remain practically

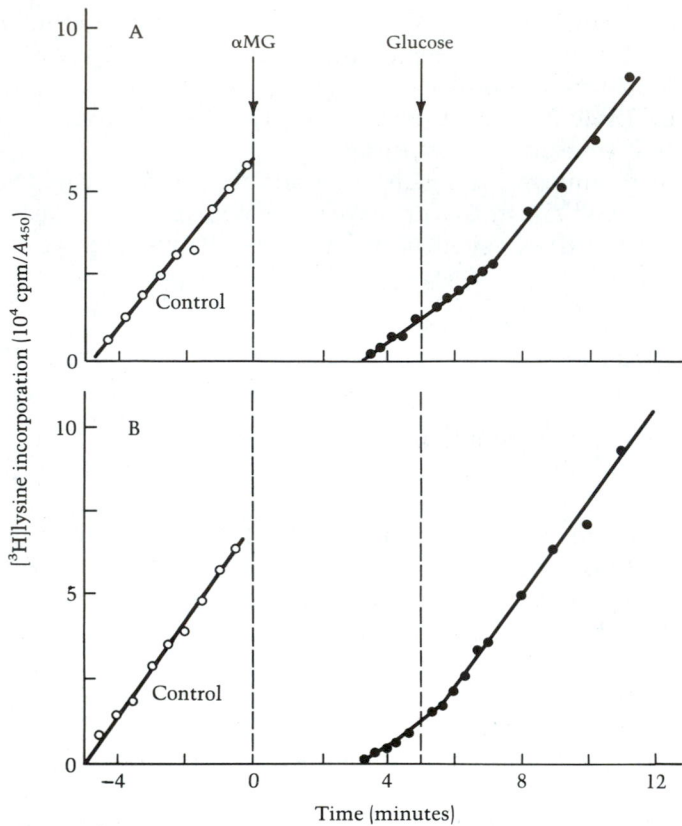

Figure 7. Protein synthesis after reversal of an αMG downshift in stringent (A) and relaxed (B) cultures. The initial conditions were as described in Figure 5, but at 5 minutes the glucose concentration was raised from 0.1 to 1.0%. Labeling with [³H]lysine was begun 3 minutes after (●) or 5 minutes before the downshift (control, ○). (After Johnsen, 1977.)

unaffected. This indicates that strong forces exist to maintain the normal ATP charging level of 0.8–0.9 (Chapter 3).

In the *energy downshift* effected by αMG, these forces are certainly challenged, but it is also clear that they adequately protect the cell. The mechanism involved can be deduced from the biochemistry of the polymerization reactions, all of which consume ATP and release AMP + PP_i (Chapter 3). The polymerization reactions are readily reversible, and one of the key enzymes in the cell is a *pyrophosphatase*, which, by splitting PP_i, drives the overall processes in the direction of polypeptide and polynucleotide formation. The immediate consequence of the downshift is to lower the capacity of the cells to regenerate ATP; the ADP and AMP pools must therefore increase at the expense of ATP. The PP_i pool may also increase; and if it does, this increase will directly inhibit polymerization and hence reduce ATP consumption. However, the same effect seems to be furthered by a specific regulatory circuit in which ADP acts as a negative effector on the pyrophosphatase (Josse, 1966). Thus, when ADP accumulates because it is not as efficiently phosphorylated as just before the shift, the concentration of PP_i will rise because its degradation is inhibited. In this way, protein synthesis, which is the main ATP-consuming reaction of the cell, and nucleic acid synthesis could very quickly be scaled down to match the reduced supply of carbon and energy. It may not be damaging to the cells if, in the process of readjustment, the ATP concentration went down considerably, providing it was quickly restored. However, this remains an open question because the relevant measurements have not been made.

Secondary effects of the energy downshift

The energy downshifts have revealed other intriguing effects. During the transition period, when the cells are semistarved of glucose, considerably more rRNA is made than is incorporated into mature ribosomes (Table 2). Parallel measurements of α_r show that the instability of part of the rRNA is not due to lack of r-protein. Actually, all the components of a ribosome are produced in amounts that would support a rate of ribosome maturation 2–3 times greater than the observed rate. The reason for this wastefulness is not known.

Another secondary phenomenon of some interest is revealed during the transition period. As shown in Table 3, the cgr_p remains low in the *relA* cultures for a considerable time after the shift. The cgr_{rRNA} on the other hand is not affected, and the leading ribosome must therefore lag more and more behind the RNA-P during transcription. As a result, an increasingly long stretch of "naked" mRNA will be exposed. It has been shown that in this situation translation of β-galactosidase cannot always be completed, presumably because the exposed segments of some of the *lac* mRNAs are severed by nucleases. A short NH_2-terminal piece of the β-galactosidase polypeptide (the auto-α fragment) can be assayed separately and has been shown to be produced in more copies than the finished chain. This phenomenon has been called INTRAGENIC POLARITY, and it has also been observed when cells grow in the presence of fusidic acid. This drug lowers the cgr_p by interfering with EF-G and thus causes the ribosomes to lag behind the transcribing RNA-P.

UPSHIFTS

Introduction

A basic difference between upshifts and downshifts should be noted: the immediate effect on dP/dt caused by adding αMG to a glucose culture (Figure 5) is essentially the result of lowering the efficiency of the PSS by reducing either cgr_p or r_{act}. Most upshift experiments cannot be expected to produce a corresponding and immediate increase in dP/dt, because the PSS has been adjusted during growth in the preshift medium to operate at near-maximal efficiency. Exceptions to this rule are shifts involving very low preshift growth rates. A clear case was illustrated in Figure 7, Chapter 6; it showed that dP/dt is increased immediately when a culture growing very slowly because of glucose limitation is supplied with excess glucose. We know the reason; at the low preshift growth rate, cgr_p is normal but relatively few of the cells' ribosomes are active. Thus, dP/dt can be raised immediately by involving more of the preexisting ribosomes in protein synthesis, that is, by increasing the efficiency of the PSS. If this efficiency is already high in the preshift culture, the growth rate will only go up if the shift causes the relative size of PSS to increase.

The mechanism by which this is achieved is often difficult to analyze. Thus, the carbon-source shifts with *S. typhimurium* discussed earlier included experiments in which succinate was replaced by glucose. This is an upshift in terms of growth rate; but the transition from $\mu = 0.8$ (succinate) to $\mu = 1.4$ (glucose) involved a long lag followed by a slow gradual increase of μ. The reason for the sluggishness of this upshift seems to be that a slow transition from the "TCA cycle," preshift carbon source to glucose was superimposed on the necessary increase in the relative size of the PSS.

Shift from minimal to rich medium

A typical immediate-response shift from minimal medium to rich medium may be performed by adding all 20 amino acids and other natural building blocks to a culture in glucose minimal medium. In *E. coli*, and probably in many bacteria, these monomers are taken up very rapidly, with the result that the respective pools swell. One obvious effect of this is that signals are generated that cause repression of many operons. A large number of experiments involving addition of individual amino acids show that repression is elicited almost instantaneously. Thus, *multiple repressions are direct effects of this simple upshift*. Note that energy and carbon metabolism can continue without interruption or change in the rich postshift medium.

We shall now examine the response of the cells in terms of the PSS and of ppGpp:

1. PSS. The indicator for a response by the PSS is the relative rate of r-protein synthesis, α_r; and we know that the preshift value of 0.14–0.15 must increase to approximately 0.25. Figure 8 shows that the transition between these boundaries begins at once, that it is quite slow, and that the time course is irregular. Note the small initial jump in α_r, which is a reproducible feature of shifts to rich media, and the gradual rise of α_r over the first few minutes to a peak value, which is reached at approximately 4 minutes and followed by a temporary drop. The subsequent slow increase of α_r does not present reproducible regularities (Gausing, 1980).

The model to be developed at the end of this chapter suggests how the α_r at a given growth rate is determined, and it explains some of the features of the transition between the boundary values of α_r in experiments like the one in Figure 8.

2. ppGpp. The response in terms of ppGpp is shown in Figure 9. Two upshifts are illustrated by their effects on the levels of ATP, GTP, and ppGpp. The filled symbols refer to an immediate-response upshift produced by adding

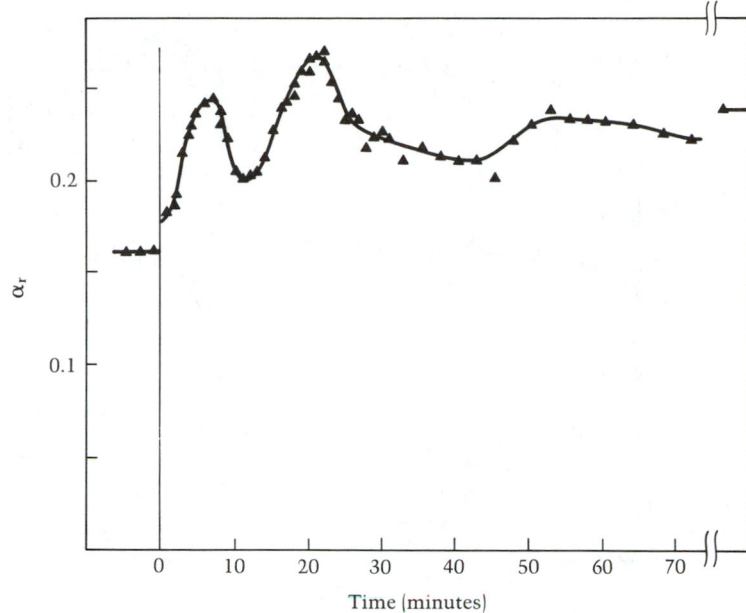

Figure 8. Measurements of α_r in an immediate response upshift. Amino acids and hypoxanthine were added at $t = 0$ to a culture of *E. coli* AS19 pregrown in glucose minimal medium at 30°C. Pulses of 30 seconds with [³H]leucine were followed by a 30-minute chase with cold leucine. The symbols refer to independent experiments. (After Gausing, 1980.)

the amino acids and other components of a rich medium to an acetate culture; the open symbols refer to a shift from acetate to acetate + glucose. Frames A and B show the results obtained with *relA*⁺ and *relA* cultures, respectively (Friesen, 1975).

Several features are notable. In all four combinations the effects on ATP and GTP levels are small and may not be significant. In contrast, the effect on the ppGpp level is dramatic in the shift to rich medium: in both cultures ppGpp levels drop precipitously to values near zero and stay there for 30–40 minutes, before slowly rising to the relatively low levels characteristic of rapid growth. What we see here is a typical immediate response; in all probability, it is caused by a rapid increase in the intracellular amino acid pools, which, in turn, brings the tRNAs up to very high levels of charging. This would be expected to reduce the rate of synthesis of ppGpp; and because this compound turns over very rapidly, the net result would be the drop in the ppGpp levels illustrated in Figure 9. In the shifts from acetate to acetate + glucose, the effect on ppGpp is modest, and the shift can be classified as sluggish.

Two corollaries of the transient disappearance of ppGpp after an immediate-response shift must be discussed. First of all, this effect is seen *in the relA*⁺ *as well as in the relA cells.* This finding suggests that, during balanced

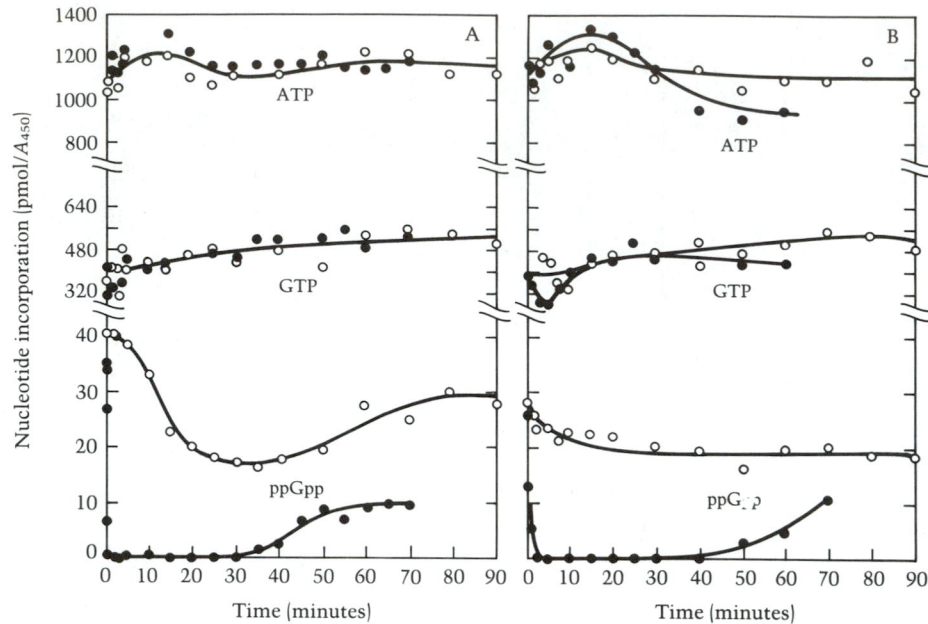

Figure 9. Responses in terms of ATP, GTP, and ppGpp to upshifts in stringent (A) and relaxed (B) cultures. ●, an immediate response shift (acetate minimal medium to rich medium); ○, a more sluggish shift (acetate minimal medium to glucose minimal). (After Friesen, 1975.)

growth, *relA*[+] and *relA* cells may synthesize their ppGpp by *the same mechanism*; this conclusion is in agreement with the fact, noted earlier, that the growth physiology of the two cell types is identical except for their different responses to starvation for amino acids or energy. The presence of some uncharged tRNA in growing cells actually may lead to production of ppGpp at a low rate, that is, of amounts of ppGpp not easily detected in vitro; production may occur without the cooperation of stringent factor. If so, the physiology of *relA* cells, including their response to upshifts, would be accounted for, but we would be left with a different problem. Why does the presence of stringent factor in *relA*[+] cells make no difference except in starvation conditions? We suggest that the factor may be inactive during normal growth and that it is activated somehow when protein synthesis is adversely affected. According to this model, activation of stringent factor would be the first step in an SOS-like response.

Inducibility and the eclipse of ppGpp after shifts to rich medium

In a number of experiments, particularly in studies of protein synthesis in vitro, a stimulating effect of ppGpp has been observed. What this means is not clear, but it has also been shown that the synthesis of β-galactosidase can be induced 10–20 minutes after an immediate-response upshift, that is, when ppGpp is virtually absent (Figure 9). Furthermore, synthesis proceeds with the normal time course and at the normal rate. It is therefore obvious that in vivo the overall process of induction *and* of synthesis does not require ppGpp.

A standard type of experiment has actually been interpreted in the opposite sense: when a pair of *relA*⁺ and *relA* cultures are semistarved of an amino acid, the *relA*⁺ culture accumulates ppGpp, and it responds to IPTG by synthesizing β-galactosidase. In the *relA* culture, the ppGpp level slowly decreases and so does the ability of the *relA* culture to respond to IPTG. The correlation between ppGpp level and inducibility is evident, but the apparent dependence of β-galactosidase synthesis on ppGpp is a consequence of the *loss* of inducibility in the *relA* caused by accumulation of excess RNA (see "Stringent factor is not necessary," earlier in this chapter).

THE BACTERIUM AS A GROWING SYSTEM

In this section we shall attempt to derive answers to one of the questions posed in the introduction to this chapter; that question concerns the remarkable efficiency of the overall system. (The second question, which asks why growth is faster in some media than in others, is best discussed at the end of the chapter.)

Growth efficiency

The efficiency of bacterial growth is to a large extent the result of regulation of ribosome synthesis, or, more precisely, of adjustment of the size and capacity of PSS so that the *input* of energy and of monomers from the fueling and biosynthetic sectors will be utilized effectively to produce proteins and nucleic acids, the main *output* of the process of growth. Evidence that such a balance establishes itself at moderate to high growth rates ($\mu >$ 0.7) was presented in Chapter 6.

Our problem can be illustrated by a "Gedanken" (imagined) experiment: picture a cell in which there is perfect balance between the input and output sectors; allow this cell to divide equally except that a little more of each of the PSS elements goes to one than to the other sister cell. The immediate effect of this imbalance would be that in the cell in which the size and capacity of the PSS are now too large relative to the capacity of the supply side, all the pools from which monomers for polymerization and energy are drawn will be reduced in size. In other words, the concentration of many effectors will be lowered; and this will generate signals for changing the degrees of repressions and inductions in many fueling and biosynthetic operons. Thus, in the cell we are considering, internal repression on the *trp* operon, for example, will be reduced; and in the sister cell where the supplies are in excess, it will be increased. In this way the whole pattern of gene expression is modified, and we assume that the *primary effects* of the uneven cell division—the multiple signals—cooperate to restore balance between input and output. It should be noted that without a response of this general nature, the whole system would be unstable. In the laboratory, analogous multiple effects are elicited in an immediate-response upshift, which may consist of adding all 20 amino acids and other natural monomers to a minimal medium culture as described earlier.

We stress the essential difference between experiments involving activation or deactivation of a single operon and situations affecting the PSS as a whole. When a single operon is involved, the repercussions of this event in the cell are small and can often be neglected. If the primary effects touch the PSS, multiple responses are generated because the activities of the PSS dominate the entire growth process. When we analyze the regulation of ribosome synthesis (and of the PSS), it is therefore necessary to take a global view: we must take into account all we know about the growth physiology of the cells, about the genetics of the network of controls that exists in the cells, and about amino acid pools, numbers of RNA-P molecules, and promoter strengths (see later). Finally, we must remember that our present knowledge is limited and that there is no way of knowing what relevant facts remain to be discovered. Therefore, at best our analysis can reveal only general features of the system and not details about the functioning of its many parts.

Amino acid pools

An extremely simple labeling experiment can give a crude estimate of the amount of an amino acid that exists free in the cell. Figure 10 shows the initial time course of incorporation of proline into protein. The extrapolation of the linear portion of the curve to background level cuts the time axis at approximately 6 seconds. If the unlabeled proline in the pool were used preferentially, no label would be incorporated into protein until 6 seconds, which then would be an estimate of the time protein synthesis could be sustained on the amount of proline present in the pool at $t = 0$. By using the volume and the protein content of the average cell described in Chapter 1 and by knowing the growth rate ($\mu = 1.25$), we find that the 6-second extrapolation lag corresponds to a concentration of free proline in the pool of approximately 10^{-4} M.

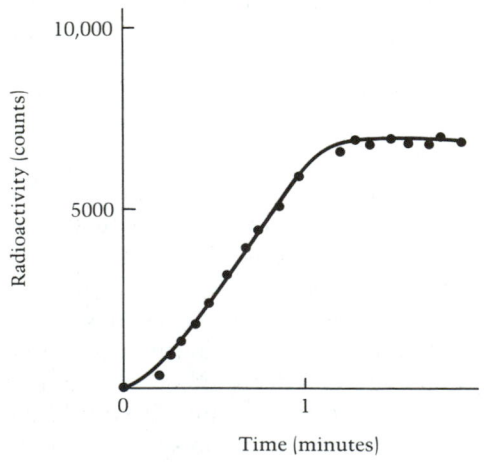

Figure 10.

Incorporation of [^{14}C]proline into protein during balanced growth. One minute after addition of label at 1.5×10^{-3} mg/ml, nonradioactive proline was added at 1 mg/ml, reducing the specific activity approximately 700 times. (After Schleif, 1967.)

Table 4. Levels of the amino acid pool

Carbon source	μ	Pool level (μg amino acid/ml cell volume)[a]			
		Met	Trp	Lys	Leu
Glucose	1.46	34 ($\cong$0.24 mM)	14 ($\cong$0.07 mM)	142 ($\cong$1.0 mM)	108 ($\cong$0.8 mM)
Glycerol	1.1	22	7.6	116	71
Succinate	0.56	25	9.6	69	75

[a]Cultures of wild-type *E. coli* in balanced growth were collected dropwise into boiling medium. The amino acid extracts obtained were tested by means of a strain requiring two amino acids, the one under study and a second that was supplied in radioactive form to the assay medium. A large inoculum of the requiring strain ensured exhaustion of the extract within approximately 10 minutes. The concentrations of the amino acids in the extract were estimated from linear standard curves obtained by plotting counts per minute against known concentrations of the cognate amino acids.

Direct measurements require sensitive methods because cultures at physiologically acceptable densities ($1-2 \times 10^8$ cells/ml) yield extracts with very low amino acid concentrations. It is also necessary to consider that pools might change during the extraction process. The result of the experiment in Figure 10 tells us that synthesis as well as consumption of proline should be stopped within a few seconds if we want to be sure no significant changes have occurred. Both requirements were met when cultures were sampled dropwise into boiling water and the extracts assayed by an extremely sensitive bioassay designed for this purpose (Maaløe, 1979). Table 4 presents the results for four amino acids, all of which appear to have pool concentrations in the range 10^{-3} to 10^{-4} M at $\mu = 1.46$. In all cases the concentrations dropped somewhat at lower growth rates.

The observed pool concentrations of 10^{-3} to 10^{-4} M should be considered in relation to the *dual role of the amino acids* as substrates for protein synthesis and as effectors. Purified aminoacyl-tRNA synthetases have been studied extensively in vitro, and the K_m values for the formation of charged tRNA range between 10^{-6} and 10^{-8} M. If these values are approximately valid for the reaction in vivo, the pool levels in growing cells would always be high enough to ensure a near-maximum rate of charging of tRNA. Because the pool levels are high compared to the K_m values for charging, an amino acid could exert its effector function without ever compromising the charging process.

However, this simple argument may not apply generally. As we have seen, the histidine operon, and a few others, appear to be regulated exclusively by attenuation (Chapter 7, Figure 10); in such instances the control is exerted through the *degree* of charging the cognate tRNA. This looks like a paradox: during rapid growth, when protein is synthesized at high rates, attenuation—say, of the *his* operon—must be at its lowest, which requires a relatively low pool concentration of histidine. Would not a correspondingly reduced degree of charging reduce the cgr_P? A possible way out is suggested by the fact that operons that are controlled by attenuation alone (in contrast to the *trp* operon) contain *many successive codons* for the effector amino acid (Chapter 7, Figure 15); in the *his* leader, the number is 7. Conceivably, "queuing effect" in a leader sequence of this kind would make attenuation hypersensitive to reduced levels of charging, and the effector function of histidine may be displayed in a concentration range that does not significantly affect the cgr_p. To test this hypothesis, mutants in the leader sequences in the *his*, *leu*, *thr*, *ilv*, and *phe* operons would have to be analyzed in as great detail as the *trp* leader has been (Chapter 7).

Ribonucleic acid polymerase content of cells

Many attempts have been made to account for the dynamics of transcription, both of messenger and of stable RNA, by implicating the concentration and/or the activity of RNA-P (for example, Bremer, 1975; Travers, 1980). We shall first consider the concentration in relation to total rates of RNA synthesis. On two-dimensional gels, the components of RNA-P (the α, β, and β' subunits) are clearly separated from each other and from contaminating proteins, and the number of core units ($\alpha_2\beta\beta'$) per milligram of total protein has been determined at several growth rates. We can now ask how many of these RNA-P molecules are engaged in RNA synthesis? This figure is obtained by dividing the number of nucleotides incorporated per second into total RNA by the cgr_{RNA} of approximately 50 at 37°C. The overall rate of RNA synthesis may be estimated as described in Chapter 6. Figure 11 shows how the total number of RNA-P core units increases with μ; it also shows that the total number of these molecules considerably exceeds the calculated number of engaged RNA-P molecules, except possibly at maximum growth rate.

The existence of large numbers of nonengaged RNA-P molecules is evident from upshift experiments. Thus, in minimal medium, approximately one-half the engaged RNA-P molecules may be synthesizing stable RNA and the other half mRNA; after a shift to rich medium, the rate of mRNA synthesis probably remains unchanged for several minutes, as indicated by the fact that dP/dt does not increase immediately. In contrast, the rate of rRNA and tRNA synthesis may go up two- or threefold, which means that the total number of engaged RNA-P molecules would have to increase by 50 to 100%. This could not happen within 1 or 2 minutes unless sufficient RNA-P were free in the cell at the time of the shift. It has been argued that the nonengaged RNA-P might be inactive, but little is known about the state and localization of these molecules; in vitro they tend to associate in a loose, nonspecific way with DNA, and the same may be true in vivo, as indicated by the very low RNA-P content in DNA-less minicells. However, a loose association is not to be confused with chemical modifications or inactivation of the RNA-P.

The state of the free RNA-P molecules, and, beyond that, the possible role of transitions between states, cannot be discussed meaningfully. This subject is extremely difficult to approach experimentally, and we know too little about it. Our immediate concern here is protein synthesis and, as em-

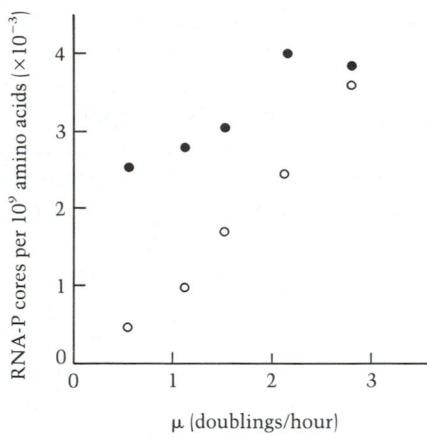

Figure 11.

Estimated numbers of RNA-P core units [total (●) and actively engaged (○)] at different growth rates. Total core units from β subunit measurements as described in Chapter 6.

phasized before, only initiations that lead to polysome formation count. However, even at this general level of understanding, an apparent excess of RNA-P makes sense: as we know, most operons are controlled by "valves" that actively regulate access to their promoters. Once a promoter becomes accessible, a high potential for initiating transcription ensures that polysome formation will be established with minimum delay. For the time being such simple reasoning must suffice. Despite all that is known about RNA-P itself and its interactions with a variety of factors and effectors, the picture we have of the complex processes by which transcription and polysome formation are initiated is incomplete, and it keeps changing. Therefore, for our purpose, initiation is treated as an event taking place in a "black box" (Chapter 6).

Promoters and promoter strength

Within the black box an RNA-P molecule interacts with a promoter, and there is now strong evidence that promoters differ greatly in strength. A promoter, p_i, is said to be strong if the maximum frequency $F_{i(max)}$ with which transcription can be initiated is high; correspondingly, a low $F_{i(max)}$ means a weak promoter.

The intrinsic strength of a promoter in the absence of accessory protein factors must be studied in vitro. In one such study, nine relatively short segments of the DNA phage fd, each carrying a promoter, were used to follow the formation of active complexes with RNA-P (Seeburg, 1977). Two observations are particularly pertinent to our discussion: first, at concentrations of RNA-P exceeding 10^{-8} M, the reaction is independent of RNA-P concentration. In a cell the size of *E. coli*, this concentration corresponds to five to ten free RNA-P molecules, that is, a small number compared to the "excess" of RNA-P indicated in Figure 11. Second, the half-time of formation of the active complex varied among the nine promoters from 20 seconds to 3 minutes. These half-times are very long compared to diffusion-limited collision frequencies in a bacterial cell, as would be expected if complicating events, such as "opening" the promoter site to form a RAPIDLY STARTING COMPLEX (Chamberlin, 1976), are rate-limiting. To quote Seeburg, "It is the relatively slow rate of (active) complex formation, which dictates the frequency of initiation events at each individual (fd) promoter. This conclusion is also supported by determination of the transcriptional activity in vivo and in vitro of various segments of the fd genome, which indicates that more RNA chains are initiated at the strong promoters (those with relatively short half-times) than at the weak ones" (Seeburg, 1977).

The comparison between in vitro and in vivo observations is always a problem, and some major differences between these experimental conditions should be emphasized. In the first place, "accessory" proteins that might take part in the formation of active complexes, by combining either with the template or with RNA-P, may not be present in the in vitro system; and second, supercoiling of the DNA in vivo may modify the efficiency of initiation.

Estimates of $F_{i(max)}$ in vivo suggest very large variations. Calculations show that, at maximum growth rate ($\mu = 2.8$, at 37°C), transcription must be initiated at the promoters for r-protein operons approximately every 4 seconds and at rRNA operons at even shorter intervals; these are very strong promoters. At the other end of the distribution we find exceedingly weak promoters. Their existence was anticipated because it seems meaningless to regulate the synthesis of a control protein such as the Lac repressor, which is produced in very small amounts, by yet another control protein. A promoter that could be activated only a few times per hour would solve the problem.

On two-dimensional gels, a set of 14 very weak spots representing proteins present in a cell in less than approximately 200 copies were selected and their fractional yield (α_i) was determined at growth rates varying from $\mu = 0.5$ to $\mu = 2.8$. The α_i values for these unidentified proteins were all approximately 10^{-4}, and eight of them were observed to *decrease* significantly as the growth rate increased. A very low α_i that decreases with increasing μ suggests constitutive synthesis from a promoter that can be activated only rarely. To see why this is so, recall that the average polysome yields the same amount of protein independently of μ and that F_i is proportional to μ (Chapter 6). To double a given quantity of protein, a fixed number of polysomes must therefore be initiated and express themselves *per doubling time*, and the relative abundance (α_i) of protein i must be proportional to the fraction of this fixed number of polysomes that contain transcripts of gene i. Finally, constitutive synthesis of protein i implies that transcription at promoter p_i always is initiated with frequency $F_{i(max)}$. Long doubling times therefore allow for more transcripts than do short times; and α_i must decrease with the decreasing doubling time accompanying increasing growth rate. Under the simple assumptions made here, *all constitutive syntheses should behave in this way*. Figure 12A illustrates that this is true for one of the arginine enzymes, for the enzymes from the *trp* operon, and for β-galactosidase under conditions of maximum induction. The basal level of β-galactosidase showed the same trend. The decrease of α_i at high growth rates was first seen for the tryptophan enzymes and described as "metabolic control" (Rose, 1972). Figure 12B shows the strikingly different behavior of three PSS proteins.

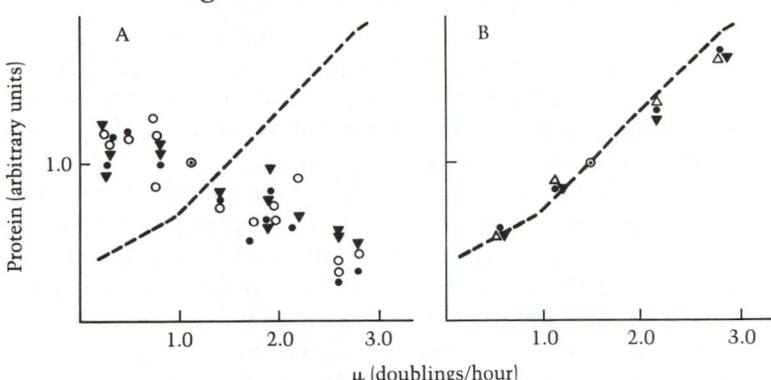

Figure 12. A. Relative synthesis rates of fully induced or constitutively produced enzymes. Activities of β-galactosidase (▼), ornithine transcarbamoylase (○), and the *trp* operon enzymes (●) were measured on samples from a culture of *E. coli* K-12 carrying the *lacUV5* promoter mutation together with *argR⁻* and *trpR⁻* mutations. B. The PSS proteins S1 (●), EF-G (△), and arginyl-tRNA synthetase (▼) were measured on cultures of *E. coli* B by two-dimensional gel electrophoresis. The dashed lines show the α_r/μ relationship (see Figure 5 in Chapter 6, and Figure 13B in this chapter); the values measured in glucose minimal media (☉) are redefined as unity for calculation of relative rates. (After Maaløe, 1979.)

It should be noted that α_i could be kept at very low levels in several ways: the half-life of the *i* messenger could be exceptionally short, its ribosome attachment site could be weak, or transcription/translation of the *i* gene could be under tight autorepression. Both a very unstable messenger and a poor ribosome binding site would increase the number of transcripts needed to generate the observed α_i, but synthesis would remain constitutive and α_i would still be expected to decrease as μ goes up. Autorepression, on the other hand, could be expected to maintain a more or less constant concentration of the *i* product (Sompayrac, 1973). This is observed for some gene products, and a μ-independent α_i suggests that the gene in question may be under autoregulation.

REGULATION OF THE SIZE OF THE PROTEIN SYNTHESIZING SYSTEM

Introduction

The many aspects of growth and regulation we shall discuss in this section can be brought into perspective by a new Gedanken experiment. Consider an *E. coli* cell with constitutive syntheses throughout and with strong and equal promoters; all gene products would then be produced in equal numbers. We shall disregard the effect of DNA replication on the relative numbers of early versus late replicating genes; under these conditions α_r is the fraction of all codons that make up the r-protein genes—approximately

0.01. In a cell with a total of 10^9 amino acids, this corresponds to 1250 ribosomes and, given adequate supplies, they would support a dP/dt of approximately 2×10^4 amino acids per second, a number that corresponds to a μ of approximately 0.1. It should not be surprising that a cell stripped of controls would not be able to grow faster.

A less obvious, but very important, consequence of our stipulations concerns the utilization of the "strong and equal promoters." We assume that the overall frequency with which polysome formation can be initiated (F_μ) is proportional to the number of ribosomes. With as few as 1250 ribosomes, only about one new polysome will be started per second from all the promoters (approximately 10^3 per genome). As we have calculated, a strong promoter can support a new start at average intervals of approximately 4 seconds; and in our hypothetical cell, the many strong promoters would be inefficiently utilized. At a given promoter, most of the time would in fact be spent waiting for a ribosome.

The efficiency of this primitive cell could have increased during evolution solely by the acquisition of the many control mechanisms that we know exist and that endow the cell with great flexibility. In any given growth medium, inductions and repressions would specifically turn on a few operons and turn off many, and thereby would establish the state of balanced growth characteristic of the medium. One obvious consequence would be that the promoters left open could be used effectively, that is, the waiting time for a ribosome to arrive would be greatly reduced.

Any state of balanced growth would be characterized exhaustively if we knew how much each gene contributed to protein synthesis. Analysis of two-dimensional gels has provided a good deal of information of this sort, but the complete DISTRIBUTION FUNCTION may never be worked out. Even so, we can also imagine a cell in which the appropriate distribution had been established, not by inductions and repressions, but by setting the different promoters at their requisite strengths. We may assume that such a cell would grow, at least for some time, equally as well as the wild type in the medium to which it has been adjusted by having few promoters at full strength and reducing the strength of most promoters to varying degrees. However, this cell would grow efficiently only in that medium, and not at all in media requiring induction of enzymes necessary for utilization of alternative carbon sources.

There is ample evidence that promoters vary greatly, and the way evolution may have worked on promoter strength is easy to visualize. First, a down-mutation in one PSS promoter would reduce the efficiency of the entire PSS; the mutant would compete unfavorably with wild type and probably would be eliminated. Second, the performance of a gene outside the PSS may be governed exclusively by the strength of its promoter, in the case of constitutive synthesis, or by a combination of promoter strength and active control. Extremely weak promoters may serve to keep the yield of certain proteins very low, and they may be the result of an accumulation of down-mutations. (Paradoxically, the end result could be a Lac-constitutive cell with a dead promoter in front of an intact *lacI* gene.) On the other hand, the performance of most operons must reflect both promoter strength and active control. In such cases, an efficient system may have evolved by combining a relatively strong promoter and effective internal control as discussed for the *ara* operon (Chapter 7). It is the interplay between these variables that we shall discuss in the following section.

Passive regulation

The model to be presented is based on the concept of PASSIVE REGULATION, the essence of which is that the operons in the PSS domain with its internal network of controls are viewed as constitutive relative to all other operons. The model is derived from the fact that the *operons of PSS are served by a small set of strong promoters, whereas a large set of weaker promoters serves the numerous fueling and biosynthetic operons*. In rapidly growing cells, these two very uneven sets contribute equally to protein synthesis (α_{PSS} being approximately 0.5).

To predict the behavior of a system with these general properties, values must be assigned to it; and because some of them are rough estimates, we shall assess how critical they are. In the preshift condition (balanced growth with glucose), the μ and α_r values are approximately 1.5 and 0.15, respectively; and in a cell with 10^9 amino acids in its proteins, approximately 10 new polysomes must be initiated per second. This frequency, F_μ, is obtained by dividing dP/dt by the average protein yield per polysome; the latter is not accurately known, and our present estimate of $F_{\mu=1.5}$ is 10 ± 2 per second at 37°C.

The cell we consider has 1000 promoters per genome-equivalent of DNA. First, we shall distinguish between promoters in the PSS and outside it. Promoters in the PSS constitute a small class from which the proteins of the PSS are derived. It contains something like 40 strong promoters and perhaps an equal number of weaker promoters per genome. Promoters outside the PSS include many virtually silent promoters, and the number to be used in our calculations is based on data indicating that *E. coli* synthesizes approximately 10^3 different proteins (O'Farrell, 1975). Assuming an average of 2.5 genes per operon (Chapters 1 and 6), we estimate that there are 40 strong PSS promoters and approximately 400 weaker ones functioning outside PSS; or $80 + 800$ in the average cell. The efficiency of the strong PSS promoters is expressed by the calculated minimum time between successive initiations of 4 seconds.

To estimate the strength of the outside promoters, several factors must be considered: first, the molecular weights of the r-proteins and of some of the factors are each approximately ⅓ the average molecular weight of other *E. coli* proteins; hence to produce the same α_i, most PSS promoters must be 2–3 times more active than the average promoter outside the PSS. Second, the 80 strong PSS promoters serve the synthesis of approximately 30% of the cell's protein, against 70% from the 800 outside promoters. Third, many of the 800 are known to be repressed to various degrees during growth, as illustrated for the *trp* operon in Chapter 7. All together, the 4-second lag attributed to strong PSS promoters must be greatly increased to obtain an average value for the outside promoters. As a minimum value we arrive at approximately

40 seconds for the lag time of outside promoters, which is close to the value previously calculated for a fully induced *lac* operon; also, to take internal repression into account, we insert the values 60, 80, and 100 seconds (Figure 13A1).

The equations to be used express the equilibrium between the loss of polysomes through mRNA decay and the gain through initiation of new ones characteristic of balanced growth:

$$S_1(k_1 - f_1) = F_\mu[f_1/(f_1 + f_2)] \tag{1a}$$

$$S_2(k_2 - f_2) = F_\mu[1 - f_1/(f_1 + f_2)] \tag{1b}$$

where k is the total number of promoters and f is the number of open promoters. The subscripts 1 and 2 refer to the PSS and to the sectors outside PSS, respectively; and S is the reciprocal average lag time of the promoters.[1]

The simultaneous equations (1a) and (1b) may be read as follows: the products on the left-hand side are the numbers of "closed" promoters $(k - f)$ in classes 1 and 2 that in 1 second make the transition to the "open" state, in which they are eligible for initiation. In equilibrium, the same number must pass from the open to the closed state due to initiation. This is expressed on the right-hand side as the products of F_μ (total number of initiations per second) and the fractions $f_1/(f_1 + f_2)$ and $1 - f_1/(f_1 + f_2)$, assigned to classes 1 and 2, respectively, on the assumption that initiation occurs at random among all open promoters $(f_1 + f_2)$.

We now introduce $\alpha_{PSS} = f_1/(f_1 + f_2)$ and obtain

$$\mathbf{F}_\mu = [(k_2/(1 - \alpha_{PSS})) - (k_1/\alpha_{PSS})][(1/S_2) - (1/S_1)]^{-1} \tag{2}$$

Figure 13A1 shows α_{PSS} as a function of F_μ, and we note that the curves representing S_2 values of 60, 80, and 100 cross a box with the sides 8 and 12 on the abscissa ($F_\mu = 10 \pm 2$). Thus, parameters that appear reasonable, readily satisfy the expected figure of 10 for $F_{\mu=1.5}$ and the corresponding α_{PSS} of 0.3, which is characteristic of balanced growth in glucose minimal medium. More striking, however, is the fact that the curves generated by equation (2) so closely resemble the experimentally defined relation between α_r and μ (Figure 13B). In conclusion, the concept of passive control, as formulated here, can account not only for the quasi-linear relation between α_r and μ at moderate to high growth rates, but also for the significant deviation from linearity at low growth rates.

It should be appreciated that the model is based to a large extent on the results of recent genetic studies. In the first place, it has become known that the r-proteins originate from approximately 20 individual transcripts, to which we arbitrarily add 20 to represent the rest of the abundant PSS proteins; it makes little difference whether this number is 10, 20, or 30. In the second place, autoregulations have been discovered that seem to link the genes of the

---

[1] Equations (1a), (1b), and (2) were formulated by M. Weis Bentzon on the basis of discussions with one of the authors (OM). Bentzon also tested whether (1a) and (1b) actually define an equilibrium by computer simulation. This work will be the subject of a separate paper by MWB and OM.

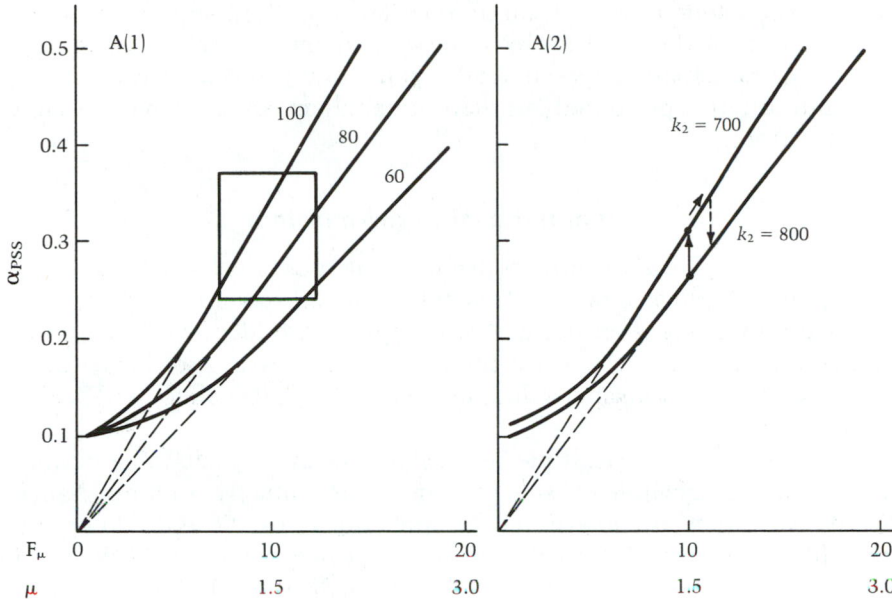

Figure 13.

A(1). Values of α_{PSS}, calculated from equation (2) for $k_1 = 80$; $k_2 = 800$; $S_1 = \frac{1}{4}$ and for S_2 values of $\frac{1}{60}$, $\frac{1}{80}$, and $\frac{1}{100}$ as indicated. The inserted frame is described in the text. A(2). Values of α_{PSS}, calculated for $k_2 = 700$ and 800 as indicated; $S_2 = \frac{1}{80}$. The arrows relate to the upshift described in the text. B. The α_r values are reproduced from Figure 5 in Chapter 6 and are compared to calculations using the parameters from the $S_2 = \frac{1}{80}$ curve in A(1). The width of the shaded zone does *not* represent calculated error limits; it points to the fact that our calculations are, at best, an approximation to biological reality.

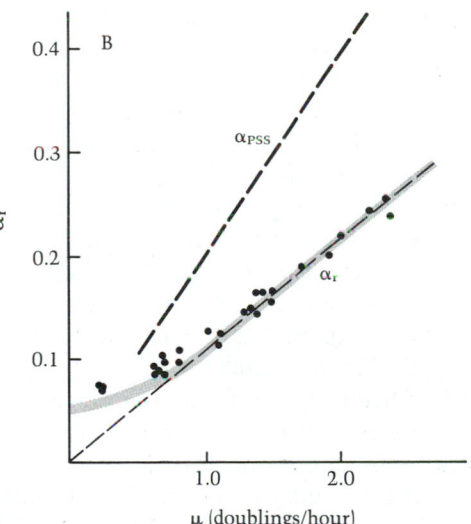

PSS domain and cause it to behave as a unit. In simple terms, the simultaneous equilibrium equations (1a) and (1b) show that α_r and α_{PSS} increase with F_μ (or μ) because high initiation frequencies favor the strong promoters.

The relationship between F_μ and α_{PSS}, expressed by equation (2), says that the cell composition, in terms of protein and RNA, is defined unambiguously by F_μ, and hence by μ. This has actually been known a long time, as illustrated by data presented in Chapter 6, Figure 4: they show that for a given value of μ, it makes no difference whether the cells grow with an unlimited supply of their carbon source or with an appropriately limited supply of glucose, the composition of the cells is the same. When these observations were made, there was no apparent reason why a pair of cultures with the same growth rate, but differing profoundly in their metabolism, should have the same gross composition.

Shifts from minimal to rich medium

We shall now consider what makes cells in balanced growth in the preshift glucose minimal medium take off toward a higher growth rate when the medium is enriched. In this upshift we identified two immediate, presumably primary responses: the multiple repressions of biosynthetic operons and the virtual disappearance of ppGpp. The latter effect will be discussed under active control.

In our model the multiple repressions mean immediate reduction of the number (k_2) of effective outside promoters, presumably with no change of F_μ. We do not know by how much the preshift k_2 of 800 goes down, but it is probably reduced by 100 to 200. According to equation (2), the values of α_{PSS} corresponding to $k_2 = 800$ and $k_2 = 700$ are 0.38 and 0.45, respectively (Figure 13A2); that is, α_{PSS} is expected to increase between 25 and 50%. This predicted response to the immediate closing of biosynthetic operons must express itself over a period corresponding to 2–3 mRNA half-lives, because the number of polysomes containing PSS mRNA can increase only *pari passu* with the disappearance of polysomes with mRNA from the now closed biosynthetic operons (as long as F_μ is unchanged). Figure 8 illustrates one of several shifts to rich medium, analyzed by Gausing (1980), and it is evident that the initial response is about as predicted. It is important that her analysis included hybridization experiments with DNA probes representing r-protein genes and covered the first 15 minutes after the shift; throughout this time, changes in α_r paralleled changes in the rate of mRNA synthesis (Gausing, 1976). The response to the shift thus parallels changes in the transcription pattern as implied in our model.

Other experiments suggest a more complex mechanism (Miura, 1981). Two different r-protein promoters were inserted in front of either the *lacZ* or the *galK* genes and introduced on λ phages integrated at the normal attachment site. Enzyme production (β-galactosidase or galactokinase) did *not* vary with μ over the range 0.3 to 1.2 (at 30°C). In contrast, an rRNA promoter in the same position led to enzyme production that paralleled the normal increase with μ of the RNA:protein ratio. No direct measurements of *lac* and *gal* mRNA were reported, and the interpretation of the results in terms of posttranscriptional autoregulation remains tentative.

The apparent oscillations that precede the establishment of the definitive α_r level cannot be accounted for in any detail. However, an obvious, but delayed response offers a clue: the initial rise of α_r tells us that the fraction of all initiations now occurring in the PSS sector has increased. This probably does not cause the cell to react until the PSS products, particularly new active ribosomes, begin to accumulate at a relative rate, $d\ln r_{act}/dt$, higher than that characterizing the preshift growth rate. It takes 2–3 minutes to assemble a ribosome at 37°C (Chapter 2); and if the relative rate of transcription of the PSS genes is suddenly raised, it will take that long before $d\ln r_{act}/dt$ changes. From then on, functional ribosomes will continue to accumulate at the new high rate for the full maturation time of 2–3 minutes. As a result, the concentration of ribosomes will increase; and in the terms of our model, this will push the system toward a higher F_μ and, with that, a higher α_{PSS}. This perturbation of the preshift system is illustrated in Figure 13A2. The initial closing of biosynthetic operons moves the system represented by a point at $F_{\mu=1.5}$ on the lower curve to the corresponding point on the upper curve (k_2 reduced from 800 to 700); and with no limitation on ribosome activity, the system would move autocatalytically toward higher values on the upper curve.

Note that several amino acid pools have been shown to be somewhat lower at low than at high growth rates (Table 4). This implies that the degrees of internal repression go down with μ and, hence, that the effect of repressing such operons must be greater in a shift from, say, succinate minimal medium to rich medium than in a similar shift from glucose minimal medium.

Obviously, the balanced preshift system cannot for long support the dP/dt corresponding to an increased ribosome concentration. We suggest that the cells react by temporarily derepressing operons in the fueling sector, thus increasing k_2 and lowering α_r and α_{PSS}, as the experimental results in Figure 8 show. Cycles of this kind may repeat themselves, but the resolution of the measurements is not sufficient to follow the protracted increase of α_r in detail. The definitive postshift α_r must be reached when the supply side cannot support further increase of dP/dt. We do not know how the new balance is maintained, but the same mechanism probably serves to stabilize any state of balanced growth.

It is clear now that the concept of passive control permits construction of a system with some of the important characteristics of a growing bacterial culture. This is in itself interesting, but we must ask whether, or to what extent, this paper work reflects biological reality. There is probably no definitive answer to this question, but we can approach an answer by examining two of the most relevant issues.

In the first place, our calculations are based on averages representing parameters with unknown distributions. This is particularly evident for the strength of the promoters outside the PSS, but it also applies to the number of genes per operon. In the case of S_2, calculations have been made for distributions of values between 1/20 and 1/2000 with a weighted average of 1/80; for $S_1 \gg S_2$, the average is a good representation of the S_2 distribution. As regards the number of genes per operon, it should be noted that the limits of ±20% assigned to F_μ allows the number of genes per operon to take values from 2 to 3. Outside this range, both slope and position of the curves on Figure 13A1 begin to change significantly.

The second issue concerns the growth characteristics the model purports to explain. Namely, are the demands sufficiently stringent to lend weight to the analysis? The merit of the model, as it stands now, is that it ties together several features of obvious importance for growth and offers a description of considerable predictive value. Most significant is the contrast between the small class of strong promoters, which governs 100 genes and is capable of supporting the synthesis of as much as 50% of the cell's protein, and the large class of relatively weak promoters, which governs at least 1000 genes. On the other hand, the model does not take into account possible changes in the degree of internal repression with μ; and, in particular, it does not in its present form explain how states of balanced growth are stabilized.

In passing we may recall that some of the newly synthesized rRNA is broken down, and the slower the cells grow the more is broken down (Chapter 6). At very low growth rates, more than half the rRNA made is never built into ribosomes. This observation is significant from the point of view of regulation; it suggests that the primary target for controlling ribosome synthesis is the r-proteins (as in our model) and that the apparent balance between the rates of synthesis of the protein and the RNA components of the ribosomes is due, in part, to instability of excess rRNA.

Active control

In analogy with the classic operon model, the adjustment of α_r and α_{PSS} to the growth conditions has traditionally been ascribed to active control, implying the existence of an effector to monitor the activity of PSS. The obvious candidate has been ppGpp, and it is also the only one that has been considered seriously. It is notable that the current picture of PSS as a subsystem with an internal network of controls makes it easy to visualize how active control might work: an effector, such as ppGpp, might control the rate of synthesis from *one* of the PSS operons directly; the rest would follow suit because they are coupled by the internal controls.

We have seen that the same low levels of ppGpp are maintained during balanced growth of *relA*$^+$ and *relA* strains (Figure 9) and that marked inhibition of RNA and protein synthesis requires much higher concentrations of ppGpp (Figure 1). Such high levels are the characteristics of the SOS-like response of stringent cells to restrictions on protein synthesis; however, relaxed cells that cannot respond in this way are indistinguishable from their stringent counterparts with respect to balanced growth and to shifts from minimal to rich media.

Even so, ppGpp may play a significant role in maintaining balanced growth and in shifts to rich media. Note that in Figure 8 α_r jumps slightly at the time of the shift; it is tempting to associate this small discontinuity with the simultaneous disappearance of ppGpp (Figure 9). The suggestion is that even at the low levels characteristic of balanced growth ppGpp exerts a modest inhibitory effect on the synthesis of the components of PSS.

When ppGpp was discovered, its negative effect on the synthesis of rRNA (and tRNA) suggested to many that it might be the regulator of ribosome synthesis. We now know that this view is too simplistic, but this conclusion does not rule out the possibility that ppGpp (or as yet unidentified effectors) serves important functions during balanced growth. One such function would be to balance the production of rRNA against that of r-protein. The problem involved here is that a single transcript from an r-protein operon yields an average of 20–30 copies of each protein encoded in that operon and that, to match these r-proteins, 20–30 whole transcripts are required from one of the seven rRNA operons in the *E. coli* chromosome. Each of the seven must therefore be transcribed a few times more frequently than the r-protein operon. The actual ratio depends on the relative gene numbers, which vary with the pattern of replication and therefore with μ.

If ppGpp is involved in this problem of balance, its role may be very elusive. The adjustments required probably would be in response to fluctuations in the rate of ribosome production analogous to those provoked by the upshift in Figure 8. During balanced growth, such fluctuations would not be synchronous; and average values, which is all we can obtain by sampling from a culture, cannot reveal what takes place in individual cells. The same arguments apply to another function for which ppGpp might be a candidate, that is, the stabilization of balanced growth. Even synchronized cultures could not reveal the fluctuations we now consider unless, as is very unlikely, they were coupled to particular events in the cell cycle. To follow random variations in a parameter such as the level of ppGpp in a bacterial cell would require techniques of unbelievable sensitivity and resolution.

Therefore, the states of balanced growth we observe in bacterial cultures and the balance between r-proteins and rRNA point to the existence of stabilizing mechanisms, not of new and different kinds but still outside the range of in vivo experiments, because the cells are too small, and such mechanisms are difficult, if not impossible, to reproduce in vitro.

What determines growth rate in a given medium?

This natural question turns up regularly in connection with bacterial growth; and as often as not, "a rate-limiting reaction" is invoked. In most cases, this is probably not the right answer. A growing bacterium is *not* like a homogeneous in vitro system in which the overall reaction rate can be increased by raising the concentration of a particular component, such as an enzyme. We can imagine cells that grow very slowly on some esoteric carbon source because an enzyme necessary to metabolize it is inefficient. During growth with this particular carbon source, "fitter" mutants may arise that grow more rapidly, either because the hypothetical key enzyme has become more efficient or perhaps because of an up-mutation in the promoter or a duplication of a gene. When we observe that balanced growth at a constant rate can be maintained for long periods, the many reactions involved probably have been "harmonized" in such a way that no single step can be said to be rate limiting.

The immediate-response upshift (Figure 8) illustrates this; if all that was needed for the cells to grow faster in the postshift medium was more ribosomes, α_r would be expected to rise steeply—perhaps to overshoot its definitive value—so as to reach the final ribosome concentration in the shortest possible time. Instead, we see α_r increasing in an irregular manner and over a long period before settling at the final postshift value. As already mentioned, we interpret this to mean that an increase in ribosome concentration necessitates *other* changes in cell composition in order to make growth at a higher rate possible.

Another relevant case is that of rRNA synthesis. At high growth rates the initiation frequency at the rRNA promoters required to produce enough rRNA is extremely high. At maximum growth rate ($\mu = 2.8$, 37°C), a new act of transcription must in fact be initiated at each rRNA promoter at intervals of approximately 1 second, that is, about as frequently as possible, considering that the promoter region must be cleared between successive initiations. Thus, at maximum growth rate, rRNA can hardly be produced in excess. These extreme frequencies of initiation at the rRNA promoters have been thought to set an upper limit to μ. However, the number of rRNA genes probably would have been amplified beyond seven if that alone had sufficed to push μ_{max} above 2.8. It seems more likely that μ_{max} reflects a state of growth in which α_{PSS} has reached a value that cannot be exceeded without concomitant reduction of biosynthetic and fueling capacities that would prevent efficient utilization of an oversized PSS.

If we discard the notion of rate-limiting step(s), what then causes growth to be faster in some media than in others? The equilibrium equations (1a) and (1b) presented earlier are relevant to the problem, but they do not contain the solution. They indicate that a relationship has evolved between PSS on the one hand and the many enzyme systems outside PSS on the other; this relationship ensures optimal balance between the two at all but the lowest growth rates. At these low growth rates, active ribosomes function with the same high efficiency as in richer media, but fewer of them are active (Chapter 6, Figure 7). This relative inefficiency of the PSS at low growth rates was first observed experimentally, and it is accounted for by our model. Qualitatively, this can be seen by extrapolating to $F_\mu = 0$ (Figure 13A1); as F_μ decreases, so does the advantage of having strong promoters and α_{PSS} must therefore approach a definite, positive value. Maximum efficiency at extremely low growth rates would demand that α_{PSS} approached zero; and according to our model, this probably would require a combination of active and passive control.

Ability to adjust the size and capacity of PSS to the growth conditions confers selective advantage (in terms of growth rate and yield) on cells that possess this ability, *whatever nutrients are available to the cells*. On the other hand, ability to utilize, say, a particular carbon source effectively has to do with the quality of certain enzymes and the chemical properties of the carbon compound considered. Hence, passive and/or active regulation of PSS does *not* explain why different media support different rates of growth of the same strain.

We therefore restate the original question with a slight change. Namely, what determines the value F_μ assumes in a given medium? Emphasis is thus on macromolecular synthesis (specifically, RNA and protein) from a particular set of nutrients and on the demands made on the fueling reactions. Three broad categories of reactions are involved: polymerization, synthesis of monomers, and maintenance; the latter is small enough to be neglected in this context (Chapter 5). To double the RNA and protein in a cell, as many monomers as are contained in these macromolecules to begin with and a fixed amount of energy (ATP) are needed for polymerization, irrespective of the actual doubling time. In addition to this fixed energy requirement, carbon and energy are needed to synthesize the monomers themselves. The energy required for these syntheses varies greatly, depending on the "quality" of the carbon source available (Chapter 3, Table 14).

As discussed earlier (Chapter 5), it has been found that several carbon sources that support very different growth rates under aerobic conditions give rise to the same O_2 consumption per unit of cell mass *and* per minute. This suggests that *the energy generated per minute is nearly constant* and that growth on a "bad carbon source" is slow because much larger amounts of energy are needed (on top of the fixed amount used for polymerization) than in the case of a good carbon source such as glucose. However, it remains to be seen to what extent the energy needed to produce the end products of the fueling reactions correlates with the rate of growth on a given carbon source. We believe that growth rate depends in a simple manner on the quality of fuel and on the type of burner used (aerobic or anaerobic).

SUMMARY

1. The global frequency of initiation of polysome formation F_μ is proportional to μ, because the average protein yield per polysome is μ independent. Our key question concerns the partitioning of F_μ between operons inside and outside the PSS.

2. Size and capacity of PSS is *not* regulated by ppGpp or other known effectors. Absence of stringent factor (*relA* deletion) has no effect on ppGpp levels in balanced growth. Stringent and relaxed strains may produce ppGpp by a common mechanism, activation of stringent factor being part of an SOS-like response to starvation.

3. Shifts between media are "sluggish" when profound enzymatic changes are elicited, for example, change of nitrogen or sulfur source. "Immediate-response" shifts elicit obvious primary effects, for example, multiple repressions and disappearance of ppGpp after shifts to rich medium. The transient loss of ppGpp does not affect inducibility.

4. Some energy downshifts immediately reduce dP/dt. Energy charging and triphosphate concentrations remain high, partly because ADP inhibits the pyrophosphatase. In $relA^+$ cultures αMG causes accumulation of ppGpp and reduces synthesis of all species of RNA.

5. Most amino acid pool levels are high relative to known K_m values for tRNA charging. Internal repression thus acts at concentrations not affecting cgr_p. Operons regulated exclusively by attenuation feature leader sequences with many repeats of the cognate codon; this may increase sensitivity to uncharged tRNA by queuing effects.

6. Cells have more RNA-P than is engaged in RNA synthesis. This ensures that transcription is initiated with minimum delay following derepression of an operon.

7. At maximum growth rate, transcription starts from r-protein and other PSS promoters every 4 seconds. These "strong" promoters (approximately 40 per genome) account for one-half the protein. The other half originates from approximately 400 intrinsically "weak" promoters, many of which are partially repressed.

8. Assuming constitutive synthesis from the PSS promoters (passive control), a pair of equilibrium equations expresses the partitioning of F_μ (the initiation frequency) between promoters inside and outside the PSS. $F_{\mu=1.5}$ is approximately 10, and the equations yield the corresponding α_r of 0.15. Varying F_μ generates the observed relation between μ and α_r.

9. Multiple repressions in a shift to rich medium reduce the number of open promoters outside the PSS and increase α_{PSS}, as observed. Balance at a higher α_{PSS} may require derepressions in the fueling sector and cause the observed transient lowering of α_{PSS}. The model does not specify how balance at the final postshift growth rate is established.

10. Our model accounts for the high efficiency of growth (constant high cgr_p), but not for medium-dependent variations of μ. We believe that growth rate depends in a simple manner on the quality of the fuel and on the type of burner used (aerobic or anaerobic).

REFERENCES

Atherly, A. G. 1979. *Escherichia coli* mutant containing a large deletion from *relA* to *argA*. J. Bacteriol. 138:530.

Bennett, P. M. and O. Maaløe. 1974. The effects of fusidic acid on growth, ribosome synthesis and RNA metabolism in *Escherichia coli*. J. Mol. Biol. 90:541.

Bremer, H. and D. G. Dalbow. 1975. Regulatory state of ribosomal genes and physiological changes in the concentration of free ribonucleic acid polymerase in *Escherichia coli*. Biochem. J. 150:9.

Cashel, M. and J. Gallant. 1969. Two compounds implicated in the function of the RC gene of *Escherichia coli*. Nature 221:838.

Chamberlin, M., W. Mangel, G. Rhodes and S. Stahl. 1976. Biochemical studies on the transcription cycle, p. 22. In Benzon Symposium IX, *Control of Ribosome Synthesis*, N. O. Kjeldgaard and O. Maaløe, eds. Munksgaard, Copenhagen.

Dennis, P. P. 1974. Synthesis and stability of individual ribosomal proteins in the presence of rifampicin. Mol. Gen. Genet. 134:39.

Fiil, N. P., K. von Meyenburg and J. D. Friesen. 1972. Accumulation and turnover of guanosine tetraphosphate in *Escherichia coli*. J. Mol. Biol. 71:769.

Friesen, J. D., N. P. Fiil and K. von Meyenburg. 1975. Synthesis and turnover of basal level guanosine tetraphosphate in *Escherichia coli*. J. Biol. Chem. 250:304.

Gausing, K. 1976. Synthesis of rRNA and r-protein mRNA in *E. coli* at different growth rates, p. 292. In Benzon Symposium IX, *Control of Ribosome Synthesis*, N. O. Kjeldgaard and O. Maaløe, eds. Munksgaard, Copenhagen.

Gausing, K. 1980. Regulation of ribosome biosynthesis in *E. coli*, p. 693. In *Ribosomes: Structure, Function, and Genetics*, G. Chambliss, G. R. Craven, J. Davies, K. Davis, L. Kahan and M. Nomura, eds. University Park Press, Baltimore.

Johnsen, K., S. Molin, O. Karlström and O. Maaløe. 1977. Control of protein synthesis in *Escherichia coli*: analysis of an energy source shift-down. J. Bacteriol. 131:18.

Josse, J. 1966. Constitutive inorganic pyrophosphatase of *Escherichia coli* II. Nature and binding of active substrate and the role of magnesium. J. Biol. Chem. 241:1948.

Maaløe, O. 1979. Regulation of the protein-synthesizing machinery—ribosomes, tRNA, factors, and so on, p. 487. In *Biological Regulation and Development*, Vol. 1, R. F. Goldberger, ed. Plenum Press, New York.

Maaløe, O. and N. O. Kjeldgaard. 1966. *Control of Macromolecular Synthesis*. Benjamin, New York, N.Y.

Miura, A., J. H. Krueger, S. Itoh, H. A. de Boer and M. Nomura. 1981. Growth-rate-dependent regulation of ribosome synthesis in *E. coli*: Expression of the *lacZ* and *galK* genes fused to ribosomal promoters. Cell 25:773.

Molin, S., K. von Meyenburg, O. Maaløe, M. T. Hansen and M. L. Pato. 1977. Control of ribosome synthesis in *Escherichia coli*: analysis of an energy source shift-down. J. Bacteriol. 131:7.

O'Farrell, P. H. 1975. High resolution two-dimensional electrophoresis of proteins. J. Biol. Chem. 250:4007.

Reeh, S., S. Pedersen and J. D. Friesen. 1976. Biosynthetic regulation of individual proteins in *relA*[+] and *relA* strains of *Escherichia coli* during amino acid starvation. Mol. Gen. Genet. 149:279.

Richmond, M. H. and O. Maaløe. 1962. The rate of growth of *Salmonella typhimurium* with individual carbon sources related to glucose metabolism or to the Krebs cycle. J. Gen. Microbiol. 27:285.

Rose, J. K. and C. Yanofsky. 1972. Metabolic regulation of the tryptophan operon of *Escherichia coli*: Repressor-independent regulation of transcription initiation frequency. J. Mol. Biol. 69:103.

Seeburg, P. H., C. Nüsslein, and H. Schaller. 1977. Interaction of RNA polymerase with promoters from bacteriophage fd. Eur. J. Biochem. 74:107.

Sompayrac, L. and O. Maaløe. 1973. Autorepressor model for control of DNA replication. Nature New Biol. 241:133.

Travers, A. A., R. Buckland and P. B. Debenham. 1980. Functional heterogeneity of *Escherichia coli* ribonucleic acid polymerase holoenzyme. Biochemistry 19:656.

Von Meyenburg, K. 1971. Transport-limited growth rates in a mutant of *Escherichia coli*. J. Bacteriol. 107:878.

Wanner, B. L. and R. McSharry. 1982. Phosphate-controlled gene expression in *Escherichia coli* K-12 using Mu*dl*-directed *lacZ* fusions. J. Mol. Biol. 158:347.

Appendix A

Visualizing the Bacterial Cell

An understanding of the molecular architecture of the bacterial cell has depended on the ability to visualize the cell. Indeed, what seemed to be purely chemical problems of macromolecular structure, including the subunit structure of certain enzymes and the structure of unit membranes, were solved by visual observation. Visualization of a bacterial cell requires the use of a microscope to produce a magnified image. Revelation of meaningful detail within the image requires that there be adequate CONTRAST between its parts and that the microscope used to produce the image have sufficient RESOLVING POWER to allow discrimination between separate points within the image. Relationships between vertically separated points of the image can only be discerned if there is adequate DEPTH OF FOCUS. These then are the chief requirements of successful visualization of bacteria: magnification, resolving power, contrast, and depth of focus.

The use of a conventional compound light microscope in visualizing subcellular detail is limited by its resolving power–a limit set by the wave nature of light. The resolving limit (the distance between two objects within an image that can just be seen to be separate) is determined by the equation

$d = 0.5\lambda/N\sin\theta$

where d is the resolving limit, λ is the wavelength of light used to illuminate the object, θ is the half-angle of the objective lens, and N is the refractive index of the material between the objective lens and the specimen. Resolving power can be increased by using a larger lens (increasing θ) or by introducing between the specimen and the objective lens a material with a higher refractive index (N) than that of air (as is done with an oil-immersion objective). In addition, a shorter wavelength (λ) of light can be used to illuminate the specimen. Therefore, microscopes that employ ultraviolet light (and hence

387

that require expensive quartz objective lenses) provide marginal improvement in resolving power; however, the use of an electron beam, as in transmission electron microscopy (TEM), provides a dramatic improvement. The limit of useful magnification (i.e., the limit of magnification that produces greater detail) depends on resolving power. The limit of useful magnification of a compound microscope is approximately 1000×, and that of a modern TEM is over 100,000×. Correspondingly, the resolving limit of the light microscope is approximately 250 nm and that of the electron microscope, 0.5 nm.

Contrast within the image produced by a conventional compound light microscope depends on differential absorption of light. Because bacteria absorb little light, the image of a bacterium produced by a conventional compound microscope is almost devoid of detail; indeed, the cell is only barely visible unless dyes are added to stain the cell completely or to differentially stain its parts. Alternatively, contrast can be generated by using a stain that remains outside the cell or that is excluded from a part of it. Such staining is termed NEGATIVE STAINING.

The phase contrast compound microscope produces a high-contrast image of unstained bacteria because this instrument generates contrast from the differences in refractive index between various parts of the cell and between the cells and the suspending medium. These differences are considerable.

Contrast is also inadequate in TEM images unless the specimen is stained. Staining specimens for observation by TEM is usually done with salts of heavy metals, such as uranium, osmium, and lead, because they are effective barriers to the passage of an electron beam.

Light microscopy (using either the conventional compound microscope, a phase-contrast microscope, or a UV microscope) and TEM depend on the same fundamental principle. A beam of electromagnetic radiation is passed through the specimen; and by a series of lens systems (made of glass, quartz, or electromagnets), an image is generated. In the last decade a totally new system of microscopy—SCANNING ELECTRON MICROSCOPY (SEM)—has become common. In SEM, an image is generated by scanning the surface of the specimen in a raster pattern. Such bombardment of the specimen generates a cloud of secondary electrons, the intensity of which is determined by the composition of the specimen at the point being irradiated and the angle between that point on the specimen and the electron beam. This cloud of electrons is collected on an anode, thereby generating an electric signal that is amplified and used to modulate the intensity of an electron beam in a cathode ray tube (TV picture tube). The electron beam scanning the specimen and that scanning the cathode ray tube are synchronized. Thus, on the surface of the cathode ray tube an image is generated that reflects the varied composition and surface topology of the specimen. SEM images have remarkable depth of focus and reveal surface images at high magnification in elaborate detail (Figure 1).

The richness of detail now known about the internal ultrastructure of the bacterial cell comes largely from TEM observations on stained thin sections (not more than 50 nm thick) cut through the cell; but interpretation of TEM images is not without its hazards. Artifacts can be generated by the process of drying, which is required before the specimen can be placed in the high

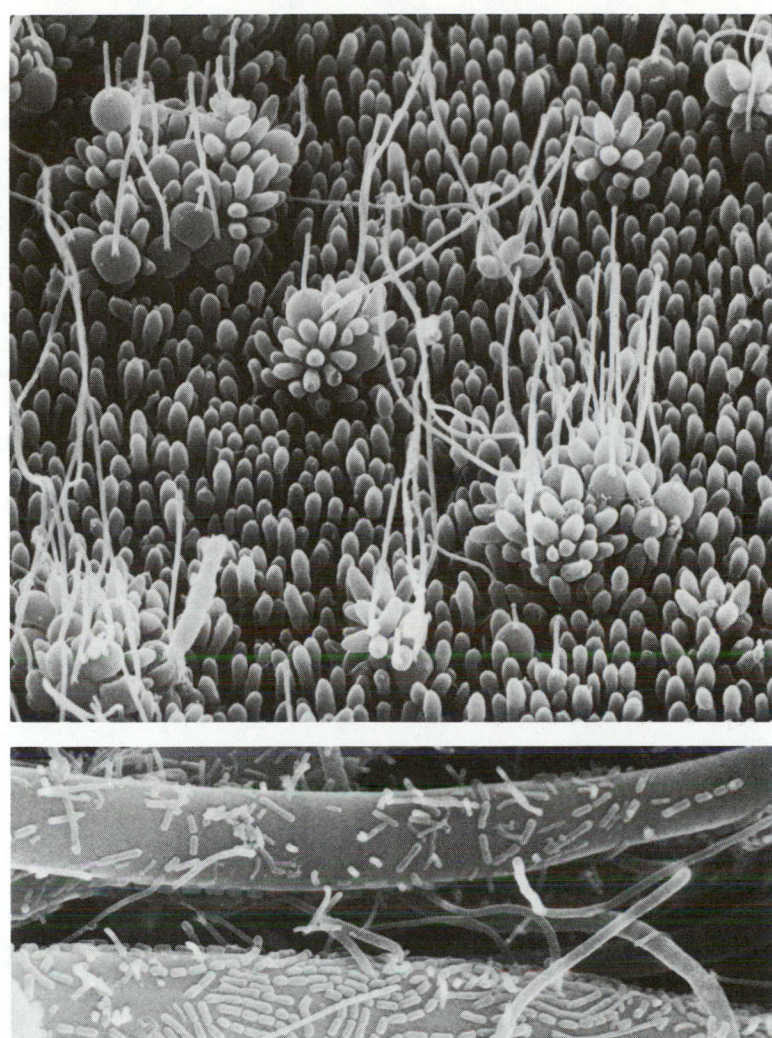

Figure 1. Scanning electron micrograph of bacteria growing in a natural environment. Mussels taken from the immediate vicinity of thermal submarine vents at the Galapagos Rift spreading zone at a depth of 2550 meters were carried to the surface by the research submersible ALVIN. Portions of the samples were rinsed with phosphate buffer, dehydrated through 25, 50, 75, and 100% ethanol or acetone, dried to the critical point, and spatter-coated with palladium–gold alloy. Samples were examined by E. Seling using an AMR Model 100 scanning electron microscope. ×6000. Top: *Hyphomicrobium*-like cells attached to the mussel surface. Bottom: Relatively slender filaments of *Beggiatoa* with rod-shaped bacteria attached to them. Courtesy of Holger W. Jannasch.

vacuum necessary for passage of the electron beam, or by one or another of the steps of sample preparation, including staining.

A number of effective techniques have been developed to improve the quality of the image of TEM micrographs and to minimize the generation of artifacts. Staining is usually accomplished by addition of the salts of heavy metals (lead, tungsten, or uranium). Negative staining with these materials is particularly effective for revealing the structure of very small objects such as viruses or proteins. METAL SHADOWING, which is the exposure of the dried specimen to a beam of platinum, palladium, or gold, reveals the shape of the specimen by the shadow that is cast. FREEZE-FRACTURING has been particularly useful in revealing the structure of membranes. In this procedure, the specimen is frozen; the frozen mass is then fractured with a knife. Because such fracture lines often run along membranes, the membranes are exposed on the fractured surface. Visualization of this surface is accomplished by shadowing it with a heavy metal, evaporating a layer of carbon onto it for support, chemically removing the specimen, and observing the replica by TEM. Artifacts deriving from the necessity of drying specimens prior to observation by TEM can be minimized by critical-point drying.

Details of light and electron microscopy and their many uses can be found in Chapters 2 and 3 of *Physical Biochemistry* by David Freifelder (W. H. Freeman and Co., 1976).

Appendix B

The Genetic Catalog—The Mapped Genes of Escherichia coli Grouped by Metabolic Function[a]

Genes	Mapped location
I. Transport (92)[b]	
A. Ions (23)	
1. Iron (6)	
exbB enterochelin	64
exbC enterochelin	58
fec iron (citrate dependent)	7
fep iron (enterochelin dependent)	13
tonA ferrichrome	3
tonB chelated iron and cyanocobalamin	27
2. Mg^{2+} (3)	
corA Mg^{2+}	85
corB Mg^{2+}	96
mgt Mg^{2+}, (System II)	92
3. K^+ (9)	
kdpA K^+ (high affinity)	15
kdpB K^+ (high affinity)	15
kdpC K^+ (high affinity)	15
kdpD regulatory gene for kdpA, B, and C	15
trkA K^+	72

Genes	Mapped location
trkB K^+	73
trkC K^+	1
trkD K^+	84
trkE K^+	28
4. Phosphate (4)	
phoS periplasmic binding protein	83
phoT phosphate	83
pit phosphate	76
pst phosphate	83
5. Sulfate (1)	
cysA sulfate	52
B. Amino Acids (20)	
argP arginine, ornithine, and lysine	62
aroP general aromatic amino acid transport	27
aroT aromatic amino acids, alanine, and glycine	27
brnQ isoleucine, leucine, and valine	9
brnR isoleucine, leucine, and valine	8
brnS isoleucine, leucine, and valine	1
brnT isoleucine	62
cycA D-alanine, D-serine, and glycine	95
dpp dipeptides	13
gltR regulatory gene for glutamate transport	92
gltS glutamate	82

[a]From Bachmann, B. J. and K. B. Low. 1980. Linkage Map of *Escherichia coli*, Edition 6. Microbiol. Revs. *44*, 1.

[b]Number in parentheses is the number of identified genes in the category.

391

II. Fueling Reactions (138)

A. Central (39)

1. EMP Pathway (19)

2. TCA Cycle (8)

3. Entner-Doudoroff Pathway (4)

4. Hexose Monophosphate Shunt (4)

5. Glyoxylate Shunt (4)

B. Catabolism of Amino Acids (12)

C. Catabolism of Carbohydrates (61)

proB proline synthesis: synthesis of L-
glutamate semialdehyde 6

proC proline synthesis: Δ-pyrroline-5-
carboxylate reductase 9

glnA glutamine synthetase (EC 6.3.1.2) 86

glnD uridylyltransferase (regulates glutamine
synthetase activity) 4

glnF regulation of glnA 69

2. Aspartate Family
(+ Isoleucine and Valine) (43)

alnR alanine regulatory gene 99

asd aspartate-semialdehyde dehydrogenase
(EC 1.2.1.11) 75

asnA asparagine synthetase (EC 6.3.1.1) 84

asnB asparagine synthetase (EC 6.3.1.1) 15

aspC aspartate aminotransferase (EC 2.6.1.1) 20

azl regulation of ilv and leu genes,
azaleucine resistance 55

flrA regulation of ilv and leu genes 100

ilvA isoleucine-valine synthesis:
threonine deaminase (EC 4.2.1.16) 84

ilvB isoleucine-valine synthesis: aceto-
lactate synthetase (EC 4.1.3.18)
(valine sensitive) 84

ilvC isoleucine-valine synthesis: ketol-
acid reductoisomerase (EC 1.1.1.86) 84

ilvD isoleucine-valine synthesis: dihydro-
acid dehydrase (EC 4.2.1.9) 84

ilvE isoleucine-valine synthesis: branched-
chain amino acid aminotransferase
(EC 2.6.1.42) 84

ilvF isoleucine-valine synthesis:
affects ilvG 54

ilvG isoleucine-valine synthesis: aceto-
lactate synthase II (EC 4.1.3.18),
valine insensitive 84

ilvH isoleucine-valine synthesis: aceto-
lactate synthase III (EC 4.1.3.18),
valine sensitive 2

ilvI isoleucine-valine synthesis: aceto-
lactate synthase (EC 4.1.3.18), valine
sensitive 2

ilvO isoleucine-valine synthesis: affects
expression of ilvG 84

ilvY isoleucine-valine synthesis: regulation
of ilvC 84

leuA leucine synthesis: 2-isopropylmalate
synthase (EC 4.1.3.12) 2

leuB leucine synthesis: 2-isopropylmalate
dehydrogenase (EC 1.1.1.85) 2

leuC leucine synthesis: α-isopropylmalate
isomerase (subunit) 2

leuD leucine synthesis: α-isopropylmalate
isomerase (subunit) 2

leuK regulation of biosynthetic enzymes for
leucine, isoleucine-valine, histidine,
and tryptophan 18

dapA diaminopimelate synthesis: dihydropico-
linate synthase (EC 4.2.1.52) 53

dapB diaminopimelate synthesis: dihydropico-
linate reductase 0

dapC diaminopimelate synthesis: tetrahydro-
picolinate succinylase 3

dapD diaminopimelate synthesis: succinate-
diaminopimelate aminotransferase 4

dapE diaminopimelate synthesis: N-succinyl-
diaminopimelate deacylase 53

lysC lysine synthesis: aspartokinase III 91

lysA lysine synthesis: diaminopimelate
decarboxylase 20

metA methionine synthesis: homoserine
acetyltransferase (EC 2.3.1.31) 90

metB cystathionine γ-synthase (EC 4.2.99.9) 88

metC cystathionine γ-lyase 65

metE methionine synthesis: tetrahydropterol-
triglutamate methyltranferase
(EC 2.1.1.14) 85

metF methionine synthesis: 5,10-methylene-
tetrahydrofolate reductase (EC 1.1.1.68) 88

metH methionine synthesis: B-12 dependent
homocysteine-N_5-methyltetrahydrofolate 90

metJ methionine synthesis: regulatory gene 88

metK methionine synthesis: methionine
adenosyltransferase (EC 2.5.1.6) 63

metL methionine synthesis: aspartokinase II 88

metM methionine synthesis: homoserine
dehydrogenase II 88

thrA synthesis of threonine: aspartokinase
I-homoserine dehdyrogenase I 0

thrB synthesis of threonine: homoserine
kinase (EC 2.7.1.39) 0

thrC threonine synthase (EC 4.2.99.2) 0

3. Glycine-Serine Family (13)

cysB regulatory gene for cysteine
biosynthesis 28

cysC adenylsulfate kinase (EC 2.7.1.25) 59

cysD sulfate adenylyltranferase (EC 2.7.7.4) 59

cysE cysteine synthesis: serine acetyltransferase (EC 2.3.1.30) 80

cysG sulfite reductase activity 73

cysH adenylsulfate reductase (EC 1.8.99.2) 59

cysI sulfite reductase activity 59

cysJ sulfite reductase activity 59

cysK cysteine synthase (EC 4.2.99.8) 52

3. Deoxynucleotides (5)

dcd dTTP biosynthesis: dCTP deaminase
   (EC 3.5.4.5)   45
dut dTTP biosynthesis: dUTPase   81
nrdA ribonucleoside diphosphate reductase,
   subunit Bl (EC 1.17.4.1)   48
nrdB ribonucleoside diphosphate reductase,
   subunit B2 (EC 1.17.4.1)   48
tdk thymidine kinase (EC 2.7.1.75)   27

C. Cofactors (35)

1. Biotin (11)

bisA reduction of biotin-d-sulfoxide   17
bisB reduction of biotin-d-sulfoxide   18
bisC reduction of biotin-d-sulfoxide   79
bisD reduction of biotin-d-sulfoxide   0
bioA 7 KAP → DAPA   17
bioB conversion of dethiobiotin to biotin   17
bioC block before pimeloyl CoA   17
bioD dethiobiotin synthetase   17
bioF pimeloyl CoA → 7 KAP   17
bioH block before pimeloyl CoA   74
bioR bio regulatory gene   89

2. Folic Acid (4)

folA dihydrofolatereductase (EC 1.5.1.3)   1
folB regulatory gene   1
pabA synthesis of p-aminobenzoate   74
pabB synthesis of p-aminobenzoate   40

3. Lipoate (2)

lip requirement for lipoate   14
lpd lipoamide dehydrogenase (EC 1.6.4.3)   3

4. Pantothenate (3)

panD aspartate l-decarboxylase   3
panC pantothenate synthetase (EC 6.3.2.1)   3
panB ketopantoate hydroxymethyl transferase
   (EC 4.1.2.12)   3

5. Pyridoxine (5)

pdxJ pyridoxine synthesis   55
pdxH pyridoxine phosphate oxidase   36
pdxC pyridoxine synthesis   20
pdxB pyridoxine synthesis   50
pdxA pyridoxine synthesis   1

6. Pyridine Nucleotides (7)

pncA synthesis: nicotinamide deaminase
   (EC 3.5.1.19)   39
nadA quinolinate synthetase (A protein)   16
nadB quinolinate synthetase (B protein)   55
nadC quinolinate phosphoribosyltransferase   3
ndh NAD dehydrogenase complex   22
pncH regulation of pncA   39

pnt pyridine nucleotide transhydrogenase
   (EC 1.6.1.1)   35

7. Thiamine (3)

thiA synthesis of thiamine   90
thiB thiamine phosphate pyrophosphorylase   90
thiC thiamine pyrimidine requirement   90

D. Components of Electron Transport Chains (20)

1. Heme Porphyrin (9)

hemA δ-aminolevulinate synthase (EC 2.3.1.37)   26
hemB 5-aminolevulinate dehydratase
   (EC 4.2.1.24)   8
hemC uroporphyrinogen I synthase (EC 4.3.1.8)   85
hemD uroporphyrinogen III cosynthase   85
hemE uroporphyrinogen decarboxylase
   (EC 4.1.1.37)   90
hemF coproporphyrinogen III oxidase
   (EC 1.3.3.3)   17
hemH ferrochelatase (EC 4.99.1.1)   11
popC δ-aminolevulinate synthesis   3
popD 5-aminolevulinate dehydratase
   (EC 4.2.1.24)   1

2. Menaquinone (3)

menA 1,4-dihydroxy-2-naphthoate →
   dimethylmenaquinone   88
menB 2-succinylbenzoate → 1,4-dihydroxy-2-
   naphthoate   48
menC chorismate → 2-succinylbenzoate   48

3. Ubiquinone (8)

ubiA 4-hydroxybenzoate-3-octapenyl 4-
   hydroxybenzoate   91
ubiB 2-octaprenylphenol → 2-octaprenyl-6-
   methoxy-phenol   85
ubiC chorismate lyase   91
ubiD 3-octaprenyl-4-hydroxybenzoate →
   2-octaprenyl-phenol   85
ubiE 2-octaprenyl-6-methoxy-1,4-benzo-
   quinone → octaprenyl-3-methyl-6-
   methoxy-1,4-benzoquinone   85
ubiF 2-octaprenyl-3-methyl-6-methoxy-1,4-
   benzoquinone → 2-octaprenyl-3-methyl-
   5-hydroxy-6-methoxy-1,4-benzoquinone   15
ubiG 2-octaprenyl-3-methyl-5-hydroxy-6-
   methoxy-1,4-benzoquinone → ubiquinone-8   48
ubiH 2-octaprenyl-6-methoxy-phenol → 2-
   octaprenyl-6-methoxy-1,4-benzoquinone   62

E. Fatty Acids and Phospholipids (10)

cls cardiolipin synthase   27
dgk diglyceride kinase   91
fabA β-hydroxy-decanoylthioester dehydrase
   (EC 4.2.1.60)   22

VI. Machinery of Protein Synthesis (164)

A. Aminoacyl-tRNA Synthetases (23)

alaS	alanyl-tRNA synthetase (EC 6.1.1.7)	58
argS	arginyl-tRNA synthetase (EC 6.1.1.19)	40
asnS	asparaginyl-tRNA synthetase	21
glnS	glutaminyl-tRNA synthetase (EC 6.1.1.18)	15
glnT	affects levels of glutamyl-tRNA[1] and glutamine synthetase	77
gltB	glutamate synthetase (EC 2.6.1.53)	69
gltE	glutamyl-tRNA synthetase (subunit)	80
gltM	glutamyl-tRNA synthetase	43
gltX	glutamyl-tRNA synthetase (catalytic subunit)	52
glyS	glycyl-tRNA synthetase (EC 6.1.1.14)	79
ileS	isoleucyl-tRNA synthetase (EC 6.1.1.5)	0
leuR	affects leuS expression	78
leuS	leucyl-tRNA synthetase (EC 6.1.1.4)	15
leuY	expression of leuS affected	10
metG	methionyl-tRNA synthetase	46
pheS	phenylalanyl-tRNA synthetase (EC 6.1.1.20) α-subunit	37
pheT	phenylalanyl-tRNA synthetase (EC 6.1.1.20) β-subunit	37
serR	level of seryl-tRNA synthetase	2
serS	seryl-tRNA synthetase (EC 6.1.1.11)	20
thrS	threonyl-tRNA synthetase (EC 6.1.1.3)	38
trpS	tryptophanyl-tRNA synthetase (EC 6.1.1.2)	74
tyrS	tyrosyl-tRNA synthetase (EC 6.1.1.1)	36
valS	valyl-tRNA synthetase (EC 6.1.1.9)	96

B. tRNA's (51)

alaT	alanine tRNA 1B	86
alaU	alanine tRNA 1B	70
aspT	aspartate tRNA 1 (rrnC)	84
asnT	asparagine tRNA	43
cca	tRNA nucleotidyltransferase	66
glnU	glutamine tRNA 2	15
glnV	glutamine tRNA 1	15
gltT	glutamate tRNA 2	89
gltU	glutamate tRNA 2	84
gltV	glutamate tRNA 2	90
glyT	glycine tRNA 2	89
glyU	glycine tRNA 1	61
glyV	glycine tRNA 3	95
glyW	glycine tRNA 3	41
hisR	histidine tRNA	84
hisT	pseudouridylate synthetase	50
ileT	isoleucine tRNA 1	86
ileU	isoleucine tRNA 1	70
leuT	leucine tRNA 1	84
leuU	leucine tRNA 2	68
leuV	leucine tRNA 1	93
leuW	a leucine tRNA	15
lysT	lysine tRNA	16
metT	methionine tRNAm	15
metY	methionine tRNA$_f$ 2	68
metZ	methionine tRNA$_f$ 1	61
newA	uridine thiolation factor A activity	9
newC	uridine thiolation factor C activity	44
serT	serine tRNA 1	16
serV	serine tRNA 3	61
supC	suppressor of UAA and UAG mutations	27
supD	suppressor of UAG mutations	43
supG	suppressor of UAA and UAG mutations	16
supH	suppressor	43
supK	a tRNA methylase	61
supN	suppressor of UAA and UAG mutations	51
supO	suppressor of UAA and UAG mutations	27
supP	suppressor of UAG mutations	96
supQ	suppressor	12
supV	suppressor of UAA and UAG mutations	84
trmA	tRNA (uracil-5)-methyltransferase (EC 2.1.1.35)	89
trmB	tRNA (guanine-7)-methyltransferase (EC 2.1.1.33)	7
trmC	deficiency of 5-methylaminoethyl-2-thio-uridine in tRNA	55
trmD	tRNA (guanine-1)-methyltransferase (EC 2.1.1.31)	58
thrT	threonine tRNA 3	89
thrU	threonine tRNA 4	89
trpT	tryptophan tRNA	84
tyrT	tyrosine tRNA 1	27
tyrU	tyrosine tRNA 2	89
tyrV	tyrosine tRNA 1	27
valT	valine tRNA 1	16

C. rRNA (22)

ksgA	methylase for rRNA	1
rrfA	5 S RNA	86
rrfB	5 S RNA	89
rrfC	5 S RNA	84
rrfD	5 S RNA	72
rrfE	5 S RNA	90
rrfF	5 S RNA	74
rrfG	5 S RNA	56
rrlA	23 S rRNA	86
rrlB	23 S rRNA	89
rrlC	23 S rRNA	84
rrlD	23 S rRNA	72
rrlE	23 S rRNA	90
rrlF	23 S rRNA	74
rrlG	23 S rRNA	56

Appendix C

Genetic Mapping

The usefulness of mutant analysis in studies of bacterial growth has been indicated in Chapter 4. One aspect of mutant analysis requires mapping—locating genes on the bacterial chromosome. Rough mapping is accomplished by techniques that involve interrupted mating (either the time of entry or the gradient of transmission of genes is measured), complementation by F-prime plasmids, and cotransfer by transduction or transformation. Fine structure mapping involves two-factor crosses (including the use of deletions), three-factor crosses, and measurement of the probability of recombination. We shall outline these various techniques in this appendix.

MAPPING BY TIME OF ENTRY

As discussed in Chapter 4, Hfr strains of enteric bacteria donate one preexisting strand of their chromosome to an F^- cell as the chromosome is being replicated. Replication and therefore entrance of the chromosome into the F^- cell occurs at a constant rate. Thus, the time required for a particular gene to enter the F^- cell is a direct measure of the gene's physical distance from the point of insertion of F into the chromosome. Chromosomes of enteric bacteria are divided into 100 theoretical units, with the zero point set arbitrarily near the genes of threonine (*thr*) and leucine (*leu*) biosynthesis. In *E. coli*, gene transfer at 37°C occurs at the rate of 1 unit/min; in *S. typhimurium*, transfer occurs at the rate of 0.77 units/min.

The time for a particular gene to be transferred and, therefore, its location on the genetic map can be determined by a technique known as an INTERRUPTED MATING. Cultures of Hfr and F^- cells are mixed to initiate the mating process. Because the union of the mating pairs is a fragile one, either the mating mixture is incubated in a culture flask, with gentle shaking, or it is filtered through a cellulose nitrate filter, which is incubated on the surface of an agar-filled petri plate. At various times the mating mixture is sampled, the sample is shaken vigorously to break apart the mating pairs, and aliquots of the sample are plated to score the number of recombinants that will develop, that is, the mated bacteria are plated on media that support the growth of recombinants but not the growth of either parent. The time of sampling that yields (on further incubation) the first recombinant of a particular gene represents the time of entry of that gene into the F⁻ cell from the Hfr cell and, therefore, is a measure of the distance between that gene and the site of insertion of the F plasmid into the Hfr chromosome. This information, along with knowledge of the point and orientation of insertion of F in the Hfr, is sufficient to map the genes. Results of an interrupted mating in which multiple samples were taken and two recombinant types were scored is shown in Figure 1. The time required for a particular gene to enter the F⁻ cell in the first mating pairs that formed can be measured by extrapolating the region of increase to the abscissa. But even then the measurement is accurate only to a unit or so—a region that contains approximately 35 genes (3500 genes / 100 units).

MAPPING BY GRADIENT OF TRANSMISSION

In interrupted mating the slope of the curve representing recombinants versus time of mating decreases progressively as the distance between the gene in question and the site of insertion of the F plasmid increases. Eventually the value of the number of recombinants reaches a maximum value, a plateau.

The plateau value of the number of recombinants for an allele decreases progressively with its distance from F in a manner consistent with the hypothesis that the decrease is a consequence of the spontaneous breaking apart of mating pairs, that is, the number of pairs that remain intact (N) over a period of time (t) would be expected to obey the formula:

$$N = N_0 e^{-kt}$$

where N_0 is the initial number of pairs and k is a rate constant describing the particular condition of incubation. Thus, the log of the maximum number of recombinants to be obtained for a particular marker would be expected to be a linear function of the time required for it to enter the F⁻. This fact can be used to map genes by GRADIENT OF TRANSMISSION.

Usually this simple, rapid technique of rough mapping is used to locate genes with respect to known genes in the same region of the chromosome. The procedure is simple (Figure 2). Hfr and F⁻ cells are mixed and allowed to mate long enough for the most distal gene of interest to enter the F⁻ in all mating pairs (i.e., for the number of recombinants to have reached their plateau value). The mating mixture is plated to select for recombinants of the most proximal marker. These are picked and scored to determine the number of more distal markers they also received. The log of the number of recombinants is plotted against the map position of previously located genes, and this standard curve is used to locate others.

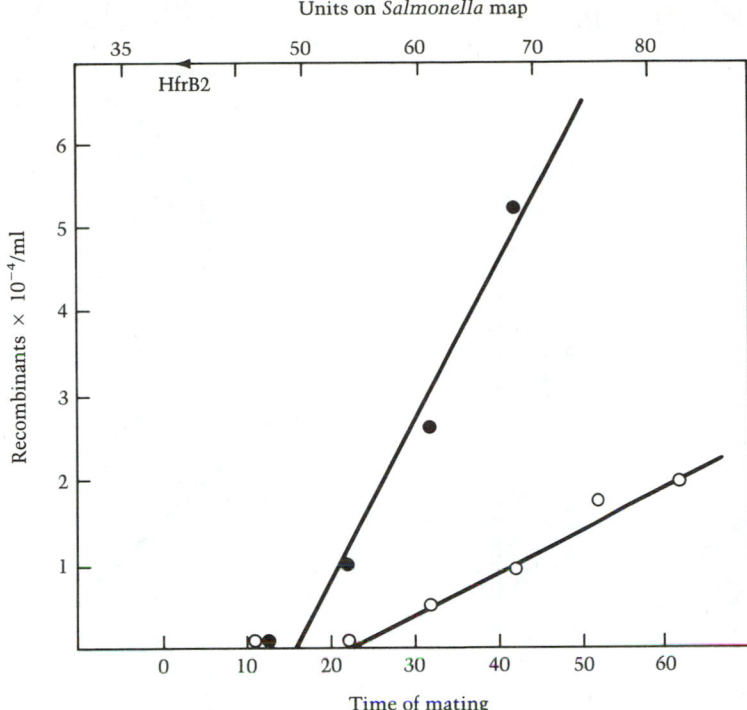

Figure 1. Kinetics of gene transfer in an interrupted Hfr × F⁻ mating. An interrupted mating between strains JL625 (Hfr B2, *arg⁻,pro⁻*) and JL435 (*F⁻ cod-8, cdd-9, gal⁻, pyrA81, pyrC1502, udk-6, udp-8, upp-101*). Mating was initiated by mixing cultures of the strains (at 0 minutes, lower abscissa scale). At various times, the mating mixture was sampled; mating pairs were broken apart by vigorous shaking of the sample; and the sample was plated, after appropriate dilution, on selective media that allowed growth of only the recombinant genotypes. The number of recombinants that were *udk⁺* (●) and *upp⁺* (○) were calculated from the number of colonies that developed on appropriate selective plates; these values were plotted according to the time at which the mating mixture was sampled (time of mating). The scale of the upper abscissa is in *Salmonella* map units (time of mating ÷ 0.77, see text); the site of insertion of the F plasmid (<), located in map units by previous experiments, is aligned with the time (0) at which the cultures were mixed. Thus, the map location of *udk* and *upp* can be determined directly from the time at which the curves start to rise (49 and 53 units, respectively).

MAPPING BY USE OF F′ STRAINS

Possibly the simplest rough mapping can be accomplished by the use of F′ strains. With a set of F′ strains that cover the entire chromosome, recessive alleles can be located quickly by a series of crosses between the F⁻ carrying the recessive allele and each of the set of F′ strains. By knowing the chromosomal segment carried by the F′ particles that complement the allele, it is roughly mapped.

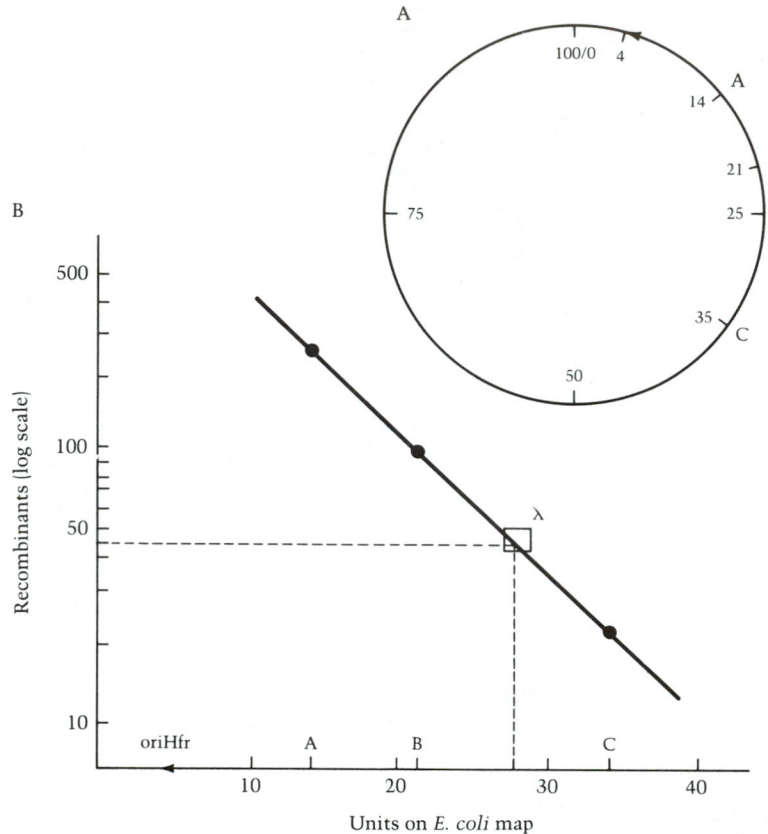

Figure 2. Mapping by gradient of transmission. A. The location of three genes
(A, B, and C) and the site (<) of insertion of the F plasmid (at 14, 21,
24, and 4 units, respectively) are shown on a map of the chromosome
of *E. coli*. A mating between this Hfr and an F⁻ using the gradient of
transmission procedure is allowed to proceed long enough for recombinants of the most distal gene (C) to have reached a maximal value
(approximately 20 minutes longer than the time at which this gene
enters F⁻ cells in the first mating pairs to form), in this case approximately 51 minutes (35 − 4 + 20; in the case of *E. coli* the transfer
replication occurs at 1 unit/min). The mating mixture is sampled and
plated on a selective medium upon which only recombinant clones
(Hfr genotype) of A grow. (This medium is constructed so that clones
bearing either allele (Hfr or F⁻) of B and C will develop. The clones
recombinant for A are then scored for their B, C, and X genotypes. B.
The logs of the number of A, B, and C recombinants are plotted
versus their known map location. The map position of X is determined by locating its number of recombinants on the curve and
reading its map position from the abscissa. Results of such experiments do not always yield the completely straight line depicted in
this idealized case, but the curve is usually sufficiently regular to
locate a gene on the linkage map with the accuracy of within a unit
or two if it lies between two genes of known location.

TWO-FACTOR CROSSES

After the approximate location of a gene on the chromosome has been established by one of the rough mapping techniques, TWO-FACTOR CROSSES are frequently done to locate it more precisely. The principle of mapping by two-factor crosses is simple. Transductional and transformational crosses are mediated by transfer of short fragments of donor DNA. If two genes are observed to be cotransferred, one can conclude that they are not separated on the chromosome by a distance greater than the length of the transferred piece. Because the pieces of DNA are random in transformational crosses and nearly so in generalized transductional crosses, frequencies of cotransfer (termed COTRANSFORMATION and COTRANSDUCTION, respectively) increase as distance between the markers decreases. Thus, cotransfer is a rough quantitative index of linkage that can be used to order a sequence of genes (Figure 3). If one assumes that the ends of transduced fragments are completely random, more precise estimates of linkage can be made from cotransduction frequencies (Table 1).

Mapping by two-point crosses is based on the assumption that the two

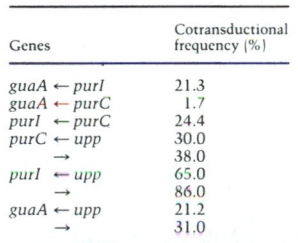

Genes	Cotransductional frequency (%)
guaA ← purI	21.3
guaA ← purC	1.7
purI ← purC	24.4
purC ← upp	30.0
→	38.0
purI ← upp	65.0
→	86.0
guaA ← upp	21.2
→	31.0

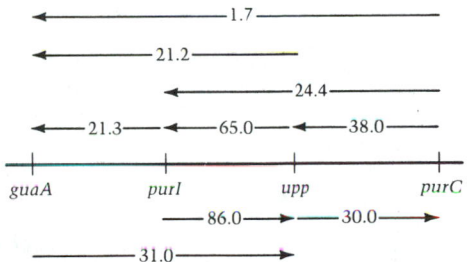

Figure 3. Mapping the relative position of genes by two-factor crosses. From a series of two-factor, phage P22-mediated crosses, cotransductional frequencies were determined between various pairs of the four *Salmonella typhimurium* genes (*guaA, purI, purC,* and *upp*), with the results shown in the box above. The arrows joining the pairs of genes that were cotransduced point in the direction of the selected marker from the donor strain. Note the significant variation of frequencies between reciprocal crosses; such differences are almost always noted, although the results from each direction of the cross are highly reproducible. In spite of these variations, the data can be used to order the genes by assuming only that frequency of cotransduction is inversely related to the physical distance between the markers. Distances are not additive. The genetic map (right) shows the arrangement of the genes indicated from cotransductional frequencies. Cotransductional frequencies (%) are shown within the arrows that point in the direction of the selected donor markers. In this case the results are internally consistent. In others they may not be, thus necessitating the employment of more precise mapping techniques such as three-factor crosses. (After Beck, 1971.)

Table 1. Theoretical relationship between physical distance between markers and their frequency of cotransduction by phages P1 and P22[a]

Percentage of cotransduction by		Distance between markers (kb)
P22	P1	
84	92	1.4
77	88	2.3
64	80	4.5
43	67	9.0
18	48	18
5	34	27
0.3	23	36
0	16	45

[a]Calculated from the formula (Kemper, 1974) $C = 1 - t + t\ln t$ in which C is the frequency of cotransduction and t is the linear distance between markers expressed as a fraction of the phage genome. The equation is based on the assumption that ends of generalized transducing particles are randomly located on the chromosome. The molecular size of phage P1 is 90 kb and of P22, 39 kb.

markers are carried on the same fragment of DNA—a good assumption in the case of transductional crosses because even in the case of efficient systems the probability of double infection (the product of the two separate probabilities) is small enough to be ignored. But in the case of transformation, coinheritance of genes derived from two separate DNA molecules (CONGRESSION) is significantly probable and must be considered in calculations of cotransformation. Frequency of congression can be determined from crosses in which one scores the frequency of inheritance of markers that are known to be unlinked by cotransformation. If the frequency of inheritance of markers of unknown linkage is significantly higher than the congressional frequency, they can be assumed to be cotransformed and, hence, to be closely linked. In most transformation crosses, frequency of congression lies in the range of a few percentage points—a value high enough to be useful in strain construction. Nonselectable mutations can be moved from one strain to another by screening clones transformed for a selectable mutation.

From the frequency of congression, the fraction of competent cells can be calculated. Because only competent cells receive donor DNA, the observed frequency of transformation (F) of a particular marker (A) is the ratio of the number of cells transformed for that marker (N_A) and the total number of cells (N_T) in the population, that is,

$$F_A = N_A/N_T \quad \text{or} \quad F_B = N_B/N_T$$

The observed frequency of congression is the product of the individual frequencies

$$F_{AB} = F_A \times F_B$$
$$N_{AB}/N_T = N_A N_B/N_T^2$$

However, only the fraction (K) of competent cells in the population participate in the cross; thus, if K is <1,

$$N_{AB} = N_A N_B/K N_T$$

or rearranging

$$K = N_A N_B/N_{AB} N_T = (N_A/N_T)(N_B/N_T)/(N_{AB} N_T) = F_A F_B/F_{AB}$$

If the recipient in a two-factor cross carries a deletion, wild-type recombinants will not be obtained unless the mutation in the donor lies outside the deletion. Thus, lack of wild-type recombinants in a cross between a strain carrying a point mutation and one carrying a deletion establishes immediately a region in which the point mutation must lie. If a set of overlapping deletions are available to be used as recipients, the location of a point mutation can be determined quickly and accurately by performing a small number of crosses (Figure 4). Such a mapping scheme is termed DELETION MAPPING.

THREE-FACTOR CROSSES

The most precise genetic technique for ordering markers is accomplished by the THREE-FACTOR CROSS. Analysis of three-factor crosses between procaryotes to determine the relevant position of mutations depends on two assumptions: (1) The immediate product of a procaryotic cross is almost always a merodiploid, and the exogenote is rarely a replicon. Therefore, an even number of crossovers (recombinational events) is required to obtain stable recombinant progeny. (2) Crossovers are somewhat infrequent. Therefore, if the three factors are closely linked, recombinant genotypes resulting from two crossovers are more frequent than those resulting from four. If all three markers can be independently scored, their relative position can be deduced from the number of crossovers required to produce a given combination of markers in the recombinant cell. Because crosses between procaryotes always involve the transfer of only a portion of the chromosome, at least two crossovers must occur for donated DNA to become incorporated into the genome of the recipient, but, in a three-point cross, for the two outside markers to be the donor genotype and the central one to be the recipient genotype, four crossovers must occur—a rarer event than two crossovers. Thus, by determining the rarest combination of markers in the recipient, the four-crossover event is identified and the gene order is established (Figure 5).

Often one wishes to order two mutations within the same gene with

A

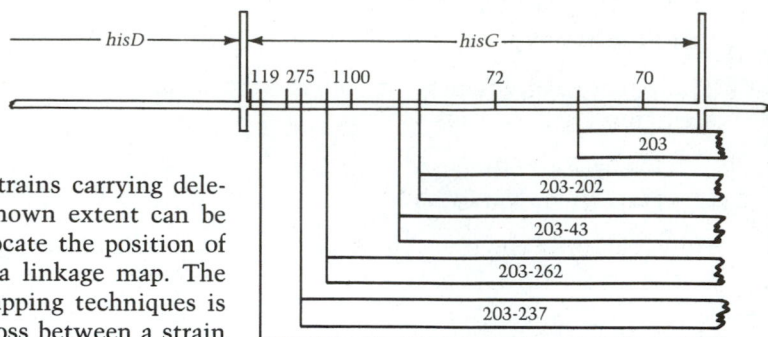

Figure 4.

Deletion mapping. Strains carrying deletion mutations of known extent can be used effectively to locate the position of other mutations on a linkage map. The principle of such mapping techniques is quite simple: in a cross between a strain that carries a certain mutation and one that carries a deletion, wild-type recombinants cannot be obtained if the mutation is in the region of the chromosome that is deleted from the other strain. A. The double horizontal line represents the *hisG* region of the chromosome of *Salmonella typhimurium*. Boxes below the line contain mutation numbers of deletion mutations, the left extent of which is indicated by the vertical line; the right-hand end extends to an unknown distance beyond *hisG*, as indicated by the jagged ends. Above the line are point mutations located by transductional crosses between

B

Prototrophic (wild-type) recombinants obtained

Deletion mutation	Point mutation				
	119	275	1100	72	70
203	+	+	+	+	−
203–202	+	+	+	−	−
203–43	+	+	+	−	−
203–262	+	+	−	−	−
203–237	+	+	−	−	−
203–9	+	−	−	−	−

strains carrying them and strains that carry deletions as summarized in B. B. Pluses and minuses indicate whether wild-type (prototrophic) recombinants were or were not obtained. The results of these 25 crosses are sufficient to construct the fine structure map of point mutations shown in A. For example, *hisG119* can be located to the left of the end of *hisG203–9* because prototrophic recombinants are obtained from crosses between them: *hisG275* lies between the ends of *hisG203–9* and *hisG203–237* because a cross between *hisG275* and *hisG203–9* does not yield prototrophic recombinants and one between it and *hisG203–237* does.

---

dogenote. Note that three of the recombinant classes require only two crossovers; one (class 4; a^+, b^-, c^+) requires four. Thus, in such a cross, class 4 would have few representatives. Alternatively, if the gene order were not known and if the rare class were found to be class 4, one could deduce that the gene order was $a-b-c$. Were a^+,b^+,c^- recombinant found to belong to the rare class, the indicated gene order would have been $a-c-b$. B. Data from a three-factor conjugational cross (Beck, 1972) are shown. The Hfr strain (JL417) carries the alleles $argE^+$, cod^+, and $strA^+$; and the recipient carries $argE^-$, cod^-, and $strA^-$. From the cross, $argE^+$ recombinants were selected; and of these, the allelic forms of *cod* and *strA* were scored. The recombinant class with the fewest representatives (4) is $argE^+$, cod^-, $strA^+$; thus, four crossovers are required to generate it and the gene order must be $argE-cod-strA$.

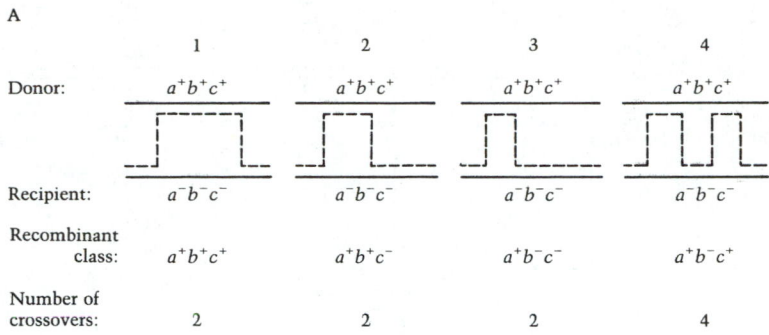

A

	1	2	3	4
Donor:	$a^+b^+c^+$	$a^+b^+c^+$	$a^+b^+c^+$	$a^+b^+c^+$
Recipient:	$a^-b^-c^-$	$a^-b^-c^-$	$a^-b^-c^-$	$a^-b^-c^-$
Recombinant class:	$a^+b^+c^+$	$a^+b^+c^-$	$a^+b^-c^-$	$a^+b^-c^+$
Number of crossovers:	2	2	2	4

B

JL417(HfrK3, *serA13*, *glpD*⁻) × JL885(*pyrC7*, *cod-101*, *argE*, *strA*⁻)

Genetic composition of class	Number of clones
argE⁺, *cod*⁺, *strA*⁺	147
argE⁺, *cod*⁻, *strA*⁺	4
argE⁺, *cod*⁺, *strA*⁻	131
argE⁺, *cod*⁻, *strA*⁻	68

C. Indicated gene order and crossover pattern of rare class

Donor:	*argE*⁺	*cod*⁺	*strA*⁺
Recipient:	*argE*⁻	*cod*⁻	*strA*⁻

Figure 5. Analysis of a three-factor cross in which each mutation can be independently scored. The analysis of three-factor crosses between procaryotes to determine the relative position of the mutations depends on two assumptions: (1) Because the immediate product of a procaryotic cross is almost always a merodiploid and because the exogenote is rarely a replicon, an even number of crossovers (recombinational events) is required to obtain stable recombinant progeny. (2) Because crossovers are somewhat infrequent, if the three factors are closely linked, recombinant genotypes resulting from two crossovers are more frequent than those requiring four. Therefore, following a three-factor cross (by transduction, conjugation, or transformation), if one first scores the number of representatives of each possible recombinant class and then identifies the class with the fewest representatives, one can deduce the gene order by determining the arrangement of genes that would require four crossovers to generate the rare class. A. The four diagrams represent portions of merodiploids resulting from a cross between a donor strain with genotype, a^+, b^+, c^+, and a recipient, a^-, b^-, c^-. The gene order is assumed to be $a-b-c$. The dashed lines follow the regions of DNA that are inherited in the four possible a^+ recombinant classes; each vertical portion of the line represents a site of recombination (crossover) between exo- and en-

A

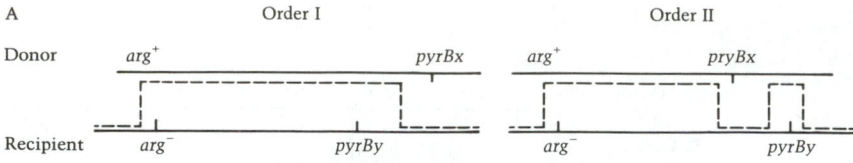

B

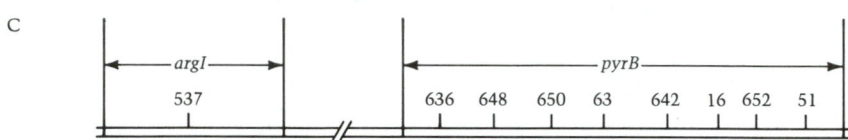

Donor *pyrB* mutation	Recipient *pyrB* mutation			
	648	650	642	652
636	18	17	10	6
63	69	65	11	18
16	69	57	54	22
51	76	65	65	64

C

Figure 6. Scheme of mapping by three-factor crosses when not all mutations can be scored independently. The analysis of three-factor crosses designed to align mutations within a single gene is complicated by the fact that the mutations usually all have the same phenotype and, therefore, cannot be easily scored. A. One approach is to make a cross in which one of the intragenic mutations (*pyrBx*) is carried in the donor; the recipient carries a different mutation (*pyrBy*) in that gene as well as a mutation in a closely linked gene (*argI*). Two crossovers are required to generate wild-type recombinants in Order I and four are required in Order II. Thus, in Order I the frequency of *argI*[+] recombinants among total *pyrB*[+] recombinants will approach the cotransductional frequency between *argI* and *pyrB* that one would observe in a two-factor cross. (The frequency will be somewhat depressed because one crossover is restricted to the region between *pyrBy* and *pyrBx*.) In Order II, the frequency of *argI*[+] recombinants will be markedly depressed because four crossovers are required. B. This system has been used to construct a fine structure map of *pyrB*. Cotransduction of *pyrB* and *argI* by phage P22 is 75%. In a set of three-factor, phage P22-mediated crosses of the type shown in A, the percentage of *argI*[+] recombinants among the total *pyrB*[+] class is shown. The values fall into two groups: one falling in the range of 54 to 76% and the other, 6 to 22%. The set of higher values (to the left of the diagonal line) can be presumed to have required two crossovers; the latter (to the right of the line) from four crossovers. Thus, for example, *pyrB63* can be located on the map (C) to the right (*argI* distal) of *pyrB650* but to the left of *pyrB642*. These data (a portion of those published by Syvanen, 1973) establish an internally consistent fine structure map.

respect to a neighboring gene. A different analysis must be used in such a case because the two mutations in the same gene usually express the same phenotype and thus cannot be independently scored. The relative position of the two markers is deduced from a cross in which one of the intragenic markers is carried by the donor and the other by the recipient. Wild-type transductants are selected and cotransduction with the other gene is scored. In one gene arrangement, two crossovers will be required for cotransduction, and in the other, four are required. In the gene arrangement requiring two crossovers, the frequency of cotransduction is higher than in the gene arrangement requiring four crossovers. Thus, by comparing the apparent cotransductional frequency in this three-factor cross with that of a two-factor cross, gene order can be deduced (Figure 6).

In practice, certain crosses of this sort will give intermediate values of linkage. In these cases it is not clear whether two or four crossovers have occurred. In such cases unequivocal placement of the mutations can be made by performing RECIPROCAL THREE-FACTOR CROSSES: in one cross *pyrBx* is on the donor and *argI⁻*, *pyrBy* is on the recipient; in the other *pyrBy* is on the donor and *argI⁻*, *pyrBx* is on the recipient. One arrangement will require two crossovers to generate *argI⁺*, *pyrB⁺*; the other will require four. The arrangement requiring four crossovers will generate a significantly lower percentage of *argI⁺* among the *pyrB⁺* class.

MAPPING BY PROBABILITY OF RECOMBINATION

The unit of genetic distance between genes on a eucaryotic chromosome is the CENTIMORGAN (RECOMBINATION UNIT), the distance over which there is a 1% frequency of crossover in a genetic cross. Knowing distances in centimorgans between various genes on the eucaryotic chromosome allows the ordering of genes because probability of recombination is roughly proportional to physical distance on the chromosome, and recombination is sufficiently rare that double or triple recombinational events do not complicate the measurement. Measured in a single cross, map distances cannot exceed 50 centimorgans because of multiple crossovers. Thus, recombinational mapping is limited to regions of less than 50 centimorgans—a long distance on the eucaryotic chromosome but quite a short one on the chromosome of enteric bacteria. Each minute or unit of chromosome of *E. coli* is approximately 20 centimorgens (Hayes, 1968), and the entire chromosome is approximately 2000 centimorgans. Conventional recombinational mapping is therefore limited to a couple of units and hence is not as useful as other techniques. However, frequency of recombination is much lower in *Streptomyces* species. Although the physical length of the chromosome of *Streptomyces coelicolor* is approximately the same as that of *E. coli*, its genetic length is approximately 200 centimorgans; therefore recombinational mapping is the principal way in which genes are located on the chromosome of this bacterium (Hopwood, 1973).

PROBLEMS

1. The following cotransductional frequencies were obtained in a series of two point crosses:

Genes	Cotransduction (%)
cmlB and *pyrD*	54
pyrD and *fabA*	50
aroA and *cmlB*	26
aroA and *pyrD*	5
aroA and *fabA*	2

What is the probable gene order?

2. A three-factor transductional cross was made between JL382 (*pyrC1502, upp-17, purC7*) and JL377 (*purI305*); *purI*$^+$ recombinants were selected and among these, *purC* and *upp* alleles were scored with the following results:

Donor	Recipient	Genotype	Number of recombinants
JL382	JL377	*purI*$^+$, *purC*$^-$, *upp*$^+$	0
		purI$^+$, *purC*$^-$, *upp*$^-$	81

What is the probable gene order of the three genes involved in the cross?

3. In a cross designed to determine the frequency of cotransduction between *fol* and *pyrA*, the donor genotype was *fol*$^-$ and the recipient was *pyrA*$^-$. The *pyrA* allele confers a growth requirement for pyrimidine base and arginine; the *fol*$^-$ allele renders a strain resistant to aminopterin, a toxic analogue of folic acid. How would you carry out the cross and score the classes of recombinants that would allow frequency of cotransduction to be calculated? Assuming that you have a basal medium that contains all salts necessary to support growth of a prototroph, what additions would you make to this medium to select and score the various genotypes?

REFERENCES

Hayes, W. 1968. *The Genetics of Bacteria and their Viruses.* John Wiley & Sons, Inc., New York.

Hopwood, D. A., K. F. Chater, J. E. Dowding and A. Vivian. 1973. Advances in *Streptomyces coelicolor* genetics. Bact. Revs. 371.

Appendix D

A Genetic Approach to Characterizing Complex Promoters in E. coli[1]

JON BECKWITH

DEPARTMENT OF MICROBIOLOGY
AND MOLECULAR GENETICS
HARVARD MEDICAL SCHOOL, BOSTON, MASSACHUSETTS

The initiation of transcription of genes or operons takes place at sites termed promoters. The analysis of a variety of regulatory systems in *E. coli* has revealed that many promoters are complex, often with several protein factors interacting to stimulate transcription. For example, the promoter of the *lac* operon comprises two well defined sites, one where RNA polymerase interacts and a second where the positive control protein CAP binds in the presence of 3',5'-cyclic AMP to enhance transcription initiation (Beckwith et al., JMB 69, 155–160, 1972). In the case of the *ara* operon, at least two positive control factors are involved in promoting initiation of transcription: the positive control protein specific to the arabinose system (the *araC* gene product), and also the CAP protein (Englesberg and Wilcox, Ann. Rev. Genet. 8, 219–242, 1974).

The suggestion that the CAP protein interacts with a separate site from RNA polymerase in the *lac* promoter came initially from the isolation and characterization of promoter mutants. It is my belief that the genetic analysis used in that case should be applicable to other regulatory systems, such as those described above. However, this approach has not been used, and, as a result, the structure of these promoters—the assignment of different regions for the interaction of different proteins—either has not been elucidated or has not been firmly established. Binding studies can be suggestive of a particular promoter structure, but they do not establish that the binding sites reflect the

[1] This article originally appeared in *Cell* 23:307–308 (1981). It is reprinted here with the permission of the author and The MIT Press, Cambridge, MA.

in vivo situation. The fact that the description of this type of genetic analysis is scattered in several articles may have prevented recognition of its utility. Furthermore, the newer techniques of gene fusion make this approach much more widely applicable. For these reasons, I wish to describe the steps that should permit the relatively rapid characterization of promoter structure.

The basic outline of the approach is as follows. First, leaky promoter mutants for a particular gene or operon are isolated (Scaife and Beckwith, CSHS *31*, 403–408, 1966). Specifically, these leaky mutants allow expression of the attached gene(s) at about 5% of the normal level. Second, the promoter mutations are mapped by use of deletions that define different regions of the promoter. Third, the promoter mutations are introduced into strain backgrounds that are mutant for one or another of the positive control factors for that particular operon. Since the mutations are only leaky, it is still possible to measure the effect of the absence of a positive control factor on the expression of the operon. This effect can be compared to the effect seen on a wild-type promoter. For example, in the *lac* operon, the expression of the wild-type promoter is reduced approximately 50 fold by the introduction of a mutation in the structural gene (*crp*) for CAP, which eliminated CAP activity (Beckwith et al., loc. cit.). When a promoter mutation that reduces the ability of RNA polymerase to interact with the *lac* promoter is combined with a crp^- mutation, there is still a 50 fold effect compared to a crp^+ background. The CAP-binding site, and therefore CAP stimulation, are still intact. However, when a promoter mutation in the CAP-binding site is combined with a crp^- mutation, there is only a 3 fold reduction in *lac* operon expression compared to a crp^+ background. Clearly, the ability of CAP to stimulate the operon has largely been lost in this second class of promoter mutants.

ISOLATION OF LEAKY PROMOTER MUTANTS

In practice, there are serious or insuperable limitations to the isolation of leaky promoter mutants in the case of most genes or operons. First, for many genes a level of 5% leakiness might permit growth of the organism at a rate essentially indistinguishable from wild-type. Second, even if the mutants are detectable, in the case of those operons that comprise only one gene, it is difficult to distinguish leaky promoter mutations from structural gene mutations that reduce the activity of the gene product (for example, those that result in a partially defective enzyme). Third, the assay for the gene product(s) of the operon may not be sensitive enough to carry out the kinds of studies with various positive control factor mutants described above.

In order to avoid these problems, I suggest that the first step in the analysis be the construction of operon fusions between the *lac* operon and the gene(s) of interest. A number of in vivo and in vitro techniques exist for the construction of such fusions (see for example Casadaban, JMB *104*, 541–555, 1976; Holowachuk et al., PNAS 77, 2124–2128, 1980). The expression of the *lac* operon provides a measure of the functioning of the promoter to which it is attached. In addition, the selective techniques available for the *lac* operon are applicable to the fusion strains. Finally, the assay for β-galactosidase provides a sensitive means of measuring effects of regulatory mutations on the expression of the particular promoter.

A combination selection and screening technique exists for detecting leaky promoter mutants in such fusions. The compound thio-orthonitrophenyl-galactoside (TONPG) is an analog of lactose, which, if accumulated in the bacterial cell, causes growth stasis. The transport of TONPG into the cell is dependent on the β-galactoside permease, product of the *lacY* gene. By including TONPG in the growth media at the proper concentration, one can select for surviving *E. coli* mutants that have as much as 10–15% *lacY* gene expression (Hopkins, JMB 87, 715–724, 1974). If the selection is done on solid media that includes the β-galactosidase indicator dye, 5-bromo-3-chloro-indolyl-β-D-galactoside (XG), one can also screen for the level of expression of the *lacZ* gene. Pale blue colonies are those that also have reduced levels of β-galactosidase activity. In this way, one can very quickly detect pleiotropic mutants, which are simultaneously reduced in the expression of *lacZ* and *lacY*. (Some of the mutants detected by this scheme may be polar mutations in a gene of the operon to which *lac* is fused. These can be eliminated by introducing into the mutant fusion strain either a mutation in the rho protein or amber or ochre suppressors, any of which will suppress polarity effects.)

The advantages of this selection and screening cannot be overemphasized. First, the direct selection permits the isolation of spontaneous mutants. Since spontaneous mutations comprise all classes of mutations including transversions, the spectrum of possible changes in the promoter should be detectable. Second, it may well be that any screening or selection procedure that relies on the very strong effects of a mutation on the expression of the gene or operon will miss all or most promoter mutants. Most of the *lac* operon promoter mutants were isolated as leaky mutations. By a different approach we were able to isolate rare *lac⁻* promoter mutants in which the expression of the *lac* operon was less than 1% of normal. The two that have so far been sequenced alter the most conserved base in the Pribnow box of the *lac* promoter (R. Johnson, W. S. Reznikoff, A. Gruyer, H. Shuman and J. Beckwith, unpublished observations). These results suggest that nearly all promoter mutants due to single base changes will be leaky mutants, and that it is only the very rare change in a particularly necessary base pair that will show up in any general screening of fully negative mutants.

This approach has been used successfully to isolate mutations in the promoter of the structural gene for a tyrosine tRNA (Berman and Beckwith, JMB *130*, 285–301, 1979; Berman and Beckwith, JMB *130*, 303–315, 1979), using fusions of the *lac* operon to the *tyrT* gene. These papers describe promoter mutants that express the *tyrT* gene at 1–11% of its normal level and give details of the isolation and characterization procedure.

CHARACTERIZATION OF PROMOTER MUTANTS

Once the promoter mutants have been isolated, it should be possible to introduce into the strains mutations in the positive control factors that are required for the expression of the particular gene or operon. As described above, assays of the effect of such mutations on β-galactosidase synthesis should indicate which component of the promoter is affected by the promoter mutation.

Finally, the mutations must be mapped to their sites within the promoter. This can be done in the fusion-promoter mutant strains themselves using the selective or screening techniques of the *lac* operon, or the promoter mutations may be recombined back into the original gene (Berman and Beckwith, op. cit., pp. 303–315), if the mapping is convenient there. Alternatively, DNA sequencing can be done with the mutations to establish their precise location.

I believe that this analysis will yield information essential to understanding the functioning of complex promoters. In addition, for those cases in which a eucaryotic promoter does function in *E. coli*, the same approach may be possible for a genetic characterization of the regulatory region.

Credits for Figures and Tables

Chapter 1

de Boer, W. E., C. Golten and W. A. Scheffers. 1975. Effects of some physical factors on flagellation and swarming of *Vibrio alginolyticus*. Nether. Sea. Res. 9:197.

Bloch, P. L., T. A. Phillips, and F. C. Neidhardt. 1980. Protein identifications on O'Farrell two-dimensional gels: locations of 81 *Escherichia coli* proteins. J. Bacteriol. 141:1409.

Bochner, B. R. and B. N. Ames. 1982. Complete analysis of cellular nucleotides by two-dimensional thin layer chromatography. J. Biol. Chem. 257:9759.

Cairns, J. 1963. The chromosome of *Escherichia coli*. Cold Spring Harbor Symposium on Quant. Biol. 28:43.

Clark, B. F. C. 1980. Structure of tRNA during protein synthesis. In *Ribosomes: Structure, Function and Genetics*. G. Chambliss, G. R. Craven, J. Davies, K. Davis, L. Kahan and M. Nomura, University Park Press, Baltimore.

Dennis, P. P. and H. Bremer. 1974. Macromolecular composition during steady-state growth of *Escherichia coli* B/r. J. Bacteriol. 119:270.

DePamphilis, M. L. and J. Adler. 1971a. Fine structure and isolation of the hook-basal body complex of flagella from *Escherichia coli* and *Bacillus subtilis*. J. Bacteriol. 105:384.

DePamphilis, M. L. and J. Adler. 1971b. Attachment of flagellar basal bodies to the cell envelope: specific attachment to the outer, lipopolysaccharide membrane and the cytoplasmic membrane. J. Bacteriol. 105:396.

DiRienzo, J. M., K. Nakamura and M. Inouye. 1978. The outer membrane of proteins of Gram-negative bacteria: biosynthesis, assembly and functions. Ann. Rev. Biochem. 47:481.

Gausing, K. 1976. Synthesis of rRNA and r-protein mRNA in *E. coli* at different growth rates. In *Control of Ribosome Synthesis. Proceedings of the Alfred Benzon Symposium IX*, Copenhagen, 1975. N. O. Kjeldgaard and O. Maaløe, eds. Academic Press, New York.

Ghuysen, J. N. 1968. Bacteriolytic enzymes in the determination of wall structure. Bacteriol. Rev. 32:425.

Kavenoff, R. and O. Ryder. 1976. Electron microscopy of membrane-associated folded chromosomes of *Escherichia coli*. Chromasoma 55:13.

Koshland, D. E., Jr. 1980. *Bacterial Chemotaxis as a Model Behavioral System*. Raven Press, New York.

Maaløe, O. 1979. Regulation of the protein synthesizing machinery—ribosomes, tRNA, factors, and so on. In *Biological Regulation and Development*, Vol. I, *Gene Expression*. R. F. Goldberger, ed. Plenum Press, New York.

Miller, O. L., Jr., B. A. Hamkalo and C. A. Thomas, Jr. 1980. Visualization of bacterial genes in action. Science 169:392.

Nikaido, H. 1973. Biosynthesis and assembly of lipopolysaccharide. In *Bacterial Membranes and Walls*. L. Leive, ed. Marcel Dekker, New York. Material on p. 137 of this reference is reprinted by courtesy of Marcel Dekker, Inc.

Roberts, R. B., R. H. Abelson, D. B. Cowie, E. T. Bolton and R. J. Britten. 1955. Studies of Biosynthesis of *Escherichia coli*. Carnegie Inst. Washington. Publ. 607.

Stanier, R. Y., E. A. Adelberg and J. L. Ingraham. 1976. *The Microbial World*, 4th Edition. Prentice-Hall, Inc., Englewood Cliffs, New Jersey.

Umbarger, H. E. 1977. A one-semester project for the immersion of graduate students in metabolic pathways. Biochem. Education 5:67.

Van Ness, J. and D. E. Pettijohn. 1979. A simple autoradiographic method for investigating long range chromosome substructure: size and number of DNA molecules in isolated nucleoids of *Escherichia coli*. J. Mol. Biol. 129:501.

Watson, S. W. and M. Mandel. 1971. Comparison of the morphology and deoxyribonucleic acid composition of 27 strains of nitrifying bacteria. J. Bacteriol. 107:563.

Zeikus, J. G., and V. G. Bowen. 1975. Fine structure of *Methanospirillum hungatii*. J. Bacteriol. 121:373.

Chapter 2

Apirion, D., B. K. Ghora, G. Plantz, T. K. Misra and D. Gegenheimer. 1980. Processing of rRNA and tRNA in *Escherichia coli*: cooperation between processing enzymes. In *tRNA: Biological Aspects*. Cold Spring Harbor Laboratory, Cold Spring Harbor, New York.

Davis, B. D., R. Dulbecco, H. N. Eisen, and H. S. Ginsberg. 1980. *Microbiology*, 3rd Edition. Harper & Row, Hagerstown, Maryland.

DePamphilis, M. L. and J. Adler. 1971a. Fine structure and isolation of the hook-basal body complex of flagella from *Escherichia coli* and *Bacillus subtilis*. J. Bacteriol. 105:384.

DePamphilis, M. L. and J. Adler. 1971b. Attachment of flagellar basal bodies to the cell envelope: specific attachment to the outer lipopolysaccaride membrane and the cytoplasmic membrane. J. Bacteriol. 105:396.

Held, W. A., B. Ballou, S. Mizushima and M. Nomura. 1974. Assembly mapping of 30S ribosomal proteins from *Escherichia coli*. J. Biol. Chem. 239:3103.

Inouye, M., J. DiRienzo, J. Maida, R. Mouva, K. Nakamura, N. Lee, R. Pirtle and I. Pirtle. 1980a. Secretion of outer membrane proteins of *Escherichia coli* across the cytoplasmic membrane. In *Precursor Processing in the Biosynthesis of Proteins*. M. Zimmerman, R. A. Mumford and R. F. Steiner, eds. Ann. N.Y. Acad. Sci. 343:362.

Inouye, M. and S. Halegoua. 1980b. Secretion and membrane localization of proteins in *Escherichia coli*. CRC Critical Reviews in Biochemistry, p. 339.

Képès, A. and F. Autissier. 1972. Topology of membrane growth in bacteria. Biochem. Biophys. Acta 265:443.

Laughrea, M. and P. B. Moore. 1978. Ribosomal components required for binding protein S1 to the 30S subunit of *Escherichia coli*. J. Mol. Biol. 122:109.

Lindahl, L. and J. M. Zengel. 1982. Expression of ribosomal genes in bacteria. Adv. Genet. 21:53.

Nierhaus, K. H. 1982. Structure, assembly and function of ribosomes. In *Current Topics in Microbiology and Immunology*, Vol. 97. W. Henle, P. H. Hofschneider, H. Koprowski, F. Melchers, R. Rott, H. G. Schweiger and P. K. Vogt, eds. Springer-Verlag, New York.

Osborn, M. J. and H. C. P. Wu. 1980. Proteins of the outer membrane Gram-negative bacteria. Ann. Rev. Biochem. 34:369.

Silverman, M. and M. Simon. 1977. Bacterial flagella. Ann. Rev. Biochem. 31:397.

Wickner, W. 1979. The assembly of proteins into biological membranes: the membrane trigger hypothesis. Ann. Rev. Biochem. 43:23.

Wu, H. C., J. J. C. Lin, P. K. Chattophadhyay, and H. Kanazawa. 1980. Biosynthesis and assembly of murein lipoprotein in *Escherichia coli*. In *Precursor Processing in the Biosynthesis of Proteins*. M. Zimmerman, R. A. Mumford and D. F. Steiner, eds. Ann. N.Y. Acad. Sci. 343:362.

Chapter 3

Alberts, B. and R. Sternglanz. 1977. Recent excitement in the DNA replication problem. Nature 269:655.

Altman, S. 1978. Biosynthesis of tRNA. In *Transfer RNA*. S. Altman, ed. MIT Press, Cambridge, Massachusetts.

Apirion, D., B. K. Ghora, G. Plantz, T. K. Misra and D. Gegenheimer. 1980. Processing of rRNA and tRNA in *Escherichia coli*: cooperation between processing enzymes. In *tRNA: Biological Aspects*. Cold Spring Harbor Laboratory, Cold Spring Harbor, New York.

Atkinson, D. E. 1968. The energy charge of the adenylate pool as a regulatory parameter. Interaction with feedback modifiers. Biochem. 7:430.

Dennis, P. P. and H. Bremer. 1974. Macromolecular composition during steady-state growth of *Escherichia coli* B/r. J. Bacteriol. 119:270.

Dills, S. S., A. Apperson, M. R. Schmidt and M. H. Saier. 1980. Carbohydrate transport in bacteria. Microbiol. Rev. 44:385.

Fersht, A. R., J. W. Knill-Jones and W.-C. Tsui. 1982. Kinetic basis of spontaneous mutation. Misinsertion frequencies, proofreading specificities and cost of proofreading by DNA polymerases of *Escherichia coli*. J. Mol. Biol. 156:37.

Gerhart, J. C. and A. B. Pardee. 1962. The enzymology of control by feedback inhibition. J. Biol. Chem. 237:891.

Haddock, B. A. and C. W. Jones. 1977. Bacterial respiration. Bacteriol. Rev. 41:47.

Hanawalt, P. C. 1972. Repair of genetic material in living cells. Endeavour 32:83.

Hinkle, P. C. and R. E. McCarthy. 1978. How cells make ATP. Sci. Am. 238, no. 3:104. Copyright © (1978) by Scientific American, Inc. All rights reserved.

Howard-Flanders, P. 1981. Inducible repair of DNA. Sci. Am. 245, no. 5:72. Copyright © (1981) by Scientific American, Inc. All rights reserved.

Ogawa, T. and T. Okazaki. 1980. Discontinuous DNA replication. Ann. Rev. Biochem. 49:421.

Pribnow, D. 1979. Genetic control signals in DNA. In *Biological Regulation and Development*. R. F. Goldberger, ed. Plenum Press, New York.

Roberts, R. B., R. H. Abelson, D. B. Cowie, E. T. Bolton and R. J. Britten. 1955. Studies of Biosynthesis of *Escherichia coli*. Carnegie Inst. Washington. Publ. 607.

Rosen, B. P. and E. R. Kashket. 1978. Energetics of bacterial transport. In *Bacterial Transport*. B. P. Rosen, ed. Marcel Dekker, Inc., New York. Material on pages 571, 573 and 574 of this reference are reprinted courtesy of Marcel Dekker, Inc.

Siebenlist, U., R. B. Simpson and W. Gilbert. 1980. *E. coli* RNA polymerase interacts homologously with two different promoters. Cell 20:269.

Steitz, J. A. 1979. Genetic signals and nucleotide sequences in messenger RNA. In *Biological Regulation and Development*. R. F. Goldberger, ed. Plenum Press, New York.

Umbarger, H. E. 1977. A one-semester project for the immersion of graduate students in metabolic pathways. Biochem. Education 5:67.

Umbarger, H. E. 1978. Amino acid biosynthesis and its regulation. Ann. Rev. Biochem. 47:533.

Wood, W. A. 1961. Fermentation of carbohydrates and related compounds. In *The Bacteria*, Vol. 2. I. C. Gunsalus and R. Y. Stanier, eds. Academic Press, New York.

Chapter 4

Alper, M. D. and B. N. Ames. 1975. Positive selection for mutants with deletions of the *gal-chl* region of the *Salmonella* chromosome as a screening procedure for mutagens that cause deletions. J. Bacteriol. 121:259.

Anderson, R. P. and J. R. Roth. 1977. Tandem genetic duplications in phage and bacteria. Ann. Rev. Microbiol. 31:473.

Bukhari, A. I., J. A. Shapiro and S. L. Adhya (eds.). 1977. *DNA: Insertion Elements, Plasmids and Episomes*. Cold Spring Harbor Laboratory, Cold Spring Harbor, New York.

Chelala, C. A. and P. Margolin. 1974. Effects of deletions on cotransduction linkage in *Salmonella typhimurium*: Evidence that bacterial chromosome deletions affect the formation of transducing DNA fragments. Molec. Gen. Genet. 131:97.

Cohen, S. N., A. C. Chang and L. Hsu. 1972. Nonchromosomal antibiotic resistance in bacteria. Genetic transformation of *E. coli* by R-factor DNA. Proc. Natl. Acad. Sci. USA 69:2110.

Crawford, J. P. 1975. Gene arrangement in the evolution of the tryptophan synthetic pathway. Bacteriol. Revs. 39:87.

Henner, D. J. and J. A. Hoch. 1980. The *Bacillus subtilis* chromosome. Microbiol. Revs. 44:57.

Hill, C. W., R. H. Grafstrom and B. S. Hillman. 1977. Chromosomal rearrangements resulting from recombination between ribosomal RNA Genes, p. 497. In *DNA Insertion Elements, Plasmids, and Episomes*. A. I. Bukhari, J. A. Shapiro and S. L. Adhya, eds. Cold Spring Harbor Laboratory, Cold Spring Harbor, New York.

Jackson, E. N., F. Laski and C. Andres. 1982. Bacteriophage P22 mutants that alter the specificity of DNA packaging. J. Mol. Biol. 154:551.

Royle, P. L., H. Matsumoto and B. W. Holloway. 1981. Genetic circularity of the *Pseudomonas aeruginosa* PAO chromosome. J. Bacteriol. 145:145.

Schmieger, H. 1972. Phage P_{22} mutants with increased or decreased transduction abilities. Molec. Gen. Genet. 119:75.

Shapiro, J. A. 1977. F, the *E. coli* sex factor, p. 671. In *DNA Insertion Elements, Plasmids, and Episomes*. A. I. Bukhari, J. A. Shapiro and S. L. Adhya, eds. Cold Spring Harbor Laboratory, Cold Spring Harbor, New York.

Chapter 5

Andersen, K. B. and K. von Meyenburg. 1980. Are growth rates of *Escherichia coli* in batch culture limited by respiration? J. Bacteriol. 144:114.

Herendeen, S. L., R. A. VanBogelen and F. C. Neidhardt. 1979. Levels of major proteins of *Escherichia coli* during growth at different temperatures. J. Bacteriol. 139:185.

Marr, A. G. and J. L. Ingraham. 1962. Effect of temperature on the composition of fatty acids in *Escherichia coli*. J. Bacteriol. 84:1260.

Marr, A. G., E. H. Nilson and D. J. Clark. 1963. The maintenance requirement of *Escherichia coli*. Ann. N.Y. Acad. Sci. 102:536.

Pirt, L. J. 1965. The maintenance energy of bacteria in growing cultures. Proc. Roy. Soc. B 163:224.

Shehata, T. E. and A. G. Marr. 1970. Synchronous growth of enteric bacteria. J. Bacteriol. 103:789.

Taketa, K. and M. Pogell. 1965. Allosteric inhibition of rat liver fructose-1,6-diphosphatase by adenosine-5'-monophosphate. J. Biol. Chem. 240:651.

Chapter 6

Bird, R. E., J. Louarn, J. Martuscelli and L. Caro. 1972. Origin and sequence of chromosome replication in *Escherichia coli*. J. Mol. Biol. 70:549.

Chandler, M., R. E. Bird and L. Caro. 1975. The replication time of the *Escherichia coli* K12 chromosome as a function of cell doubling time. J. Mol. Biol. 94:127.

Cooper, S. and C. E. Helmstetter. 1968. Chromosome replication and the division cycle of *Escherichia coli* B/r. J. Mol. Biol. 31:519.

Donachie, W. D. 1968. Relationship between cell size and time of initiation of DNA replication. Nature 219:1077.

Ecker, R. E. and M. Schaechter. 1963. Bacterial growth under conditions of limited nutrition. Ann. N.Y. Acad. Sci. 102:549.

Engbaek, F., N. O. Kjeldgaard and O. Maaløe. 1973. Chain growth rate of β-galactosidase during exponential growth and amino acid starvation. J. Mol. Biol. 75:109.

Helmstetter, C. E. and S. Cooper. 1968. DNA synthesis during the division cycle of rapidly growing *Escherichia coli* B/r. J. Mol. Biol. 31:507.

Jacobsen, H. 1974. Thesis, Copenhagen University.

Maaløe, O. and N. O. Kjeldgaard. 1966. *Control of Macromolecular Synthesis*. W. A. Benjamin, New York.

Maaløe, O. 1979. Regulation of the protein synthesizing machinery—ribosomes, tRNA, factors, and so on. In *Biological Regulation and Development*, Vol. I. *Gene Expression*. R. F. Goldberger, ed. Plenum Press, New York.

Molin, S. 1976. Ribosomal RNA chain elongation rates in *Escherichia coli*. In *Control of Ribosome Synthesis. Proceedings of the Alfred Benzon Symposium IX*, Copenhagen, 1975. N. O. Kjeldgaard and O. Maaløe, eds. Academic Press, New York.

Pato, M. L. and K. von Meyenburg. 1970. Residual RNA synthesis in *Escherichia coli* after inhibition of initiation of transcription by rifampicin. Cold Spring Harbor Symp. Biol. 35:497.

Pato, M. L., P. M. Bennett and K. von Meyenburg. 1973. Messenger ribonucleic acid synthesis and degradation in *Escherichia coli* during inhibition of translation. J. Bacteriol. 116:710.

Prescott, D. M. and P. L. Kuempel. 1972. Bidirectional replication of the chromosome in *Escherichia coli*. Proc. Natl. Acad. Sci. USA 69:2842.

Ryter, A. and A. Chang. 1975. Localization of transcribing genes in the bacterial cell by means of high resolution autoradiography. J. Mol. Biol. 98:797.

Chapter 7

Barry, G., C. L. Squires and C. Squires. 1979. Control features within the *rplJL-rpoBC* transcription unit of *Escherichia coli*. Proc. Natl. Acad. Sci. USA 77:7084.

Fiil, N. P., J. D. Friesen and P. P. Dennis. 1980. *Genetics and Evolution of RNA Polymerase, tRNA and Ribosomes*. University of Tokyo Press, Tokyo.

Jacob, F. and J. Monod. 1961. Genetic regulatory mechanisms in the synthesis of proteins. J. Mol. Biol. 3:318.

Johnsen, M., T. Christensen, P. P. Dennis and N. P. Fiil. 1982. Autogenous control: ribosomal protein L10–L12 complex binds to the leader sequence of its RNA. EMBO J. 1:999.

Maaløe, O. 1979. Regulation of the protein synthesizing machinery—ribosomes, tRNA, factors, and so on. In *Biological Regulation and Development*, Vol. I. *Gene Expression*. R. F. Goldberger, ed. Plenum Press, New York.

Nomura, M., J. L. Yates, D. Dean and L. E. Post. 1980. Feedback regulation of ribosomal protein gene expression in *Escherichia coli*: structural homology of ribosomal RNA and ribosomal protein mRNA. Proc. Natl. Acad. Sci. USA 77:7084.

Ogden, S., D. Haggerty, C. M. Stoner, D. Kolodrubetz and R. Schleif. 1980. The *Escherichia coli* L-arabinose operon: binding sites of the regulatory proteins and a mechanism of positive and negative regulation. Proc. Natl. Acad. Sci. USA 77:3346.

Schleif, R. 1982. Personal communication.

Squires, C., A. Krainer, G. Barry, W.-F. Shen and C. L. Squires. 1981. Nucleotide sequence at the end of the gene for the RNA polymerase β' subunit (rpoC). Nucleic Acids Res. 9:6827.

Stent, G. S. and R. Calendar. 1978. *Molecular Genetics*. Freeman, San Francisco.

Yanofsky, C. 1981. Attenuation in the control of expression of bacterial operons. Nature 289:751.

Zengel, J. M., D. Mueckl, and L. Lindahl. 1980. Protein L4 of the *E. coli* ribosome regulates an eleven gene r protein operon. Cell 21:523.

Chapter 8

Fiil, N. 1969. A functional analysis of the *rel* gene in *Escherichia coli*. J. Mol. Biol. 45:195.

Fiil, N. P., K. von Meyenburg and J. D. Friesen. 1972. Accumulation and turnover of guanosine tetraphosphate in *Escherichia coli*. J. Mol. Biol. 71:769.

Gausing, K. 1980. Regulation of ribosome biosynthesis in *E. coli*. In *Ribosomes: Structure, Function and Genetics*. G. Chambliss, G. R. Craven, J. Davies, K. Davis, L. Kahan and M. Nomura, eds. University Park Press, Baltimore.

Hansen, M. T., K. L. Pato, S. Molin, N. P. Fiil, and K. von Meyenburg. 1975. Simple downshift and resulting lack of correlation between ppGpp pool size and ribonucleic acid accumulation. J. Bacteriol. 131:18.

Johnsen, K., S. Molin, O. Karlström and O. Maaløe. 1977. Control of protein synthesis in *Escherichia coli*: analysis of an energy source shift-down. J. Bacteriol. 131:18.

Maaløe, O. 1979. Regulation of the protein synthesizing machinery—ribosomes, tRNA, factors, and so on. In *Biological Regulation and Development*, Vol. I, *Gene Expression*. R. F. Goldberger, ed. Plenum Press, New York.

Molin, S., K. von Meyenburg, O. Maaløe, M. T. Hansen and M. L. Pato. 1977. Control of ribosome synthesis in *Escherichia coli*: analysis of an energy source shift-down. J. Bacteriol. 131:7.

Pedersen, S., P. L. Bloch, S. Reeh and F. C. Neidhardt. 1978. Patterns of protein synthesis in *E. coli*: a catalog of the amount of 140 individual proteins at different growth rates. Cell 14:179.

Schleif, R. 1967. Ph.D. Thesis, University of California, Berkeley.

Appendix C

Beck, C. F. and J. L. Ingraham. 1971. Location on the chromosome of *Salmonella typhimurium* of genes governing pyrimidine metabolism. Molec. Gen. Genet. 111:303.

Beck, C. F., J. L. Ingraham and J. Neuhard. 1972. Location on the chromosome of *Salmonella typhimurium* of genes governing pyrimidine metabolism. II. Uridine kinase, cytosine deaminase, and thymidine kinase. Molec. Gen. Genet. 115:208.

Kemper, J. 1974. Gene order and co-transduction in the *leu-ara-fol-pyrA* region of the *Salmonella typhimurium* linkage map. J. Bacteriol. 117:94.

Syvanen, J. M. and J. R. Roth. 1973. Structural genes for catalytic and regulatory subunits of aspartic transcarbamylase. J. Mol. Biol. 76:363.

Index

α protein, 100
Abortive transductant, 209
Absorbance, dry weight and, 232–233
Acetate, 150, 151
Acetyl CoA
  biosynthetic requirements for, 128
  formation of, 138–142
  structure, 123
N-Acetylglucosamine, 17, 113
N-Acetylmuramic acid, 17, 28, 64, 113
O-Acetylserine sulfhydrylase, 133
Acinetobacter species, 43, 202, 205
ACP, *see* Acyl carrier protein
Acridine orange, 196
Actinomyces antibioticus, 18
Actinomyces streptomycini, 18
Acyl carrier protein, 116, 117
Adaptation, 34
Adenine, tautomeric forms, 197
Adenosine triphosphate
  energy charge concept, 166–168
  generation of, 143–154
Adenosine triphosphatase, 144
Adenylate cyclase, 327
Adenylate kinase, 166
Adhesin, 36
Adhesion, 24, 36
Aerobe, strict, 148
Aerobic growth on glucose
  cost of fueling reactions, 143
  pathways used, 139
Aerobic growth on malate
  cost of fueling reactions, 143
  pathways used, 139
Agrobacterium tumefaciens, 206
Alarmones, 50
Alkaline phosphatase, 154

Alkylating agent, mutagenic effects, 196, 198
Allosteric inhibition, 164, 165–166
Allosteric protein, 164
Ameboid motion, 21
Amino acids
  amounts required in protein synthesis, 110, 124
  biosynthetic costs, 129–132
  codons for, 195
  number of kinds of, 19
  pools, 368–370
  transport, 160–163
Aminoacyl-tRNA synthetase, 104, 369
2-Amino purine, 196
Ammonia oxidizers, 147
Ammonium ion, 19, 20
  assimilation into cell constituents, 133
  biosynthetic requirements for, 128
Amphibolic pathways, 52, 137
Anabolic reduction charge, 168
Anaerobe, 148
Anaerobic growth on glucose, pathways used, 140
Anaerobic respiration, 148–149
Anapleurotic reactions, 137, 139, 140
Antibiotics, use, 273–275
Anticodon, 104
Antiport, 157, 158
Arabinose operon, 328–332
ara promoter, 101, 329, 330, 331, 417
Archebacteria, 16, 25–27, 103
Arrhenius plot, 251, 252, 253
Arthrobacter pyridinolis, 160
A site, 104, 105
Assembly, 50, 52–53, *see also* *specific cell structure*
  directed, 50
Assimilation reactions, 52
Assimilatory nitrate reductase, 128

Attenuation, 100, 319
  growth physiology and, 337–338
  in S10 operon, 341
  in *trp* operon, 334–335
Attractant, 34
Autoregulation, 339, 341, 350
Autorepression, 329
Auxotroph, 181, 182
Azoferredoxin, 129
Azotobacter vinelandii, 132

β operon, 341–345
β protein, 89
Bacillus subtilis, 18, 65, 103, 206, 274, 305
  ultrastructure, 25
  genetic map, 178–179
Bacteriophage, *see* Phage
Bactoprenol, *see* Undecaprenol phosphate
Balanced growth, 5, *see also* Exponential phase; Growth; Growth rate
  amino acid incorporation, 368
  experimental considerations, 268–270
  parameters of, 275–289, 302
  replication during, 292–295
  theory, 229–230
Base analogue, mutagenic effects, 196–197
Base analogue selection method, 185, 186
Base pair mutation, 194
Bayer's junctions, 55–56, 61, 66
Bent loop model, 59, 60, 68–69
Binding proteins, 30, 156–157
Bioassay, 237
Biosynthetic pathways, 50, 123, 125–132, 134
  coordination among, 163–168
Biosynthetic reactions, 50
5-Bromo-3-chloro-indolyl-β-D-galactosidase, 419

427

FREDERICK C. NEIDHARDT is Professor of Microbiology and Immunology at the University of Michigan. Born in Philadelphia in 1931, he received his Ph.D. in 1956 from Harvard University where he worked with Boris Magasanik. Before joining the University of Michigan in 1970 as chairman of the Department of Microbiology, he taught at Harvard Medical School and Purdue University. For his research in bacterial physiology, Dr. Neidhardt has received the Eli Lilly Award in Bacteriology and Immunology (1966) and the Alexander von Humboldt Senior U.S. Scientist Award (1979) from the Federal Republic of Germany. He is an editor of *Biochemical and Biophysical Research Communications* and has served on the editorial boards of the *Journal of Bacteriology* and the *Journal of Biological Chemistry*. He has been a member of the National Board of Medical Examiners and in 1981–82 was president of the American Society for Microbiology.

OLE MAALØE is a professor at the University Institute of Microbiology, Copenhagen, Denmark. Born in 1914, he received his Ph.D. in 1946 from the University of Copenhagen. He has been the Director of the World Health Organization's Center on Biological Standardization at the State Serum Institute in Copenhagen, a visiting professor in the Molecular Biology and Virus Laboratory at Berkeley, California, and an invited lecturer in the People's Republic of China in 1973 and 1980. Dr. Maaløe's major research interest is bacterial growth, in several of its experimental and theoretical aspects. He is a member of the Royal Danish Academy of Sciences and Letters and a science advisor to the Basel Institute of Immunology, Switzerland, the European Molecular Biology Laboratory, and the Max Planck Institute of Molecular Genetics in Berlin, Germany.

JOHN L. INGRAHAM is Professor of Bacteriology at the University of California, Davis. Born in Berkeley, California in 1924, he received his Ph.D. in 1951 from the University of California, Berkeley, under the joint direction of Roger Y. Stanier and Ralph Emerson. After employment with the DuPont Company and the U.S. Department of Agriculture, he joined the staff of the University of California, Davis, where he has been ever since. Dr. Ingraham has studied the physiology and genetics of the malo-lactic fermentation, the biosynthesis of fusel oil, the factors precluding growth of bacteria at low temperature, pyrimidine nucleotide metabolism, and, currently, denitrification. He has coauthored two texts, *The Microbial World* and *Introduction to the Microbial World*. He was a Guggenheim Fellow and is an honorary member of the Spanish Microbiology Society.